Pollutants and Water Management

Pollutants and Water Management

Resources, Strategies and Scarcity

Edited by

Pardeep Singh
PGDAV College, University of Delhi
New Delhi
India

Rishikesh Singh
Banaras Hindu University
Varanasi, Uttar Pradesh
India

Vipin Kumar Singh
Banaras Hindu University
Varanasi, Uttar Pradesh
India

Rahul Bhadouria
University of Delhi
New Delhi
India

WILEY Blackwell

This edition first published 2021

Registered Offices
John Wiley & Sons, Inc., 111 River Street, Hoboken, NJ 07030, USA
John Wiley & Sons Ltd, The Atrium, Southern Gate, Chichester, West Sussex, PO19 8SQ, UK

Editorial Office
9600 Garsington Road, Oxford, OX4 2DQ, UK

For details of our global editorial offices, customer services, and more information about Wiley products visit us at www.wiley.com.

Wiley also publishes its books in a variety of electronic formats and by print-on-demand. Some content that appears in standard print versions of this book may not be available in other formats.

Library of Congress Cataloging-in-Publication Data is applied for
Hardback: 9781119693611

Cover Design: Wiley
Cover Image: © Daxiao Productions/Shutterstock

Set in 9.5/12.5pt STIXTwoText by SPi Global, Pondicherry, India

10 9 8 7 6 5 4 3 2 1

Contents

List of Contributors

Saami Ahmed
Department of Chemistry, Zakir Husain Delhi College, University of Delhi, New Delhi, India

Zeenat Arif
Department of Chemical Engineering and Technology, IIT (BHU), Varanasi, Uttar Pradesh, India

B.S. Bhau
Department of Botany, Central University of Jammu, Samba, Jammu and Kashmir, India

Swati Chaudhary
Department of Applied Sciences, M.S.I.T., GGSIP University, New Delhi, India

Meenakshi Chaurasia
Department of Botany, University of Delhi, New Delhi, India

Sunil Dhar
Department of Environmental Sciences, Central University of Jammu, Samba, Jammu and Kashmir, India

Kajol Goria
Department of Environmental Sciences, Central University of Jammu, Samba, Jammu and Kashmir, India

Ankita Gupta
Institute of Environment and Sustainable Development, Banaras Hindu University, Varanasi, Uttar Pradesh, India

R.N. Jadejaa
Department of Environmental Studies, Faculty of Science, The Maharaja Sayajirao University of Baroda, Vadodara, Gujarat, India

S. Jayakumar
Environmental Informatics and Spatial Modelling Lab (EISML), Department of Ecology and Environmental Sciences, School of Life Sciences, Pondicherry University, Pondicherry, Puducherry, India

Jayant Karwadiya
Institute of Environment and Sustainable Development, Banaras Hindu University, Varanasi, Uttar Pradesh, India

Mahima Kaushik
Nano-bioconjugate Chemistry Lab, Cluster Innovation Centre, University of Delhi, New Delhi, India

Richa Kothari
Department of Environmental Sciences, Central University of Jammu, Samba, Jammu and Kashmir, India

Agam Kumar
Environmental Informatics and Spatial Modelling Lab (EISML), Department of Ecology and Environmental Sciences, School of Life Sciences, Pondicherry University, Pondicherry, Puducherry, India

Lawrence Kumar
Department of Nanoscience and Technology, Central University of Jharkhand, Ranchi, Jharkhand, India

Mohan Kumar
Department of Chemistry, Shri Varshney College, Aligarh, Uttar Pradesh, India

Pawan Kumar
Department of Physics, Mahatma Gandhi Central University, Motihari, Bihar, India

Pradeep Kumar
Department of Chemical Engineering and Technology, IIT (BHU), Varanasi, Uttar Pradesh, India

Ravishankar Kumar
Department of Environmental Science and Technology, Central University of Punjab, Bathinda, Punjab, India

Kavita Kumari
Department of Nanoscience and Technology, Central University of Jharkhand, Ranchi, Jharkhand, India

Avinash K. Kushwaha
Department of Botany, BHU, Varanasi, Uttar Pradesh, India

P.K. Mishra
Department of Chemical Engineering and Technology, IIT (BHU), Varanasi, Uttar Pradesh, India

Virendra Kumar Mishra
Institute of Environment and Sustainable Development, Banaras Hindu University, Varanasi, Uttar Pradesh, India

Sunil Mittal
Department of Environmental Science and Technology, Central University of Punjab, Bathinda, Punjab, India

Indica Mohan
Department of Environmental Sciences, Central University of Jammu, Samba, Jammu and Kashmir, India

Anam Naheed
Institute of Environment and Sustainable Development, Banaras Hindu University, Varanasi, Uttar Pradesh, India

Supriya Nath
Central Water and Power Research Station, Pune, Maharashtra, India

Ramesh Oraon
Department of Nanoscience and Technology, Central University of Jharkhand, Ranchi, Jharkhand, India

Kajal Patel
Department of Botany, University of Delhi, New Delhi, India

Sanjeet Kumar Paswan
Department of Nanoscience and Technology, Central University of Jharkhand, Ranchi, Jharkhand, India

Deepak Pathania
Department of Environmental Sciences, Central University of Jammu, Samba, Jammu and Kashmir, India

K.S. Rao
Department of Botany, University of Delhi, New Delhi, India

Krishna Rawat
School of Environment and Sustainable Development, Central University of Gujarat, Gandhinagar, Gujarat, India

Prafulla Kumar Sahoo
Department of Environmental Science and Technology, Central University of Punjab, Bathinda, Punjab, India

M. Sathya
Environmental Informatics and Spatial Modelling Lab (EISML), Department of Ecology and Environmental Sciences, School of Life Sciences, Pondicherry University, Pondicherry, Puducherry, India

Naresh K. Sethy
Department of Chemical Engineering and Technology, IIT (BHU), Varanasi, Uttar Pradesh, India

Vhatkar Shashikant Shivaji
Department of Nanoscience and Technology, Central University of Jharkhand, Ranchi, Jharkhand, India

Reetika Shukla
Institute of Environment and Sustainable Development, Banaras Hindu University, Varanasi, Uttar Pradesh, India

Sushil Kumar Shukla
Department of Transport Science and Technology, Central University of Jharkhand, Ranchi, Jharkhand, India

Ajeet Kumar Singh
Environmental Informatics and Spatial Modelling Lab (EISML), Department of Ecology and Environmental Sciences, School of Life Sciences, Pondicherry University, Pondicherry, Puducherry, India

Anubhuti Singh
Institute of Environment and Sustainable Development, Banaras Hindu University, Varanasi, Uttar Pradesh, India

Gurudatta Singh
Institute of Environment and Sustainable Development, Banaras Hindu University, Varanasi, Uttar Pradesh, India

Priyanka Singh
Institute of Environment and Sustainable Development, Banaras Hindu University, Varanasi, Uttar Pradesh, India

Ram Kishore Singh
Department of Nanoscience and Technology, Central University of Jharkhand, Ranchi, Jharkhand, India

Sukhendra Singh
School of Biochemical Engineering, IIT (BHU) Varanasi, Varanasi, Uttar Pradesh, India

Rupika Sinha
Department of Biotechnology, MNNIT, Prayagraj, Uttar Pradesh, India

K.S. Sista
Research and Development, Tata Steel, Jamshedpur, CIndia

Swati
Department of Botany, BHU, Varanasi, Uttar Pradesh, India

Indu Tripathi
Department of Botany, University of Delhi, New Delhi, India
Department of Environmental Studies, University of Delhi, New Delhi, India

Shashank Tripathi
Institute of Environment and Sustainable Development, Banaras Hindu University, Varanasi, Uttar Pradesh, India

B. Verma
Department of Chemical Engineering and Technology, IIT (BHU), Varanasi, Uttar Pradesh, India

Satyam Verma
Environmental Informatics and Spatial Modelling Lab (EISML), Department of Ecology and Environmental Sciences, School of Life Sciences, Pondicherry University, Pondicherry, Puducherry, India

Jitesh Narottam Vyas
Central Water and Power Research Station, Pune, Maharashtra, India

Amit Kumar Yadav
School of Environment and Sustainable Development, Central University of Gujarat, Gandhinagar, Gujarat, India

Deepak Yadav
Chemical Engineering Department, Harcourt Butler Technical University, Kanpur, Uttar Pradesh, India

Monika Yadav
Department of Environmental Studies, Faculty of Science, The Maharaja Sayajirao University of Baroda, Vadodara, Gujarat, India

Part I

Water Pollution and Its Security

1

Water Security and Human Health in Relation to Climate Change

An Indian Perspective

Ravishankar Kumar, Prafulla Kumar Sahoo, and Sunil Mittal

Department of Environmental Science and Technology, Central University of Punjab, Bathinda, Punjab, India

1.1 Introduction

The capacity of a population to maintain sustainable access to sufficient quantities of acceptable quality water to ensure human well-being, livelihood, socio-economic development, protection against water-borne and water-related disasters, and to preserve ecosystems is termed as water security (UN Water 2013). Water demand is increasing with time due to the booming population, rapid industrialization, rampant urbanization, and extensive agricultural practices. In the world, nearly 785 million people lack a safe drinking water service, including 144 million people dependent on surface water (WHO 2019). Nearly, 1.8 billion people use feces contaminated drinking water sources and have a high risk of contracting cholera, dysentery, typhoid, and polio (WHO 2019). It has been estimated that the world population will be around 9 billion by 2050 and water availability will be less than the current availability (UN WWDR 2015). As per a World Health Organization (WHO) estimation, by 2025, 50% of the global population will be living in water scarcity areas (WHO 2019). By 2050, the global water demand is expected to increase by 20–30% as compared with the current scenario, due to growing demand in the domestic and industrial sectors (UN WWDR 2019). The estimation of the United Nations World Water Development Report (2016) indicated that more than 40% of the global population could be living in severe water stress areas by 2050.

Presently, the world's two most populous countries, India and China, are facing severe water security problems. However, the conditions are more critical in India both in terms of quantity and quality due to a lack of required infrastructure, health services, and management. India has only 4% of the world's freshwater but accounts for 16% of the global population. India ranked 120th out of 122 nations in water quality index and 133rd among 180 nations in water availability (NITI Aayog 2018). Approximately 21% of diseases are related to water among all diseases of the country (Snyder 2020). As per UNICEF and WHO (2012) estimates, approximately 97 million Indians do not have access to safe water. Further, the

Pollutants and Water Management: Resources, Strategies and Scarcity, First Edition. Edited by Pardeep Singh, Rishikesh Singh, Vipin Kumar Singh, and Rahul Bhadouria.

findings of the 2011 census revealed that 138 million rural households had access to safe drinking water, whereas 685–690 million people lacked access to safe drinking water. An ironic fact is that more than 41% of the rural population (out of 833 million people) of India own mobile phones but have no access to potable water which is a basic need. Only 18% of the rural population have access to treated water (Unitus Seed Fund 2014; Forbes India, 2015).

The NITI Aayog report (2018) also said that India is facing its worst water crisis in history, which is only expected to become worse as the country's water demand is projected to be twice the available supply by 2030. The report said that 600 million currently face high to extreme water shortage, with around two lakh people dying every year due to inadequate access to potable water. The increasing water shortage will also affect the gross domestic product (GDP) of the nation, with the country suffering a loss of up to 6% of GDP in 2030 (NITI Aayog 2018).

The quality of both river and groundwater is deteriorating at a rapid pace, making water scarcity more severe. Even toxic heavy metals like uranium, lead, cadmium, selenium, and so on are also reported in groundwater samples from various states (Chowdhury et al. 2016; Kumar et al. 2018, 2020; Sharma et al. 2020). This may lead to severe consequences for water resources. According to the IDSA report (2010), it has been reported that India is expected to become "water-stressed" by 2025 and "water-scarce" by 2050.

Further, climate change is also affecting the water security of India as rising temperature affects the Himalayan glaciers as well as altering the monsoon pattern. The combination of these two factors affects the level of river water due to the melting of glaciers and intense rainfall. Further, groundwater resources are also affected directly and indirectly by the alteration of these factors. High water temperature, changes in timing, intensity, and duration of precipitation are the significant consequences of climate change which can further affect the water quality. The alternate pattern of precipitation leads to floods and droughts, which play an important role in the degradation of water quality by adding a quantum of concentrated pollutants. As per the World Bank report (2018), climate change can affect 6% GDP of some regions due to water security, resulting in migration and conflict. As per the United Nations Convention to Combat Desertification (UNCCD), by 2030, due to climate change impacts on water scarcity, 24–700 million people may be displaced from some arid and semi-arid places.

The achievement of water security in the future will be a very challenging task. This chapter describes in detail the current situation and future challenges regarding water security along with prospective health changes. Further, the impact of climate change on water security and health has been analyzed. The available opportunities are also discussed to manage future challenges related to water security.

1.2 Quantity of Available Water Resources in India

The annual precipitation (rainfall+snowfall) is estimated as 4000 billion cubic meters (BCM). Out of total annual precipitation, 3000 BCM falls during the monsoon season (Jun to September) (Central Water Commission 2014). Around 53.3% of total annual precipitation is lost due to evapotranspiration, which leaves a balance of 1986.5 BCM. The total annual utilizable water resources of India are 1123 BCM, which consists of 690 BCM surface water and 433 BCM of groundwater (Central Water Commission 2014). The National Commission on Integrated Water Resources Development (NCIWRD) projected that total water demand to expect 973 (low demand scenario) to 1180 BCM (high demand scenario). The water used for agriculture is the highest projected demand (70%), followed by households (23%) and industries (7%) (NCIWRD 1999). The per capita average water availability in India in the year 2001 was 1816 m^3, and it is expected to reduce to 1140 m^3 in 2050 (MoWR 2015). The people of the Indian state of Andhra Pradesh have the highest access to safe treated water, i.e., 36%, and it is lowest for Bihar (2%) (Forbes India 2015). The annual surface water availability of India has decreased since the year 1950 (Table 1.1).

Rivers are the primary sources of surface water in India and are considered as the lifeline of Indian cities. There are 15 large, 45 medium, and 120 minor rivers in India (Raj 2010). The rivers are either rainfed and/or based on the Himalayan glacier. The annual water potential in the major river basins of India is 1869.35 BCM, but the utilizable potential is 690 BCM. The Ganga basin has the highest utilizable potential, i.e., 250 BCM. The detailed account of surface water potential of Indian rivers is depicted in Table 1.2.

Table 1.1 Annual surface water availability of India.

S. no	Year	Annual surface water availability (m^3/capita/year)
1	1951	5177
2	1991	2209
3	2001	1820
4	2025	1341
5	2050	1140

Source: Govt. of India (2009).

Table 1.2 Overview of surface water potential of Indian rivers.

S. no	River basin	Catchment area (sq km)	Average water resources potential (BCM)	Utilizable surface water resources (BCM)
1	Indus (up to border)	321 289	73.31	46
2	(a) Ganga	861 452	525.02	250
	(b) Brahmaputra	194 413	537.24	24
	(c) Barak and others	41 723	48.36	
3	Godavari	312 812	110.54	76.3
4	Krishna	268 948	78.12	58
5	Cauvery	81 155	21.36	19
6	Subarnarekha	29 196	12.37	6.8
7	Brahmani and Baitarani	51 822	28.48	18.3

(Continued)

Table 1.2 (Continued)

S. no	River basin	Catchment area (sq km)	Average water resources potential (BCM)	Utilizable surface water resources (BCM)
8	Mahanadi	141 589	66.88	50
9	Pennar	55 213	6.32	6.9
10	Mahi	34 842	11.02	3.1
11	Sabarmati	21 674	3.81	1.9
12	Narmada	98 796	45.64	34.5
13	Tapi	65 145	14.88	14.5
14	West flowing rivers from Tapi to Tadri	55 940	87.41	11.9
15	West flowing rivers from Tadri to Kanyakumari	56 177	113.53	24.3
16	East flowing rivers between Mahanadi and Pennar	86 643	22.52	13.1
17	East flowing rivers between Pennar and Kanyakumari	100 139	16.46	16.5
18	West flowing rivers of Kutch and Saurashtra including Luni	321 851	15.1	15
19	Area of inland drainage of Rajasthan	36 202	0	NA
20	Minor river basins draining into Myanmar and Bangladesh		31	NA
	Total		1869.35	690

Source: Central Water Commission, http://cwc.gov.in/water-info.

India is the largest and fastest consumer of groundwater, which fulfills the demands of nearly 80 and 50% of the rural and urban population, respectively (Shankar et al. 2011). The groundwater resources of the country are estimated to be 433 BCM, which is 39% of the total water resources of India (CGWB 2017). The net groundwater availability is 396 BCM, while the available for potential use is 245 BCM. The stage of groundwater development is 61% (CGWB 2017). The Indian state Uttar Pradesh has the highest net annual groundwater availability (~72 BCM) and Delhi has the least (0.29 BCM) (CGWB 2014). Around 85% of the rural population uses groundwater for drinking purposes. The volume of groundwater is inadequate to fulfill the demand of the large population, agricultural practices, rampant industrialization, and urbanization. The overall account of groundwater resources assessment 2004–2017 is presented in Table 1.3.

The per capita average water availability in India is continuously decreasing. India has a huge potential in river and precipitation water (rainfall+snowfall), but currently, not even

Table 1.3 Groundwater resources assessment from 2004–2017.

Year	Annual replenishable groundwater resources (BCM)	Net annual groundwater availability (BCM)	Annual groundwater draft for irrigation, domestic, and industrial uses (BCM)	Stage of groundwater development
2004	433	399	231	58%
2009	431	396	243	61%
2011	433	398	245	62%
2013	447	411	253	62%
2017	432	393	249	63%

Source: CGWB (2017).

50% of the potential is being used. Due to the lack of use of the water potential of river and precipitation, groundwater resources are under tremendous pressure and the water table is continuously increasing in most parts of the country over time.

1.3 Quality of Available Water Resources in India

Water quality of both available surface and groundwater resources does not satisfy the criteria for potable water in most parts of the country. The Ministry of Jal Shakti report revealed that 70% of water resources in India are polluted by untreated sewage and industrial effluents. The monitoring report of the Central Pollution Control Board (CPCB 2011), based on biological oxygen demand (BOD) and coliform bacteria count, indicated that organic pollution is predominant in aquatic bodies. The groundwater of around 600 districts (i.e. almost one-third of India) is nonpotable. On the other hand, the Central Groundwater Water Board (CGWB) has reported the presence of contaminants like fluoride, nitrate, arsenic, iron, and other heavy metals in the groundwater of many regions (Table 1.4). As and F^- contamination of groundwater is a significant public health risk concern for Indian people. As and F^- contamination of groundwater is a health threat for approximately 100 and 66 million Indian people, respectively (Bindal and Singh 2019; Kadam et al. 2020). Other major groundwater contaminants like U, NO_3^-, Fe, HCO_3^-, etc. have also been reported in several parts of India. High nitrate content in water is another grave concern in many states (Ministry of Water Resources 2014; Kaur et al. 2019). Apart from governmental organizations, various studies/reports on groundwater and surface water quality have confirmed the presence of other contaminants like uranium, cadmium, lead, copper, sulfate, pesticides, and organic pollutant in the water resources of India (Bacquart et al. 2012; Mittal et al. 2014; Chowdhury et al. 2016; Kumar et al. 2016; Bajwa et al. 2017).

Both the groundwater and surface water quality are not qualifying criteria for potable water in most parts of the country. Surface water is continuously facing quality issues due to the discharge of sewage and industrial and agricultural wastes. Groundwater in India is affected by heavy metals (As, Fe, Pb, U) and anions (F^-, NO_3^-, SO_4^{2-}) in different parts of the country.

Table 1.4 Number of states and districts affected by geogenic contamination in groundwater.

Contaminants	No of affected states	No of affected districts
Arsenic (As)	10	68
Fluoride (F^-)	20	276
Nitrate (NO_3^-)	21	387
Iron (Fe)	24	297

Source: CGWB (2019).

1.4 The Impact of Climate Change on the Quantity of Water Resources

Climate change affects water resources through warming of the atmosphere, alterations in the hydrologic cycle, glacier melting, rising sea levels, and changes in precipitation patterns (amount, timing, and intensity). In the Indian scenario, due to the alteration of monsoon patterns, rainfall becomes more intense and cumbersome, and it is concentrated on fewer rainy days. Climate change influences the quantity of water resources of India through the impact on glaciers, groundwater, and flood events. The probable climate change impacts on water resources of India are depicted through the flow diagram in Figure 1.1.

1.4.1 Rainfall

Using decade-wise average rainfall annual data of 116 years of data (1901–2019), no significant trend was observed for annual rainfall on a national basis (Figure 1.2). However, a decreasing trend in annual rainfall was observed across India since the year 2000. This data set is based on more than 2000 rain gauge data spread over the country.

Climate change has affected the rainfall pattern of India in the form of fewer rainy days, but more extreme rainfall events. This is resulting in an increased amount of rainfall in each event, leading to significant flooding. Most of the global models suggest that Indian summer monsoons will intensify. The timing of seasonal variation may also shift, causing a drying during the late summer growing season. There has been a significant change in precipitation and temperature pattern in India from 2000 to 2015. This could indicate a signature of climate change in India (Goyal and Surampalli 2018).

1.4.2 Glaciers

Around 9040 glaciers have been reported in India, covering nearly 18 528 km^2 in the Indus, Ganges, and Brahmaputra basins (Sangewar et al. 2009; Sharma et al. 2013). Any changes in a glacier can affect river run-off and the water availability in the Himalayan rivers (Indus, Ganges, and Brahmaputra) and agricultural practices in India. The annual rate of glacial shrinkage is reported to be nearly 0.2–0.7% in the Indian Himalayan region for 11 river basins during the period 1960–2004 with a mean extent of 0.32–1.40 km^2 (Kulkarni

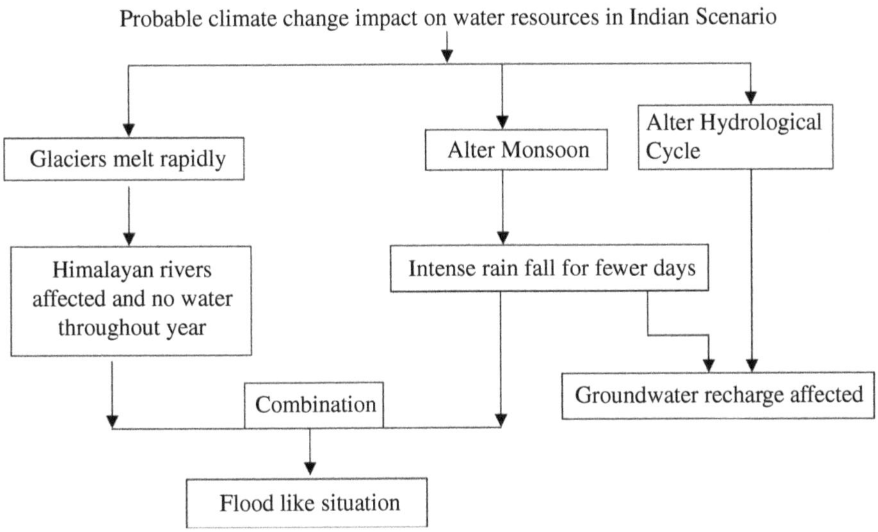

Figure 1.1 Impact of climate change on water resources.

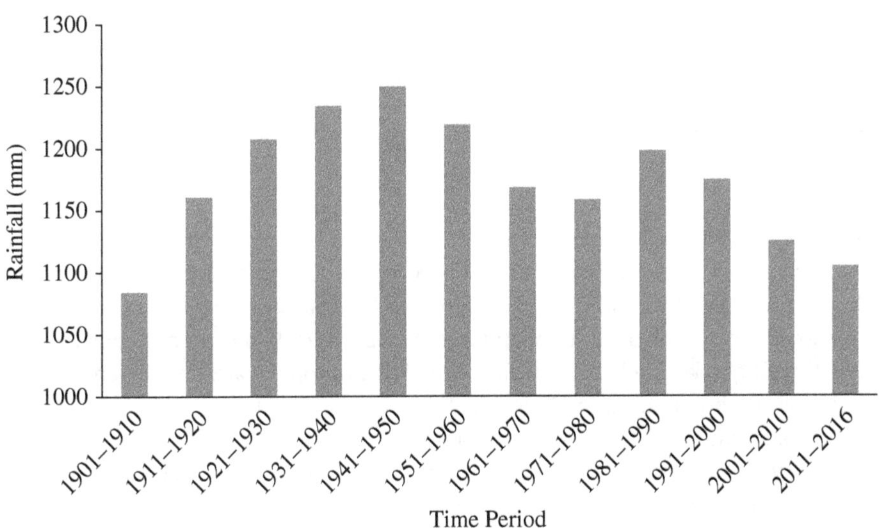

Figure 1.2 Decade-wise average rainfall annual data of India. (*Source:* Envi Stats India 2018; https://data.gov.in/keywords/annual-rainfall.)

et al. 2011; Bolch et al. 2012). Ramanathan (2011) reported the mass balance of Chhota Shigri glacier (15.7 km^2), located in the Chandra River basin of Himachal Pradesh, showed a net loss of about 1000 m from 2002–2009. The flow diagram demonstrating the impact of climate change on glaciers is depicted in Figure 1.3.

In India, climate change is expected to affect Himalayan rivers (Ganges and Brahmaputra) due to the faster rate of melting of Himalayan glaciers. Himalayan glaciers are known as the "Water Tower of Asia," a major source of water in all major Asian rivers (Shiva 2009).

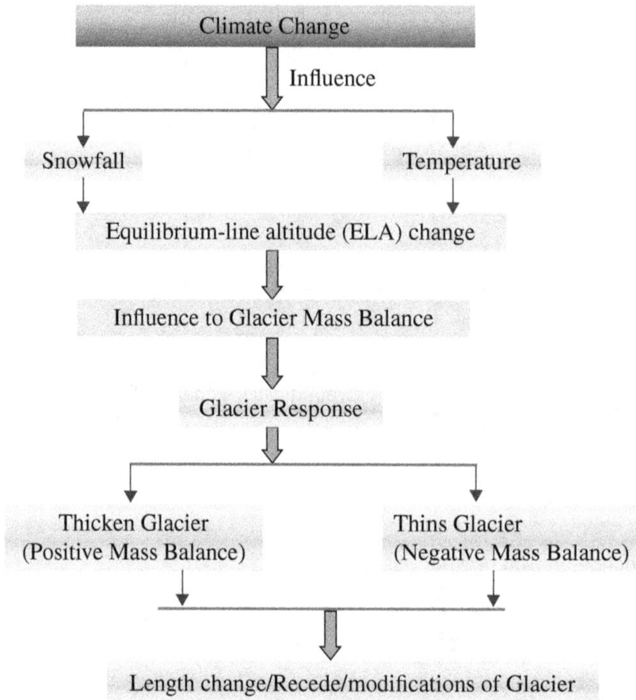

Figure 1.3 The flow diagram of the impact of climate change on glaciers. (*Source:* Pandey and Venkataraman 2012.)

As per the Intergovernmental Panel on Climate Change (IPCC), these glaciers are receding faster than any other part of the world (IPCC 2007). The Gangotri glacier (source of the river Ganga), receded 20–23 miles/year, whereas other glaciers can retreat more than 30 miles/year as a result of rising temperatures (Shiva 2009). If the conditions continue, glaciers will melt quicker and no glaciers will be left to supply water for the entire year, then rivers like Brahmaputra and Ganges will become seasonal rivers. In the monsoon season, the combination of the heavy melting of glaciers and intense heavy rainfall for fewer days may create a flash flood-like situation. On the other hand, reduced rainfall in the rest of the year may lead to drought in some regions. Chevaturi et al. (2016) illustrated the climate change impact on the northern region of Ladakh. The Ladakh area is unique due to its location in high altitude, dry desert with cold temperatures, and water flows to the mountains. Research showed a warming trend with reduced seasonal precipitation, making it highly sensitive to temperature changes.

1.4.3 Sea Level

Rising sea levels and flooding are the biggest threats of climate change. As temperature rises, ice melts and water level rises. This threatens to engulf coastal areas and cause mass displacement and loss of life. Initial predictions expected a sea-level rise of over 59 cm by 2100, but current rates will likely exceed this by a wide margin. According to Pandve (2010), a sea-level rise of 1 m would inundate up to 5763 km of India, as many cities lie only a few feet above sea level, making severe coastal floods.

1.4.4 Groundwater

Groundwater resources are affected due to an inadequate amount of water percolating down to aquifers due to reduced rainfall. The increased atmospheric temperature also increases the rate of evapotranspiration, which leads to a reduction in the actual amount of groundwater available for human use. India extracts 1000 km^3 of groundwater annually, which is 25% of groundwater at a global level (Mukherji 2019).

Climate change affects Indian water resources through warming of the atmosphere, alterations in the hydrologic cycle, melting of glaciers, rising sea levels, and changes in precipitation patterns (amount, timing, and intensity). The alteration of monsoon patterns decreases rainy days but increases the amount of rainfall. Himalayan glaciers are receding faster than any other part of the world. Further, the combined impacts of changes in precipitation patterns, glaciers melting, and sea-level rise has caused flood-like situations in different parts of the country. One noticeable thing, if the conditions continue, glaciers will melt quicker and no glaciers will be left to supply water for the entire year, then rivers like Brahmaputra and Ganges will become seasonal rivers.

1.5 Impact of Climate Change on the Quality of Water Resources

The impact of climate change on water quality has not gained much concern as an emerging topic in water research to date. However, possible effects are discussed with the association of health as depicted in Figure 1.4. Floods and droughts also affect the surface water qualitatively (in terms of pollutant concentration) and quantitatively. Whenever drought condition persists, the groundwater resources are depleted and the concentration of the pollutants are elevated in the residual water (IPCC 2007). Changes in precipitation or hydrological pattern and increased run-off can result in the rise of pathogens and contaminants in water bodies. Increased frequency and intensity of rainfall may cause more water pollution due to run-off water. The decrease in dissolved oxygen in water due to the increase in the temperature of the water is the direct consequence of climate change on water quality. Further, the concentration of dissolved carbon, phosphates, nitrates, and micropollutants are also directly altered as a consequence of climate change and they produce an adverse impact on health (Delpla et al. 2009).

Climate change is not only expected to influence the quantity of groundwater but also to influence the quality of groundwater (Dragoni and Sukhija 2008). Water recharges during an arid period contain a high concentration of salts and increases total dissolved solids (TDS). However, in a wet period, the reverse phenomena can occur. Climate change increases sea surface temperatures and results in rising sea levels. Further, rising sea levels may lead to saltwater intrusion into coastal aquifers, which influences groundwater quality and contaminates drinking water sources whenever salty water percolates into the freshwater system. It is very difficult to reverse the process. Climate change influences the

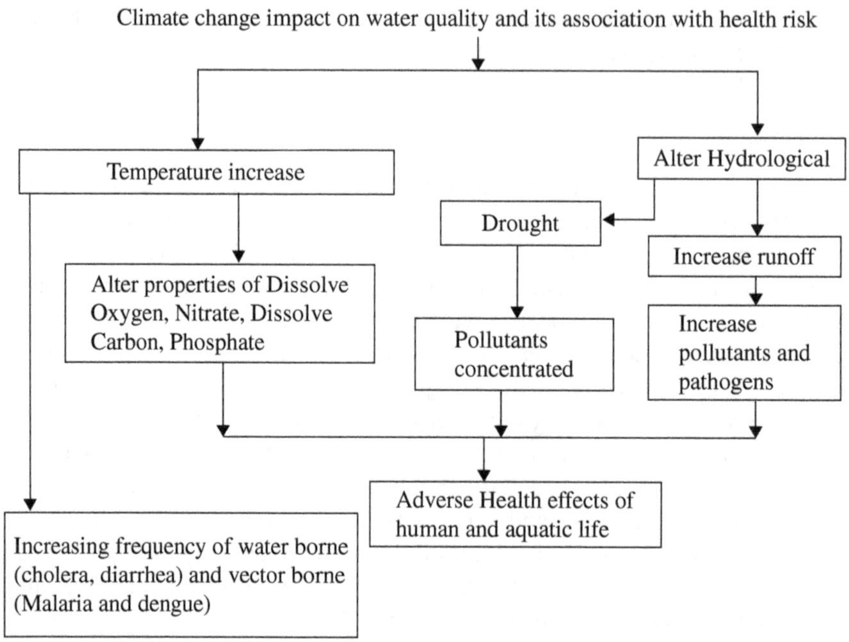

Figure 1.4 Impact of climate change on water quality and its association with health.

amount or pattern of precipitation, resulting in a flood-like situation and affects groundwater quality through the release of agrochemicals/industrial wastes from soil to groundwater.

Climate change affects water quality through the decrease of dissolved oxygen due to the rise of temperature, while alternations to the hydrological cycle increase pathogens and contaminants in surface water. Groundwater quality has been indirectly affected by climate change due to increases in TDS, salts, and other contaminants. Further, rising sea levels may lead to saltwater percolation in coastal aquifers, which influences groundwater quality.

1.6 The Health Perspective in Association with Water Security and Climate Change

As per the WHO (2018), in the period between 2030 and 2050, climate change could be the reason for approximately 250 000 additional deaths per year by malnutrition, malaria, diarrhea, and heat stress. The additional health costs by 2030 are estimated to project USD 2–4 billion/year. Climate change affects health through polluted air, unsafe drinking water, insufficient food, and shelter safety. Extreme high air temperatures directly affect cardiovascular and respiratory systems, particularly to older adults. In Europe, more than 70 000 deaths were recorded under the influence of a summer heatwave during 2003 (Robine et al. 2008). High temperature also increases ozone levels and other pollutants in the air, leading to cardiovascular and respiratory diseases. The levels of pollen and other aerial allergens are high in extreme temperature/heat. This can trigger asthma, which affects nearly 300 million people in the world (WHO 2018). Apart from this, climate change has a high impact on water-related diseases. The nonuniform rainfall patterns are likely to affect freshwater and make it unsafe for humans. This water can compromise hygiene and increase the risk of diarrheal disease, which kills over 500 000 children aged under five years, every year (IPCC 2014).

India is one of the major countries that suffers from water-related diseases. The security of drinking water ensures the prevention and control of water-borne diseases. As per the WHO assessment, around 37.7 million people in India are affected by water-borne diseases every year, and among them, 75% are children (Khurana and Sen 2009). The World Bank has also estimated that 21% of communicable diseases in India are related to unsafe water. The impact of climate change increases the risks of water-borne diseases like cholera, malaria, and dengue by warming of the climate and intense rainfall. A UN report stated that more than one lakh people die annually from water-borne diseases and 73 working days are lost due to water-borne diseases. Another report stated that 1.5 million children die annually from diarrhea (Khurana and Sen 2009). Apart from water-borne diseases, cancer, cardiovascular diseases, mental disorders, and other diseases are reported due to probable contaminants found in water (Kaur et al. 2019). A resulting economic burden of $600 million has been estimated per year due to water-borne diseases. Further, climate change makes the situation more critical. Rising temperatures often bring negative impacts to human health and life. The incidences of water-borne diseases like cholera, diarrhea, and so on,. become more prevalent in warmer climates (Figure 1.4). Vector-borne diseases like malaria can thrive when the temperature increases as a result of global warming. It is also estimated that up to 2050, the malaria vector will shift away from central regions towards southwestern and northern states due to the variation of rainfall (Kiszewski et al. 2004). Malaria kills over 400 000 people every year on the global level.

Vector-borne diseases like dengue also increase in warm and rainy climate due to the increasing mosquito population. The *Aedes* mosquito vector of dengue is also highly sensitive to climate conditions, and studies suggest that climate change is likely to increase exposure to dengue. Apart from the risks caused by increased temperature, intense rainfall could result in floods and waterlogging in several places. Waterlogged areas will then become the potential grounds for mosquitoes breeding. In India, especially in the Ganges basin, poor habitats have no choice for drinking and cooking other than using the polluted water of rivers. This results in numerous diseases. Among these diseases, stomach infections like diarrhea and dysentery are common. People living in rural areas and urban slums will be more vulnerable to diseases and infections because they do not have access to piped water and cannot afford to buy clean water. Water shortages have an enormously devastating impact on human health, including malnutrition, pathogen or chemical loading, and infectious diseases from water contamination. In the future, this cycle of diseases will place an enormous burden on the government, who will have to scramble to provide health care for all those affected and have to take preventive measures to control the situation from worsening.

Climate change affects health through polluted air, unsafe drinking water, insufficient food, and shelter safety. The nonuniform rainfall patterns are likely to affect freshwater in India and make it unsafe for humans. This water can compromise hygiene and increase the risk of diarrheal disease, in these cases, children are the main sufferers. Further, the impact of climate change also increases the risks of water and vector-borne diseases like cholera, malaria, and dengue by warming of the climate and intense rainfall.

1.7 Major Challenges to Water Security

1.7.1 Water Demand for the Future

There are several reports published by national and international agencies on the current and future demand of water (Tables 1.1 and 1.5) for India. Based on these reports, it can be analyzed that meeting the water supply-demand of India will be a serious challenge. The most serious concern is the growing population, which is likely to increase to 1.4 billion by 2050. To meet food security, the agricultural sector also needs a huge amount of water.

1.7.2 Overexploitation of Groundwater

The water table in India is depleting at a rate of 0.4–0.6 m per year. Out of the total assessment units (blocks/taluks/mandals/districts/firkas/valleys), nearly 17.5, 4.5, 14, and 64% units have been categorized as overexploited, critical, semi-critical, and safe, respectively (CGWB 2017). So, preventing the overexploitation of groundwater will be another challenge.

1.7.3 Management of Water Resources

- **Water availability:** The water resources of India have a large gap between potential and availability. The potential of water resources has been estimated at 1869 BCM and annual precipitation is 4000 BCM. Out of a total potential 1869 BCM, India uses 1123 BCM of water. The topographical and large temporal variability and regional mismatch between water availability and demands are the major reasons for the difference between potential and availability (Jain 2019).
- **Flood management:** The large variability of rainfall in space and time in India causes flooding in different parts of the country. Indian rivers carry more than 70% of their annual flow in four months during the monsoon period. There is an essential need to

Table 1.5 International reports on current and future demands of water of India.

World Bank Report 1999			
Year	**Expected demand**	**Year**	**Per capita water availability**
1997	552 BCM	1947	5000 m^3 per year
2025	1050 BCM	1997	2000 m^3 per year
		2025	1500 m^3 per year
The Mckinsey Report 2009			
2009	740 billion m^3		
2030	1.5 trillion m^3		

Source: IDSA (2010).

conserve flood water and flows for the growing demands of water in the country. Flood management can also play a key role in groundwater recharge and drought management. Nearly 500 BCM of water has been estimated through flood flows in Indian rivers (Jain 2019). In the current scenario, the management of storage flood water is not sufficient. The management of storage flood water can be used to meet growing demands throughout the year. It will also help in water-related disasters like floods and droughts.

- **Water transfer between water enriched and water-stressed regions:** India has large temporal and geographical variability about water availability. The transfer of water between water surplus regions to deficit regions could be a very effective approach in meeting the demand of the entire country.
- **Recycle and reuse:** In the current scenario, less of the urban water supply is recycled and reused, and a large quantity of water is wasted. Around 40% of the water in some cities in India is wasted due to leakage or theft. For instance, the Arab states treat 55% of wastewater, and 15% is reused, which is used in farm irrigation, environmental protection, and industrial cooling (Jain 2019).
- **Impact of climate change:** Warming of the lower atmosphere affects rainfall, snowfall, and glaciers, and raises sea levels, which all interfere with the quantity of water resources. Rising sea levels increase flooding in coastal areas and the intrusion of seawater alters water quality in rivers, lakes, and groundwater.
- Maintain water quality of resources and provide safe drinking water for rural areas.
- Hydro-diplomacy with neighboring countries to solve water conflicts.

1.7.4 Health Prospective

The prevention and control of water- and vector-borne diseases can be a difficult task due to the association with poor water quality and warming of the climate. Apart from that, the presence of arsenic, uranium, lead, cadmium, etc. leads to an increase in health problems due to their probable correlation with cancer and cardiovascular, neurological, and skin diseases.

Projected water demand is continuously increasing day by day due to the rising demand for water by agriculture, industry, and households, as well as the growing population. Groundwater resources are under tremendous pressure and the water table in India is depleting at the rate of 0.4–0.6 m per year. India is not using the full potential of river water, precipitation, and floodwater.

1.8 Government Initiatives to Ensure Water Security

Recently, the Indian government formed the Ministry of Jal Shakti in May 2019 by merging two ministries: the Ministry of Water Resources, River Development, and Ganga Rejuvenation and the Ministry of Drinking Water and Sanitation. The Government of India had also established the National Water Mission, which is one of the eight National Missions under the National Action Plan on Climate Change 2008. Now, National Water Mission is operating under the Ministry of Jal Shakti and the main objective is "conservation of water, minimizing wastage and ensuring its more equitable distribution both across and within States through integrated water resources development and management." The National Water Mission is working towards five goals as follows:

a) Building a comprehensive water database in the public domain and an assessment of the impact of climate change on water resources
b) Promotion of citizen and state actions for water conservation, augmentation, and preservation
c) Focused attention to vulnerable areas including overexploited areas
d) Increasing water use efficiency by 20%
e) Promotion of basin level integrated water resource management

In the 12th five-year plan (2012–2017) of India, more emphasis has been given on aquifer mapping, watershed development, and the involvement of nongovernmental organizations (NGOs) in developing irrigation capacity. Previously, the National Democratic Alliance (NDA) government established a separate ministry on "River Development and Ganga Rejuvenation" to accelerate the development of rivers and approved a 20 000 crores budget to the Namami Ganges scheme for the historical river Ganga. Further, the NDA government made it mandatory that 50% of work under the Mahatma Gandhi National Rural Employment Guarantee Act (MGNREGA) 2005 should be for the improvement of water conservation work like the construction of check dams and de-silting of water bodies. Recently, in the union budget 2016–2017, 60 000 crore rupees for a groundwater recharge project, 259.6 crore rupees for river basin management, and 660.27 crore rupees for water resources management were allocated and particular emphasis was given to the National Rural Drinking Water Program. Several water-related projects such as rainwater harvesting, artificial groundwater recharge, watershed management, etc. are already being run by central and state governments. Further, a substantial amount has been allocated for groundwater recharge projects in drought-hit areas to combat the challenges of climate change. The national adaptation fund was established to analyze climate change threats. The government also paid specific attention to arsenic-affected areas and constructed specially designed new wells for the mitigation of arsenic pollution in groundwater.

The Indian government formed the Ministry of Jal Shakti in May 2019 as the main regulating body of water resources in the country. For improved water quality and quantity of water resources, the Indian government launched several schemes, namely, Namami Ganges, the National Rural Drinking Water Program, the national adaptation fund (for climate change threats), the National Water Mission, etc.

1.9 Managing Water Resources Under Climate Change

India has the potential to transform the increasing number of challenges in water security into opportunities. Based on the available potential of the water sector, it can be concluded that India is not a water-deficit country. In India, 90% of water resources are suitable for growing crops. Some of the reasons against water security in the Indian context are water resource mismanagement, inadequate use of water potential, lack of required government attention, and lack of the willingness to adopt the latest technologies. Hence, fulfilling these lacunae can combat current and future water security problems. India has the opportunity to establish, as a nation, water security for a vast population. Some of the efforts required are as follows:

- **Government priories:** The success of any project or mission is largely dependent on government policies and attention. Hence, water security should be the primary agenda of the government.
- **Strict actions as well as rules and regulations:** Stringent regulation is needed and strict action should be taken against those causing water pollution and wastage.
- **Potential to use surface water:** The surface water used is 690 BCM (55.6%) out of a potential of 1869 BCM (Central Water Commission 2014). The use of the rest of the water is restricted due to a high level of pollution. The Ganges-Brahmaputra-Meghna (GBM) and Indus river systems have an average annual potential of water of 1110.62 BCM and 207.7 BCM, respectively (Central Water Commission: Indus Water Commission). These two river systems have two-thirds of the water potential of India. The need of the hour is to use the potential of surface water.
- **Investments in worthwhile water projects:** The current need is to accelerate and extend successful water projects to the entire country and make success stories like rainwater harvesting, watershed management, groundwater mapping, and other government initiatives. To accelerate government projects, monitoring should be carried out by officials from civil societies, NGOs, and others. Specialized grievance cells should be established.
- **Management of water:** It is estimated that 40–50% of the supplied water is lost due to leakage of pipes and connections. Hence, technology is required to instantly detect leakage. Recently, Danish technology was used in some municipalities, which is capable of detecting even minor leakages that are invisible to the eye. This type of technology is needed to be spread to the entire country.
- **Use of the potential of seawater:** India has 7516.6 km of coastal area and a huge potential for fulfilling the growing water demand. The use of desalination of seawater would be another excellent approach for fulfilling the demand for future needs.
- **Management of rainfall:** Only 18% of rainwater is used effectively, whereas 48% enters into rivers and the rest percolates in the ocean (Hegde 2012). Thus, enormous potential exists to use rainwater to fulfill future demand.
- **Use of wastewater in agriculture and other sectors:** These practices are ongoing but more is needed from sewage treatment, desalination, and other innovative technologies due to the huge amount of water released from domestic and industrial activities.
- **Flood management:** In every monsoon, certain parts of India are affected by floods and a huge amount of water flows is wasted. Therefore, there is a need to turn this into an opportunity by managing this huge amount of water.
- **Hydro-diplomacy:** Many river water conflicts are ongoing between India and its adjacent countries like Nepal, Bhutan, Pakistan, China, and Bangladesh. There is a need for extreme hydro-diplomacy to solve conflicts with these countries.
- **Deficit irrigation:** In this strategy, less water is supplied to crops. No significant reduction of growth yield is estimated by the systematic use of this method. A study carried out on a North China plain on winter wheat saved 25% water with no significant loss of yield

(UN WWDR 2015). In India, a study carried out using this strategy in the vegetative phase for groundnut gave positive results. More research is required regarding deficit irrigation on Indian crops for water conservation strategies.

- **Good groundwater governance:** A Netherlands funded APFAMGS (Andhra Pradesh Farmer Managed Groundwater Systems) project is an excellent example of the governance of groundwater resources. This project has been applied in 638 groundwater overexploited villages of Andhra Pradesh. The officials of this project adapted appropriate cropping systems based on available groundwater resources. The governance acted as pressure to adapt suitable water saving and harvesting projects. Low investment organic agriculture was promoted, and the rules were formulated to ensure the sustainability of groundwater resources.

India is becoming a water-deficient country and climate change is making the situation more critical. The use of the maximum potential of river water, seawater, precipitation, wastewater, and good water governance can minimize the impact of climate change on water resources.

1.10 Conclusion and Recommendations

Water security has been a grave issue in India due to a lack of proper management, the slow rate of establishing water projects, inadequate water monitoring, and a lack of appropriate preventive measures. The degradation of water quality results in increased water-borne and vector-borne diseases. Apart from this, contaminants such as arsenic, fluoride, uranium, nitrate, cadmium, and lead found in water are also responsible for various serious diseases like cancer, cardiovascular, mental disorders, and others. The effects of climate change, including the increase of temperature, changes in regional precipitation patterns, floods, droughts events, etc., make the situation more critical in respect of water quantity, water quality, and water-related diseases. India has the potential to resolve future challenges by the use of surface water to accelerate the establishment of water projects, adopting new technologies, hydro-diplomacy with adjacent countries, and making stringent rules and regulations.

Some recommendations are given as follows:

- Use of the maximum potential of surface water, seawater, and rainfall
- Turning flooding incidents to opportunities by managing huge amounts of water
- Promotion of water-efficient irrigation systems like drip irrigation, sprinkler systems, etc.
- Hydro-diplomacy and solving conflicts with neighboring countries
- Applying strict regulations and taking action against those causing water pollution and wastage

References

Bacquart, T., Bradshaw, K., Frisbie, S. et al. (2012). A survey of arsenic, manganese, boron, thorium, and other toxic metals in the groundwater of a West Bengal, India neighbourhood. *Metallomics* 4 (7): 653–659.

Bajwa, B.S., Kumar, S., Singh, S. et al. (2017). Uranium and other heavy toxic elements distribution in the drinking water samples of SW-Punjab, India. *Journal of Radiation Research and Applied Sciences* 10 (1): 13–19.

Bindal, S. and Singh, C.K. (2019). Predicting groundwater arsenic contamination: regions at risk in highest populated state of India. *Water Research* 159: 65–76.

Bolch, T., Kulkarni, A., Kääb, A. et al. (2012). The state and fate of Himalayan glaciers. *Science* 336 (6079): 310–314.

Central Water Commission (2014). Government of India Ministry of Water Resources, Annual Report 2013–14. http://mowr.gov.in/sites/default/files/AR_CWC_2013-14_2.pdf (accessed 30 May 2020).

CGWB (2014). Ministry of Water Resources. Groundwater year book – India, year 2013–14. www.cgwb.gov.in/documents/Ground%20Water%20Year%20Book%202013-14.pdf (accessed 25 January 2020).

CGWB (2017). National Compilation on Dynamic Ground Water Resources of India, 2017. http://cgwb.gov.in/GW-Assessment/GWRA-2017-National-Compilation.pdf (accessed 30 May 2020).

CGWB (2019). Ground water quality scenario in India. http://cgwb.gov.in/WQ/GROUND%20WATER%20QUALITY%20SCENARIO%20IN%20INDIA.pdf (accessed 12 March 2019).

Chevaturi, A., Dimri, A., and Thayyen, R. (2016). Climate change over Leh (Ladakh), India. *Theoretical and Applied Climatology* 1: 1–15.

Chowdhury, S., Mazumder, M.J., Al-Attas, O., and Husain, T. (2016). Heavy metals in drinking water: occurrences, implications, and future needs in developing countries. *Science of the Total Environment* 569: 476–488.

CPCB (2011). Status of water quality in India. http://cpcb.nic.in/upload/NewItems/NewItem_198_Status_of_WQ_in_India_2011.pdf (accessed 25 January 2020).

Delpla, I., Jung, A.V., Baures, E. et al. (2009). Impacts of climate change on surface water quality in relation to drinking water production. *Environment International* 35 (8): 1225–1233.

Dragoni, W. and Sukhija, B.S. (2008). Climate change and groundwater: a short review. *Geological Society, London, Special Publications* 288 (1): 1–12.

Envi Stats India (2018). EnviStats India 2018. www.indiaenvironmentportal.org.in/files/file/EnviStats_India_27sep18.pdf (accessed 25 January 2020).

Forbes India (2015). Ten facts about drinking water in India that may make you sick. www.forbesindia.com/m/blog/economy-policy/ten-facts-about-drinking-water-in-india-that-may-make-you-sick (accessed 31 March 2020).

Govt. of India (2009). Background note for consultation meeting with policy makers on review of National Water Policy. New Delhi: Ministry of Water Resources.

Goyal, M.K. and Surampalli, R.Y. (2018). Impact of climate change on water resources in India. *Journal of Environmental Engineering* 144 (7): 04018054.

Hegde, N. G. (2012). Water scarcity and security in India. In BAIF-Indian Science Congress. www.indiawaterportal.org/articles/water-scarcity-and-security-india (accessed 31 March 2020).

IPCC (2007). *Fourth Assessment Report. Intergovernmental Panel on Climate Change.* Cambridge: Cambridge University Press.

IPCC Working Group III (2014). Climate Change 2014. Mitigation of Climate Change: Summary for Policymakers. Geneva: Intergovernmental Panel on Climate Change.

IDSA (2010). Water Security for India: The External Dynamics. Institute for Defence Studies and Analyses. www.indiaenvironmentportal.org.in/files/book_WaterSecurity.pdf (accessed 25 March 2020).

Jain, S.K. (2019). Water resources management in India–challenges and the way forward. *Current Science* 117 (4): 569–576.

Kadam, A., Wagh, V., Umrikar, B., and Sankhua, R. (2020). An implication of boron and fluoride contamination and its exposure risk in groundwater resources in semi-arid region, Western India. *Environment, Development and Sustainability* 22: 7033–7056.

Kaur, G., Kumar, R., Mittal, S. et al. (2019). Ground/drinking water contaminants and cancer incidence: a case study of rural areas of South West Punjab, India. *Human and Ecological Risk Assessment: An International Journal* https://doi.org/10.1080/10807039.2019.1705145.

Khurana, I., and Sen, R. (2009). Drinking water quality in rural India: Issues and approaches: www.indiawaterportal.org/articles/issues-and-approaches-drinking-water-quality (accessed 25 January 2020).

Kiszewski, A., Mellinger, A., Spielman, A. et al. (2004). A global index representing the stability of malaria transmission. *The American Journal of Tropical Medicine and Hygiene* 70 (5): 486–498.

Kulkarni, A.V., Rathore, B.P., Singh, S.K., and Bahuguna, I.M. (2011). Understanding changes in the Himalayan cryosphere using remote sensing techniques. *International Journal of Remote Sensing* 32 (3): 601–615.

Kumar, R., Kumar, R., Mittal, S. et al. (2016). Role of soil physicochemical characteristics on the present state of arsenic and its adsorption in alluvial soils of two agri-intensive region of Bathinda, Punjab, India. *Journal of Soils and Sediments* 16 (2): 605–620.

Kumar, R., Vaid, U., and Mittal, S. (2018). Water crisis: issues and challenges in Punjab. In: *Water Resources Management* (eds. P.S. Vijay and Y. Shalini), 93–103. Singapore: Springer.

Kumar, R., Mittal, S., Sahoo, P.K., and Sahoo, S.K. (2020). Source apportionment, chemometric pattern recognition and health risk assessment of groundwater from southwestern Punjab, India. *Environmental Geochemistry and Health* https://doi.org/10.1007/s10653-020-00518-1.

Ministry of Water Resources (2014). Groundwater quality scenario. http://wrmin.nic.in/forms/list.aspx?lid=327&Id=4 (accessed 25 February 2020).

Mittal, S., Kaur, G., and Vishwakarma, G.S. (2014). Effects of environmental pesticides on the health of rural communities in the Malwa Region of Punjab, India: a review. *Human and Ecological Risk Assessment: An International Journal* 20 (2): 366–387.

MoWR (2015). A year of inclusive development in water resources sector. National Water Mission, Ministry of WR, RD&GR, Govt of India. www.nationalwatermission.gov.in/sites/default/files/COP%2021%20MoWR%20Booklet.pdf (accessed 17 January 2020).

Mukherji, A (2019). Climate change, groundwater & agriculture in India. www.chinawaterrisk.org/opinions/climate-change-groundwater-agriculture-in-india (accessed 25 March 2020).

NCIWRD (1999). GoI, Integrated Water Resource Development: A Plan for Action. Report of the National Commission for Integrated Water Resource Development, Vol 1, Government of India (GoI), Sept 1999. New Delhi: Government of India.

NITI Aayog (2018). Composite Water Management Index. A tool for water management. National Institution for Transforming India, GOI. New Delhi: Government of India. www.niti.gov.in/writereaddata/files/document_publication/2018-05-18-Water-index-Report_vS6B.pdf (accessed 25 March 2020).

Pandey, P. and Venkataraman, G. (2012). Climate change effect on glacier behavior: A case study from the Himalayas. *Earthzine: Climate Consensus*. https://earthzine.org/climate-change-effect-on-glacier-behavior-a-case-study-from-the-himalayas-2/.

Pandve, H. (2010). Climate change and coastal mega cities of India. *Indian Journal of Occupational and Environmental Medicine* 14 (1): 22–23.

Raj, A. (2010). Environmental economics. Water security in India: The coming challenges. http://futuredirections.org.au/wp-content/uploads/2011/05/1286517242-FDI%20Strategic%20Analysis%20Paper%208%20October%202010.pdf (accessed 25 March 2020).

Ramanathan, A.L. (2011). *Status Report on Chhota Shigri Glacier (Himachal Pradesh)*. New Delhi, India: Dept. of Science and Technology, Ministry of Science and Technology.

Robine, J.M., Cheung, S.L.K., Le Roy, S. et al. (2008). Death toll exceeded 70,000 in Europe during the summer of 2003. *Comptes Rendus Biologies* 331 (2): 171–178.

Sangewar, C.V., Shukla, S.P., and Singh, R.K. (2009). *Inventory of the Himalayan Glaciers*. Special Publication 558. Calcutta, India: Geological Survey of India.

Shankar, P.V., Kulkarni, H., and Krishnan, S. (2011). India's groundwater challenge and the way forward. *Economic and Political Weekly*: 37–45. www.jstor.org/stable/27918012.

Sharma, A.K., Singh, S.K., and Kulkarni, A.V. (2013). Glacier inventory in Indus, Ganga and Brahmaputra basins of the Himalaya. *National Academy Science Letters* 36 (5): 497–505.

Sharma, S., Kumar, R., Sahoo, P.K., and Mittal, S. (2020). Geochemical relationship and translocation mechanism of arsenic in rice plants: a case study from health prone south West Punjab, India. *Groundwater for Sustainable Development* https://doi.org/10.1016/j.gsd.2020.100333.

Shiva, V. (2009). Climate change in the Himalayas. Bonn: Federal Ministry for the Environment, Nature Conservation and Nuclear Safety.

Snyder, S. (2020). Water in crisis-India. https://thewaterproject.org/water-crisis/water-in-crisis-India (accessed 1 June 2020).

UN Water (2013). What is water security? Infographic. www.unwater.org/publications/water-security-infographic (accessed 1 May 2020).

UN WWDR (2015). The United Nations World Water Development Report. Water for a sustainable world. www.unesco.org/new/fileadmin/MULTIMEDIA/HQ/SC/images/WWDR2015Facts_Figures_ENG_web.pdf (accessed 25 May 2020).

UN WWDR (2019). World Water Development Report. www.unwater.org/publications/world-water-development-report-2019 (accessed 25 May 2020).

UNICEF and WHO (2012). Progress on drinking water and sanitation 2012 update. www.unicef.org/media/files/JMPreport2012.pdf (accessed 15 May 2020).

United Nations World Water Development Report (2016). Water and jobs: facts and figures. https://unesdoc.unesco.org/ark:/48223/pf0000244041 (accessed 7 June 2020).

Unitus Seed Fund (2014). 10 facts about drinking water in India that may just make you sick. http://usf.vc/entrepreneur-info/10-facts (accessed 7 June 2020).

WHO (2018). Climate change and health. www.who.int/news-room/fact-sheets/detail/climate-change-and-health (accessed 9 June 2020).

WHO (2019). Drinking water facts sheet. www.who.int/news-room/fact-sheets/detail/drinking-water (accessed 5 June 2020).

World Bank report (2018). High and dry: Climate change, water, and the economy. www.worldbank.org/en/topic/water/publication/high-and-dry-climate-change-water-and-the-economy (accessed 8 June 2020).

2

Assessment of Anthropogenic Pressure and Population Attitude for the Conservation of Kanwar Wetland, Begusarai, India

A Case Study

Ajeet Kumar Singh, M. Sathya, Satyam Verma, Agam Kumar, and S. Jayakumar

Environmental Informatics and Spatial Modelling Lab (EISML), Department of Ecology and Environmental Sciences, School of Life Sciences, Pondicherry University, Pondicherry, Puducherry, India

2.1 Introduction

Wetlands are a ubiquitous ecosystem, found in all regions and climatic conditions across the globe. But they have long been a misunderstood ecosystem due to the misconception that they are "wastelands, disease-infested, dark and dirty swamps," and therefore, considered worthless and large numbers of them have been converted for the other land-use activities (Shah 2012). Later, the advancement of wetland research made the importance of these ecosystems clear, as they interlink land and aquatic ecosystems, and therefore, understanding developed from the *conversion* to the *conservation* of wetlands (Shah 2012). They are considered as the "transitional zone between terrestrial and aquatic ecosystems," where the water table is usually at or near the surface or the land is covered by shallow water (Mitsch and Gosselink 1986). These are the most productive and biologically diverse ecosystems, but they are also very fragile in nature (Mitsch and Gosselink 2000). They were defined for the first time during the world convention on wetlands at Ramsar city, Iran, on 2 February 1971 as:

> Wetlands are area[s] of marsh, fen, peatland or water, whether natural or artificial, permanent or temporary, with water that is static or flowing, fresh, brackish or salt, including areas of marine water, the depth of which at low tide does not exceed six meters.

The wetlands contribute approximately 40–45% of the total global estimate of natural productivity and ecosystem services, as reported in the various wetland ecosystem assessment reports.

Pollutants and Water Management: Resources, Strategies and Scarcity, First Edition. Edited by Pardeep Singh, Rishikesh Singh, Vipin Kumar Singh, and Rahul Bhadouria.

2.1.1 Distribution of Wetlands

2.1.1.1 Global Distribution

Wetlands cover a small proportion (only 2–6% area) of the Earth's land surface. They are distributed in every continent and country except Antarctica (Tiner 2009). The total estimated area of wetlands over the globe was recorded in a range of 747–885 million hectares (Finlayson et al. 2001). Approximately, half of the global wetland area occurs between 50°N and 70°N latitude, one-third between 20°N and 30°S latitude, and the remaining in the temperate zones (Tiner 2009). A total of 2386 wetlands are identified as Ramsar sites in 171 countries, which cover 253 771 669 ha area.

2.1.1.2 Distribution of Wetlands in India

The total estimated area of wetland over the Asian continent was recorded in a range of 211–224 million hectares (Watkins and Parish 1999). In India, it is estimated as 15 260 572 ha (NWIA 2011; Panigrahy et al. 2012), which is about 4.63% of the total geographical area of the country (NWIA 2011). Most of the natural wetlands in India are found at high altitudes in the Himalayas, followed by wetlands of major rivers in their flood plains and coastal wetlands (Prasad et al. 2002). These are distributed from the cold arid trans-Himalayan zone to wet Terai regions of the Himalayan foothills, including the floodplains of the rivers Ganga and Brahmaputra. In the Deccan peninsula, the river floodplains of Cauvery, Krishna, Godavari, and Tapti contribute to different wetlands. The vast Indian coastline of 7500 km, contributes different types of wetlands in the state of West Bengal, Orissa, Andhra Pradesh, Tamil Nadu, Kerala, Karnataka, Gujrat, Lakshadweep, and Andaman and Nicobar states (Figure 2.1). In addition, a large number of artificial wetlands also contribute to the total estimated wetland area (Panigrahy et al. 2012).

A total of 115 wetlands were identified in different states of India, as "wetlands of national importance" for conservation under the National Wetland Conservation Program in 2009. Wetlands in India have been classified under different categories as shown in Table 2.1 (NWIA 2011).

Currently, there are 37 wetlands of international importance in India that are designated as Ramsar sites, covering 1 067 939 ha.

2.1.1.3 Distribution of Wetlands in Bihar

In Bihar, the total estimated area of wetlands is 403 209 ha (NWIA 2011), which is approximately 2.64% of the total wetland area of India. Wetlands in Bihar are locally called *Chaur*, *Maun*, *Taal*, *Jheel*, and *Pats*. The natural inland wetlands dominate in Bihar, which comprises about 92% of the total estimated area of the wetlands of Bihar, while the artificial area accounts for 3.5% and the small wetlands area of 1–2.5 ha accounts for 4.5% (NWIA 2011).

As Bihar is a land-locked state from all sides in the Gangetic flood plain, limited categories of wetlands exist here (Figure 2.2). The district-wise wetland area in Bihar is shown in Figure 2.3 (NWIA 2011).

The major rivers which contribute to different wetlands in Bihar are Ganga, Kosi, Gandak, Bagmati, Mahananda, and Kamla-Balan (Ghosh et al. 2004). The three important wetlands of Bihar, i.e. Kanwar, Barilla, and Kusheshwar Asthan were included in the list of NWCP (2009).

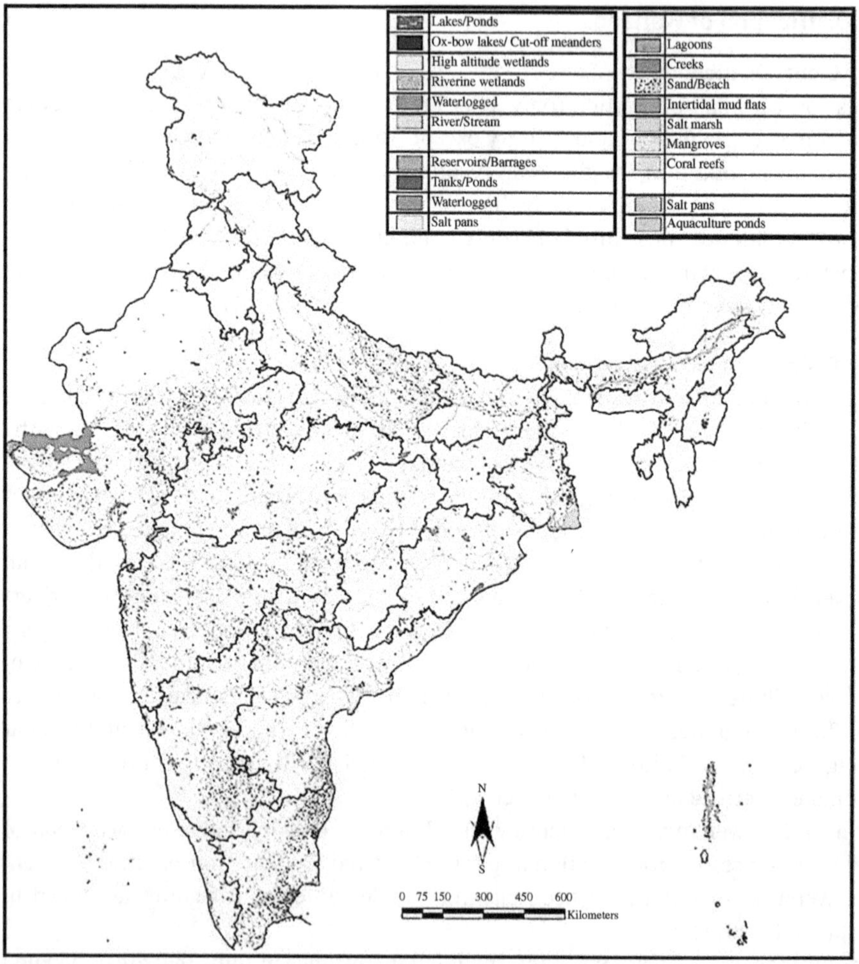

Figure 2.1 A map showing different types of wetlands in India. (Source: National Wetland Atlas 2011, Space Application Centre, Indian Space Research Organization.)

Table 2.1 Category wise distribution of wetlands in India.

S. no	Wetland category	Total wetland area (ha)	Wetland area (%)
1	Lake/pond	729 532	4.78
2	Oxbow lake	104 124	0.68
3	High altitude wetland	124 253	0.81
4	Riverine wetland	91 682	0.6
5	Waterlogged (natural)	315 091	2.06
6	River/stream	5 258 385	34.46
7	Reservoir/barrage	2 481 987	16.26

(Continued)

Table 2.1 (Continued)

S. no	Wetland category	Total wetland area (ha)	Wetland area (%)
8	Tank/pond	1 310 443	8.59
9	Waterlogged (artificial)	135 704	0.89
10	Salt pan (inland)	13 698	0.09
11	Lagoon	246 044	1.61
12	Creek	206 698	1.35
13	Sand/beach	63 033	0.41
14	Intertidal mudflat	2 413 642	15.82
15	Salt Marsh	161 144	1.06
16	Mangrove	471 407	3.09
17	Coral reef	142 003	0.93
18	Salt pan (coastal)	148 913	0.98
19	Aquaculture pond	287 232	1.88
20	Wetlands (<2.25 ha)	555 557	3.64
	Total	**15 260 572**	**100**

Source: National Wetland Atlas (2011), Space Application Centre (ISRO), Ahmedabad.

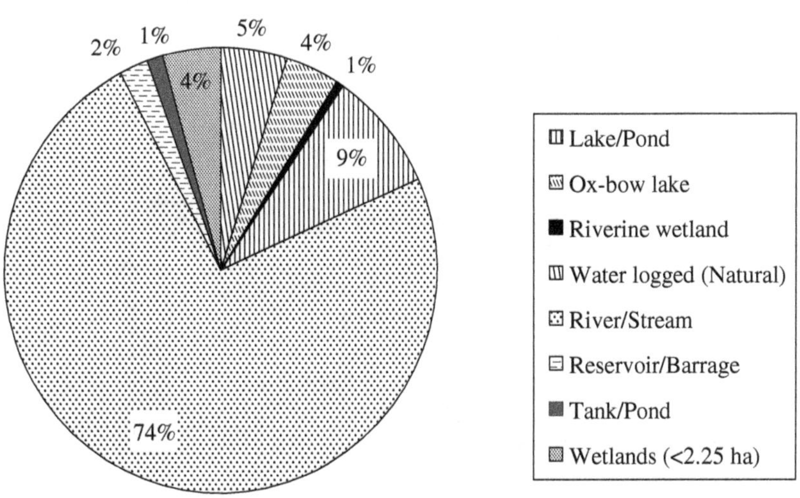

Figure 2.2 Percentage of wetlands area under different categories in Bihar.

2.1.2 Significance of Wetlands

Wetlands are a highly productive ecosystem, providing many important benefits, which can be described as "Goods and Services."

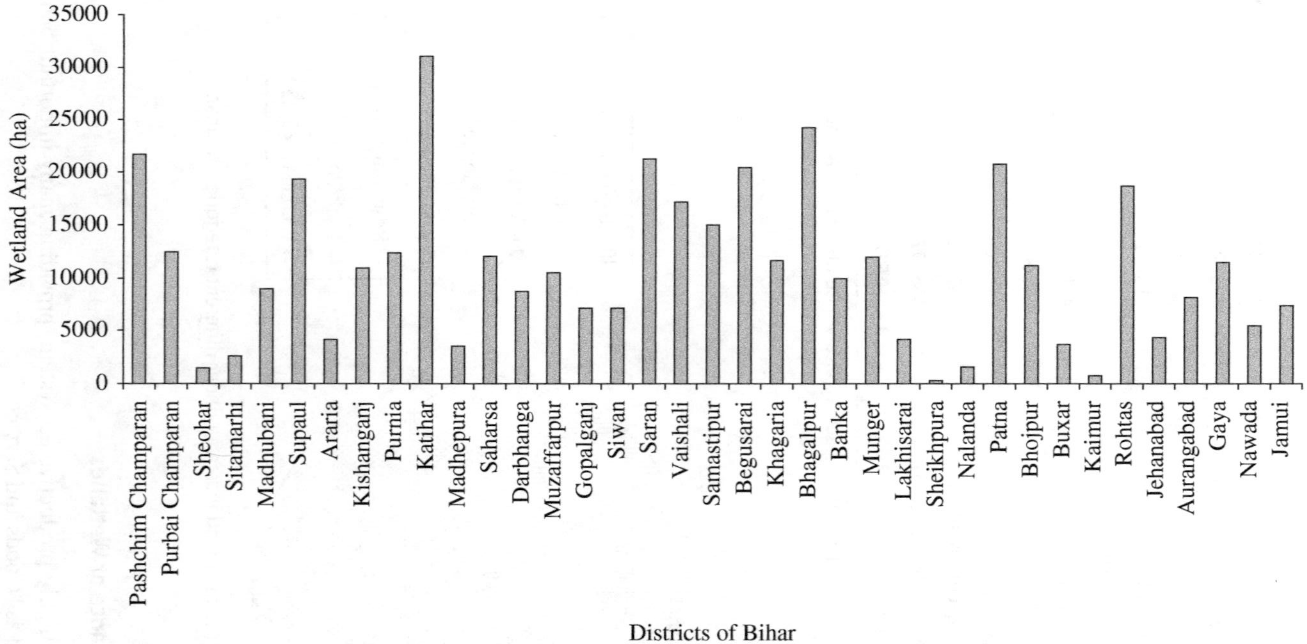

Figure 2.3 District-wise wetland area in Bihar. (Source: National Wetland Atlas 2011, Ministry of Environment and Forest, Government of India.)

2.1.2.1 Ecological Significance

The ecological importance of various wetlands toward biodiversity contribution is well documented by different agencies. These areas provide suitable habitats for fish, micro- and macro-flora and fauna, and act as a wintering ground for a variety of migratory and residential birds (Gilver and Mclnnes 1994; Gopal and Chauhan 2006; Alongi 2009). In addition, wetlands trap a large amount of carbon, silts, and various pollutants, and maintain the quality of water (Manson et al. 2005).

2.1.2.2 Hydrologic and Climatic Significance

Wetlands play an effective role in flood control by reducing sediment load, recharging aquifers, and carbon sequestration (Carter 1986; Ford and Bedford 1987; Bullock 1992; Hilman 1998). These areas also have a significant role in deciding local weather and topographic, edaphic, and hydrological variations (Kusler and Riexinger 1986; Bullock and Acreman 2003).

2.1.2.3 Socio-Economic Significance

Apart from providing various ecosystem services, the wetland ecosystem is vital for socio-economic development, as the majority of the human population is dependent on these ecosystems for their livelihood. Wetlands are the sources of highly valued commercial products and fisheries, and they are the wintering grounds for a large number of avifauna and some other wildlife species. In addition, different wetlands have significant historical and cultural importance, which attracts visitors and provides revenue through eco-tourism.

2.1.3 Background of the Current Study

Wetlands have been the center-stage of research since 1971, after the first international convention on wetlands, popularly called the Ramsar Convention. Different studies have revealed that wetlands have a significant contribution to biodiversity, microbial biogeochemistry, nutrient cycling, greenhouse gas emission, and safeguarding natural and anthropogenic hazards (Dobermann et al. 1996; Craft and Casey 2000; Mathew et al. 2002; Chandra et al. 2009; Tiner 2009; Prusty et al. 2010; Nasir and Harikumar 2011; Niraula 2012; Mohan et al. 2013; Giri and Singh 2014; Singh and Jayakumar 2015).

Several studies have been conducted on the Kanwar wetland since it was notified as a protected area in 1986 and the status of water quality, aquatic flora and fauna, and the socio-economy of the dependent population were documented (Singh and Roy 1989, 1990a, b, 1991a, b; Rakshi and Sanghi 1996). Later, the land use land cover (LULC) dynamics, resource inventory and socio-economy, contamination status of water, soil/sediments, and vegetation were estimated in different research studies (Ghosh et al. 2004; Ambastha et al. 2007a, b; Roy et al. 2008; Shardendu et al. 2012; Anand and Joshi 2013; Kumar 2013a; Singh and Jayakumar 2015, 2016, 2017; Singh et al. 2018). All these studies were conducted to meet their specific objectives and indicated the degrading status of the wetland. The total avifauna population visiting the site was reported to have decreased by 56.63%, including 69.49% of migratory birds and 53.27% of residential birds in 2011, as compared with 2002. Approximately, 3961 ha (62.8%) of the notified land area of the wildlife sanctuary is under dispute between the administration and the local population. Only 2350 ha (37.2%) of the sanctuary area is undisputed as per the forest range officer of Begusarai district. The wetland area supports the livelihoods of a large population, and the increasing human population around this wetland and their encroachments have posed serious challenges for both ecological diversity and quality of resources. These studies have also indicated the degradation of wetland status and suggested an inclusive research approach for the conservation and management of this wetland. In the past one and half decades, the wetland has been under severe threat of degradation due to heavy siltation, water shortage, invasive weeds, and anthropogenic encroachments including illegal land acquisition, agriculture, bird trapping, cattle grazing, tree harvesting, waste dumping, and the establishment of small-scale industries like a brick kiln in and around the sanctuary.

As it is one of the more remotely located wetlands of national importance, not only does it support huge natural biodiversity, it also influences local hydrology, weather, and the socio-economy of the inhabiting population. However, this wetland still only has a limited number of scientific studies, which is an indication that it has not been given much attention by the scientific community for its proper monitoring and conservation. Therefore, this work was conducted to identify the key factors that adversely affect the wetland, as well as human health and the societal status of this area. This study is also aimed to find solutions for the identified factors.

2.2 Materials and Method

2.2.1 Study Area

The study was carried out in Kanwar wetland (Figure 2.4), one of the largest inland freshwater lakes in the Indo-Gangetic biogeographic zone. It was formed naturally by the meandering of the Burhi Gandak River in the geologic past (Rodger and Panwar 1988a, b).

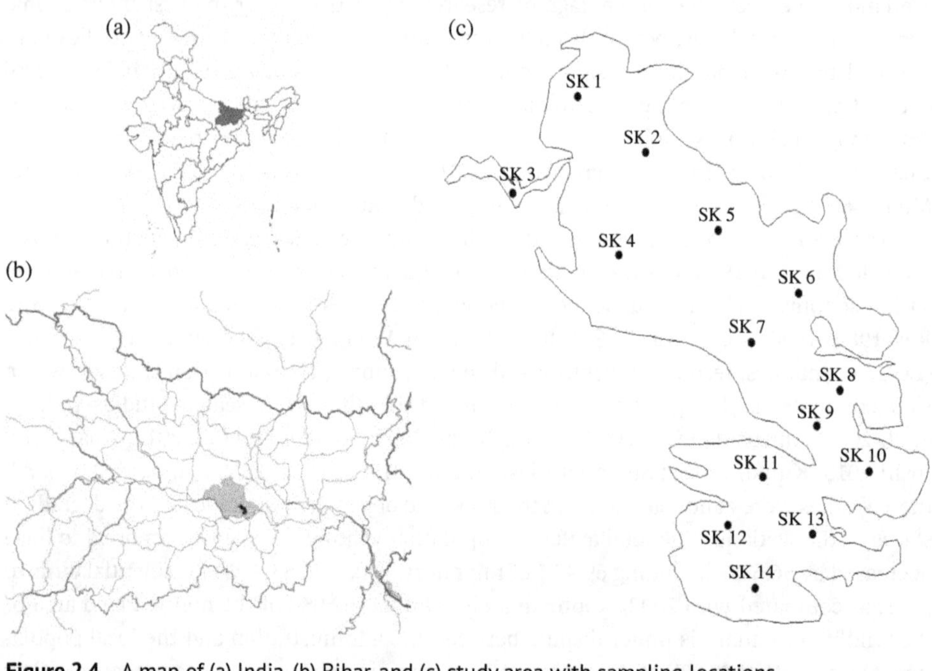

Figure 2.4 A map of (a) India, (b) Bihar, and (c) study area with sampling locations.

It was declared as a Wildlife Sanctuary in 1989 by the Ministry of Environment and Forest, Government of India. Geographically, it is located between (25°35′00″ – 25°40′00″ N and 86°05′00″ – 86°10′00″ E) and covering an area of 6311 ha. It is classified as wetland type 19, because of the type of wetland habitat present, i.e. paddies (Scott 1989). It is a flat terrain with an average elevation of 44 m above mean sea level. This area is found in the tropical wet zone of India (Sarthi and Singh 2013), and temperature varies in a range of 7–38 °C. It mainly experiences four distinct seasons of winter (December–February), summer (March–May), monsoon (June–August), and post-monsoon (September–November). The average annual rainfall is 1100 mm mainly received from the southwest monsoon between July to September. It often gets recharged either from rainwater or excess water from the river Burhi Gandak during the monsoon period. It was identified as a lake of national importance and included in the National Wetland Conservation Programme by the Ministry of Environment and Forest, Government of India in 2009.

The wetland supports a large variety of micro- and macro-flora and fauna in and around its boundary. The vegetation around the wetland area is characterized by tropical dry mixed forest, tropical seasonal swamp forests, and wooded grasslands (Champion and Seth 1968). The major tree species Peepal (*Ficus religiosa*), Indian Banyan (*Ficus benghalensis*), Cluster fig (*Ficus racemosa*), White fig (*Ficus infectoria*), Babool (*Acacia nilotica*), Sissoo (*Dalbergia sissoo*), Arjuna (*Terminalia arjuna*), and so on support the shoreline and raised mounds or *rahies*. Common reed (*Phragmites karka*), water hyacinth (*Eichhornia crassipes*), lotus (*Nelumbo nucifera*), Hydrilla (*Hydrilla verticillata*), and *Sesbania* species are the most prominent aquatic species of this wetland (Rakshi and Sanghi 1996).

The wetland is an important waterfowl habitat of different residential (107 species) and migratory (59 species) birds (Ramakrishna et al. 2002). The white-rumped vulture (*Gyps bengalensis*), the Indian vulture (*Gyps indicus*), the greater spotted eagle (*Aquila clanga*), and the Sarus crane (*Grus antigone*) are the most threatened bird species in this lake (Collar 2005). Different wild animals like blue bulls (*Boselaphus tragocamelus*), bats (*Chiroptera*), squirrels (*Sciuridae*), and jackals are often visible in the late monsoon seasons (Rakshi and Sanghi 1996).

More than 16 hamlets are situated around the wetland. These hamlets were less populated and mainly dominated by the local community during the late 1980s. They were completely dependent on the lake for fishing, bird capture, fuelwood, livestock grazing, and wetland paddy cultivation for their livelihoods (Rodger and Panwar 1988a; Rakshi and Sanghi 1996). Nowadays, the population has drastically increased because of the movement of other communities. During the late monsoon season, when the water recedes, the people cultivate different crops on the open areas of the wetland. The four crop seasons are Kharif (August–November), Rabi (November–February), Aghani, and Garma (March–May). Major crops such as paddy, sugarcane, wheat, maize, pulses, and different vegetables are commonly cultivated. In addition, a practice of double cropping pattern is also common.

2.2.2 Data Collection

First, secondary data including toposheets from the Survey of India (SOI), Patna branch, population data of 2001 and 2011 from the census report, the Government of India website, and rainfall and temperature data from the European Centre for Medium-Range Weather Forecasts were collected for this area.

The LULC change dynamics were estimated by using Landsat imageries and geographic information system (GIS) software. The samples of water and soil/sediments were collected from 14 different locations (Figure 2.4). The samples of aquatic species of flora and fauna, and the cultivated vegetation species, especially crops, were also collected and estimated for the general parameters and the content of specific contaminants, i.e. heavy metals, by following APHA procedures (APHA 2005). Further, the water quality index (WQI) and the health risk index (HRI) was calculated by adopting standard procedures. For WQI, the modified weighted arithmetic mean procedures were used (Shah and Joshi 2017; Singh et al. 2020). The HRI was calculated by using the formula given by Khan et al. (2010).

$$\text{Health risk index}\left(\text{HRI}\right)=\frac{\text{Daily intake of metal}\left(\text{DIM}\right)}{\text{Oral reference dose}}$$

Where,

$$\text{Daily intake of metal}\left(\text{DIM}\right)=\frac{\text{C}_{\text{metal}} * \text{D}_{\text{food intake}}}{\text{Average body weight}}$$

Further, with the help of toposheets, ten villages were identified in close vicinity of the wetland, which were also documented in the Gazette of Bihar, while this wetland was declared as a wildlife sanctuary. These villages are Manjhaul, Jaimangalpur, Jaimangalgarh, Ekamba, Parora, Narayanpiper, Sakra, Rajaur, Kanausi, and Manikpur. A rapid survey was conducted among these villages from June 2013 to August 2013 to understand the overall wetland resource use pattern and its dynamics, to identify different resource user groups (RUGs) and their dependency and socio-economic status, and the attitude for the conservation of this wetland. The objective-oriented questionnaire was prepared after the rapid survey, with a total of 47 key questions, which were open-ended (with options), closed-ended (yes/no), and suggestive types. Later, an intensive survey was conducted through a specific questionnaire among the identified RUGs in these villages from July 2015 to July 2017. A total of 147 households were interviewed to gather information on various aspects. In addition, the local representatives of government agencies, leaders, social workers, and regular visitors were also interviewed.

2.3 Results

2.3.1 Climatic Variation

The monthly average rainfall and temperature over the study area are shown in Figures 2.5 and 2.6, respectively. The average monthly rainfall was 107.25 mm in 2012, 128.44 mm in 2013, 92.61 mm in 2014, 102.16 mm in 2015, and 101.24 mm in 2016, while the average monthly temperatures for the years 2013, 2014, 2015, and 2016 were 20.1, 20, 20.1, 20.6, and 20.94 °C, respectively.

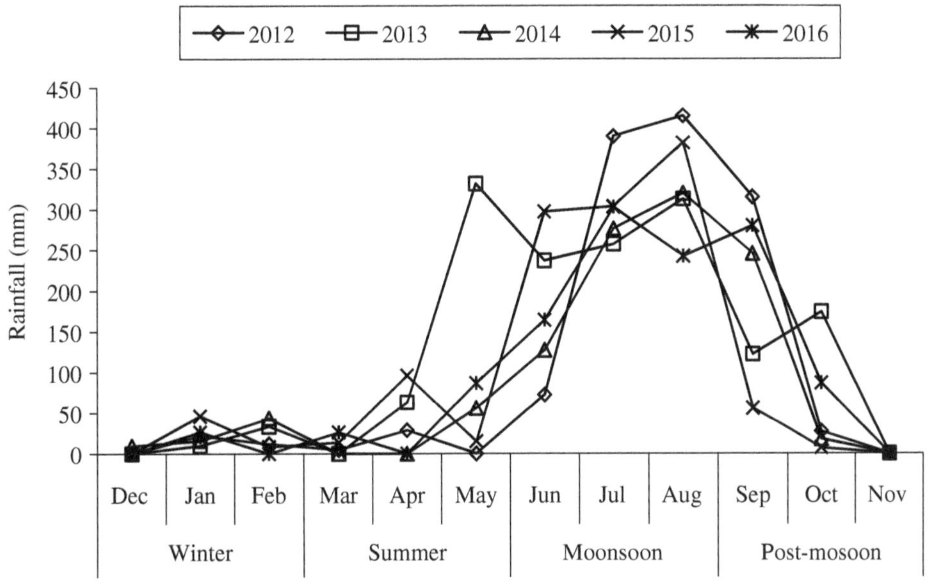

Figure 2.5 Monthly rainfall (mm) variations.

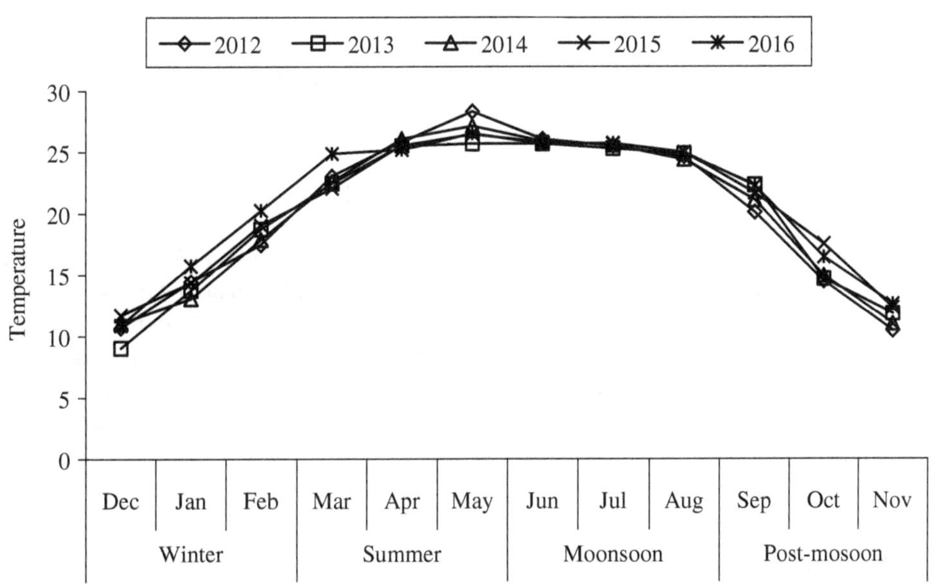

Figure 2.6 Monthly temperature variations.

2.3.2 Hydrological Variation

The hydrological regime of the wetland is highly variable throughout the year. Therefore, a contour line map of water availability in different months, i.e. October, December, March,

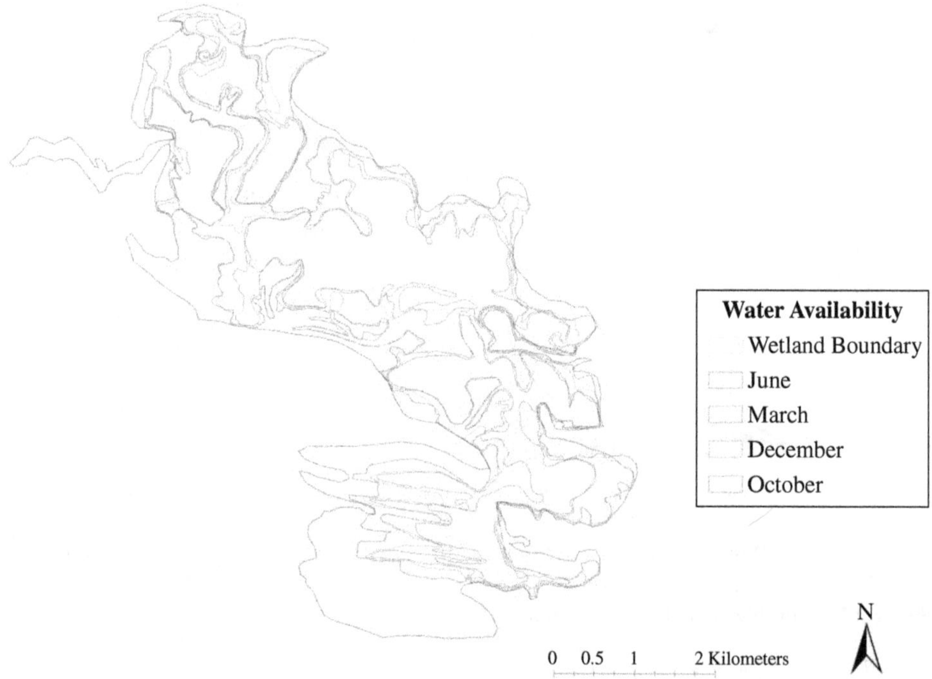

Figure 2.7 Contours lines based on water availability in different months in the Kanwar wetland by using Landsat 8 data from 2014 to 2017 for October, December, March, and June.

and June during the study period was prepared and depicted in Figure 2.7. The average water spread was recorded to the maximum extent in October. Later, it started decreasing in December, followed by June, and the minimum water spread was in March. This is an indication of un-uniform water availability and a pattern of siltation across the wetland. It also clarifies that in the lower stretch of this wetland, the pattern of siltation is more frequent, followed by the upper and middle stretches. The most frequent pattern of silt deposition in the lower stretch could be due to its existence toward the outlet channel of the wetland, which is not appropriately managed, and the runoffs from the whole catchment are drained to this part.

2.3.3 Land Use Land Cover Change

The natural land cover in the wetland area was identified into four classes, i.e. waterlogged, open/scrub, agriculture, and settlements. In this study, the change in the LULC pattern under the classes was analyzed for the years 1988 and 2016, which is depicted in Figure 2.8. In 1988, the maximum area (59.58%) of this wetland was covered with water, followed by scrub/open (39.79%), agriculture (0.39%), and settlement (0.25%), but in 2016, waterlogged and scrub/open area were reduced to 35.84 and 29%, respectively. The average rate of the LULC pattern is shown in Table 2.2.

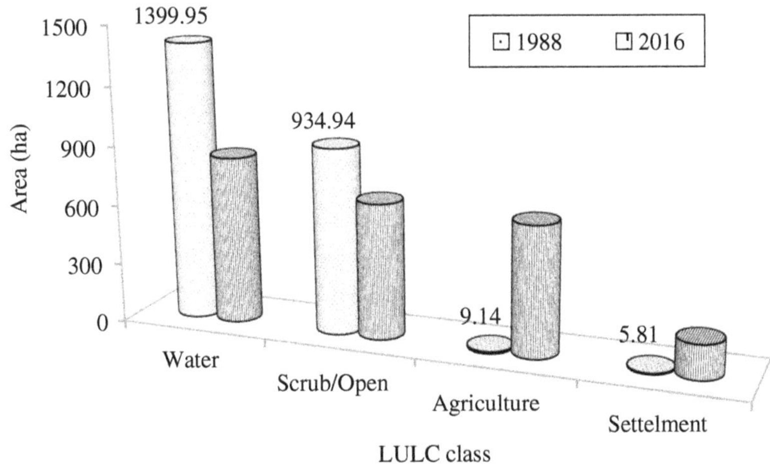

Figure 2.8 Area under four land use land cover change class in 1988 and 2016.

Table 2.2 Land use land cover change analysis in the Kanwar wetland between 1988 and 2016.

LU/LC class	Feb. 1988	Feb. 2016	Change	% Change	Yearly change (ha/year)
Water (ha)	1399.95	842.29	−557.66	−39.83	−19.92
Scrub/Open (ha)	934.94	681.67	−253.27	−27.09	−9.05
Agriculture (ha)	9.14	653.02	643.88	7044.64	22.99
Settlement (ha)	5.81	172.89	167.08	2875.73	5.97
Total area (ha)	2349.84	2349.87	–	–	–

2.3.4 Water Quality

The water samples were analyzed for 17 parameters and compared with standard permissible limits. Some of the parameters with the previously reported values for the same wetland in this study are shown in Table 2.3.

Further, an overall WQI was calculated by adopting standard procedures proposed in various studies (Tiwari and Mishra 1985; Dwivedi et al. 1997; Trivedi 2010; Shah and Joshi 2017) and is shown in Figure 2.9. The overall WQI was recorded in the medium

Table 2.3 Water quality parameters of the Kanwar wetland.

Parameters	Current study				Permissible limits	Year 2002[a]
Year	2014	2015	2016	Average ± SD	IS:2296[b]	
pH	7.89	8.41	8.25	8.18 ± 0.26	6.5–8.5	7.83
EC (μS/cm)	126.32	290.93	323.93	247.06 ± 105.86	1000–2250	204.24

(Continued)

Table 2.3 (Continued)

Parameters	Current study				Permissible limits	Year 2002[a]
Year	2014	2015	2016	Average ± SD	IS:2296[b]	
DO (mg/l)	7.24	7.24	3.92	6.13 ± 1.91	4–6	6.04
Alkalinity (mg/l)	298.28	244.39	218.89	253.85 ± 40.53	(200–600)[c]	122.9
Hardness (mg/l)	303.89	290.36	193.67	262.64 ± 60.11	300–600	130.28
Phosphate (mg/l)	0.07	0.01	0.10	0.06 ± 0.04	NA	0.0007
Nitrate (mg/l)	2.00	2.26	3.49	2.58 ± 0.79	20–50	0.0007
Sulfate (mg/l)	26.53	33.00	33.23	30.92 ± 3.80	200–400	1.19
Chloride (mg/l)	12.59	10.21	25.74	16.18 ± 8.37	250–600	13.12
Calcium (mg/l)	37.24	6.71	41.07	28.34 ± 18.83	75–200	25.73
Magnesium (mg/l)	9.53	4.27	5.22	6.34 ± 2.80	30–100	10.23
Cr (mg/l)	0.09	0.10	0.11	0.10 ± 0.01	0.05	NA
Mn (mg/l)	1.15	1.17	1.34	1.22 ± 0.10	0.5	NA
Cu (mg/l)	5.65	5.95	7.08	6.23 ± 0.76	1.5	NA
Zn (mg/l)	1.34	1.55	1.57	1.49 ± 0.13	15	NA
Pb (mg/l)	1.55	1.66	1.82	1.67 ± 0.14	0.1	NA
Cd (mg/l)	0.02	0.02	0.04	0.03 ± 0.01	0.01	NA

NA: not available.
[a] Ramakrishna et al. (2002).
[b] BIS (1991).
[c] IS:10500.

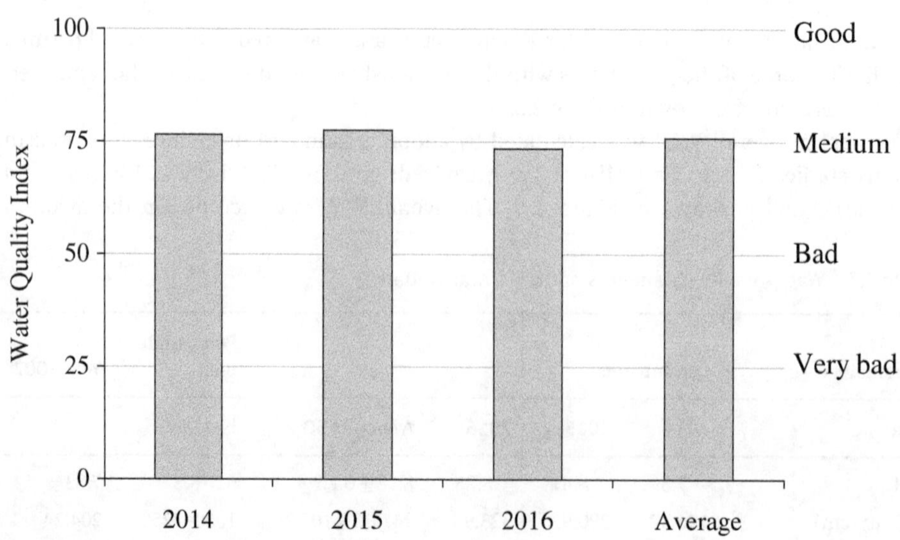

Figure 2.9 Water quality index of the Kanwar wetland.

range, which is not safe for direct consumption without disinfection as per the guidelines of the Water Environment Partnership in Asia (WEPA n.d.).

2.3.5 Sediment Quality

The sediment samples were analyzed for 15 parameters and compared with different sediment quality guidelines established by various agencies, which are summarized in Table 2.4.

2.3.6 Heavy Metal Contamination and Health Risk Index

The health risk assessment is directly associated with the level of human exposure of the various contaminants, especially heavy metals through different mediums including air, water, and food items. In this study, the heavy metals were also estimated in the aquatic species and most frequently cultivated vegetation species, i.e. sorghum, wheat, grains, and

Table 2.4 Sediment quality parameters of the Kanwar wetland.

Parameters	Current study					
Physical parameters						
WHC (%)	49.6–65.5					
Bulk density (g/cc)	0.77–0.86					
pH	6.5–8.76					
Electrical conductivity (μS/Cm)	115–136					
Macronutrients						
Organic carbon (%)	0.14–2.7					
Available N (mg/kg)	1110–5831					
Available P (mg/kg)	79.1–665					
Available K (mg/kg)	89–1017					
Available S (mg/kg)	82.2–226.5					
Micronutrients, i.e. heavy metals		Sediment quality guidelines				
		GRS[a]	GDS[b]	UCC[c]	Ontario MoE[d]	USEPA
Cr (mg/kg)	34–234	150	97	83	26–110	25–75
Mn (mg/kg)	217–1108	1715	1075	600	460–1100	NA
Cu (mg/kg)	14.6–118	55	25	25	16–110	25–50
Zn (mg/kg)	119–608	106	76	71	120–820	90–200
Pb (mg/kg)	15.6–223	23	NA	17	31–250	NA
Cd (mg/kg)	0–1.6	0.55	NA	0.098	0.6–10	NA

GRS: Ganga river sediments; GDS: Ganga delta sediment; UCC: upper continental crust; MoE: Ministry of Environment; USEPA: United States Environmental Protection Agency.

[a] Singh et al. (2003).

[b] Datta and Subramanian (1998).

[c] McLennan (2001).

[d] McDonald et al. (2000).

Table 2.5 Heavy metal content in various species of the Kanwar wetland.

Species	Cr (mg/kg)	Mn (mg/kg)	Cu (mg/kg)	Zn (mg/kg)	Pb (mg/kg)	Cd (mg/kg)
Hydrilla	111–164	116–148	38.5–55	9.6–12.85	133–158	0.89–0.98
Eichhornia	172–224	72–88	48–72	8.0–18	212–252	1.8–2.8
Lotus	28–46	26–42	12–31	6.0–22	114–163	1.02–1.8
Fish	22–36	12.0–26	5.3–21.5	6.24–20	146–172	0.97–1
Sorghum	58–92	12.6–23	22.5–52	4.0–15	217–264	1.01–1.26
Wheat	38.3–42	24–25.7	27.2–39	8.52–9.63	95.9–101	0.98–1.25
Maize	32–34.41	16.5–19	15.2–16.9	7.15–8.27	89–111	0.97–1.1
Rice	42–46.63	8.14–8.93	23–25.44	5.5–6.2	101–105	0.9–1.15
Permissible limit	1[a]	1[b]	30[b]	50[b]	5[c]	0.3[c]
Indian Standard[d]	20	NA	30	50	2.5	1.5

NA: not available.
[a] GB2762 (2012).
[b] CODEX (2001).
[c] FAO/WHO (2011).
[d] Awashthi (2000).

rice, and compared with the standard permissible limits and the previously reported values for the same crops (Table 2.5). Further, the HRI due to heavy metal contamination was calculated as per the standard formula (Khan et al. 2010; Singh et al. 2018) and the guidelines of the United States Environmental Protection Agency (USEPA 2006) for the cultivated crop grains of wheat, maize, and rice to understand the potential human health issues of the dependent population (Figure 2.10).

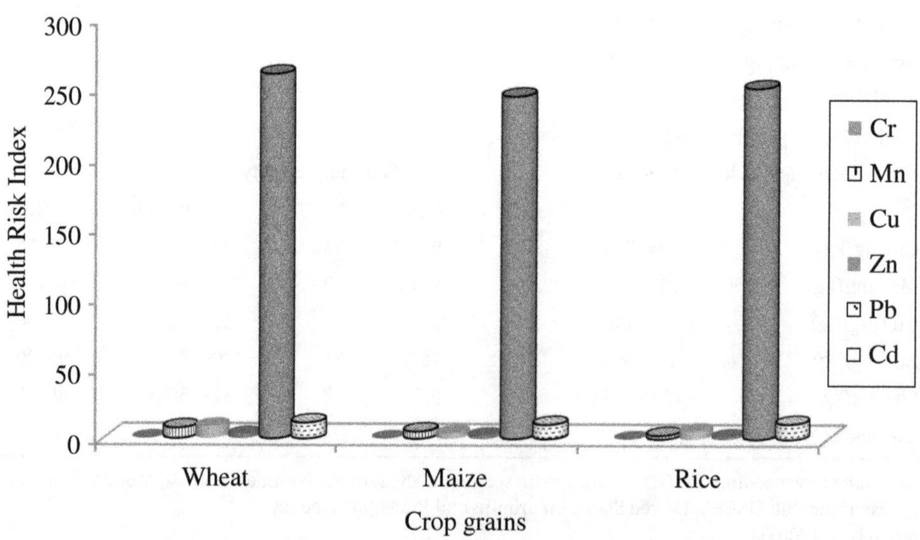

Figure 2.10 Health risk index of heavy metals via cultivated crop grains in the Kanwar wetland.

2.3.7 Socio-Economic Dependence and Anthropogenic Pressure

The economy of the nearby population is significantly dependent on the wetland resource. It was recorded that 70.1% of the population was directly involved in agricultural activities inside the wetland area during the lean period. Sometimes, their activity extended throughout the year if there was no flooding. The major crops cultivated were sugarcane, sorghum, maize, wheat, and paddy. During agriculture, a large number of chemical fertilizers of nitrogen, phosphorous, and potassium (NPK), biofertilizers, and different chemicals including pesticides/insecticides and herbicides were commonly used. The range of chemical fertilizer application varied from 170.97 to 864.85 kg/ha, while biofertilizer application was recorded between 350 and 635 kg/ha. Chemical fertilizers were used two to three times for each cultivated crop, whereas biofertilizers were used once at the time of cultivation. However, the maximum cultivators (61.22%) were not satisfied with the yield of crops.

The activity of bird poaching, practices to drain out the water, young fish dying due to pollution, and overexploitation of other resources like water, fish, and land were commonly recorded during the survey. The range of vegetation biomass collection was in the range of 75–210 kg/day/head, while the fish catch was highly variable with the season, ranging from 20–60 kg/day/head.

2.3.7.1 Demographic Status and Literacy

The change in village-wise population and number of households among ten villages between 2001 and 2011 is summarized in Table 2.6. The increase in population and number of households were largest at Kanausi and smallest at Manikpur villages. Whereas, the annual rate of increase in population and number of households among the ten villages in a decade were largest at Manjhaul and least at Jaimangalgarh (Table 2.6). The literacy rate (%) among the villages is shown in Figure 2.11. The maximum (51.75%) and minimum (39.15%) literacy rates were in Narayanpiper and Jaimangalpur, respectively.

2.3.7.2 Resource User Groups and their Socio-Economic Status and Dependency

All the responses of the questionnaire survey were summed up and RUGs were decided on a ranking basis (Fisher 2004). The Kanwar wetland resource users were categorized into three groups, namely Landowner, Laborer, and Business. The maximum (53%) respondents belonged to the labor group, followed by the landowner (27%) and small-scale business (20%) groups. The landowners had their registered land of a good amount (>20 katha, i.e. 0.3 ha), whereas the other two groups did not have registered land. These groups used to get land on lease from landowners. The labor group population was deprived in education, pakka housing, and land (Table 2.7). Their annual income was in the range of INR 12 500–25 000 household/year, which was less than the landowners (INR 25 000–125 000 household/year) and business (INR 15 000–50 000 household/year). The source of income of all RUGs was mainly based on agricultural yields. Most of the

Table 2.6 Changes in village-wise population and number of households.

S. no.	Village	Population					Number of households				
		2001	2011	Change	% change	Rate (per year)	2001	2011	Change	% change	Rate (per year)
1	Manjhaul	31 261	35 905	4644	14.86	465	5336	7372	2036	38.16	204
2	Jaimangalpur	1774	2309	535	30.16	53.5	373	507	134	35.92	14
3	Jaimangalgarh	538	731	193	35.87	20	130	175	45	34.62	5
4	Ekamba	4803	6923	2120	44.14	212	793	1471	678	85.50	68
5	Parora	4251	6353	2102	49.45	211	792	1275	483	60.98	49
6	Narayanpiper	5396	7451	2055	38.08	206	1015	1541	526	51.82	53
7	Sakra	3354	4416	1062	31.66	107	591	973	382	64.64	39
8	Rajaur	4420	6075	1655	37.44	166	832	1391	559	67.19	56
9	Kanausi	2179	3831	1652	75.81	166	454	857	403	88.77	41
10	Manikpur	2849	3150	301	10.57	31	507	611	104	20.51	11
	Total	**60 825**	**77 144**	**16 319**	**26.86**	**1632**	**10 823**	**16 173**	**5350**	**49.43**	**535**

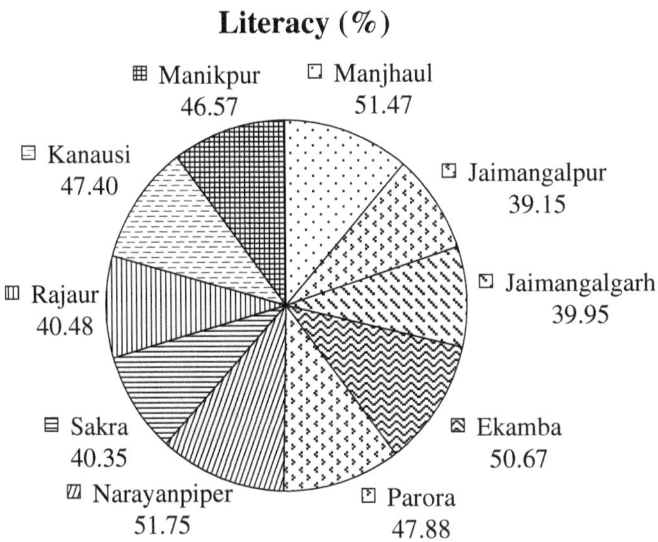

Figure 2.11 Village-wise literacy.

Table 2.7 Socio-economic characteristics of three resource user groups based on a questionnaire survey (N = 147).

S. no.	Characteristics	Labor	Landowner	Business
1	Households (%)	53.06	26.53	20.41
2	Average household size	7.2	6.4	6.6
3	Literacy (%)	31	46	42
4	Pakka houses (%)	31	81	41
5	Electrified households (%)	13	81	31
6	Cattle owning households (%)	82.4	76.34	41.65
7	Landholding (mean ± SD) (ha)	0.015 ± 0.003	1.6 ± 0.72	0.4 ± 0.08
8	Involved in agriculture (%)	92	100	85
9	Engaged in other services (%)	–	21	–
10	Unemployment (%)	78	–	–
11	Average household income (INR/household/year)	18 500 ± 2532	60 000 ± 2632	37 000 ± 1236

RUG population was unemployed and directly dependent for their livelihood on different materials from the wetland (Table 2.8).

2.3.7.3 Attitude for Conservation

The responses regarding the conservation of this wetland are summarized in Table 2.9. Most of the population (71.43%) was not aware of the protected status of this wetland, but they were very positive in their attitude to conserve it, with no compromises in access to the resources.

Table 2.8 Dependence of resource user groups (in %) on wetland resources and their use.

S. no.	Resources	Labor	Landowner	Business	Common uses
1	Biomass	86	15	62	Cattle fodder, fuel
2	Open land	72	92	76	Agriculture
3	Water	80	56	82	Irrigation
4	Construction material	60	43	60	House surface fill

Table 2.9 Attitude of people for conservation.

		Responses (in %); N = 147	
S. no.	Questions	Yes	No
1	About the awareness of the protected status of the wetland	28.57	71.43
2	Making arrangements to retain water throughout the year	61.22	38.78
3	Allowing cultivation on open land after the water recedes	89.8	10.2
4	Developing it as a tourist site	69.39	30.61
5	Whether any change has been observed in the agriculture pattern in the wetland	92.8	7.2
6	Complete draining of wetland and allowing the land for agriculture	10.2	89.8
7	Complete ban on the access to wetland resources	0	100

2.4 Discussion

Many of the wetlands in India support the livelihood of thousands of people every day, however, an increasing population and their encroachment into the wetlands either directly or indirectly pose a threat of degradation. In addition, the variations in climatic conditions, especially rainfall and temperature, which are crucial for maintaining water levels, is another important reason in the degradation of wetlands (Sarthi and Singh 2013). The average monthly precipitation over this area was in the range of 92.61–128.44 mm during 2012–2016, which is less than the expected precipitation (140 mm), and clearly indicates a water shortage. This site is a declared wildlife sanctuary and bird sanctuary with an area of 6311 ha by MoEF, the Government of India, and the Government of Bihar, however, there is still no clear-cut outer boundary demarcation, and only the core waterlogging area of 2350 ha is monitored by the agencies. This wetland is used for the water supply of irrigation and domestic purposes, fishing, netting of waterfowls, the harvesting of wild fruits like singhara (*Trapa natans*) and makhana (*Euryale ferox*), edible mollusks (*Pila globosa*) (Ambastha et al. 2007a, b; Kumar 2013a), and the cultivation of wild rice on open lands (Scott 1989). When this wetland was declared a protected area in 1986, all the landholders who lost their land were promised compensation under the provisions of the Land Acquisition Act (1894) by the government, however, many people are still claiming land ownership rights in the wetland area. There have been more than 15 legal petitions and approximately 1600 public objections are still pending before the local administration and court regarding these issues as per different reports. The rights for the access of different wetland resources were given to some of the local communities for fishing, vegetation biomass collection, and wild paddy cultivation in the wetland under the provisions of the Indian Wildlife (Protection) Act 1972, while land deals, bird trapping, and wild animal poaching were banned (Rakshi and Sanghi 1996; Kumar 2013b).

As per the report of the Centre for Monitoring Indian Economy (CMIE 2008), the villages surrounding the Kanwar wetland are highly populated and underdeveloped. The wetland has lost its connectivity to the main river and is now considered an oxbow and a rain-fed wetland in some of the literature (Kumar 2013a). The reduced rainfall over this region is the reason for water scarcity and the heavy siltation is the reason for the decrease in the water logging area. The relatively elevated areas of the wetland, locally called *Rahis*, and the low-lying areas, called *Dove*, were intensively encroached by people, which have altered the land-use dynamics of this wetland. The water and sediment quality is mainly becoming compromised with harmful heavy metals from different sources (Figure 2.12), which are posing a severe threat not only for the resources of the wetland but also to the health of the dependent population. The heavy siltation in different stretches, the premature death of fish, the intensive growth of weed biomass like *Eichhornia*, and the reporting of frequent human diseases of skin allergies, indigestion, blood pressure, and kidney and nervous disorders are very common among the population (Singh et al. 2018).

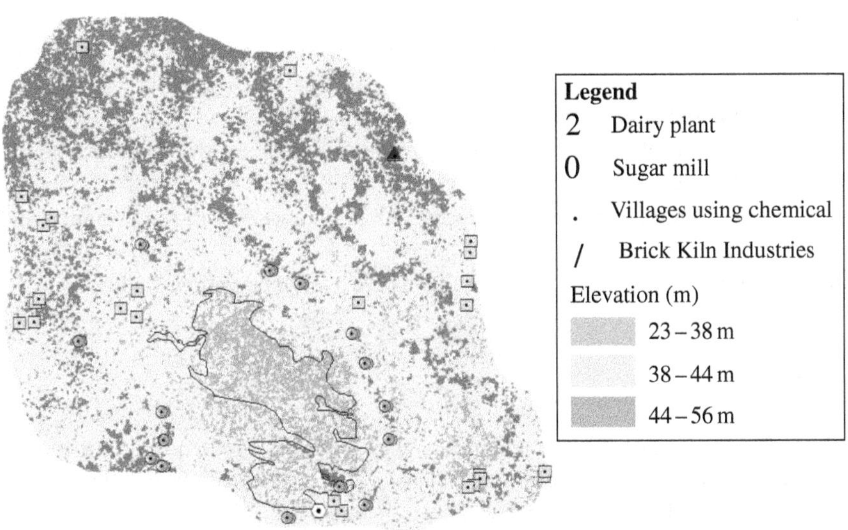

Figure 2.12 Elevation map of the Kanwar watershed showing possible sources of heavy metals. (Source: Based on Struhsaker et al. 2005.)

Even though most of the population around the wetland was highly dependent on its resources, they were still very positive and ready to cooperate with the administration in the co-management and conservation of the wetland. Therefore, a collaborative approach needs to be initiated to address the issues of this wetland and dependent population for its effective management and conservation as suggested by Struhsaker et al. (2005).

2.5 Conclusion

The conclusion of this study is that the Kanwar wetland is facing severe problems of water scarcity, heavy siltation, and enormous weed growth due to climatic variations and human encroachments. Water was being localized in patches during lean periods, i.e. summer, and open/scrub areas were used for other purposes by the local population. A significant change in LULC has taken place in the wetland. Different resources of this wetland like water, sediment, aquatic, and cultivated species of flora and fauna are contaminated with harmful heavy metals like lead (Pb) and cadmium (Cd).

Most of the nearby population is dependent on wetland resources for their livelihoods and they are willing to conserve it in cooperation with the administration. Therefore, an inclusive and collaborative approach should be initiated to conserve this important wetland, without compromising the interests of locals with the following ***recommendations***:

- The inlet channel of the wetland should be cleaned to allow water from the Basahi barrage to maintain the water level for the maximum time.
- A barrage across the outlet channel can help maintain the water level.
- An immediate de-siltation and weed removal program should be initiated.
- A clear-cut boundary demarcation should be made for the Kanwar wildlife sanctuary.
- Illegal land acquisition, agriculture, waste disposal, land digging, and bird trapping in the wetland should be stopped.
- The land disputes among different community groups in the wetland area should be resolved.
- The over-extraction of water from the wetland should be avoided.
- Conflicts among different sections of society for the utilization of the Kanwar wetland resources should be resolved.
- Awareness programs should be initiated to make people aware of the importance and values of wetlands in society.
- Eco-tourism should be promoted to provide alternative livelihood options for local people.

References

Alongi, D.M. (2009). Paradigm shifts in mangrove biology. In: Coastal Wetlands: An Integrated Ecosystem Approach (eds. G.M.E. Perillo, E. Walanski, D.R. Cahoon and M.M. Brinson), 615–640. Amsterdam: Elsevier.

Ambastha, K., Hussain, S.A., and Badola, R. (2007a). Resource dependence and attitudes of local people toward conservation of Kabartal wetland: a case study of Indo-Gangetic plains. *Wetlands Ecology and Management* 15: 287–302.

Ambastha, K., Hussain, S.A., and Badola, R. (2007b). Social and economic considerations in conserving wetlands of Indo-Gangetic plains: a case study of KabarTal wetland, India. *Environmentalist* 27: 261–273.

Anand, S. and Joshi, P.K. (2013). Remote sensing to quantify wetland loss. 14th Esri India User Conference, New Delhi.

APHA (2005). Standard Methods for the Examination of Water and Waste Water, 21ste. Washington, D.C: American Water Works Association/Water Environment Federation.

Awashthi, S.K. (2000). Prevention of Food Adulteration Act 37, 1954. Central and State Rules as Amended for 1999. Ashoka Law House, New Delhi.

BIS: Bureau of Indian Standards. IS: 2296 and IS: 10500 (1991). *Indian Standards for Surface Water*, New Delhi.

Bullock, A. (1992). Dambo hydrology in Southern Africa: review and reassessment. *Journal of Hydrology* 134: 373–396.

Bullock, A. and Acreman, M. (2003). The role of wetlands in the hydrological cycle. *Hydrology and Earth System Sciences* 7 (3): 358–389.

Carter, V. (1986). An overview of the hydrologic concerns related to wetlands in United States. *Canadian Journal of Botany* 64: 364–374.

Champion, H.G. and Seth, S.K. (1968). A Revised Survey of the Forest Types of India. New Delhi, India: Government of India Press.

Chandra, R., Prusty, B.A.K., and Azeez, P.A. (2009). Impact of prosopis juliflora on herbaceous diversity in Keoladeo National Park, Bharatpur. *Proceedings of the National Symposium on Prosopis: Ecological, Economic Significance, and Management Challenges* (20–21 February). Bhuj: GIDE.

CMIE. (2008). *Monthly reports, June 2008 to July 2009*. Mumbai: Economic Intelligence Services, Centre for monitoring Indian Economy.

CODEX Alimenttarius. CAC/GL 39.c (2001) *Maximum levels of heavy metals in cereals, pulses and legumes*. Joint FAO/WHO standards. Rome: FAO.

Collar, N.J. (2005). Threatened birds of Asia – The birdlife international Red Data Book. Cambridge: Bird Life International.

Craft, C.B. and Casey, W.P. (2000). Sediment and nutrient accumulation in the floodplain and depressional freshwater wetlands of Georgia, USA. *Wetlands* 20 (2): 323–332.

Datta, D.K. and Subramanian, V. (1998). Distribution and fraction of heavy metals in the surface sediments in the Ganges, Brahmaputra, Meghna river system in the Bengal basin. *Environmental Geology* 36: 93–101.

Dobermann, A., Sta, P.C.C., and Cassman, K.G. (1996). Nutrient cycling. *Agroecosystem* 46: 1–10.

Dwivedi, S., Tiwari, I.C., and Bhargava, D.S. (1997). Water quality of river Ganga at Varanasi. *Management Science* 6: 1–4.

FAO/WHO (2011). *Joint FAO/WHO food standards programme codex committee on contaminants in foods, 5th session*. Rome: FAO.

Finlayson, C.M., Davidson, N.C., and Stevenson N.J (2001). Wetland inventory, assessment and monitoring: Practical techniques and identification of major issues. *Proceedings of Workshop 4, 2nd International Conference on Wetlands and Development*, Dakar, Senegal (8–14 November 1998). Darwin: Environment Australia, Supervising Scientist.

Fisher, M. (2004). Household welfare and forest dependence in southern Malawi. *Environment and Development Economics* 9: 135–154.

Ford, J. and Bedford, B.L. (1987). The hydrology of Alaskan wetlands, U.S.A. *Arctic and Alpine Research* 19: 209–229.

GB 2762-2012. (2012). *Maximum Levels of Contaminants in Foods. National Food Safety Standards*. Beijing: National Standards of the People's Republic of China. www.chinesestandard.net.

Ghosh A.K., Bose, N., Singh, K.R.P., and Sinha, R.K. (2004). Study of spatio-temporal changes in the wetlands of north Bihar through remote sensing. ISCO 2004-13th *International Soil Conservation Organisation Conference* (July 2004), Brisbane. Braeside: Australian Society of Soil Science Incorporated and the International Erosion Control Association

Gilver, D.J. and Mclnnes, R.J. (1994). Wetland hydrological vulnerability and the use of classification procedures: a Scottish case study. *Journal of Environmental Management* 42: 403–414.

Giri, S. and Singh, A.K. (2014). Assessment of surface water quality using heavy metal pollution index in Subarnarekha River, India. *Water Quality, Exposure & Health* 5: 173–182.

Gopal, B. and Chauhan, M. (2006). Biodiversity and its conservation in the Sundarban mangrove ecosystem. *Aquatic Sciences* 68: 338–354.

Hilman, G.R. (1998). Floodwave attenuation by a wetland following a beaver dam failure on second order boreal stream. *Wetlands* 18: 21–34.

Khan, S., Rehman, S., Khan, A.Z. et al. (2010). Soil and vegetables enrichment with heavy metals from geological sources in Gilgit, Northern Pakistan. *Ecotoxicology and Environmental Safety* 73 (7): 1820–1827.

Kumar, M. (2013a). Resource inventory analysis of Kabartal wetland. *Research on Humanities and Social Sciences* 1 (8): 13–26.

Kumar, R. (2013b). Land deals in Kanwar bird sanctuary area banned. *The Times of India*, 13 January.

Kusler, J.A. and Riexinger, P. (1986). National wetland assessment symposium. Association of state wetland managers, New York.

Manson, F.J., Loneragan, N.R., Harch, B.D. et al. (2005). A broad scale analysis of links between coastal fisheries production and mangrove extent: a case study for north-eastern Australia. *Fisheries Research* 74: 69–85.

Mathew, M., Sathishkumar, M., and Azeez, P.A. (2002). Sediment quality of wetlands in Coimbatore, Tamilnadu, India. *Bulletin of Environmental Contamination and Toxicology* 68: 389–393.

McDonald, D.D., Ingersoll, C.G., and Berger, T.A. (2000). Development and evaluation of consensus based sediment quality guidelines for freshwater ecosystems. *Archives of Environmental Contamination and Toxicology* 39 (1): 20–31.

McLennan, S.M. (2001). Relationship between the trace element composition of sedimentary rocks and upper continental crust. *Geochemistry, Geophysics, Geosystems* 2 (4) https://doi.org/10.1029/2000GC0001092000GC000109.

Mitsch, W.J. and Gosselink, J.G. (1986). Wetlands, 539. New York: Van Nostrand Reinhold.

Mitsch, W.J. and Gosselink, J.G. (2000). The value of wetlands: importance of scale and landscape setting. *Ecological Economics* 35: 25–33.

Mohan, M., Augustine, T., Jayasooryan, K.K. et al. (2013). Fractionation of selected metals in the sediments of Cochin estuary and Periyar River, south-west coast of India. *Environmentalists* 32: 383–393.

Nasir, U.P. and Harikumar, P.S. (2011). Ecotoxicity and ecosystem health of a Ramsar wetland system of India. *Journal of Environment Protection* 2: 710–719.

National Wetland Atlas (2011). SAC/EPSA/ABHG/NWIA/ATLAS/34/2011. Ahmedabad: Space Applications Centre (ISRO).

Niraula, R. (2012). Evaluation of the limnological status of Beeshazar Lake, a Ramsar site in central Nepal. *Journal of Water Resource Protection* 4: 256–263.

NWCP (2009) *National Wetland Conservation Programme. Guidelines for Conservation and Management of Wetlands in India. Conservation and Survey Division.* New Delhi: Ministry of Environment and Forests, Government of India.

NWIA (2011). Information Brochure: National Wetland Inventory and Assessment. New Delhi: Ministry of Environment and Forest, Government of India.

Panigrahy, S., Murthy, T.V.R., Patel, J.G., and Singh, T.S. (2012). Wetlands of India: inventory and assessment at 1:50,000scale using geospatial techniques. *Current Science* 102 (6): 852–856.

Prasad, S.N., Ramachandra, T.V., Ahalya, N. et al. (2002). Conservation of wetlands of India: a review. *Tropical Ecology* 43 (1): 173–186.

Prusty, B.A.K., Chandra, R., and Azeez, P.A. (2010). Macronutrients along the sediment profile in a subtropical monsoonal wetland in India. *Wetlands Ecology and Management* 18: 91–105.

Rakshi and Sanghi (1996). Documentation of People's Priorities for Biodiversity and Conservation (BCPP). Remda, Daltonganj, India: Nature conservation Society.

Ramakrishna, Muley, E.V., Sddiqui, S.Z., and Vasanth, M. (2002). Wetland Ecosystem Series – 4, fauna of Kanwar Lake wetland, Bihar. Zoological Survey of India, 117–129.

Rodger, W.A. and Panwar, H.S. (1988a). Bio-Geographical Classification of India. New Forest, Dehradun: Wildlife Institute of India.

Rodger, W.A. and Panwar, H.S. (1988b). Planning Wildlife Protected Area Network in India, vol. 2. Dehradun: WII.

Roy, S.P., Ramanand, R., Prabhakar, A.K., and Singh, J.P. (2008). *Sustainable development of Kanwar Lake, Begusarai (North-Bihar, India)*, 2008, 1578–1585. In: Sengupta M, Dalwani R, editors. Kusatsu: Proceedings of Taal 2007: The 12th World Lake Conference.

Sarthi, P.P. and Singh, A.K. (2013). A simple approach about the characteristics of available surface water in Bihar state of India. *Geosciences* 3 (2): 68–76.

Scott, D.A. (1989). A Directory of Asian Wetlands. Cambridge, UK: IUCN.

Shah, R. (2012). *Wetlands: Refuge to existence.* India Water Week 2012 – Water, Energy and Food Security (10–14 April 2012). New Delhi: Ministry of Water Resources, Government of India.

Shah, K.A. and Joshi, G.S. (2017). Evaluation of water quality index for river Sabarmati. Gujarat, India. *Applied Water Science* https://doi.org/10.1007/s13201-015-0318-7.

Shardendu, S., Sayantan, D., Sharma, D., and Irfan, S. (2012). Luxuary uptake and removal of phosphorous from water column by representative aquatic plants and its implication for wetland management. *Soil Science* https://doi.org/10.5402/2012516947.

Singh, A.K. and Jayakumar, S. (2015). A review on methods to estimate CH_4 and N_2O flux in terrestrial ecosystems. *Climate Change & Environmental Sustainability* 3 (2): 104–113.

Singh, A.K. and Jayakumar, S. (2016). Water quality assessment of Kanwar Lake, Begusarai, Bihar, India. *Imperial Journal of Interdisciplinary Research* 2 (4): 793–803.

Singh, A.K. and Jayakumar, S. (2017). Heavy metal contamination in sediments of Kanwarjheel Wildlife Sanctuary, Bihar, India. *Pollution Research* 36 (1): 48–55.

Singh, J.P. and Roy, S.P. (1989). Faunistic composition and community structure of Kanwar Lake, Begusarai (Bihar). *Biology Bulletin of India* 11 (1): 1–8.

Singh, J.P. and Roy, S.P. (1990a). *Investigations on the Limnological Profiles of the Kanwar Lake, Begusarai, Bihar. Recent Trends in Limnology*. Muzaffarnagar: Society of Biosciences.

Singh, J.P. and Roy, S.P. (1990b). Some aspect of ecology of birds of Kanwar Lake, Begusarai, Bihar. *Journal of Freshwater Biology* 2 (3): 175–188.

Singh, J.P. and Roy, S.P. (1991a). Seasonal variation in the macrophytic biomass production and its effect on the water chemistry of Kanwar Lake, Begusarai, Bihar. *Environmental Ecology* 9 (1): 72–76.

Singh, J.P. and Roy, S.P. (1991b). Interaction between macrophytic biomass and macroinvertibrate abundance in Kanwar Lake, Begusarai, Bihar. *Journal of Freshwater Biology* 3 (3): 229–234.

Singh, M., Muller, G., and Singh, I.B. (2003). Geogenic distribution and baseline concentration of heavy metals in sediments of the Ganga river, India. *Journal of Geochemical Exploration* 80: 1–17.

Singh, A.K., Sathya, M., Verma, S., and Jayakumar, S. (2018). Health risk assessment of heavy metals in the crop grains grown on open soils of Kanwar wetland, India. *Euro-Mediterranean Journal for Environmental Integration* https://doi.org/10.1007/$41207-018-0073-x.

Singh, A.K., Sathya, M., Verma, S., and Jayakumar, S. (2020). Spatio-temporal variation of water quality index in Kanwar wetland, Begusaria, India. *Sustainable Water Resources Management* (Accepted).

Struhsaker, T.T., Struhsaker, P.J., and Siex, K.S. (2005). Conserving Africa's rain forests: problems in protected areas and possible solutions. *Biological Conservations* 123: 45–54.

Tiner, R.W. (2009). Global distribution of wetlands. In: Encyclopedia of Inland Waters, vol. 3 (ed. G.E. Likens), 526–530. Oxford: Elsevier.

Tiwari, T.N. and Mishra, M. (1985). A preliminary assignment of water quality index to major rivers. *Journal of Indian Water Resources Society* 1 (3): 34–38.

Trivedi, R.C. (2010). Water Quality of the Ganga River – An Overview. *Aquatic Ecosystem Health and Management* https://doi.org/10.1080/14634988.2010.528740.

US-EPA IRIS (2006) Integrated Risk Information System. www.epa.gov/iris/substS (accessed 24 September 2020).

Watkins, D. and Parish, F. (1999). Review of wetland inventory information in Asia. In: Global Review of Wetland Resources and Priorities for Wetland Inventory (eds. C.M. Finlyson and A.G. Spiers), 41. Canberra, Australia: Supervising Scientist report, 144.

WEPA (n.d.). Water Environment Partnership in Asia to raise well-being by catalyzing knowledge and action. http://wepa-db.net/3rd/en/index.html (accessed 24 September 2020).

3

Grossly Polluting Industries and Their Effect on Water Resources in India

Zeenat Arif[1], *Naresh Kumar Sethy*[1], *Swati*[2], *Pradeep Kumar Mishra*[1], *and Bhawna Verma*[1]

[1] *Department of Chemical Engineering and Technology, IIT (BHU), Varanasi, Uttar Pradesh, India*
[2] *Department of Botany, BHU, Varanasi, Uttar Pradesh, India*

3.1 Introduction

Water is one of the most essential resources for the existence of animal and plant life. The existence of plants, animals, and humans on Earth depends on nature playing an important role in providing water (Sharma et al. 2016). Rivers, lakes, and reservoirs are a major source of water in India. Some 80% of India's population is dependent on 14 major rivers for food and their livelihood. A growing population, accelerating industrialization, intensification of agriculture, and expanding urbanization exert heavy pressure on our vast but limited water resources. Despite so many priceless advantages, it is a poorly managed resource (Nwankwoala and Nwagbogwu 2012) because of its easy availability at a low cost. Both surface and groundwater resources are under stress. The increase in the growth rate of industrialization and urbanization created an unconstructive and adverse impact on water resources. The main reasons of water pollution are: (i) soil weathering; (ii) decomposition of organic materials; and (iii) untreated or not efficiently treated discharge effluent from industries, sewage, and municipal waste. From these mentioned causes, industrial pollution is considered as one of the factors causing the degradation of water quality (Järup 2000). Water discharge from industries and households carries chemicals and biological matter, which leads to an increase in demands on oxygen (DO) present in water, thereby reducing the level of DO due to biological oxygen demand (BOD) and chemical oxygen demand (COD) (Afzal et al. 2018).

In developing countries like India, the major contributors to water pollution are the end-points of effluents released from industries. Contaminated wastewater released from industrial areas creates negative impacts on human health and also makes the soil quality unsuitable for irrigation purposes and crops. So, it becomes crucially important to access the quality of water before directly discharging it into the environment (Bharti et al. 2013). Based on the industry type, many levels of pollutants are directly or indirectly discharged through public sewer lines into water bodies (Kumar et al. 2018. 2019). According to the

Pollutants and Water Management: Resources, Strategies and Scarcity, First Edition. Edited by Pardeep Singh, Rishikesh Singh, Vipin Kumar Singh, and Rahul Bhadouria.

State of India's Environment (SoE) 2019 report (www.downtoearth.org.in/news/pollution/grossly-polluting-industries-more-than-doubled-in-8-years-soe-in-figures-64962), there has been a substantial increase in the number of grossly polluting industries (GPIs) between 2011 and 2018. There was a 136% increase in the number of GPIs over the period for the demands of the increasing population in the country.

GPIs are defined as industries that discharge more than 100 000 L of wastewater and/or hazardous chemicals into rivers or water bodies. The different types of GPIs are pulp and paper mills, distilleries, sugar mills, textile units, tanneries, thermal power plants, food, dairy, and beverage industries, chemical units, slaughterhouses, etc. As a result of this, the freshwater that Indian rivers carry is now so severely polluted that rivers threaten the very life they once nurtured. There are 992 GPIs in the river Ganga and its tributaries, i.e. Kali and Ramganga, and, from these, 851 industrial units are found in Uttar Pradesh, 61 in Uttarakhand, 43 in West Bengal, and 40 in Bihar (www.downtoearth.org.in/news/pollution/faith-to-filth-thanks-to-kumbh-prayagraj-sinks-in-solid-waste-64579). The major water pollution contributors in Kosi, Ramganga, and Kali river catchments are tanneries, sugar and distilleries, and pulp and paper units. Panigrahi and Pattnaik (2017) in their report mentioned that the Gomati River near Lucknow urban center is mostly polluted by heavy metals. Malik et al. (2014) mentioned that many industries are situated near the bank of the Yamuna River, therefore, it is considered the most polluted river in India. Ahirwar et al. (2015) state the major cause of water pollution in the River Son is due to effluents discharged from thermal power plants and pulp and paper industries. Therefore, it can be said that water body pollution due to industrial discharge affects the food web and hampers public health (Ahirwar et al. 2015). It becomes crucially important to access and control the quality of water bodies by proper treatment of the wastewater at the source itself before directly discharging it to the environment (Bharti et al. 2013).

This chapter examines the major contributing wastewater industries, their present status, and the major pollutants released from different types of industries situated in India. The impact of pollutant discharge from these industries into rivers, drains, and so on, are also discussed in detail. The current status and quality of rivers including the River Ganga and its tributaries, Yamuna, Cauvery, Narmada, and Bandi are also highlighted. The role of different legal authorities and monitoring bodies to control the water pollution caused by industries and their rules, regulations, and drawbacks are explained in the last section of the chapter.

3.2 Industrialization in India

Over the last century, industrial production in India has risen more than 50-fold. India has made rapid industrialization strides and is one of the world's ten most developing nations. However, poor monitoring and control of emissions combined with unplanned growth have resulted in significant harm to the environment.

Industrialization is the period of social and economic transition that transforms a human civilization from an agrarian community into an industrial culture, including the comprehensive rearrangement of an economy into an industrial culture (Bhandari and Garg 2015). Industrialization in a country leads to increases in urbanization, extraction of natural resources, industrial growth, and environmental pollution, and a decrease in environmental suitability.

Industrialization is on the expansion as is environmental pollution because of emissions and waste produced from these industries. Industrial pollution because of its higher production can possibly cause irreversible responses in nature and henceforth represents a significant risk to sustainable development (Baek et al. 2009).

3.3 Categorization of Industries

The Central Pollution Board (CPCB) listed 17 groups of the most polluting industrial sectors that contribute to environmental degradation in terms of particulate matter, gases, and effluents. Currently, 77% of industries contribute to water pollution, while 15% contribute to air pollution and the remaining 8% to air and water contamination (Annual Report 2004–2009).

The industries that rely on natural resources are the most polluting and are rising rapidly. Depending upon the pollution load generated and the water extracted by industries, they are classified into three major groups: (i) major water-intensive industries, including tanneries, slaughterhouses, pulp and paper, textile, breweries and soft drinks, etc.; (ii) moderate water-consuming industries, including canning industries, roller and flour mills, and food products; (iii) industries using water for sanitation, including electrical and electronic industries, woods, rubber, glass, wax, paint, and plastic (http://14.139.116.20:8080/jspui/bitstream/10603/33831/4/chapter4.pdf).

The Government of India, Ministry of Environment Forestry and Climate Change (MoEFCC) classifies all industries as red, orange, and green based on their ability to cause pollution: red –high polluting; orange – less polluting; and green – nonpolluting industries. The MoEFCC has developed industrial sector categorization criteria based on the Pollution Index which is a feature of emissions (air pollutants), effluents (water pollutants), produced hazardous waste, and resource consumption. They are classified as follows (MOI 2019).

- Industrial sectors having a Pollution Index score of 60 and above – red category
- Industrial sectors having a Pollution Index score of 41–59 – orange category
- Industrial sectors having a Pollution Index score of 21–40 – green category
- Industrial sectors having a Pollution Index score ≤20 – white category

3.4 Criteria for Determination of Grossly Polluting Industries

GPIs are classified as effluent discharging industries into a waterway and (i) handling hazardous substances, (ii) effluent with a BOD load of 100 kg per day or more, or (iii) a combination of (i) and (ii) according to the CPCB Annual Report (2004–2009).

According to the latest data from 36 states/UTs obtained by the CPCB, 2743 industries have been listed as GPIs for routine monitoring by CPCB/SPCBs/PCCs. Of those, 2497 units are still in service and 246 units have been closed by themselves. Of the operating units, it is stated that 2222 units comply with environmental standards. There has been a 136% increase in the number of GPIs. In 2011, only 22 states/UTs were covered to identify industries, whereas, in 2019, the data received belonged to 36 states, this implies that there has been increased industrialization/development over this time in India. Also, compliance status (based on discharge limit parameters and the validity of consent letters) increased to 89% in 2018 – it was only 79.8% in 2011 (Groundwater Year Book – India 2018–2019).

3.4.1 State Distribution of Grossly Polluting Industries in India

The graph shown in Figure 3.1 represents the state presentation of GPIs in India in 2019. Table 3.1 depicts four major states comprising a large number of GPIs as per the CPCB report 2019.

The study of state and category distribution shows that the states of Maharashtra, Uttar Pradesh, Gujarat, Andhra Pradesh, and Tamil Nadu have significantly large numbers of industries in the sectors previously listed and that the tannery, textile, and sugar sectors have the maximum number of industries, followed by pharmaceuticals, distillery, cement, and fertilizer. It also indicates that agro-based and chemical industries have major shares of 47% and 37%, respectively, of the total number of industries. Table 3.1 shows that Uttar Pradesh comprises nearly 43.7% of GPIs in India followed by Haryana, Gujarat, and Andhra Pradesh.

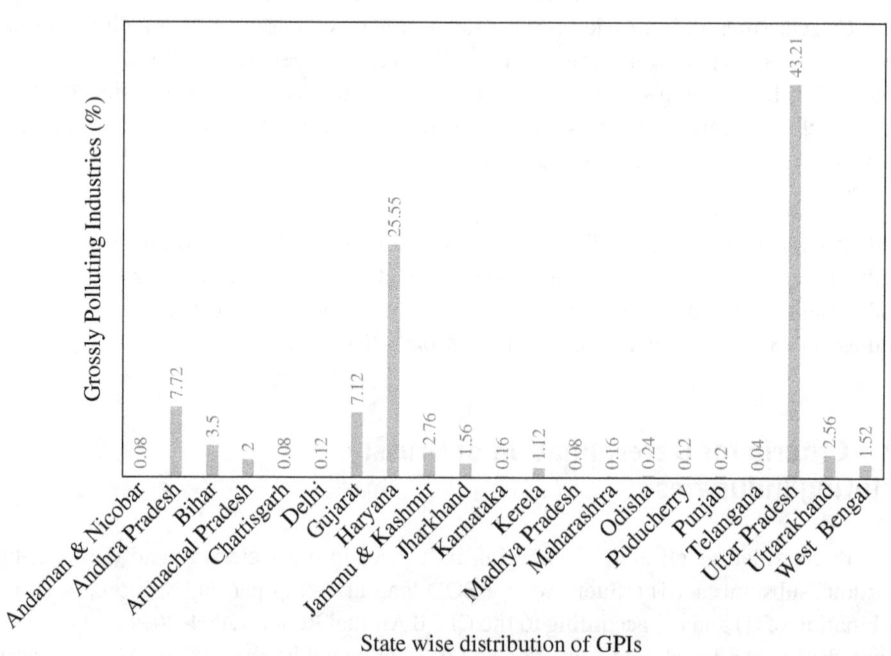

Figure 3.1 State distribution of grossly polluting industries in India.

Table 3.1 Four major states comprising large numbers of grossly polluting industries as per the Central Pollution Control Board report 2019.

States	Total no. of industries	No. of operational industries	No. of complying industries	No. of noncomplying industries
Andhra Pradesh	198	193	189	4
Gujarat	191	178	139	39
Haryana	660	638	625	13
Uttar Pradesh	1218	1079	950	129
Total GPIs in India	2743	2497	2222	275

3.5 Different Type of Grossly Polluting Industries and their Impact on Water Bodies

MoEFCC notified 17 categories of GPIs engaged in different production systems (https://cpcbenvis.nic.in/cpcb_newsletters/PollutingIndustries.pdf):

- Chlor alkali/copper smelting
- Integrated iron industry/dyes and intermediates
- Sugar
- Distilleries
- Fertilizers
- Cement plants
- Petrochemicals
- Pulp and paper
- Pesticides
- Tanneries
- Textile and dye industries
- Oil refineries
- Pharmaceutical
- Zinc smelting
- Copper smelting
- Aluminum smelting
- Thermal power plants

Most of these industries are situated near riverbanks where the untreated/partially treated wastewater is discharged directly into river bodies thereby deteriorating water quality (Paul 2017). Wastewater generated from industries results from the production process and the final effluent includes both residual wastewater and final treated effluent. The report of the Ministry of Water Resource (MoWR) concluded that around 62% of wastewater is generated from large and medium-scale industries whereas small-scale industries generate 32% of wastewater. Water consumption in the Indian industrial sector is 2–3.5 more per unit of production compared with similar plants operating in other countries (www.downtoearth.org.in/news/pollution/grossly-polluting-industries-more-than-doubled-in-8-years-soe-in-figures-64962). Some major industries and their discharged pollutants are discussed in the following sections.

3.5.1 The Textile Industry

In India, the textile industry ranks in second place for providing employment, with 45 million people, and according to estimates, it will increase to 55 million by 2020 (Invest India 2019a); 2.3% contribution was made by the textiles industry and 13% to industrial production of gross domestic product (GDP) contribution. During the manufacturing process, a huge quantity of water is utilized, while, at the same time, water effluent is generated that consists of dyes, heavy metals, organic and inorganic wastes, detergents, and surfactants. According to Manikandan et al. (2015), the quality of water released from textile units are turbid, loaded with high organic and inorganic constituents, and possess a low BOD/COD ratio, which means that a large proportion of the water is nonbiodegradable and is not fit for any application if discharged directly. According to the study carried out by Kumar et al. (2020), it was concluded that for an average production capacity of 8000 kg of fabric/day, the amount of freshwater utilized is around 1.6 million liters/day, which generates 200–350 m^3 of wastewater/ton of finished products, thus introducing 100 kg COD/ton of fabric to the water body.

3.5.2 The Leather Industry

India's contribution to the leather industry is 12.9% of the world's total leather manufacture, which produces around 3 billion square feet of leather yearly and provides employment to more than 4 million individuals (Invest India 2019b). The major pollutants present in wastewater discharged from the tanneries are chromium, salts, solids, sulfides, alkalinity, lime, etc. The characteristics of tannery wastewater are high chromium (Cr) concentration (toxic element having an adverse effect on flora and fauna and to the human beings), high BOD and COD, high total dissolved solids (TDS), and high total suspended solids (TSS). Kanpur's leather tanning industries are among the major contributors to river pollution and are classified by the MoEFCC as "Red Category" industries in India (MoEFCC 2017). Untreated or potentially processed effluents have risen to 20 times the number of contaminants, such as cyanide and chromium, in 22 highly contaminated regions of the world. It increases the BOD from 10–190 times, which can destroy human life in the affected region in severe situations. The research conducted by Dandira and Madanhire in 2013 also reported that COD in wastewater discharged via tanneries is in the range of 1000–43000 mg/l, which is 25–275 times greater than normal COD (Chaudhary and Walker 2019).

3.5.3 Food-Related Industries

The growth rate of the food industry in India was 322 billion dollars in 2016 and is estimated to increase to 543 billion dollars in 2020. Different types of food processing industries are grain, beverage, dairy, and edible oil industries (Invest India 2020). Different wastes released from food-processing industries are poultry, dairy, meat, sugar processing, etc.

All these wastes are highly degradable and when they enter water bodies, they decrease the DO level to a significant extent. Major industrial discharges are organic contaminants rich in fats, proteins, and pathogens. The quality of effluents from food processing units have moderate to high BOD, high COD, TSS, and TDS, and large nutrient contents such as nitrogen, phosphorous, heavy grease, and oil.

3.5.4 The Metal Industry

The major processes involved in the metal industry are mining, ore refining, alloying, casting, and sheet/folding. The most widely used techniques for the manufacture of metals are open-hearth furnaces, oxygen incinerators, blast furnaces, and electric arc boilers (MOI 2015). An excess amount of water is generated from metal processing; the water is highly acidic and rich in toxic elements, for example, cyanogens, phenols, coke, metal, soluble bases, limestone, oils, etc.

3.5.5 The Paper and Pulp Industry

Rice husk, wood, cellulose, fibers, and waste papers are the main materials used for paper production. The paper production process is a water exhaustive process as large quantities of water are required in the production of paper. As a result, the water discharged from the paper industry is contaminated with trichlorophenol, trichloroguicol, dichlorophenol, dichloroguicol, pentachlorophenol, dibenzo-p-dioxin, and benzofuran, which are responsible for increases in BOD and COD, and suspended solid (SS) level, which, if entering the human body through the food chain, can lead to reproductive problems and cancer (PrintWeek India 2018).

3.5.6 The Sugar Industry

In India, the total production of sugar is 4.1 million tons of sugar from 380 factories. Molasses, alcohol, other by-products, and liquid wastes are the effluents generated from different steps in the sugar process which cause pollution to local rivers and lakes. The use of phosphoric acid and sulfur dioxide during the production of sugar subsequently increases BOD, COD, SS, unpleasant odors, and heavy metals (Singh et al. 2018).

3.5.7 Thermal Power Plants

According to Singh et al. (2018), in India, there are more than 200 power plants (116 coal-based, 65 gas-based, and 19 diesel-based). The discharge of hot water from these power plants causes subsequent changes in the DO level. Other pollutants released from thermal power plants are coal dust, fly ash, and chlorinated water, and contamination of water due to heavy metals like Hg, Bi, As, and Cr are the main constituents (Sonwane 2015).

3.5.8 The Petroleum and Oil Industry

In India, the demand for oil is set to increase by 4% during 2016–2030. It was reported that petroleum product production increased from 243.5 million metric tons (MMT) in 2016–2017 to 262.3 MMT in 2018–2019 (Invest India 2019c). Pollutants that are released at each step, from starting manufacturing to transportation, are oil spillage, saline water, and phenolic mixtures. Their discharge causes a poisonous effect on plants, aquatic life, and human life, and may lead to tumor formation, cancer, and mutation.

3.6 Major Water Body Pollution Due to Grossly Polluting Industries

3.6.1 The Status of Water Resources in India

The country's inland water resources are listed as rivers and canals; reservoirs; dams and ponds; bells, oxbow lakes; water that is unsafe; and brackish water. Aside from rivers and canals, total water bodies occupy an area of approximately 7 million hectares (Chaudhary and Walker 2019; Gagan 2015). There are few flowing rivers in the desert and they are lost in the desert. There are whole arid regions where evaporation is equal to rainfall, and thus, no airflow. Coastal rivers have medium and small river basins (Chaudhary and Walker 2019).

Together, the rivers Brahmaputra, Ganga, Indus, and Godavari occupy more than half of the country's territory. The entire west coast, extending 1500 km between Surat in Gujarat and Cape Comorin in Tamilnadu, is fed by 14 medium and 18 minor river basins, leaving major river basins including Mumbai, Panji, Cochin, and Trivandrum. On the east coast of Peninsular India, three places have large river basins (Gagan 2015). These three pockets are the area south of the River Cauvery from Madurai to Cape Comorin; the area between the basin of Penner and Cauvery where Chennai and Pondicherry are located; and the area between the basins of Mahanadi and Godavari in Orissa. There is a tremendous variation both in the quantity of discharge from a major basin to a minor one and in the quality of discharge from region to region.

3.6.2 The Quality of Some Major Rivers in India Due to Grossly Polluting Industries Wastewater Discharge and their Impact

The Himalaya, often called as "The water tower of India," is situated at the Indian north subcontinent and is the main source for the largest rivers. These are the Indus, the Ganges, and the Brahmaputra (Panigrahi and Pattnaik 2017). The water quality of these rivers and other major rivers of India are discussed in detail in the following section.

The Ganga: The river covers (26.3%) of the total geographical area of the country. It is considered as the sacred and holy river in India and covers the states of Uttar Pradesh, Uttarakhand, Delhi, some portions of Haryana and Punjab, Himachal Pradesh, Madhya Pradesh, Rajasthan, Bihar, and West Bengal (Agarwal 2015). The Gamukh, the ice cave of Gangotri, is considered as the origin of the River Ganga. Water quality of the Ganga with respect to levels of coliform bacteria was found to be in excess (greater than two lakh). Approximately 260 million liters per day of industrial wastewater is directly discharged into the river (Srivastava et al. 2016). Season-wise, COD ranges between 12.5 and 65 mg/l, chloride concentration ranges from 14 to 25 mg/l, and sulfate concentration ranges from 25 to 38 mg/l (Srivastava et al. 2016). Total hardness lies in the range of 114–184 mg/l (Srivastava et al. 2016). Kanpur, a city in Uttar Pradesh, is the hub of tanneries and textile units where the Ganga is found to be highly polluted due to industrial discharge. High levels of coliform of up to 100 000 000 per 100 ml and an average BOD of 40 mg/l in Varanasi were reported in a study by Agarwal (2015).

The Brahmaputra: By discharge, the Brahmaputra is considered as the largest river in the world (Panigrahi and Pattnaik 2019). Guwahati city is situated on the banks of the river. According to a study by Kotoky and Sarma (2017), the physicochemical characteristics of the river are mean a pH of 9.37 and mean a DO of 10.76 mg/l. The main cause of river pollution is the discharge of crude oil along with agriculture and municipal run-off (Kotoky and Sarma 2017).

Yamuna River: The origin of the Yamuna River is the Yamnotri glacier and the different states of India covered by the river are Rajasthan, Uttar Pradesh, Himachal Pradesh, Delhi, Haryana, and Madhya Pradesh (Panigrahi and Pattnaik 2019). There are different industries situated on the banks of the Yamuna River, including electroplating industries, thereby directly increasing the Cr concentration of the river, causing a threat to human health when entering the body through the food chain (Malik et al. 2014). There are almost 359 factories that discharge their effluent directly or indirectly at the Yamuna, further intensifying pollution loads in the Ganga (Maheshwari 2016).

Kali River: The origin of the Kali River is from the village of Antwada, in the Muzaffarnagar district in western Uttar Pradesh. Before meeting the Ganga, it covers the districts of Meerut, Ghaziabad, Bulandshahr, Aligarh, Etah, Farrukhabad, and Kannauj. Different types of effluent are discharged from sugar, distillery, and textile industries. Slaughterhouses and chemical units have made the Kali River almost dead. The water has been deprived of oxygen and has turned into a black color. The concentration of heavy metal is currently exceeding the permissible limit. For example, the Cr level in Muzaffarnagar is 30 times higher than the permissible value, while in Meerut, the Pb concentration is almost 100 times higher than the standard value (Mishra et al. 2015).

Son River: Son River, a tributary of the River Ganga, originates from the Amarkantak plateau. Metal industries are the main cause of Son River pollution (Ahirwar et al. 2015). As per the study analysis carried out by Ahirwar et al. (2015), the TDS (898.5 mg/l) and hardness of the Son River was found to be more than the permissible limit and the other mean values of parameters were a pH of 7.19, Fe of 0.2484 mg/l, Cr of 0.0026, Mg of 14.5 mg/l, and Na of 76.24 mg/l were also higher (Ahirwar et al. 2015).

Gomti River: This river mainly carries the industrial waste pollution from eastern Uttar Pradesh. The water samples of the Gomti River were found to be rich in heavy metals like Cr (2.22–19.13 mg/l), Fe (1606–3412 mg/l), Mn (82.6–263.9 mg/l), Pb (6.3–75.3 mg/l), and Ni and Zn (3.1–101.7 mg/l) (Panigrahi and Pattnaik 2019).

Brahmani River: The characteristics of the Brahmani River area are as follows: a pH of 7.6; mean fluorine of 0.49 mg/l; nitrate of 4.18 mg/l; and mean alkalinity of 48.48 mg/l (Nath et al. 2018). Nitrate concentration is more than the permissible limit of this river which adversely affects fish diversity (Das et al. 2016).

Mahanadi River: The river originates from the Chhattisgarh basin. Cuttack city is located on the banks of the Mahanadi delta. Rourkela steel plant effluents and biomedical waste (Panigrahi and Patra 2013) are the main causes of Mahanadi River pollution. The water of the Mahanadi River is turning toward a condition of eutrophication (Panigrahi and Pattnaik 2019).

Cauvery River: The river originates from Brahmagiri hill. The pH value of the river varies from 6.5 to 9.4. In Salem district, the high-level discharge of industrial effluents badly affects the planktonic population (Panigrahi and Pattnaik 2019). Different heavy metals have been found in the Cauvery River in excess amounts, such as Cr, Co, Cu, Mn, Ni, Zn, and Pb, which contaminate the water body.

Bandi River: The discharge effluents of textile dyeing units are the main cause of pollution in Bandi River, which is situated at Pali. Water quality deterioration of the Bandi River has an adverse effect on human beings and aquatic ecosystems. Textile dyeing industries have production capacities ranging between 725 and 3625 kg of cloth/day, while abstracting water that comes from this river is in the range of 30–275 m^3/day and daily discharging wastewater has a high BOD and COD ranging from 41–76 l/kg.

Narmada River: The importance of this river can be identified by the fact that pilgrims performed a holy pilgrimage called Narmada Parikrama there. Along with industrial discharge, the other reasons for the deterioration of water quality of the Narmada River are human activities like the disposal of dead bodies and municipal wastewater. People in the adjacent area of the Narmada River suffer from health problems like gastric ulcers, skin problems, respiratory problems, diarrhea, and stomach problems (Katakwar 2016). The other major rivers of Gujarat, namely Sabarmati, Tapi, Mahisagar, and Damanganga, have been found to have a COD level in an extremely critical state (Pimparkar et al. 2016).

Damodar River: This river is the most polluted river after the Ganga. The level of coliform is higher due to wastewater discharged from cement plants, coal washing plants, and other industries. This leads to an increase in COD value, dissolved and suspended solids, and hardness. The other characteristics of water are pH in the range of 7.4–8.9, hardness ranges between 133 and 327 mg/l, Cr concentration between 30 and 160 μg/l, Cu concentration was found in the range of 10–56 μg/l, and Pb was between 38 and 146 μg/l (Panigrahi and Pattnaik 2019).

The status of other rivers and their tributaries contaminated from different types of industrial effluent in different states is explained in Table 3.2.

Impacts on rivers due to industrialization (Edokpayi et al. 2010; Gagan et al. 2016; Mali et al. 2015; Taskeena et al. 2017):

- Industries are dumping toxic waste into water bodies, as a result of which, the DO of surface waters is falling at an incredible rate, endangering aquatic plants and animals.
- Apart from the DO, the chemical composition of water is often modified, which deteriorates the water quality making it hazardous.
- This has resulted in significant ecological degradation, such as a decline in water quality and availability, extreme floods, habitat loss, and changes in the distribution and structure of aquatic biota rendering surface watercourses unsustainable in the provision of goods and services.
- Using poor-quality water in agriculture can affect the crop yield and also causes food insecurity.
- Toxic substances in water bodies may lead to a high population growth of pathogens in the water and the use of this contaminated water for domestic and other purposes is dangerous for human health and society at large.
- This also affects wildlife groups that use surface water for drinking or as habitat.
- Degradation of water quality, water supply degradation, and the destruction of marine biodiversity are influential environmental ecosystem characteristics that need immediate action at both global and regional levels.

Table 3.2 Impact of different types of industries on water bodies.

Type of industry	Effects
1) Chemicals (dyes, paints, fertilizers, etc.)/ Ankleshwar, Gujarat state	Amla khadi (a rivulet flowing through the industrial estate) in Ankleshwar carries extremely toxic, often acidic, dark brown or black effluent throughout the year
2) Foundry/Howrah, West Bengal state	Air heavily polluted by particulate matter, water contaminated by heavy metals viz. lead, cadmium, chromium, iron, manganese, and lead. Pollution potentially affecting about 1 million people in Howrah city and 4 million in Howrah district.
3) Chemicals (organic and inorganic, dyes, and pharmaceuticals)/ Nandesari, Gujarat state	The Mahi River is contaminated by industrial effluents. Due to soil and groundwater contamination, both crop quantity and quality have declined tremendously, contributing to poverty in the region and forcing the local population to rely more heavily on industry for economic subsistence. This is harmful to humans because the polluted water leads to skin allergies, breathing and circulatory disorders, kidney problems, gastrointestinal disease, chronic stomach problems, and diarrhea, which have all been noted in communities.
4) Chemicals/Panipat, Haryana state	The concentration of mercury in the sample taken from a tube well near an industrial area in Panipat was 0.2683 mg/l, more than 268 times the permissible limit of 0.001 mg/l set by the World Health Organization for drinking water, according to the nongovernmental organization, the Centre for Science and Environment.
5) Mainly tanneries and textile dyeing/Palar River Basin, Tamil Nadu, India	According to a study sponsored by the Asian Development Bank, pollution loads in the Palar river are extremely high: total suspended solids, 29 938 kg/day; total dissolved solids, 400 302 kg/day; chloride, 101 434 kg/day; sulfide, 3818 kg/day; BOD, 23 496 kg/day; COD, 70 990 kg/day; total chromium, 474 kg/day; and cyanide: 22 kg/day
6) Chemicals/Vapi, Gujarat, India	Sediment and effluent in a channel bypassing the common effluent treatment plant and flowing into the river are heavily contaminated with cadmium, chromium, copper, lead, mercury, nickel, and zinc. Greenpeace declared Vapi a Global Toxic Hotspot in 1999.

Source: Rajaram and Das (2008).

3.6.3 Quality of Groundwater Resources in India Due to Wastewater Discharge by Grossly Polluting Industries and Its Impact

Groundwater, which is a precious and essential source of drinking water, must be cautiously maintained to maintain its quality under normal limits. In many parts of India, water available for industrial activities has reached a level of crisis. The Central Ground Water Board (CGWB) reports states that out of the total replenishable groundwater (431.42 billion cubic meters), the percentage of total replenishable groundwater in the Ganga Basin is 39.6% and the other four basins are the Godavari at 9.42%, Brahmputra (6.15%), Indus (6.14%), and Krishna at 6.12% having more than 5% of total replenishable

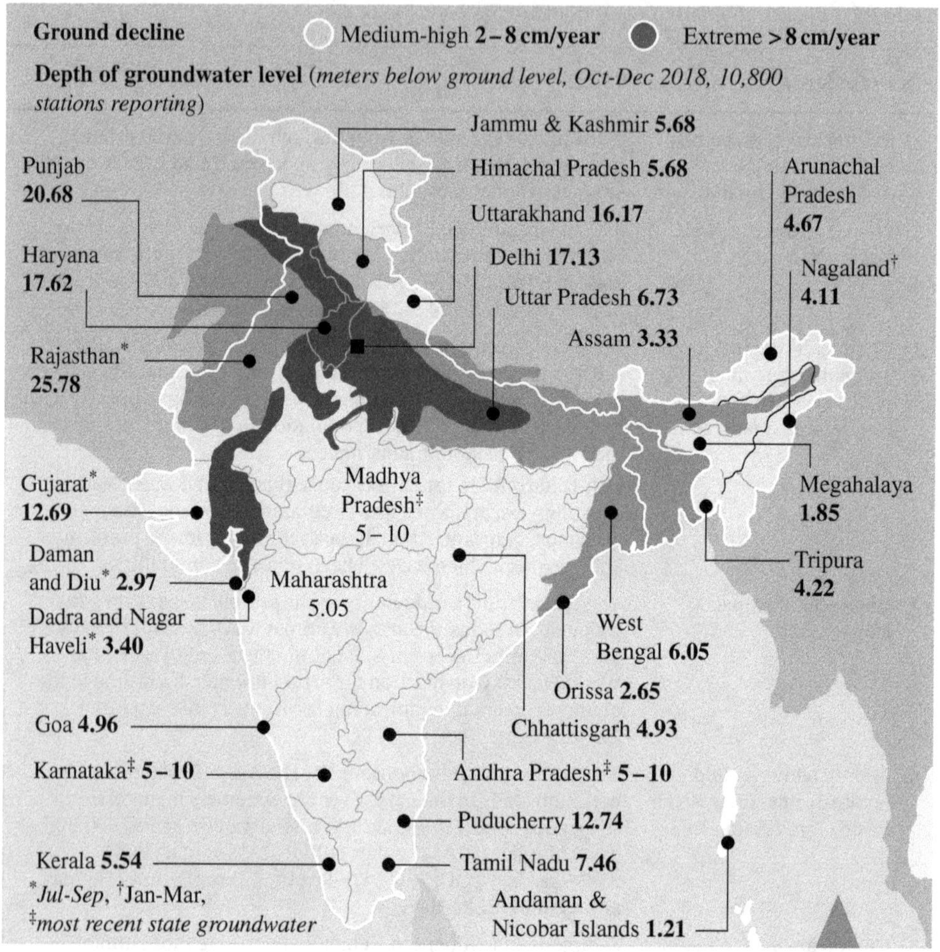

Figure 3.2 Status of groundwater level in India in 2018. (*Source:* Groundwater Year Book – India 2017–2018.)

groundwater resources in the country (Parmar 2017). India observed a 40% decline of groundwater level in recent years (2017–2018) – an unsustainable rate. If this continues, it is estimated that nearly 60% of groundwater resources will be in a critical state by the year 2030. The status of the groundwater level in different states of India is shown in Figure 3.2 for 2018.

Groundwater depletion happens when parameters (BOD, COD, TSS, TDS, nitrates, chlorides, fluoride, etc.) are altered by the addition or withdrawal of other compounds beyond their normal variations. The contact of a bore well or any other groundwater source with industrial effluent discharge can alter the quality of water. The effects of groundwater contamination by industry are outlined as follows (Gagan et al. 2016; Mali et al. 2015):

- Contaminated water with heavy metals or any other toxic chemical may cause health effects to consumers, such as arsenic poisoning, mercury poisoning, lead toxicity causes cancer, respiratory damages, liver damages, and kidney failure.

- Contaminated water also reduces crop yield and quality due to toxic chemicals, as various industries discharge partially treated or untreated effluent into agricultural fields which degrades soil quality.
- Consumption of these contaminated foods creates human health issues and food insecurity.
- Tannery effluents with improper treatment add chromium into the groundwater and surface water. This causes various diseases and environmental degradation.
- Sugar mill effluents have a higher number of suspended solids, dissolved solids, BOD, COD, chloride sulfate, nitrates, calcium, and magnesium. The continuous use of these effluents harms crops when used for irrigation. As a result, a higher number of various elements are deposited in the soil, causing pollution.
- Dumping of industrial discharge on any site causes leaching of heavy metals and toxic chemicals. This infiltrates through the soil profile into groundwater and hence contaminates groundwater quality.

3.7 Environmental Infrastructure in Grossly Polluting Industries and its Performance

Industrial pollution management in many developing countries is complicated by the fact that a large number of small-scale units lack the capacity to mitigate emissions and follow regulations properly (Bansal 2018). Not only are these units restricted by their size, they often have inadequate technological expertise, face financial bottlenecks, and have limited managerial capacities. These features mean that effluent treatment is not feasible by individual units (Kathuria 2001). In addition, the costs associated with individual care will be prohibitive, thus making it impractical. Effluent treatment plants (ETPs) and common effluent treatment plants (CETPs) have been advocated as a cost-effective alternative for meeting the requirements for small-scale polluting units in industrial estates/ sectors (Bansal 2018). A list of operational CETPs, with an online continuous effluent monitoring system (OCEMS) and zero liquid discharge (ZLD) units in India are listed in Table 3.3.

The first big move toward regulating industrial discharges was implemented in India in 1993–1994 with a priority action plan to classify polluting industries along rivers; they were ordered to install ETPs within three months or face closure (Rajaram and Das 2008). The outcome of this program was believed to be a considerable success.

Different bodies with different functions were also created at a central and state level related to groundwater for the conservation and management of water (Garg 2012). These included:

- MoWR
- River Development and Ganga Rejuvenation
- Central Pollution Control Board (CPCB)
- Central Water Commission (CWC)
- Central Ground Water Authority (CGWA)
- CGWB
- Ministry of Industry (MoI)

Table 3.3 Lists of operational effluent treatment plants, with an online continuous effluent monitoring system and zero liquid discharge units in some states.

S. no.	State	Operational CETPs	OCEMS connectivity with CPCB	No. of CETPs with ZLD
1)	Andhra Pradesh	0	0	0
2)	Delhi	6	6	0
3)	Gujarat	0	0	0
4)	Haryana	0	0	0
5)	Himachal Pradesh	0	0	0
6)	Jammu and Kashmir	0	0	0
7)	Jharkhand	0	0	0
8)	Karnataka	0	0	0
9)	Kerala	0	0	0
10)	Madhya Pradesh	13	4	9
11)	Maharashtra	0	0	0
12)	Punjab	33	16	17
13)	Rajasthan	19	15	4
14)	Tamil Nadu	1	0	1
15)	Telangana	2	0	2
16)	Tripura	1	1	0
17)	Uttar Pradesh	10	9	1
18)	Uttarakhand	6	4	2
19)	West Bengal	0	0	0
	Total	191	127	63

Source: https://cpcbenvis.nic.in/cpcb_newsletters/PollutingIndustries.pdf.

However, it must be understood that the mere installation of ETPs in medium or large units without regular monitoring of millions of small-scale units does not guarantee clean discharges. This is evident from the pollution of surface and groundwater around industrial areas. Specific factors including both legal and illegal processes that are responsible for the pollution of water bodies, even after ETPs and CETPs were installed (Maheshwari 2016), are shown in Figure 3.3.

In order to prevent these dangerous circumstances, it is advised that the authorities take legal or collective action on defaulters or handle drainage using creative approaches. To improve water resources quality, there are some legal actions and steps for enhancing existing treatment systems as follows (Afzal et al. 2018; Maheshwari 2016; Sharma et al. 2016):

- Strict enforcement – closure or warning of units that do not meet the requirements or norms.
- Restriction of water consumption/restriction of wastewater production.
- Upgrade of individual ETPs.

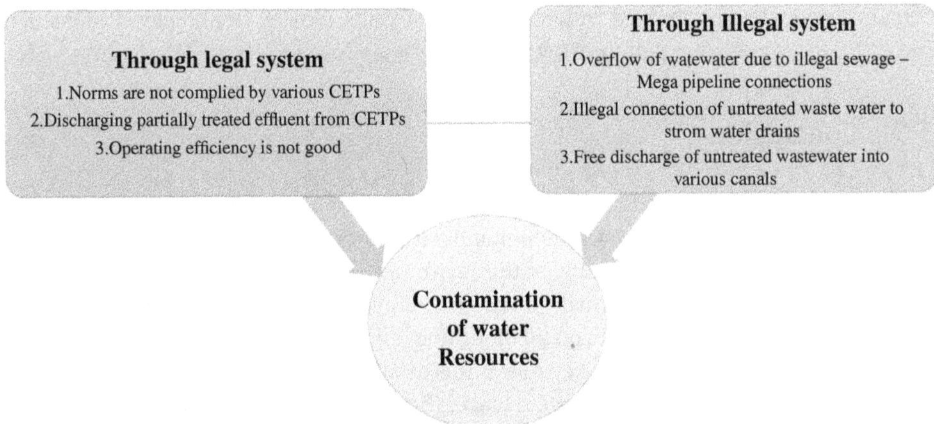

Figure 3.3 Factors responsible for the contamination of water resources. (*Source:* Based on Maheshwari 2016.)

- Updating of CETPs and stricter CETP inlet/outlet specifications.
- Most importantly, ETPs and CETPs should be operated by technicians or professionals.
- National Environmental Quality Standards (NEQS) should be strictly enforced for industrial effluents.
- Avoiding malpractice by authorities.
- Periodic monitoring and evaluation processes should be there to ensure the proper functioning of CETPs and ETPs.

3.8 Challenges Faced in Industrial Water Regulations

Poor coordination and regulation between different regulatory bodies lead to the improper management of industrial wastewater. Also, poor or sometimes ignored water tariffs and the provision of very rare incentives to industry owners if they are reuse water do not attract industry owners to undertake water abstraction minimization and water reuse maximization. This conflict between industry and local communities will become unmanageable if not addressed. Water control measures have not shown significant improvement because although different water bodies are made to regulate and monitor this issue, the powers assigned to these regulating bodies are not framed properly. For example, the MoWR is the principal agency responsible for water but no authority has been given to take action related to water pollution or for the industrial use of water. Similarly, the MoI focuses on the planning and development of water resources for industrial purposes, but it has no power to regulate and monitor water use by GPIs.

To meet discharge norms, wastewater treatment has been installed by industry. However, a common practice adopted by industry is to measure the pollution level in a given quantity of water to meet the required standard by diluting the effluent in freshwater. Second, reusing water from different production steps is not commonly followed, therefore, clear water is contaminated when discharged with effluent, but if water reusing technology is adopted, it is possible to reduce water pollution if water pricing encourages industries to conserve water.

3.9 Conclusion

With heavy industrialization and globalization, an awful impact on the environment arises. It is an unpleasant reality that even though industrialization increases employment opportunities, they also degrade the quality of the environment. Because of this increasing pollution, the entire spectrum of the environment including air, water, and the land is damaged, and it also has a cascading impact on all biotic and abiotic factors. However, in the process of industrialization, water plays a key role as it is used in almost all types of processes such as manufacturing, storage, energy, etc. Hence, the load on all water resources increases, both on humans and other living creatures. Industrialization also degrades the quality of water by discharging partially treated or untreated wastewater effluent into water bodies. Highly polluting industries like GPIs contribute more to water pollution. As a result of a lack of infrastructure and management, these industries release a lot of untreated wastewater into nearby water bodies where it mixes with harmful pollutants such as heavy metals, organic nonbiodegradable pollutants, inorganic pollutants, and so on. CPCB and state pollution control boards provide certain standards of discharged water that should be followed by all these GPIs, however, most of these industries do not comply with standards and they discharge partially treated polluted wastewater into various water bodies. Today, Ganga, Yamuna, Godavari, and so on, are all major sources where water is turned into wastewater. The government and NGOs undertake various policies and actions to combat this situation and to clean water resources. They apply different treatment processes in these industries such as the ZLD method, ETPs, CETPs, the polluter pays principle, and OCEMS. Henceforth, if the previously mentioned actions are implemented and executed fairly, they may help to improve the environment and water resources. While this may seem like baby steps away from environmental degradation, these actions can help if implemented properly.

References

Afzal, M.S., Ashraf, A., and Nabeel, M. (2018). Characterization of industrial effluents and groundwater of Hattar industrial estate, Haripur. *Advances in Agriculture and Environmental Science* 1 (2): 70–77.

Agarwal, P.K. (2015). A review of Ganga river pollution-reasons and remedies. *Journal of Indian Water Resources Society* 35 (3): 46–52.

Ahirwar, M.K., Gupta, G.S., Kirar, N., and Ahirwar, P. (2015). Study of heavy metal pollution from effluent of Orient Paper Mill, Amalia. *International Journal of Innovative Research & Development* 3 (11): 28–38.

Annual Report, 2004–2005, Central Pollution Control Board (CPCB). https://nmcg.nic.in/pdf/plloution%20assessment.pdf (accessed February 2020).

Baek, J., Cho, Y.S., and Koo, W.W. (2009). The environmental consequences of globalization: a country specific time-series analysis. *Ecological Economics* 68 (8–9): 2255–2264.

Bansal N. Industrial Development and Challenges of Water Pollution in Coastal Areas: The Case of Surat, India, IOP Conf. Series: Earth and Environmental Science 120(2018) 01200. doi: 10.1088/1755-1315/120/1/012001.

Bhandari, D. and Garg, R.K. (2015). Effects of industrialization on environment (Indian scenario). *Global Journal for Research Analysis* 4 (12): 2277–8160.

Bharti, P.K., Kumar, P., and Singh, V. (2013). Impact of industrial effluents on ground water and soil quality in the vicinity of industrial area of Panipat city, India. *Journal of Applied and Natural Science* 5 (1): 132–136.

Chaudhary, M. and Walker, T.R. (2019). River Ganga pollution: causes and failed management plans. *Environment International* 126: 202–206.

Das, M.K., Samanta, S., Sajina, A.M. et al. (2016). Fish diversity, community structure and ecological integrity of river Brahmani. *Journal of the Inland Fisheries Society of India* 48 (1): 1–13.

Edokpayi, J.N., Odiyo, J.O., and Durowoju, O.S. (2010). Impact of wastewater on surface water quality in developing countries: a case study of South. *International Journal of Environmental Sciences* 1 (4): 401–416.

Gagan, M. (2015). Evaluation and prediction of deviation in physico chemical characteristics of River Ganga. *International Journal of Advancements in Research and Technology* 4 (6): 14–30.

Gagan, M., Amit, C., Avinash, K., and Kumar, A. (2016). Impact of industrial effluent on ground water and surface water quality: a case study of Dhampur region (U.P.), India. *Journal of Chemical and Pharmaceutical Sciences* 9 (2): 709–713.

Garg, M. (2012). Water pollution in India: causes and remedies. *International Journal of Physical and Social Sciences* 2 (6): 555–567.

DownToEarth (2019). Grossly polluting industries more than doubled in 8 years: SOE in Figures. www.downtoearth.org.in/news/pollution/grossly-polluting-industries-more-than-doubled-in-8-years-soe-in-figures-64962 (accessed 12 May 2020).

Groundwater Year Book – India (2017–2018). https://vajiramias.com/current-affairs/ground-water-year-book-india-2017-18/5d5261ee1d5def64dbdba43f (accessed 12 May 2020).

Groundwater Year Book – India (2018–2019). www.indiaenvironmentportal.org.in/files/file/GW%20Year%20Book_2018-19.pdf (accessed 12 May 2020).

Invest India (2019a). Snapshot: India – Knitting the future: India is among the world's largest producers of textiles and apparel. www.investindia.gov.in/sector/textiles-apparel (accessed 10 May 2020).

Invest India (2019b). Snapshot: World's tannery. www.investindia.gov.in/sector/leather (accessed 10 May 2020).

Invest India (2019c). Snapshot: World's fastest-growing energy market: India is the 3rd largest energy and oil consumer in the world after China and the US. www.investindia.gov.in/sector/oil-gas (accessed 10 May 2020).

Invest India (2020). Snapshot: Food processing – A sunrise sector. www.investindia.gov.in/sector/food-processing (accessed 10 May 2020).

Järup, L. (2000). Cadmium overload and toxicity. *Oxford Journals, Medicine & Health Nephrology Dialysis Transplantation* 17: 35–39.

Katakwar, M. (2016). Narmada river water: pollution and its impact on human health. *International Journal of Chemical Studies* 4 (2): 66–70.

Kathuria V. Pollution: Prevention vs Control: Is EOP Treatment the Solution? Economic and Political Weekly 36 (2001) 2745–2748.

Kotoky, P. and Sarma, B. (2017). Assessment of water quality index of the Brahmaputra River of Guwahati City of Kamrup District of Assam, India. *International Journal of Engineering Research & Technology* 6: 536–540.

Kumar, V., Chauhan, R.K., Srivastava, S. et al. (2018). Contamination, enrichment and translocation of heavy metals in certain leafy vegetables grown in composite effluent irrigated soil. *Archives of Agriculture and Environmental Science* 3 (3): 252–260.

Kumar, V., Kumar, P., Singh, J., and Kumar, P. (2019). Potential of water fern (Azolla pinnata R. Br.) in phytoremediation of integrated industrial effluent of SIIDCUL, Haridwar, India: removal of physicochemical and heavy metal pollutants. *International Journal of Phytoremediation* 22: 1–12.

Kumar, V., Kumar, P., Singh, J., and Kumar, P. (2020). Current status of water pollution by integrated industrial hubs (IIHs) in India. In: Environmental Degradation: Causes and Remediation Strategies, vol. 1 (eds. V. Kumar, J. Singh and P. Kumar), 104–119. Haridwar: Agro Environ Media, Agriculture and Environmental Science Academy.

Maheshwari, R. (2016). Impact of industrial estates on water resources. *International Journal of Environmental Science and Development* 7 (12): 933–939.

Mali, S., Sanyal, S., Bhatt, B., and Pathak, H. (2015). Water pollution and agriculture. *National Academy of Agricultural Sciences*: 39–47.

Malik, D., Singh, S., Thakur, J. et al. (2014). Heavy metal pollution of the Yamuna River: an introspection. *International Journal of Current Microbiology and Applied Science* 3 (10): 856–863.

Manikandan, P., Palanisamy, P.N., Baskar, R. et al. (2015). Physicochemical analysis of textile industrial effluents from Tripur city, TN, India. *International Journal of Advanced Research in Science and Engineering* 4: 93–104.

Mishra, S., Kumar, A., Yadav, S., and Singhal, M.K. (2015). Assessment of heavy metal contamination in Kali river, Uttar Pradesh, India. *Journal of Applied and Natural Science* 7: 1016–1020.

MoEFCC (Ministry of Environment and Forest and Climate Change) 2017. Ministry of Environment and Forest and Climate Change, Government of India, Unstarrred Question No. 354. "Industrial garbage in major rivers." To be Answered on 6.06.2017: Minister of State (Independent Charge) for Environment, Forest and Climate Change (Shri Anil Madhav Dave). www.ippapublicpolicy.org/file/paper/594418ed7f130.pdf.

MOI (2015). Maps of India: Metal Industry in India. https://business.mapsofindia.com/india-industry/basic-metals.html (accessed 5 May 2020).

MOI (2019) Parivesh – A newsletter from ENVIS center – Central Pollution Control Board (https://cpcbenvis.nic.in/cpcb_newsletters/Polluting%20Industries.pdf.

Nath, T.K., Tripathy, B., and Das, A. (2018). A study of water quality of river Brahmani, Odisha (India) to assess its potability. *International Journal of Engineering Research & Technology* 7: 301–311.

Nwankwoala, H. and Nwagbogwu, C. (2012). Characteristics and quality assessment of groundwater in parts of Akure, South-Western Nigeria. *Journal of Environmental Science and Water Resources* 1 (4): 67–73.

Panigrahi, S. and Patra, A.K. (2013). Studies on seasonal variations in phytoplankton diversity of river Mahanadi, Cuttack city, Odisha. India. *Indian Journal of Scientific Research* 4 (2): 211–217.

Panigrahi, A.K. and Pattnaik, S. (2017). A review on pollution status of river Bhagirathi-Hooghly in the stretch of West Bengal, India. *Indian Journal of Applied Research* 9 (8): 53–57.

Panigrahi, A.K. and Pattnaik, S.A. (2019). Review on consequences of pollution of some Indian Major Rivers and their remedial measures. *International Journal of Research & Review* 6 (7): 373–383.

Parmar, S.K. (2017). Present status of groundwater in India. *IOSR Journal of Mechanical and Civil Engineering* 14: 93–95.

Paul, D. (2017). Research on heavy metal pollution of river Ganga: a review. *Annals of Agrarian Science* 15: 278–286.

Pimparkar, M., Tyagi, S., Khatri, N., and Rawtani, D. (2016). Development of critically index to assess water quality in majors rivers of Gujarat. *Environmental Claims Journal* 28: 320–345.

PrintWeek India (2018). Indian paper industry moving out of the woods. www.printweek.in/news/indian-paper-industry-moving-woods-29617 (accessed 5 May 2020).

Rajaram, T. and Das, A. (2008). Water pollution by industrial effluents in India: discharge scenarios and case for participatory ecosystem specific local regulation. *Future* 40: 56–69.

Sharma, S., Sharma, G., and Yadav, R.K. (2016). Impact of industrial development on groundwater quality- a case study of impact of effluent from Vishwakarma industrial area, Jaipur on ground water. *Bulletin of Environment, Pharmacology and Life Sciences* 5: 12–15.

Singh, U., Singh, S., Tiwari, R.K., and Pandey, R.S. (2018). Water pollution due to discharge of industrial effluents with special reference to Uttar Pradesh, India – a review. *International Archive of Applied Sciences and Technology* 9: 111–121.

Sonwane, S.M. (2015). Impact of effluent discharge from thermal power station on status of fish species of river Tapi at Bhusawal, district Jalgaon Maharashtra, IOSR. *Journal of Environmental Science and Toxicology in Food Technology* 9: 63–72.

Srivastava, V., Prasad, C., Gaur, A. et al. (2016). Physico-chemical and biological parameters investigation of river Ganga: from source to plain of Allahabad in India. *European Journal of Experimental Biology* 6: 1–5.

Taskeena, H., Saltanat, P. et al. (2017). Seasonal variations in water quality parameters of river Yamuna, India. *International Journal of Current Microbiology and Applied Sciences Volume* 6 (5): 694–712.

Part II

Phytoremediation of Water Pollution

4

Phytoremediation

Status and Outlook

Kajal Patel[1], Indu Tripathi[1,2], Meenakshi Chaurasia[1], and K.S. Rao[1]

[1] *Department of Botany, University of Delhi, New Delhi, India*
[2] *Department of Environmental Studies, University of Delhi, New Delhi, India*

4.1 Introduction

The economy and development of any nation depend on ever-increasing urbanization, automation, agriculture, and other anthropogenic activities. These processes lead to global concerns, of which environmental pollution is a serious threat to living beings. Pollution can be defined as the contamination of biotic and abiotic components of the ecosystem to such an extent that natural processes are adversely influenced (Krishna et al. 2017). Based on contaminated components of the environment, pollution can be divided into three categories, i.e. air, water, and soil. The key factors that cause contamination are "pollutants," which can be categorized in broad classes, i.e. solid, liquid, and gaseous pollutants, and can include various substances like metals, chemicals, exhaust, smog, solid particles, etc. (Krishna et al. 2017). Among all of the mentioned risk factors, heavy metals (HMs) pollution occupies a growing concern, as unlike many other compounds, HMs are persistent and cannot be degraded from the environment. Moreover, HM pollution could lead to biomagnification in food chains which consequently becomes a hazard for humans and wildlife (Begum et al. 2009; Mohammed et al. 2011; Mahurpawar 2015). The promulgation and deployment of HMs in reservoirs of the ecosphere are caused by various anthropogenic activities such as municipal wastewaters, automobiles, power plants, mining, construction, metallurgy, petrochemical heavy industry effluents, and others. The burden of deaths attributable to modifying environmental factors is around 24.3% globally, while it is almost 30% for India (Prüss-Üstün et al. 2016). Environmental risk factors such as air, water, and soil pollution, chemical exposure, climate change, and UV-rays contribute to more than 100 diseases and injuries. These diseases include infectious, parasitic, neonatal, and nutritional diseases and noncommunicable diseases such as stroke, heart diseases, cancer, and respiratory diseases account for two-thirds of total deaths (Prüss-Üstün et al. 2016).

Among all the risk factors, water pollution is of greatest concern because of its greater repercussions on ecosystems. Water pollution has been identified as the leading cause of

Pollutants and Water Management: Resources, Strategies and Scarcity, First Edition. Edited by Pardeep Singh, Rishikesh Singh, Vipin Kumar Singh, and Rahul Bhadouria.

disease and death around the world (Haseena et al. 2017). Thus, remediation of HM contaminants from different natural resources is a must and it could be accomplished by using several physicochemical and biological techniques. Soil washing, vitrification, solidification, excavation, encapsulation, and so on are examples of physicochemical methods. These techniques are very expensive, labor intensive, and cannot be applied to extensive contaminated land. They eventually cause damage or alterations in environmental components (Truong et al. 2010; Barbosa et al. 2015). Thus, biological methods, also called bioremediation techniques, are an economically possible way to overcome all the disadvantages of physical and chemical methods. Under bioremediation, the use of plants to remediate HM contamination, i.e. phytoremediation, is a more viable and sustainable approach due to cost-effectiveness, less damage to the environment, large-scale use, and so on. But, like other techniques, it also has disadvantages such as the time it takes and incomplete decontamination. Several plants have a great capacity to accumulate and mitigate pollutants from the atmosphere. Plants possess numerous mechanisms to abate HMs, which include resistance, exclusion, compartmentalization, phytochelatin production, metal uptake and translocation, and so on. Based on these mechanisms, phytoremediation techniques can also be categorized as extraction, stabilization, degradation, filtration, and volatilization (Adiloğlu 2017). There are many plants reported to have the potential to tolerate or accumulate different metals at different levels (Prasad 2007; Saraswat and Rai 2009; Fulekar 2016). Based on the process, they can be classified into excluders, indicators, accumulators, and hyperaccumulators (Ali et al. 2013).

Since phytoremediation is a promising field with immense potential, many researchers are contributing toward its advancement to make it more efficient. These advancements are achieved on the grounds of natural and synthetic approaches that use gene-induced and chemical-induced strategies. This chapter comprehensively describes various aspects related to HM pollution and its remediation. Phytoremediation techniques have been found to be interdisciplinary or integrative in nature because of the involvement of various components and approaches such as soil, water, air, plants, anthropogenic activities, natural sources, chemicals, and others. Thus, to completely understand and study phytoremediation techniques, it is essential to have background knowledge in fields like ecology, soil biology, soil ecology, environmental engineering, microbiology, anthropology, chemistry, plant biology, plant physiology, molecular biology, and plant genetic engineering and breeding. The integration of scientific knowledge and basic knowledge in these fields can help in the understanding of the complete mechanism, potentially making this approach more successful and also answering many questions about the commercial application of the technique.

4.2 The Status of Heavy Metal Pollution: Global and Indian Scenarios

HMs represent a class of major inorganic pollutants with a nonbiodegradable and persistent nature. They are the special group of metals and metalloids with high atomic numbers and weights, and a density of more than 3.5 g/cm^3. There are some HMs that are required as macro/micro-nutrients for organisms, and thus, their deficiency or surplus can affect

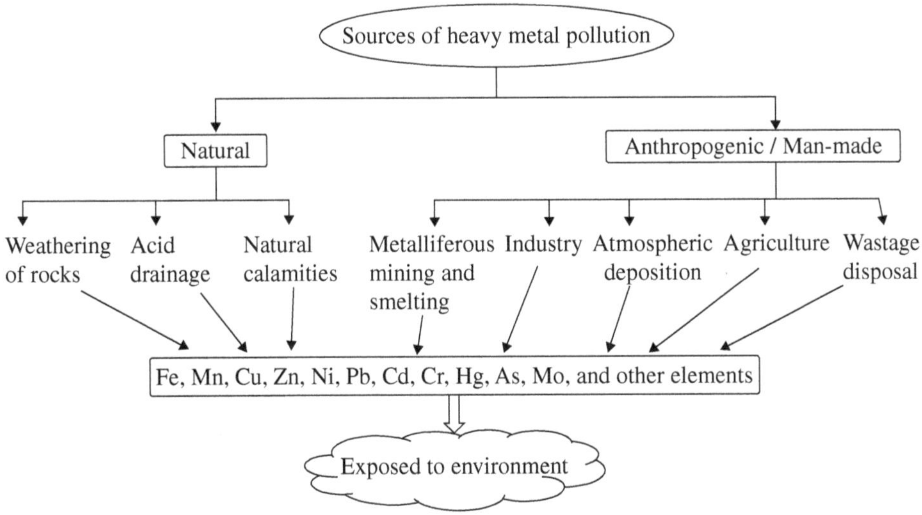

Figure 4.1 Classification of various causes of heavy metal pollution. (*Sources:* Ross 1994; Ali et al. 2013.)

living beings. Among various listed HMs pollutants, Cd, Cr, Cu, Hg, Pb, and Zn are the most common (USEPA 2000; Sharma et al. 2016). Pollution caused by various HMs in ecosystems could be due to both biotic and abiotic factors resulting from both natural (like weathering of rocks) as well as anthropogenic or man-made (like industries) causes (Figure 4.1) (Ross 1994; Ali et al. 2013).

Soil and water are the major sources of HM pollutants, and from these sources, they enter into the food chain, affecting both abiotic (soil, air, water) and biotic (plants, animals including human beings and microorganisms) components of the ecosystem (Table 4.1) (Mohammed et al. 2011; Singh and Kalamdhad 2011; Mahurpawar 2015). The world's most contaminated areas pose health risks to more than 10 million people in various countries (Shakoor et al. 2013). The rise in anthropogenic activities results in the gradual rise and types of HMs in the soil and water, and HM poisoning is a major cause of public health issues (Fernandez-Luqueno et al. 2013; Su 2014). On a global scale, the ever-growing population and exponential rise in anthropogenic activities like urbanization, thermal power plants, agricultural activities, and industrialization have escalated the emission of various toxic HMs, fly ash, particulate matter, and waste into the atmosphere (Suvarapu et al. 2014). According to Zhang et al. (2015), different HM pollution rates contributed by different sources include nonmetallic mining and smelting activities (NMMSA), industries, urban areas, nonmetal areas, irrigation areas, and remote areas (Figure 4.2). They concluded that cadmium (Cd) showed the highest pollution rates in all areas, ranging from 10 to 65%, and NMMSA contributed the highest for almost each HM, ranging from 5 to 65%. About 42 rivers in India were found to have at least two toxic HMs beyond the permissible limits (CWC 2019). Kumar et al. (2019a) mentioned that HMs such as Zn and Pb in Indian soils exceeded the limits for natural soil guidelines. The permissible limits for soil, water, and plants for most common HMs are documented in Table 4.2 (FAO 1985; Awashthi 2000; EPA 2002; WHO/FAO 2007; Singh et al. 2010).

Table 4.1 Hazards of heavy metals on various components of ecosystems.

Components	Effects	Reference
Soil	1) Decrease in organic matter and nutrient imbalance	Kuperman and Carreiro (1997);
	2) Change in physiochemical properties	Mora et al. (2005)
	3) Reduction in soil enzymatic activities and soil biota	
	4) Degraded soil health and reduced soil fertility	
Water	1) Contamination of water bodies	Mohammed et al. (2011);
	2) Degraded water quality	Siddiqui and Pandey (2019)
	3) Disturbed water properties like pH, CEC, BOD, TDS, etc.	
Air	1) Increase in atmospheric particulate matter	Li et al. (2013);
	2) Increase in atmospheric pollution and degraded air quality	Oucher et al. (2015)
	3) Sublimation to other natural resources via the air	
	4) Entrance to the food chain	
Plants	1) Imbalance in nutrients	Nagajyoti et al. (2010)
	2) Changes in plant characteristic features	
	3) Loss in plant health thus decline in growth	
	4) Reduction in yield	
Humans	1) Quelling of growth and development	Mohammed et al. (2011);
	2) Reduced reproductive capacity	Mahurpawar (2015)
	3) Loss of tissue resurgence and oxygen dissipation	
	4) Disturbed biochemical reactions	

CEC, cation exchange capacity; BOD, biological oxygen demand; TDS, total dissolved solids.
Sources: Singh and Kalamdhad (2011); Mohammed et al. (2011); Mahurpawar (2015).

Some of the relevant reasons for the need for remediation of HMs (Belluck et al. 2006; Neilson and Rajakaruna 2015) are as follows:

i) Negative impacts on soil resources, plant growth, human health, and the environment.
ii) Reduced future toxic liability.
iii) Translocation from the sources of origin for HMs (like industry, agricultural, sewage wastes, and others) to soil resources and water bodies.
iv) Limitation of environmental liability in terms of current and future legislation.

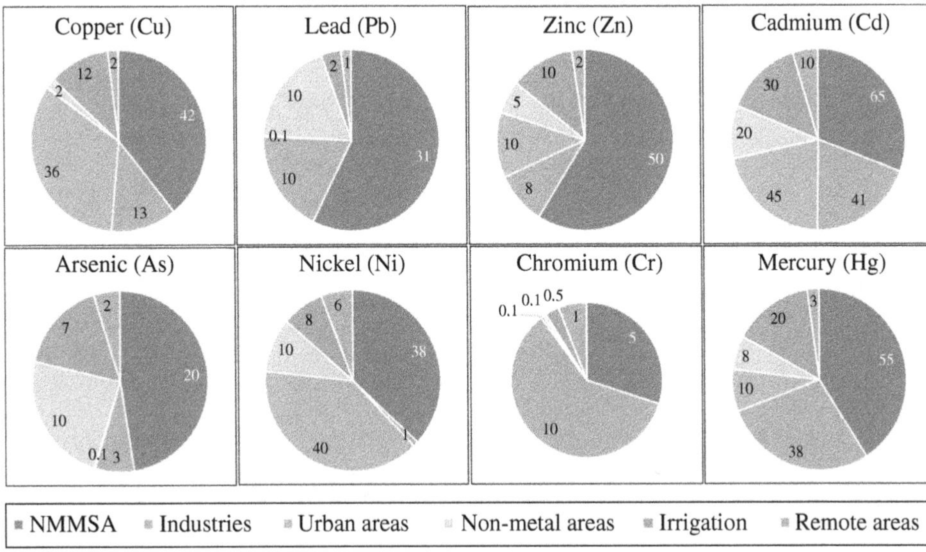

Figure 4.2 Pollution rates of various heavy metals from diverse sources including activities form both urban and rural areas (NMMSA, nonmetallic mining and smelting activities).

Table 4.2 Permissible limits for common heavy metals in soil, water, and plants (Indian and global standards).

	Soil (μg/g)		Water (μg/ml)		Plants (μg/g)	
Heavy metals	**Indian**	**EPA**	**Indian**	**FAO**	**Indian**	**WHO/FAO**
Cd	3–6	3.0	0.01	0.01	1.5	0.2
Cu	135–270	140	0.05	0.20	30.0	40.0
Pb	250–500	300	0.10	5.0	2.5	5.0
Zn	300–600	300	5.0	2.0	50.0	60.0
Mn	–	–	0.10	0.20	–	–
Ni	75–150	75	0.20	0.20	1.5	–
Cr	–	150	0.05	0.10	20.0	–

Sources: Awashthi (2000); FAO (1985); Singh et al. (2010).

4.3 Status of Phytoremediation

For the management of hazards and health risks posed by HM contamination, a wide range of remediation techniques have evolved. Based on functioning, these remediation techniques can be physical, chemical, electrical, thermal, and biological. Based on the site of operation, these techniques can be broadly categorized as ex situ and in situ techniques. Depending on technique-based classifications and the inclusion of characteristic

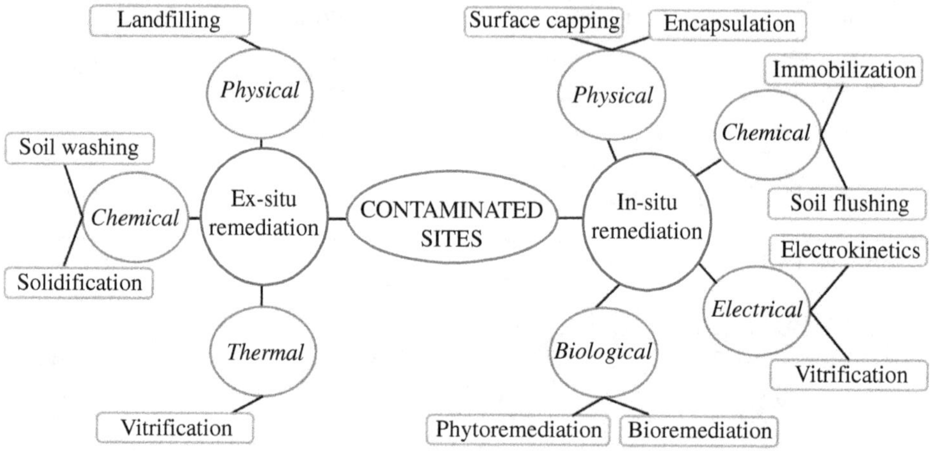

Figure 4.3 Schematic diagram of the categorization of remediation techniques used for heavy metal decontamination. (*Sources:* Ghosh and Singh 2005; Liu et al. 2018.)

mechanisms, several techniques have been developed (Figure 4.3) (Ghosh and Singh 2005; Liu et al. 2018). Under the class of biological in situ techniques, phytoremediation is emerging as a highly efficient, sustainable, economical, socially accepted, and eco-friendly approach for the decontamination of polluted areas. This is not a new technology as plants were prospected and utilized in wastewater treatment almost 300 years ago. The term of "phytoremediation" is composed of two words: phyto (Greek) means plants and remedium (Latin) means "to restore or recover," which is also commonly known as "green remediation." It involves the utilization of plants to eliminate, destruct, metabolize, immobilize, and stabilize a variety of pollutants for on-site decontamination (Fulekar et al. 2009; Purakayastha and Chhonkar 2010). Some characteristic features which make a plant ideal for phytoremediation are a high growth rate, production of more above-ground biomass, more accumulation potential, better translocation of toxic metals from roots to shoots, tolerant and resistant toward toxicity of HMs, a resilient and adaptative capability, easy to cultivate and harvest, and resistant to pests, pathogens, and herbivory (Ali et al. 2013). Proper utilization of the harvested biomass post-remediation also makes this technique a more preferable and reliable option. The practice of phytoremediation depends on the following factors:

i) The plant's capacity to uptake, store, and metabolize contaminants.
ii) Plant and microorganism interaction in the rhizosphere.
iii) Properties and chemical reactivity of metals.
iv) Plant growth and their density on the target site.

While phytoremediation has advantages as mentioned earlier, it also has some disadvantages such as it is only applicable for low to moderately contaminated sites, it is metal-specific, time consuming, low efficiency, and has been limited to pilot demonstrations until now (Table 4.3) (USEPA 2000; Liu et al. 2018). The subcategorization of this technique can be done on the basis of the mode of action of plants used to remove metals and the utilization of post-remediated plant wastes in the form of timber, fuel, and so on (Figure 4.4). The knowledge and utilization of these techniques for remediation purpose are summarized in Table 4.3.

Table 4.3 Overview of the current status of phytoremediation.

Mechanism	Media	Targeted contaminants	Plants utilized	Application status
Phytoextraction	Soil, sediment, sludges	Ag, Cd, Co, Cr, Cu, Hg, Mn, Mo, Ni, Pb, Zn, U	Indian mustard, pennycress, sweet alyssum, sunflower, hybrid poplar	Laboratory, pilot, and field
Rhizofiltration	Groundwater, surface water	Ag, Cd, Co, Cr, Cu, Hg, Mn, Mo, Ni, Pb, Zn, U	Sunflower, Indian mustard, water hyacinth	Laboratory, pilot
Phytostabilization	Soil, sediment, sludges	As, Cd, Co, Cr, Cu, Hs, Pb, Zn	Indian mustard, hybrid poplar, grasses	Field
Rhizodegradation	Soil, sediment, sludges, groundwater	Organic compounds	Red mulberry, grasses, hybrid poplar, cattail, rice	Field
Phytodegradation	Soil, sediment. Sludges, water	Organic compounds	Algae, stonewort, hybrid poplar, black willow, bald cypress	Field demo
Phytovolatilization	Soil, sediment, sludges, groundwater	Hg, Se, As, and chlorinated solvents	Poplar, alfalfa, black locust, Indian mustard	Laboratory and field
Vegetation cover	Soil, sediment, sludges	Organic and inorganic compounds	Poplar and grasses	Field

Sources: Adapted from USEPA (2000); Liu et al. (2018).

There are six subcategories of phytoremediation techniques:

i) *Phytodegradation*: Involves the use of plants to degrade organic pollutants by using the plant's metabolism and also microbial activities in the rhizosphere. It is based on the activity of plants named the "green liver of biosphere" which accumulates toxic substances and detoxifies them. *Brassica campestris*, *Festuca arundinacea*, and *Helianthus annuus* have been reported to degrade the petroleum hydrocarbons (Park et al. 2011; Ali et al. 2013).

ii) *Phytoextraction*: This mechanism works on the principle of extraction of HMs using plants by translocating the contaminants into the harvestable parts of the plants such as the wood, leaves, etc. For phytoextraction, plants with specific characteristics are required, for instance, the ability to accumulate metal, tolerance to high metal concentrations, rapid growth and a high potential for the accumulation of biomass, and can easily be grown in the ecosystem. Based on its mode of working, it is also known as phytoaccumulation, phytosorption, or phytosequestration. Canola (*Brassica napus*) and radish (*Raphanus sativus*) were found to decontaminate multi-contaminated soils through the phytoextraction process (Marchiol et al. 2004; Ali et al. 2013).

iii) *Phytovolatilization*: The most controversial technique among all the techniques of phytoremediation because it partially removes the pollutants and translocates them to another compartment of the ecosystem, which is not safe. In this technique, the plant uptakes the pollutants and converts them to a volatile or vapor form. It is generally used for the removal of organic pollutants and HMs that occur in gaseous form in the atmosphere like arsenic, mercury, and selenium. For example, *Pteris vittata* is used to remediate arsenic-contaminated soil (Sakakibara et al. 2010; Ali et al. 2013).

iv) *Phytofiltration*: The passing of polluted water through plant biomass or through roots helps in their trapping, accumulation and then their degradation takes place using the phytofiltration technique. Phytofiltration can be performed by different parts of plants: it is called rhizofiltration when plant roots are used, blastofiltration when plant seedlings are used, and caulofiltration when excised plant shoots are used for the decontamination process. Generally, aquatic plants either floating, submerged, or emergent are used, e.g. aquatic ferns including *Salvinia* sp. and *Azolla* sp. (Olguín and Sánchez-Galván 2012; Ali et al. 2013).

v) *Phytostabilization*: This technique directly affects the mobility of metal ions in the contaminated site. Here, plants are used to stabilize or immobilize the HMs through adsorption by roots, precipitation, and complexation of metal ions. It prevents the percolation of toxic metals in groundwater and thus in the ecosystem. Restrained by its temporary effect, it is not considered as any permanent solution for the detoxification of HMs. This technique particularly involves different mechanisms like the allocation of litter and vegetation cover and by increasing the water-holding capacity of soil to hinder leaching. For example, root systems of Norway spruce (*Picea abies*) and poplar (*Populus tremula*) are used for the phytostabilization of HMs (Brunner et al. 2008; Ali et al. 2013).

vi) *Rhizospheric bioremediation*: A plant microbe interaction method of remediation of HMs in plant rhizospheric or the root region. The quantitative estimation of the "rhizopsheric effect" can be done by calculating the ratio of microbes in rhizospheric soil to the number of microbes in nonrhizospheric soil, i.e. R/S ratio. The composition of the microbial community in the rhizosphere depends on the root type, plant species, plant age, and soil properties. Some plants like corn, cotton, and bean plants can be utilized (Anderson et al. 1993; Fulekar et al. 2009).

From the information provided by USEPA (2002), it can be observed that all the above subcategories of phytoremediation techniques are still limited to laboratory and pilot experiments. Although most of the toxic HMs can be removed from sewage, soil, air, and water, it still needs robust investigation and large-scale field experiments to support them. In addition to this, some species like Indian mustard, poplar, willow, algae, and some of the grasses are the most prominent types of plants utilized on a large scale for field experiments. For better waste resource management, innovative technology is highly essential. Phytoremediation as a much cheaper and sustainable technology has emerged as a promising technology for water treatment and the conservation of water and soil. Hence, the following section deals with the metal hyperaccumulators that have been identified during various studies and research. The identification of these hyperaccumulators is considered to be the first step for the advancements of a plant's capability to be utilized for phytoremediation, which will be discussed in the coming sections.

4.4 Metal Hyperaccumulators for Phytoremediation

Some plants have a good capacity for phytoremediation and these can thrive in high metal-contaminated sites and are known as metallophytes. They could be "obligate metallophytes" (essentially needing metals for their survival) or "facultative metallophytes" (tolerating high metal concentrations but are not confined to them). The natural sensitivity or tolerance of any plant directly depends on plant species and its genotypes. On the basis of criteria of sensitivity or tolerance, plants can be divided into four groups: indicators (wheat, oats, maize, etc.), excluders (members of Gramineae, e.g. Sudangrass), and accumulators and hyperaccumulators (members of the Brassicaceae and Asteraceae family, e.g. spinach, lettuce) (Tlustoš et al. 2006; Ali et al. 2013). A certain number of plants show an extreme accumulation of HMs in their aerial parts that are almost 100-fold higher than normal plants from contaminated sites, these are termed as hyperaccumulators. The process of hyperaccumulation also includes the healthy growth of these plants to maintain self-sustainability in their natural habitats and the quantity of biomass produced by particular plants. The term "hyperaccumulators" was first coined by Brooks et al. (1977) to define Ni hyperaccumulating plants. A particular ratio, i.e. shoot to root metal concentration, is termed as the "translocation factor" (TF), which is always found to be >1 for hyperaccumulators. Some of the examples of hyperaccumulators are *Thlaspi caerulescens* and *Alyssum bertolonii*, which accumulate Zn, Cd, and Ni. Additionally, these hyperaccumulators can also be used for the phytomining of high-valued HMs such as silver, palladium, and platinum (Tlustoš et al. 2006). Ojuederie and Babalola (2017) listed families of hyperaccumulators for specific metals, e.g. Asteraceae (Ni, Pb, Cd, Zn), Brassicaceae (Ni, Zn, Cd), Caryophyllaceae (Zn, Pb, Cd), Crassulaceae (Pb), Euphorbiaceae (Cu, Fe, Pb, Zn), Fabaceae (Se, Pb), Poaceae (Cr), Pteridaceae (As, Hg), and Violaceae (Pb, Zn Cd). Some of the best-known hyperaccumulators utilized so far are listed in Table 4.4.

Several reports regarding tree cultivation on polluted lands suggested a number of diverse tree species for the purpose of the bioremediation of land, which included *Betula* (Birch), *Acer* (Sycamore), *Populus* (Poplar), *Alnus* (Alder), and *Salix* (Willow). Among all of them, Willow proved to be the most promising player due to its fast growth, bushy canopy, and short rotation coppice system in which it can be harvested every three to five years, ultimately leading to the rapid uptake of toxic metals and their elimination after harvesting. Trees species are considered better because of the by-products obtained after their harvesting such as their wood which can be used to produce renewable bioenergy (Figure 4.4) (Punshon and Dickinson 1997; Pulford et al. 2002; Pulford and Watson 2003). Some studies revealed that leaves of some tree species can also play a significant role in HM accumulation followed by their removal viz. *Aesculus hippocastanum* (horse chestnut), *Tilia* sp. (linden), *Salix alba* (willow), *Populus tremula* (European aspen), *Robinia pseudoacacia* (black locust), *Quercus infectoria* (Aleppo oak), and *Pinus nigra* (black pine), and thus, can also be utilized as biomonitors for HM pollution in the environment (Cicek and Koparal 2004; Tomašević et al. 2004).

Table 4.4 Various hyperaccumulator plants for heavy metals.

Scientific name	Family	Common name	Metals accumulated	Sources
Agrostis sp.	Poaceae	Highland bentgrass	As, Cd, Cu, Pb, and Zn	Boshoff et al. (2014)
Salix sp.	Salicaceae	Osier or willow	Cd, Zn	Punshon and Dickinson (1997)
Amanita strobiliformis	Amanitaceae	European pine cone lepidella	Ag	Borovička et al. (2007)
Bacopa monnieri	Plantaginaceae	Brahmi	Hg, Cr, Cu, Cd	Prasad (2007)
Brassica juncea	Brassicaceae	Indian mustard	Ni, Cr	Saraswat and Rai (2009)
Vallisneria americana	Hydrocharitaceae	Tape grass	Cd, Cr, Cu, Ni, Pb, Zn	Prasad (2007)
Helianthus annuus	Asteraceae	Sunflower	Cd, Pb, Zn	Fulekar (2016)
Hydrilla verticillata	Hydrocharitaceae	Hydrilla	Hg, Fe, Ni, Pb	Prasad (2007)
Pteris vittata	Pteridaceae	Ladder brake	As	Ma et al. (2001)
Salvinia sp.	Salviniaceae	Karib or water ferns	Cr, Fe, Pb, Zn, Cd, Mn	Prasad (2007)
Spirodela polyrhiza	Lemnaceae	Giant duckweed	Ni, Cu, Cd	Duman et al. (2009)
Thlaspi caerulescens	Brassicaceae	Alpine pennycress	Cd, Pb, Zn	Banasova et al. (2008)
Azolla pinnata	Salviniaceae	Mosquito-fern	Cu, Cr	Jain et al. (1990)
Arabidopsis halleri	Brassicaceae	Rockcress	Cd, Zn	Zhang et al. (2017)
Lemna minor	Lemnaceae	Duckweed	Cu, Cr	Jain et al. (1990)
Pistia stratiotes	Araceae	Water lettuce	Ag, Cd, Cr, Cu, Hg, Ni, Pb, Zn	Odjegba and Fasidi (2004)
Sedum plumbizincicola	Crassulaceae	Stonecrops	Cd, Zn	Liu et al. (2017)
Euphorbia macroclada	Euphorbiaceae	Spurge	Cu, Pb, Zn	Nematian and Kazemeini (2013)
Alyssum sp.	Brassicaceae	Sweet alyssum	Ni	Robinson et al. (1997)
Arabis paniculata	Brassicaceae	Rockcress	Pb, Zn, Cd	Tang et al. (2009)
Sesbania drummondii	Fabaceae	Poison-bean	Pb	Sahi et al. (2002)
Arabidopsis thaliana	Brassicaceae	Thale cress	Zn, Cd	Saraswat and Rai (2009)

Table 4.4 (Continued)

Scientific name	Family	Common name	Metals accumulated	Sources
Cynodon dactylon	Poaceae	Bermuda grass	Ni, Cr	Saraswat and Rai (2009)
Viola baoshanensis	Violaceae	Pansy	Pb, Zn, and Cd	Zhuang et al. (2005)
Typha domingensis	Typhaceae	Cumbungi	Cd, Cu, Pb, Zn	Prasad (2007)
Eichhornia crassipes	Pontederiaceae	Water hyacinth	As, Cd, Co, Cr Cu, Al, Ni, Pb, Zn, Hg, P, Pt, Pd, Rh	Prasad (2007)

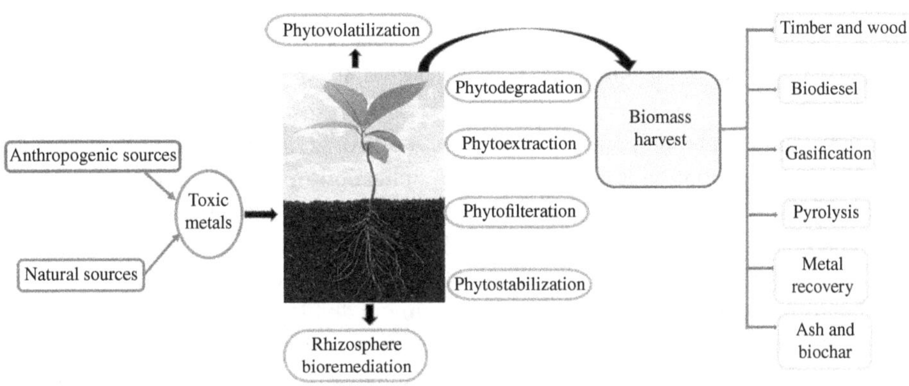

Figure 4.4 Flow chart for the phytoremediation of heavy metals and the utilization of plant waste in different forms that have economic value in terms of capital returns. (*Sources:* Punshon and Dickinson 1997; Pulford et al. 2002; Pulford and Watson 2003.)

4.5 Advancements in Techniques for the Improvement of the Phytoremediation Ability in Plants and the Status of Genetically Modified Organisms

Gene-induced strategies: The power of phytoremediation in plants relies on several features that depend on various metabolic processes mostly controlled by the network of regulating genes. Therefore, implanting specific genes from hyperaccumulators to other plants using genetic technologies may improve the potential of plants to decontaminate polluted sites. For this purpose, transgenic plant technology provides many advantages over conventional methods. A strategic approach used for the production of genetically modified plants (GMOs) with superior phytoremediation ability is given in Figure 4.5 (Jabeen et al. 2009). Recent studies show that the engineering of new transgenic plants with a high phytoremediation potential will have enhanced synthesis of eligible products such as enzymes, proteins, and various polymers (Macek et al. 2006). The manipulation of

Scrutinization of specific with high metal accumulating properties governed with special features.

↓

Examination of the scrutinized plant species under various metal stress treatment and make a comparison among them and also with other plants

↓

Analyse the particular gene which is responsible for characteristic feature which make it superior in remediation of metals from the selected plant species

↓

Isolation and cloning of the particular tolerating species.

↓

Cloning followed by examination of the cloned gene product in prokaryotic system.

↓

Insertion of the target gene to enhance the synthesis of specific gene product into the target plant.

↓

Observation of the target plant for improved metal accumulating capacity.

Figure 4.5 Schematic chart of procedures followed to develop GMOs (genetically modified organisms) for phytoremediation. (*Source:* Based on Jabeen et al. 2009.)

gene controlling metal transporters, vacuolar targeting of metals for their mitigation, and high biomass production will find a potential application in the development of plants with the augmented power of phytoremediation (Table 4.5). This approach can also introduce invasive species as hyperaccumulators of metals during the scrutinization process, which requires additional management and care for their handling and implementation. According to Ruiz et al. (2003), there is a procedure of "biological encapsulation" which can help to overcome the issues of using invasive species (like the spread of these nonnative species to native species in ecosystems) as genetically improved phytoremediation species. Fulekar et al. (2009) suggested various levels in molecular mechanisms which can be targeted for engineering genes:

1) Gene for enhanced biomass production
2) Gene for the enhanced synthesis of metal ion transporter proteins
3) Gene for enhanced production of metal sequestration proteins (phytochelators)
4) Genes for biodegradative enzymes
5) Genes to change the oxidative states of metals
6) Gene to escalate the growth rate
7) Genes to develop better root systems, i.e. more branching, deep-growing, and fine roots

Chemical-induced strategies: Naturally, plants can synthesize some specialized compounds on HM aggregation, generally known as chelating compounds. These compounds help plants to bind and translocate the HM complexes to plant cells where it is sequestered. Therefore, to improve the power of HMs remediation, the production of high amounts of chelating agents is an essential requirement. Some of the examples of chelating agents are EDTA (ethylenediaminetetraacetic acid), NTA (nitrilotriacetic acid), citrate, oxalate, malate, succinate, tartrate, phthalate, salicylate, and acetate, which can be used for chemical-based strategies for the improvement of plants. Blaylock et al. (1997) reported that use of some synthetic and natural zeolites and synthetic organic chelating agents such as

Table 4.5 Utilization and development of various transgenic plants (GMOs) for the purpose of phytoremediation.

S.no.	Gene	Source	Target	Effect	References
1	MT2	Human	*Nicotiana tabacum*	Cd tolerance	Misra and Gedamu (1989)
2	PsMTA	*Pisum sativum*	*Arabidopsis thaliana*	Cu accumulation	Evans et al. (1992)
3	MT-1	Mouse	*Nicotiana tabacum*	Cd tolerance	Pan et al. (1994)
4	MerAPe9, merA 18	*Shigella*	*Liriodendron tulipifera*	Maximum Hg evolution	Rugh et al. (1998)
5	γ-glutamylcysteine synthase	*Escherichia coli*	*Brassica juncea*	Cd tolerance	Zhu et al. (1999)
6	Ferritin	*Arabidopsis thaliana*	*Nicotiana tabacum, Oryzae sativa*	Fe accumulation enhancement	Goto et al. (1999)
7	APS1	*Arabidopsis thaliana*	*Brassica juncea*	Se accumulation enhancement	Pilon-Smits et al. (1999)
8	Glutathione-S-transferase	*Nicotiana tabacum*	*Arabidopsis thaliana*	Al, Cu, and Na tolerance	Ezaki et al. (2000)
9	RCS1	*Oryzae sativa*	*Nicotiana tabacum*	Cd tolerance	Harada et al. (2001)
10	CAL2	*Enterobacter cloacae*	*Brassica napus*	As tolerance	Nie et al. (2002)
11	MerAPe9, merA 18	Gram (−) ve bacteria	*Arabidopsis thaliana*	Hg volatilization	Dhankher et al. (2002)
12	SMTA	*Arabidopsis bisculatus*	*Brassica juncea*	Se volatilization	Ellis et al. (2004)
13	CUP1	*Saccharomyces cerevisiae*	*Helianthus annuus*	Cd tolerance	Watanabe et al. (2005)
14	PCS1	*Thlaspi caerulescens*	*Nicotiana glauca*	Cu, Zn, and Pb accumulation increased	Martínez et al. (2006)
15	PCS	*Arabidopsis thaliana*	*Brassica juncea*	Increased Cd, As accumulation and tolerance	Gasic and Korban (2007)
16	SK2-type dehydrin	*Brassica juncea*	*Nicotiana tabacum*	Cd and Zn tolerance	Xu et al. (2008)
17	CAT	*Brassica juncea*	*Nicotiana tabacum*	Cd tolerance	Guan et al. (2009)
18	SPDS	*Malus domestica*	*Pyrus communis*	Cd, Pb, and Zn tolerance	Wen et al. (2010)
19	YCF1	*Saccharomyces cerevisiae*	*Brassica juncea*	Enhanced Pb and Cd accumulation	Bhuiyan et al. (2011)

(*Continued*)

Table 4.5 (Continued)

S.no.	Gene	Source	Target	Effect	References
20	Cystathionine-γ synthase	*Oryzae sativa*	*Nicotiana tabacum*	Na and Zn tolerance	Singh et al. (2012)
21	GST and CYP2E1	*Medicago sativa*, human	*Medicago sativa*	Hg tolerance	Zhang et al. (2013)
22	AIP	*Panax ginseng*	*Arabidopsis thaliana*	Enhanced Al tolerance	Jang et al. (2014)
23	γ-GCS-GS	*Streptococcus thermophilus*	*Beta vulgaris*	Increased Cd, Zn, and Cu accumulation	Liu et al. (2015)
24	MTP	*Oryzae sativa*	*Nicotiana tabacum*	Increased Cd accumulation and As tolerance	Das et al. (2016)
25	HMA3	*Sedum plumbizincicola*	*Saccharomyces cerevisiae*	Cd tolerance	Liu et al. (2017)
26	MYB	*Raphanus sativus*	*Nicotiana tabacum and Petunia*	Cu, Zn tolerance	Ai et al. (2018)
27	ATP-sulfurylase	*Arabidopsis thaliana*	*Medicago sativa*	V and Pb accumulation and tolerance	Kumar et al. (2019b)

EDTA, DTPA (diethylene triamino-pentacetic acid), HEDTA (N-hydroxy ethylene triamine triacetic acid), EGTA (ethylene glycol bis [β-amino-ethyl ether]-N, N′, N′-tetraacetic acid), and NTA can significantly enhance the uptake of toxic metals by raising their availability through increasing solubility and mobility of metals in agriculture. In addition, the application of some synthetic cross-linked polyacrylate and hydrogels hinders the entry of HMs, and thus, guards plant roots against HM poisoning. Thus, labile metal contaminants are liberated through increased solubility in soil which is followed by complexation of chelate and HM, and thus may result in hyperaccumulation. Consequently, harvesting plants with sufficient biomass leads to phytoextraction but the possibility of leaching of some synthetic compounds to the nearby environment limits this chemical-based strategy of the remediation process (Purakayastha and Chhonkar 2010).

4.6 Physiological Mechanisms for the Sequestration of Metals

The key mechanisms and the pathways that plants follow for HM sequestration are the absorption of toxic metals from contaminated sites followed by their translocation, degradation, and stabilization over the roots and finally accumulation in the harvestable parts (Fulekar et al. 2009). The physiological and molecular processes (Figure 4.6) of HMs remediation/sequestration are explained in the following section.

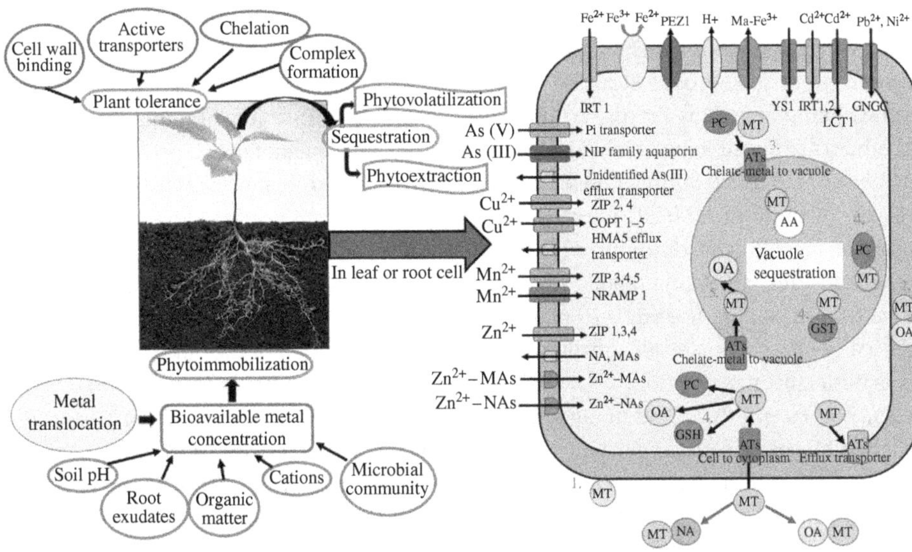

Figure 4.6 Schematic diagram for physiological and molecular mechanisms used by plants for the phytoremediation of heavy metals. At the physiological level, phytoimmobilization takes place in plant roots and plant tolerance results by several mechanisms such as cell wall binding, active transporters, chelation, and complex formation and sequestration of heavy metals done by either phytovolatilization or by phytoextraction (on the left-hand side). At a molecular level, several transporters and ion channels present in the cell wall help in metalloid (metal ions) efflux or influx for sequestration via different pathways as shown in (1). Binding of metalloids to the cell wall (2). Chelation of metal-organic acid on the outside of cell membranes (3). Active metalloid efflux from the cell via efflux transporters (4). Metal-chelator complexes (5). Vacuolar compartmentalization or sequestration inside the cell. (ATs, active transporters; ZIP/IRT, zinc/iron regulated protein; COPT, copper putative transporter; NRAMP, natural resistance-associated macrophage protein; HMA, heavy metal P-type ATPase; H^{+}ATP, HA-proton efflux transporter; FRO, ferric reductase/oxidase; PEZ1, phenolics efflux zero1; TOM1, efflux transporter; YS1, yellow stripe-like1; NA, nicotinamine; MA, family of mugineic acids; LCT1, low-affinity cation transporter; GNGC channels, putative transition metal transporter, MTs, metal(loids); OA, organic acid; AA, amino acids; GSH, glutathione; PC, phytochelatines). (*Source:* Modified from Gajić et al. 2018.)

Metal uptake mechanisms: The first step in the direction of phytoremediation, i.e. the absorption of HMs, depends on its bioavailability and various physicochemical and biological processes of the soil. The efficient metal uptake by plants can be accomplished by cell wall binding, active transporters present in root cells, chelation, and complex formations. Different exudates from plant roots which include chelating molecules (e.g. mugenic and aveic acids) and phytosiderophores that help to bind iron are the major agents. The bioavailability of metals in soil depends on various factors such as chemical properties of metals (charge, ion size, reactivity, etc.), soil properties (pH, cation exchange capacity, texture, etc.), organic matter content, root exudates, and the microbial community present in the rhizosphere (Fulekar et al. 2009).

Metal translocation: Once the metals have been absorbed by plant roots, they enter into the root cell and from there they are translocated to various aerial parts of the plant either by symplastic or by apoplastic pathways. This translocation is affected by the concentration and type of metal ions. Reports suggest that phytochelatin and various chelators (such as citrate, free histidine, malate, and others) assist in binding and translocation of metal ions through xylem sap (Brooks et al. 1977; Prezemeck and Haase 1991; Kramer et al. 1996). Several channels such as Zrt and Irt-like proteins, cation diffusion facilitators, antiporters, and others are found to be involved in the translocation of metal ions.

Mechanism to resist: Simply known as the mechanism of tolerance which helps plants to resist the toxicity of HMs. The tolerance is signified by a network of physiological and molecular mechanisms of plants which consequently results in the exclusion of toxic elements, and thus, shows metabolic tolerance to particular HMs. This mechanism follows the path of providing tolerance to plants by interrupting the free diffusion or passage of electrically charges metalloids through lipophilic cellular membranes into the cytosol of plant cells (Jabeen et al. 2009).

Exclusion mechanism: The uptake and translocation of metalloids by plant tissues are mediated by specific channels, which are known as transporters and antiporters, for the influx and efflux of metalloids. The specificity of the channel used by ions depends on their electrical charges. Sometimes these charges obstruct the free passage of ions into the cell, so excluding them delimits them to a particular area in the cell later on (Jabeen et al. 2009).

Vacuolar compartmentalization: The entry of metals into the cell cytoplasm via different channels located in the cell wall leads to their compartmentalization to the vacuole for sequestration. Sometimes this compartmentalization needs the binding of metal ions with specific compounds such as phytochelatins (PCs) and anthocyanin. This process stabilizes the free concentration of metalloids and impels them to a limited vacuolar space from where they are directly or indirectly extracted from the plant (Jabeen et al. 2009).

Phytochelatins: These are specific compounds belonging to a family of thiol-rich peptides (rich in n-Glu-Cys) that are produced by plants on HM accumulation. It has been found that some of the HMs like Cd and As have a good potential to stimulate the synthesis of PCs while others like Zn and Ni do not have that potential. The process is started by activating the PC-synthase enzyme which is responsible for phytochelatin synthesis (Grill et al. 1989). These compounds bind and chelate the HMs and are transported to vacuoles where they are stabilized by forming a complex with sulfides or organic acids.

Metallothionein: These are another group of compounds that are low molecular weight, cysteine-rich metal-binding proteins providing thiols for metal chelation. The metallothionein compounds are reported to increase the capability of plants to accumulate HMs by many folds (Pan et al. 1994; Hesegawa et al. 1997). The metallothionein metal complex can be glutathioned, so they may be carried to vacuoles for long-term sequestration. As the synthesis of the metallothionein compound is also observed under other abiotic stresses, it has been concluded that metallothionines are part of the stress response machinery in plants.

Molecular mechanism for sequestration: Binding of metal ions to the cell wall and thus their uptake is facilitated by several active transporters such as zinc iron (ZIP/IRT) regulated proteins, low-affinity cation transporters (LCT1), GNGC channels, copper putative transporters (COPT), putative transition metal transporters, and others. Then, chelation of metal outside of the cell membrane takes place via complex formation of metalloids (MTs) with organic acid (OA), nicotinamine (NA), glutathione (GSH), and phytochelatines (PC). Simultaneously, active metalloids efflux from the cell takes place via efflux transporters like the HA-proton efflux transporter (H^+ATP), ferric reductase/oxidase (FRO), phenolics efflux zero1 (PEZ1), efflux transporter (TOM1), and others to maintain ionic balance. Lastly, the vacuolar compartmentalization or sequestration inside cells also takes place via chelating to form complexes as happened outside the membrane. The whole process from uptake to sequestration helps plants to accumulate more HMs and causes the decontamination of the targeted medium. At the same time, plants maintain ionic balance to improve their sustainability and metal accumulation and sequestration efficiency (Gajić et al. 2018).

4.7 Socio-Economic Costs and Benefits of Phytoremediation

The progress in phytoremediation technology has attracted the attention of researchers and different industrial and governmental sectors due to its social and economic benefits over conventional remediation techniques. To summarize the economic value for the the phytoremediation technique, there is a case study in China for the calculation of cost and benefits, which was completed for the implementation of a phytoremediation protocol to clean up highly contaminated land under a two-year project (Wan et al. 2016). They concluded that the total cost for the project was US$ 75 375 per hm^2 or US$ 37.7 per m^3 which comprises the initial capital cost of almost 46% and operational costs of 54% of the total expenditure. There was also a benefit of around US$ 26 098 which was the sum of benefits accrued during and post the phytoremediation period, which was almost 35% of the total cost. We compared various remediation techniques to assess the economic advantage of the phytoremediation technique relative to other physical and chemical techniques (Table 4.6). It shows that solidification costs almost 2–5 folds higher, while physical and chemical extraction costs 7–8 times higher than that of phytoremediation. Off-site disposal and high-temperature thermal desorption costs around 13–22 times and 2–7 times higher than that of phytoremediation, respectively. On-site

Table 4.6 Comparison of costs of various remediation techniques.

S. no.	Remediation techniques	Cost (US$/$m^3$)
1	Phytoremediation	37.7
2	Plant extraction	19–78
3	Solidification	87–190
4	Extraction	240–290
5	Off-site disposal	480–813
6	Off-site high temp. Thermal desorption	81–252
7	On-site biopiling	130–260
8	On-site landfilling	100
9	Soil washing	71.4
10	Bioremediation	59.9
11	Excavation	47.8

Source: Adapted from Wan et al. (2016).

techniques like biopiling is 3.5–7 times and landfill is 2–3 times costlier, while other techniques like soil washing, excavation, and bioremediation cost 2–2.5 fold than that of the phytoremediation technique. Thus, it can be inferred that the financial costs of the phytoremediation technique were lower than most other remediation technologies. By improving the mechanization level of the technology, the costs can be further reduced. Also, the resulted benefits and reduced loss caused due to environmental pollution supplement it in a way that can offset the total expenditure in a few years. The use of different plant species for the removal of toxic HMs is a sustainable, economic, and environmentally friendly approach. Some of the advantageous features of phytoremediation are as follows (Ali et al. 2013; Mahar et al. 2016):

1) Minimal disturbance to ecosystems, i.e. an eco-friendly approach
2) Low cost requirements, i.e. economically effective
3) Easy application due to the amendments of a variety of compounds
4) Reduction in the generation of waste
5) Abatement of the spreading out of contaminants via air and water
6) High acceptance at a social level due to the economic benefits
7) On a large scale, potential energy can be stored and utilized to generate thermal energy
8) Economically viable (provide benefits in terms of capital return during and after phytoremediation)

These benefits make phytoremediation a socio-economically more feasible technique. Despite many advantages, phytoremediation also has limitations and challenges listed as follows (Ali et al. 2013; Mahar et al. 2016):

1) Time consuming as it needs a long time to clean up
2) The potential of remediation by hyperaccumulators is limited by the slow growth rate and low biomass
3) Appropriate management of the polluted biomass (toxic waste) is a must after phytoextraction
4) A limited bioavailability of metal ions due to compact bonding with soil
5) Suitable for low to moderately contaminated sites
6) The mismanagement and deficit of appropriate actions can cause jeopardy in the food chain
7) The HM accumulation ability of some plants can be affected by pests and diseases
8) Soil amendments and various agricultural practices may negatively impact the mobility of pollutants
9) Climatic and weather conditions may also have an impact on phytoremediation

To overcome these obstacles or limitations and to contribute toward the achievement of Sustainable Development Goals, phytoremediation can be considered as a superior approach.

4.8 Conclusion and Recommendations

Although phytoremediation is an emerging technology, it can provide a great opportunity as a green technology for treating contaminated water and soil resources, and hence, can contribute significantly to the management of natural resources. Furthermore, owing to many of its benefits, it is now emerging as a socio-economically viable technology. There are many steps that can be taken forward to upgrade and promote its utilization. Some of the recommendations are listed to make phytoremediation a suitable option for the future (Jabeen et al. 2009; Purakayastha and Chhonkar 2010; Ali et al. 2013):

- Collaboration at increased levels among universities, research institutes, local public, government, and private bodies to create a multidisciplinary unit to debate queries.
- As phytoremediation is truly integrative and multidisciplinary in nature, the collaboration of researchers from different fields is a prerequisite.
- Encourage and incentivize local authorities, private companies, and the public to apply phytoremediation.
- Affiliate the biotechnology to further improve metal uptake and its remediation properties by gene cloning, manipulation, and development of the multigene approach for the simultaneous transfer of several genes into the most efficient candidates to develop "superbug" plants.
- Formulation of data on the basis of large-scale field conditions (which are different from greenhouse and laboratory conditions) for genetically transformed plant species.
- Before the utilization of GMOs for phytoremediation in fields, there should be a comprehensive and decisive risk assessment.
- Formulation of elementary, efficient, and robust techniques to improve the understanding of metalloids in plants, which can ultimately update the knowledge of metal accumulation.

References

Adiloğlu, S. (2017). Heavy metal removal with phytoremediation. In: *Advances in Bioremediation and Phytoremediation* (ed. N. Shiomi), 115–126. Rijeka: IntechOpen https://doi.org/10.5772/intechopen.70330.

Ai, T.N., Naing, A.H., Yun, B.W. et al. (2018). Overexpression of RsMYB1 enhances anthocyanin accumulation and heavy metal stress tolerance in transgenic petunia. *Frontiers in Plant Science* 9 (1388): 1–15. https://doi.org/10.3389/fpls.2018.01388.

Ali, H., Khan, E., and Sajad, M.A. (2013). Phytoremediation of heavy metals-concepts and applications. *Chemosphere* 91 (7): 869–881. https://doi.org/10.1016/j.chemosphere.2013.01.075.

Anderson, T.A., Guthrie, E.A., and Walton, B.T. (1993). Bioremediation in the rhizosphere. *Environmental Science & Technology* 27 (13): 2630–2636. https://doi.org/10.1021/es00049a001.

Awashthi, S.K. (2000). Prevention of Food Adulteration Act No 37 1954. Central and State Rules as Amended for 1999. New Delhi: Ashoka Law House.

Banasova, V., Horak, O., Nadubinska, M. et al. (2008). Heavy metal content in *Thlaspi caerulescens* J. et C. Presl growing on metalliferous and non-metalliferous soils in Central Slovakia. *International Journal of Environment and Pollution* 33 (2–3): 133–145. https://doi.org/10.1504/ijep.2008.019388.

Barbosa, B., Boléo, S., Sidella, S. et al. (2015). Phytoremediation of heavy metal-contaminated soils using the perennial energy crops *Miscanthus* spp. and *Arundo donax* L. *Bioenergy Research* 8 (4): 1500–1511. https://doi.org/10.1007/s12155-015-9688-9.

Begum, A., Ramaiah, M., Khan, I. et al. (2009). Analysis of heavy metals concentration in soil and lichens from various localities of Hosur road, Bangalore, India. *Journal of Chemistry* 6 (1): 13–22. https://doi.org/10.1155/2009/943695.

Belluck, D., Benjamin, S., and David, S. (2006). Why remediate? In: *Phytoremediation of Metal-Contaminated Soils*, NATO Science Series, vol. 68 (eds. J.L. Morel, G. Echevarria and N. Goncharova), 1–23. Dordrecht: Springer https://doi.org/10.1007/1-4020-4688-x_1.

Bhuiyan, M.S.U., Min, S.R., Jeong, W.J. et al. (2011). Overexpression of a yeast cadmium factor 1 (YCF1) enhances heavy metal tolerance and accumulation in *Brassica juncea*. *Plant Cell, Tissue and Organ Culture (PCTOC)* 105 (1): 85–91. https://doi.org/10.1007/s11240-010-9845-y.

Blaylock, M.J., Salt, D.E., and Dushenkov, S. (1997). Enhanced accumulation of Pb in Indian mustard by soil-applied chelating agents. *Environmental Science and Technology* 31 (3): 860–865. https://doi.org/10.1021/es960552a.

Borovička, J., Řanda, J., Jelínek, E. et al. (2007). Hyperaccumulation of silver by *Amanita strobiliformis* and related species of the section *Lepidella*. *Mycological Research* 111 (11): 1339–1344. https://doi.org/10.1016/j.mycres.2007.08.015.

Boshoff, M., De Jonge, M., Scheifler, R. et al. (2014). Predicting As, Cd, Cu, Pb and Zn levels in grasses (*Agrostis* sp. and *Poa* sp.) and stinging nettle (*Urtica dioica*) applying soil–plant transfer models. *Science of the Total Environment* 493: 862–871. https://doi.org/10.1016/j.scitotenv.2014.06.076.

Brooks, R.R., Lee, J., Reeves, R.D. et al. (1977). Detection of nickeliferous rocks by analysis of herbarium specimens of indicator plants. *Journal of Geochemical Exploration* 7: 49–57. https://doi.org/10.1016/0375-6742(77)90074-7.

Brunner, I., Luster, J., Günthardt-Goerg, M.S. et al. (2008). Heavy metal accumulation and phytostabilisation potential of tree fine roots in a contaminated soil. *Environmental Pollution* 152 (3): 559–568. https://doi.org/10.1016/j.envpol.2007.07.006.

Cicek, A. and Koparal, A.S. (2004). Accumulation of sulfur and heavy metals in soil and tree leaves sampled from the surroundings of Tuncbilek Thermal Power Plant. *Chemosphere* 57 (8): 1031–1036. https://doi.org/10.1016/j.chemosphere.2004.07.038.

CWC (2019). Status of Trace & Toxic Metals in Indian Rivers. River Data Compilation-2, Directorate Central Water Commission. New Delhi, 110066, Department of Water Resources, River Development and Ganga Rejuvenation, Ministry of Jal Shakti, Government of India.

Das, N., Bhattacharya, S., and Maiti, M.K. (2016). Enhanced cadmium accumulation and tolerance in transgenic tobacco overexpressing rice metal tolerance protein gene OsMTP1 is promising for phytoremediation. *Plant Physiology and Biochemistry* 105: 297–309. https://doi.org/10.1016/j.plaphy.2016.04.049.

Dhankher, O.P., Li, Y., Rosen, B.P. et al. (2002). Engineering tolerance and hyperaccumulation of arsenic in plants by combining arsenate reductase and γ-glutamylcysteine synthetase expression. *Nature Biotechnology* 20 (11): 1140. https://doi.org/10.1038/nbt747.

Duman, F., Leblebici, Z., and Aksoy, A. (2009). Bioaccumulation of nickel, copper, and cadmium by *Spirodela polyrhiza* and *Lemna gibba*. *Journal of Freshwater Ecology* 24 (1): 177–179. https://doi.org/10.1080/02705060.2009.9664279.

Ellis, D.R., Sors, T.G., Brunk, D.G. et al. (2004). Production of Se-methylselenocysteine in transgenic plants expressing selenocysteine methyltransferase. *BMC Plant Biology* 4 (1): 1–11. https://doi.org/10.1186/1471-2229-4-1.

EPA (2002). Heavy Metals in Wastes, European Commission on Environment. http://ec.europa.eu/environment/waste/studies/pdf/heavy metal ls report.pdf.

Evans, K.M., Gatehouse, J.A., Lindsay, W.P. et al. (1992). Expression of the pea metallothionein-like gene PsMTA in *Escherichia coli* and *Arabidopsis thaliana* and analysis of trace metal ion accumulation: implications for PsMTA function. *Plant Molecular Biology* 20: 1019–1028. https://doi.org/10.1007/bf00028889.

Ezaki, B., Gardner, R.C., Ezaki, Y. et al. (2000). Expression of aluminium-induced genes in transgenic *Arabidopsis* plants can ameliorate aluminium stress and/or oxidative stress. *Plant Physiology* 122 (3): 657–666. https://doi.org/10.1104/pp.122.3.657.

FAO (1985). Water Quality for Agriculture. Paper No. 29 (Rev. 1). Rome: UNESCO. www.waterboards.ca.gov/water_issues/programs/tmdl/records/state_board/1985/ref2648.pdf.

Fernandez-Luqueno, F., López-Valdez, F., Gamero-Melo, P. et al. (2013). Heavy metal pollution in drinking water – a global risk for human health: a review. *African Journal of Environmental Science and Technology* 7 (7): 567–584. https://doi.org/10.5897/AJEST12.197.

Fulekar, M.H. (2016). Phytoremediation of heavy metals by *Helianthus annuus* in aquatic and soil environment. *International Journal of Current Microbiology and Applied Sciences* 5 (7): 392–404. https://doi.org/10.20546/ijcmas.2016.507.043.

Fulekar, M.H., Singh, A., and Bhaduri, A.M. (2009). Genetic engineering strategies for enhancing phytoremediation of heavy metals. *African Journal of Biotechnology* 8 (4): 529–535. https://academicjournals.org/journal/AJB/article-abstract/F8EEF568043.

Gajić, G., Djurdjević, L., Kostić, O. et al. (2018). Ecological potential of plants for phytoremediation and ecorestoration of fly ash deposits and mine wastes. *Frontiers in Environmental Science* 6: 124. https://doi.org/10.3389/fenvs.2018.00124.

Gasic, K. and Korban, S.S. (2007). Transgenic Indian mustard (*Brassica juncea*) plants expressing an *Arabidopsis* phytochelatin synthase (AtPCS1) exhibit enhanced As and Cd tolerance. *Plant Molecular Biology* 64 (4): 361–369. https://doi.org/10.1007/s11103-007-9158-7.

Ghosh, M. and Singh, S.P. (2005). A review on phytoremediation of heavy metals and utilization of it's by products. *Asian Journal of Energy and Environment* 6 (4): 214–231. https://doi.org/10.15666/aeer/0301_001018.

Goto, F., Yoshihara, T., Shigemoto, N. et al. (1999). Iron fortification of rice seed by the soybean ferritin gene. *Nature Biotechnology* 17 (3): 282–286. https://doi.org/10.1038/7029.

Grill, E., Loffler, S., Winnacker, E.L. et al. (1989). Phytochelatins, the heavy-metal binding peptides of plants, are synthesized from glutathione by a specific gamma-glutamylcysteine dipeptidyl transpeptidase (phytochelatin synthase). *Proceedings of the National Academy of Sciences* 86 (18): 6838–6842. https://doi.org/10.1073/pnas.86.18.6838.

Guan, Z., Chai, T., Zhang, Y. et al. (2009). Enhancement of Cd tolerance in transgenic tobacco plants overexpressing a Cd-induced catalase cDNA. *Chemosphere* 76 (5): 623–630. https://doi.org/10.1016/j.chemosphere.2009.04.047.

Harada, E., Choi, Y.E., Tsuchisaka, A. et al. (2001). Transgenic tobacco plants expressing a rice cysteine synthase gene are tolerant to toxic levels of cadmium. *Journal of Plant Physiology* 158 (5): 655–661. https://doi.org/10.1078/0176-1617-00314.

Haseena, M., Malik, M.F., Javed, A. et al. (2017). Water pollution and human health. *Environmental Risk Assessment and Remediation* 1 (3): 16–19. https://doi.org/10.4066/2529-8046.100020.

Hesegawa, I., Terada, E., Sunairi, M. et al. (1997). Genetic improvement of heavy metal tolerance in plants by transfer of the yeast metallothionein gene (CUPI). *Plant and Soil* 196: 277–281. https://doi.org/10.1007/978-94-009-0047-9_117.

Jabeen, R., Ahmad, A., and Iqbal, M. (2009). Phytoremediation of heavy metals: physiological and molecular mechanisms. *The Botanical Review* 75 (4): 339–364. https://doi.org/10.1007/s12229-009-9036-x.

Jain, S.K., Vasudevan, P., and Jha, N.K. (1990). *Azolla pinnata* R. Br. and *Lemna minor* L. for removal of lead and zinc from polluted water. *Water Research* 24 (2): 177–183. https://doi.org/10.1016/0043-1354(90)90100-k.

Jang, M.G., Kim, Y.J., Jang, G.H. et al. (2014). Ectopic overexpression of the aluminium-induced protein gene from *Panax ginseng* enhances heavy metal tolerance in transgenic *Arabidopsis*. *Plant Cell, Tissue and Organ Culture (PCTOC)* 119 (1): 95–106. https://doi.org/10.1007/s11240-014-0516-2.

Kramer, U., Cotter-Howels, J.D., Charnock, J.M. et al. (1996). Free histidine as a metal chelator in plants that accumulate nickel. *Nature* 379 (6566): 635–638. https://doi.org/10.1038/379635a0.

Krishna, I.M., Manickam, V., Shah, A. et al. (2017). *Environmental Management: Science and Engineering for Industry*. BS Publication, Butterworth-Heinemann, An imprint of Elsevier, U.K.

Kumar, V., Sharma, A., Kaur, P. et al. (2019a). Pollution assessment of heavy metals in soils of Indian and ecological risk assessment: a state of the art. *Chemosphere* 216: 449–462. https://doi.org/10.1016/j.chemosphere.2018.10.066.

Kumar, V., AlMomin, S., Al-Shatti, A. et al. (2019b). Enhancement of heavy metal tolerance and accumulation efficiency by expressing *Arabidopsis* ATP sulfurylase gene in alfalfa. *International Journal of Phytoremediation* 21 (11): 1112–1121. https://doi.org/10.1080/15226514.2019.1606784.

Kuperman, R.G. and Carreiro, M.M. (1997). Soil heavy metal concentrations, microbial biomass and enzyme activities in a contaminated grassland ecosystem. *Soil Biology and Biochemistry* 29 (2): 179–190. https://doi.org/10.1016/s0038-0717(96)00297-0.

Li, H., Qian, X., and Wang, Q.G. (2013). Heavy metals in atmospheric particulate matter: a comprehensive understanding is needed for monitoring and risk mitigation. *Environmental Science and Technology* 47: 13210–13211. https://doi.org/10.1021/es404751a.

Liu, D., An, Z., Mao, Z. et al. (2015). Enhanced heavy metal tolerance and accumulation by transgenic sugar beets expressing *Streptococcus thermophilus* StGCS-GS in the presence of Cd, Zn and Cu alone or in combination. *PLoS One* 10 (6): 1–15. https://doi.org/10.1371/journal.pone.0128824.

Liu, H., Zhao, H., Wu, L. et al. (2017). Heavy metal ATPase 3 (HMA3) confers cadmium hypertolerance on the cadmium/zinc hyperaccumulator *Sedum plumbizincicola*. *New Phytologist* 215 (2): 687–698. https://doi.org/10.1111/nph.14622.

Liu, L., Li, W., Song, W., and Guo, M. (2018). Remediation techniques for heavy metal-contaminated soils: principles and applicability. *Science of the Total Environment* 633: 206–219. https://doi.org/10.1016/j.scitotenv.2018.03.161.

Ma, L.Q., Komar, K.M., Tu, C. et al. (2001). A fern that hyperaccumulates arsenic. *Nature* 409 (6820): 579–579. https://doi.org/10.1038/35054664.

Macek, T., Francova, K., Sura, M. et al. (2006). Genetically modified plants with improved properties for phytoremediation purposes. In: *Phytoremediation of Metal-Contaminated*

Soils, vol. 68 (eds. J.L. Morel, G. Echevarria and N. Goncharova), 85–108. Dordrecht: Springer https://doi.org/10.1007/1-4020-4688-x_4.

Mahar, A., Wang, P., Ali, A. et al. (2016). Challenges and opportunities in the phytoremediation of heavy metals contaminated soils: a review. *Ecotoxicology and Environmental Safety* 126: 111–121. https://doi.org/10.1016/j.ecoenv.2015.12.023.

Mahurpawar, M. (2015). Effects of heavy metals on human health. *International Journal of Research Granthaalayah* 530: 1–7. https://doi.org/10.12691/jephh-5-3-2.

Marchiol, L., Assolari, S., Sacco, P. et al. (2004). Phytoextraction of heavy metals by canola (*Brassica napus*) and radish (*Raphanus sativus*) grown on multi-contaminated soil. *Environmental Pollution* 132 (1): 21–27. https://doi.org/10.1016/j.envpol.2004.04.001.

Martínez, M., Bernal, P., Almela, C. et al. (2006). An engineered plant that accumulates higher levels of heavy metals than *Thlaspi caerulescens*, with yields of 100 times more biomass in mine soils. *Chemosphere* 64 (3): 478–485. https://doi.org/10.1016/j.chemosphere.2005.10.044.

Misra, S. and Gedamu, L. (1989). Heavy metal tolerant transgenic *Brassica napus* L. and *Nicotiana tabacum* L. plants. *Theoretical and Applied Genetics* 78 (2): 161–168. https://doi.org/10.1007/bf00288793.

Mohammed, A.S., Kapri, A., and Goel, R. (2011). Heavy metal pollution: source, impact, and remedies. In: *Biomanagement of Metal Contaminated Soils* (eds. M. Khan, M. Zaidi Khan, A. Zaidi, et al.), 1–28. Dordrecht: Springer https://doi.org/10.1007/978-94-007-1914-9_1.

Mora, A.P., Calvo, J.J.O., Cabrera, F. et al. (2005). Changes in enzyme activities and microbial biomass after "in situ" remediation of a heavy metal-contaminated soil. *Applied Soil Ecology* 28: 125–137. https://doi.org/10.1016/j.apsoil.2004.07.006.

Nagajyoti, P.C., Lee, K.D., and Sreekanth, T.V.M. (2010). Heavy metals, occurrence and toxicity for plants: a review. *Environmental Chemistry Letters* 8 (3): 199–216. https://doi.org/10.1007/s10311-010-0297-8.

Neilson, S. and Rajakaruna, N. (2015). Phytoremediation of agricultural soils: using plants to clean metal-contaminated arable land. In: *Phytoremediation*, vol. 1 (eds. A.A. Ansari, S.S. Gill, R. Gill, et al.), 159–168. Cham: Springer International Publishing https://doi.org/10.1007/978-3-319-10395-2_11.

Nematian, M.A. and Kazemeini, F. (2013). Accumulation of Pb, Zn, Cu and Fe in plants and hyperaccumulator choice in Galali iron mine area, Iran. *International Journal of Agriculture and Crop Sciences* 5 (4): 426–432. https://pdfs.semanticscholar.org/f099/1c015a613dc9536b77c06c048191d6b52fbf.pdf.

Nie, L., Shah, S., Rashid, A. et al. (2002). Phytoremediation of arsenate contaminated soil by transgenic canola and the plant growth-promoting bacterium *Enterobacter cloacae* CAL2. *Plant Physiology and Biochemistry* 40 (4): 355–361. https://doi.org/10.1016/s0981-9428(02)01375-x.

Odjegba, V.J. and Fasidi, I.O. (2004). Accumulation of trace elements by *Pistia stratiotes*: implications for phytoremediation. *Ecotoxicology* 13 (7): 637–646. https://doi.org/10.1007/s10646-003-4424-1.

Ojuederie, O.B. and Babalola, O.O. (2017). Microbial and plant-assisted bioremediation of heavy metal polluted environments: a review. *International Journal of Environmental Research and Public Health* 14 (12): 1504. https://doi.org/10.3390/ijerph14121504.

Olguín, E.J. and Sánchez-Galván, G. (2012). Heavy metal removal in phytofiltration and phycoremediation: the need to differentiate between bioadsorption and bioaccumulation. *New Biotechnology* 30 (1): 3–8. https://doi.org/10.1016/j.nbt.2012.05.020.

Oucher, N., Kerbachi, R., Ghezloun, A. et al. (2015). Magnitude of air pollution by heavy metals associated with aerosols particles in Algiers. *Energy Procedia* 74 (C): 51–58. https://doi.org/10.1016/j.egypro.2015.07.520.

Pan, A., Yang, M., Tie, F. et al. (1994). Expression of mouse metallothionein-I gene confers cadmium resistance in transgenic tobacco plants. *Plant Molecular Biology* 24 (2): 341–351. https://doi.org/10.1007/bf00020172.

Park, S., Kim, K.S., Kim, J.T. et al. (2011). Effects of humic acid on phytodegradation of petroleum hydrocarbons in soil simultaneously contaminated with heavy metals. *Journal of Environmental Sciences* 23 (12): 2034–2041. https://doi.org/10.1016/s1001-0742(10)60670-5.

Pilon-Smits, E.A., Hwang, S., Lytle, C.M. et al. (1999). Overexpression of ATP sulfurylase in Indian mustard leads to increased selenate uptake, reduction, and tolerance. *Plant Physiology* 119 (1): 123–132. https://doi.org/10.1104/pp.119.1.123.

Prasad, M.N.V. (2007). Aquatic plants for phytotechnology. In: *Environmental Bioremediation Technologies* (eds. S.N. Singh and R.D. Tripathi), 259–274. Berlin, Heidelberg: Springer https://doi.org/10.1007/978-3-540-34793-4_11.

Prezemeck, E. and Haase, N.U. (1991). The binding of manganese, copper and cadmium to peptides of the xylem sap of plant roots. *Water, Air, and Soil Pollution* 57 (1): 569–577. https://doi.org/10.1007/bf00282920.

Prüss-Üstün, A., Wolf, J., Corvalán, C. et al. (2016). *Preventing Disease through Healthy Environments: A Global Assessment of the Burden of Disease from Environmental Risks.* Switzerland: World Health Organization (WHO), Geneva https://doi.org/10.1016/j.toxlet.2016.07.028.

Pulford, I.D. and Watson, C. (2003). Phytoremediation of heavy metal-contaminated land by trees – a review. *Environment International* 29 (4): 529–540. https://doi.org/10.1016/s0160-4120(02)00152-6.

Pulford, I.D., Riddell-Black, D., and Stewart, C. (2002). Heavy metal uptake by willow clones from sewage sludge-treated soil: the potential for phytoremediation. *International Journal of Phytoremediation* 4 (1): 59–72. https://doi.org/10.1080/15226510208500073.

Punshon, T. and Dickinson, N.M. (1997). Acclimation of *Salix* to metal stress. *The New Phytologist* 137 (2): 303–314. https://doi.org/10.1046/j.1469-8137.1997.00802.x.

Purakayastha, T.J. and Chhonkar, P.K. (2010). Phytoremediation of heavy metal contaminated soils. In: *Soil Heavy Metals*, vol. 19 (eds. I. Sherameti and A. Varma), 389–429. Berlin, Heidelberg: Springer https://doi.org/10.1007/978-3-642-02436-8_18.

Robinson, B.H., Chiarucci, A., Brooks, R.R. et al. (1997). The nickel hyperaccumulator plant *Alyssum bertolonii* as a potential agent for phytoremediation and phytomining of nickel. *Journal of Geochemical Exploration* 59 (2): 75–86. https://doi.org/10.1016/s0375-6742(97)00010-1.

Ross, S.M. (1994). *Toxic Metals in Soil-Plant Systems*, vol. 469. Chichester: Wiley https://doi.org/10.2307/2261650.

Rugh, C.L., Gragson, G.M., Meagher, R.B. et al. (1998). Toxic mercury reduction and remediation using transgenic plants with a modified bacterial gene. *Hortscience* 33 (4): 618–621. https://doi.org/10.21273/hortsci.33.4.618.

Ruiz, O.N., Hussein, H.S., Terry, N. et al. (2003). Phytoremediation of organomercurial compounds via chloroplast genetic engineering. *Plant Physiology* 132 (3): 1344–1352. https://doi.org/10.1104/pp.103.020958.

Sahi, S.V., Bryant, N.L., Sharma, N.C. et al. (2002). Characterization of a lead hyperaccumulator shrub, *Sesbania drummondii*. *Environmental Science & Technology* 36 (21): 4676–4680. https://doi.org/10.1021/es020675x.

Sakakibara, M., Watanabe, A., Inoue, M. et al. (2010). Phytoextraction and phytovolatilization of arsenic from As-contaminated soils by *Pteris vittata*. *Proceedings of The Annual International Conference on Soils, Sediments, Water and Energy* 12 (1): 267–272. https://scholarworks.umass.edu/soilsproceedings/vol12/iss1/26.

Saraswat, S. and Rai, J.P.N. (2009). Phytoextraction potential of six plant species grown in multimetal contaminated soil. *Chemistry and Ecology* 25 (1): 1–11. https://doi.org/10.1080/02757540802657185.

Shakoor, M.B., Shafaqat, A., Farid, M. et al. (2013). Heavy metal pollution, a global problem and its remediation by chemically enhanced phytoremediation: a review. *Journal of Biodiversity and Environmental Sciences* 3 (3): 12–20.

Sharma, A., Kaur, M., Katnoria, J.K. et al. (2016). Heavy metal pollution: a global pollutant of rising concern. In: *Toxicity and Waste Management Using Bioremediation* (eds. A.K. Rathoure and V.K. Dhatwalia), 1–26. Pennsylvania, U.S.: IGI Global https://doi.org/10.4018/978-1-4666-9734-8.ch001.

Siddiqui, E. and Pandey, J. (2019). Assessment of heavy metal pollution in water and surface sediment and evaluation of ecological risks associated with sediment contamination in the Ganga River: a basin-scale study. *Environmental Science and Pollution Research* 26 (11): 10926–10940. https://doi.org/10.1007/s11356-019-04495-6.

Singh, J. and Kalamdhad, A. (2011). Effects of heavy metals on soil, plants, human health and aquatic life. *International Journal of Research in Chemistry and Environment* 1 (2): 15–21. www.ijrce.org.

Singh, A., Sharma, R.K., Agrawal, M. et al. (2010). Risk assessment of heavy metal toxicity through contaminated vegetables from waste water irrigated area of Varanasi, India. *Tropical Ecology* 51 (2): 375–387. https://doi.org/10.1080/10807039.2015.1017875.

Singh, A.K., Kumar, R., Pareek, A. et al. (2012). Overexpression of rice CBS domain containing protein improves salinity, oxidative, and heavy metal tolerance in transgenic tobacco. *Molecular Biotechnology* 52 (3): 205–216. https://doi.org/10.1007/s12033-011-9487-2.

Su, C. (2014). A review on heavy metal contamination in soil worldwide: situation, impact and remediation techniques. *Environmental Skeptics and Critics* 3 (2): 24–38.

Suvarapu, L.N., Seo, Y.K., and Baek, S.O. (2014). Heavy metals in Indian atmosphere: a review. *Research Journal of Chemistry and Environment* 18 (8): 99–111. https://doi.org/10.1177/0748233716654827.

Tang, Y.T., Qiu, R.L., Zeng, X.W. et al. (2009). Lead, zinc, cadmium hyperaccumulation and growth stimulation in *Arabis paniculata* Franch. *Environmental and Experimental Botany* 66 (1): 126–134. https://doi.org/10.1016/j.envexpbot.2008.12.016.

Tlustoš, P., Pavlikova, D., Szakova, J. et al. (2006). Plant accumulation capacity for potentially toxic elements. In: *Phytoremediation of Metal-Contaminated Soils*, vol. 68 (eds. J.L. Morel, G. Echevarria and N. Goncharova), 53–84. Dordrecht: Springer https://doi.org/10.1007/1-4020-4688-x_3.

Tomašević, M., Rajšić, S., Đorđević, D. et al. (2004). Heavy metals accumulation in tree leaves from urban areas. *Environmental Chemistry Letters* 2 (3): 151–154. https://doi.org/10.1007/s10311-004-0081-8.

Truong, P.N., Foong, Y.K., Guthrie, M. et al. (2010). Phytoremediation of heavy metal contaminated soils and water using vetiver grass. In: *Environmental Bioengineering*, vol. 11 (eds. L. Wang, J.H. Tay, S. Tay, et al.), 233–275. Totowa, NJ: Humana Press https://doi.org/10.1007/978-1-60327-031-1_8.

USEPA (2000). Introduction to phytoremediation (EPA/600/R-99/ 107). Washington, DC: Environmental Protection Agency. https://nepis.epa.gov/Exe/ZyPDF.cgi/30003T7G.PDF?Dockey=30003T7G.PDF.

Wan, X., Lei, M., and Chen, T. (2016). Cost–benefit calculation of phytoremediation technology for heavy-metal-contaminated soil. *Science of the Total Environment* 563: 796–802. https://doi.org/10.1016/j.scitotenv.2015.12.080.

Watanabe, M., Shinmachi, F., Noguchi, A. et al. (2005). Introduction of yeast metallothionein gene (CUP1) into plant and evaluation of heavy metal tolerance of transgenic plant at the callus stage. *Soil Science & Plant Nutrition* 51 (1): 129–133. https://doi.org/10.1111/j.1747-0765.2005.tb00016.x.

Wen, X.P., Ban, Y., Inoue, H. et al. (2010). Spermidine levels are implicated in heavy metal tolerance in a spermidine synthase overexpressing transgenic European pear by exerting antioxidant activities. *Transgenic Research* 19 (1): 91–103. https://doi.org/10.1007/s11248-009-9296-6.

WHO/FAO (2007). Joint FAO/WHO Food Standard Programme Codex Alimentarius Commission 13th Session. Report of the Thirty-Eight Session of the Codex Committee on Food Hygiene, ALINORM 07/30/13, Houston. www.fao.org/input/download/report/671/al30_13e.pdf.

Xu, J., Zhang, Y.X., Wei, W. et al. (2008). BjDHNs confer heavy-metal tolerance in plants. *Molecular Biotechnology* 38 (2): 91–98. https://doi.org/10.1007/s12033-007-9005-8.

Zhang, Y., Liu, J., Zhou, Y. et al. (2013). Enhanced phytoremediation of mixed heavy metal (mercury)–organic pollutants (trichloroethylene) with transgenic alfalfa co-expressing glutathione S-transferase and human P450 2E1. *Journal of Hazardous Materials* 260: 1100–1107. https://doi.org/10.1016/j.jhazmat.2013.06.065.

Zhang, X., Zhong, T., Liu, L. et al. (2015). Impact of soil heavy metal pollution on food safety in China. *PLoS One* 10 (8): 0135182. https://doi.org/10.1371/journal.pone.0135182.

Zhang, Z., Wen, X., Huang, Y. et al. (2017). Higher accumulation capacity of cadmium than zinc by *Arabidopsis halleri* ssp. Germmifera in the field using different sowing strategies. *Plant and Soil* 418 (1–2): 165–176. https://doi.org/10.1007/s11104-017-3285-y.

Zhu, Y.L., Pilon-Smits, E.A., Tarun, A.S. et al. (1999). Cadmium tolerance and accumulation in Indian mustard is enhanced by overexpressing γ-glutamylcysteine synthetase. *Plant Physiology* 121 (4): 1169–1177. https://doi.org/10.1104/pp.121.4.1169.

Zhuang, P., Ye, Z.H., Lan, C.Y. et al. (2005). Chemically assisted phytoextraction of heavy metal contaminated soils using three plant species. *Plant and Soil* 276 (1–2): 153–162. https://doi.org/10.1007/s11104-005-3901-0.

5

Phytoremediation of Heavy Metals from the Biosphere Perspective and Solutions

Indica Mohan[1], Kajol Goria[1], Sunil Dhar[1], Richa Kothari[1], B.S. Bhau[2], and Deepak Pathania[1]

[1] *Department of Environmental Sciences, Central University of Jammu, Samba, Jammu and Kashmir, India*
[2] *Department of Botany, Central University of Jammu, Samba, Jammu and Kashmir, India*

5.1 Introduction

The increase in the human population and anthropogenic activities has resulted in substantial assimilation of heavy metals among soil and water ecosystems. Lead (Pb), mercury (Hg), copper (Cu), zinc (Zn), cadmium (Cd), nickel (Ni), and chromium (Cr) are important pollutants present in high amounts. There are many cases of different metals causing problems and alterations in biotic systems. Heavy metals in their biological forms, generated through anthropogenic activities, may cause damage to both manmade and natural ecosystems. The presence of heavy metals in water, soil, and air in trace amounts can cause observable damage to organisms. Bioaccumulation of heavy metal in the food chain poses dangers to human beings. Humans are exposed to heavy metal contamination commonly due to the intake of contaminated air, water, and food (Kabata-Pendias and Pendias 2001; Pivetz 2001).

The contamination of organic pollutants like polychlorinated biphenyls (PCBs), biocides, dichlorodiphenyltrichloroethane (DDT), etc. and heavy metals in ecosystems can cause an imbalance in the environment and pose serious problems among flora and fauna. The natural sources of metals into ecosystems involve the weathering of rocks and soil, erosion, and volcanic explosions (Chekroun and Baghour 2013). Anthropogenic activities such as fertilizers and pesticides, smelting, mining, sludge dumping, electroplating, industrial effluent toxic material, etc. result in the contamination of soil and water ecosystems.

In the environment, about 60% of elements are referred to as heavy metals, having a density greater than 5 g/cm^3. These heavy metals are mainly metalloids, transition metals, basic metals, lanthanides, and actinides, and generally have a large atomic number and atomic mass, and may form compounds and complexes at variable oxidation numbers. There are a number of redox processes used abiotically (chemical and photochemical) or as

Pollutants and Water Management: Resources, Strategies and Scarcity, First Edition. Edited by Pardeep Singh, Rishikesh Singh, Vipin Kumar Singh, and Rahul Bhadouria.

biochemical transformation, including plants and microorganisms in the environment (Wuana and Okieimen 2011). There are around 82 heavy metals, among these, at least 20 are classified as toxic. Depending on the requirements and functions in living systems, heavy metals are categorized into two different groups: essential and nonessential heavy metals. Essential heavy metals are required for physiological functions, development, and body growth in the living organism. On the other hand, nonessential heavy metals are not needed for any physiological function in the living organism, and these include As, Cd, Pb, Hg, and Sn (Pilon-Smits 2005; Tangahu et al. 2011).

Heavy metals in small quantities are nutritionally essential for a healthy upbringing but their high doses disturb the biochemistry and normal physiology of living systems. Heavy metals responsible for human health-related issues and poisoning mainly include key environmental pollutants like chromium, mercury, lead, zinc, arsenic, cadmium, copper, and nickel, most probably in areas with high anthropogenic activities (Van der Ent et al. 2013). Many of these elements are highly toxic in both elemental and soluble salt forms. Heavy metals like Cd, Sn, Cr, Cu, Zn, As, Hg, and Pb are a threat to human health (Navarro-Aviñó et al. 2003; Shen et al. 2002; Singh and Verma 2007; Yadav et al. 2010). Today, environmental pollution with toxic heavy metals has become a global crisis. These heavy metals are a serious problem because they are difficult to degrade biologically and their toxicity level can only be reduced by transforming them from one oxidation state to another (Göhre and Paszkowski 2006). Therefore, heavy metal pollution produces potential risks to human health and causes ecological disturbances by bioaccumulation and biomagnifications in the food chain.

5.2 Heavy Metals

5.2.1 Sources

Intensive agricultural practices and an increasing population tied with industrialization has caused serious environmental problems, comprising the generation and discharge/emission of substantial quantities of toxic wastes, organic contaminants, and heavy metals to water and soil environments (Danh et al. 2009; Younes and Siegers 1984).

Inorganic pollutants like heavy metals are uninterruptedly added to the environment via the dumping of industrial wastes and urban sewage slurry in contiguous fields, water resources, and agricultural soils (Khan et al. 2007). Developmental activities like nuclear fuel mining and milling, occasional nuclear disasters, nuclear weapons testing, etc. lead on releasing these pollutants along with other organic contaminants. These are eventually accumulated in air, soils, and water resources, and are thought to be easily transferred to humans and other living organisms via the food chain due to the transfer of metal or organic contaminants through water to plant or soil to plant (Khan et al. 2007; Sheoran et al. 2010). Soil contamination mainly occurs due to organic pollutants such as pesticides, polycyclic aromatic hydrocarbons, artificial fertilizers, petroleum hydrocarbon, etc. (Danh et al. 2009).

Heavy metals are used in many forms of in the daily workplace, such as in cars, plastics, solar panels, mobile phones, antiseptics, golf clubs, power generation plants, mining, pharmaceutical productions, and industries (Chekroun and Baghour 2013). Different heavy metals are accountable on behalf of different heavy metals discharges, although many of these are released in association as a result of certain chemical reactions and subsequent emissions

(Brooks et al. 1998; Fulekar et al. 2009; Ghosh 2010; Khan 2005; Lindqvist et al. 1991; Lone et al. 2008; Meharg 2005; Popek et al. 2017; Prasad 2003; Rafati et al. 2011; Revathi et al. 2011; Sangeeta and Maiti 2010' Shenbagavalli and Mahimairaja 2010; Tak et al. 2013; Vamerali et al. 2010; Willey 2007; Wright 2007; Wu et al. 2010; Zacchini et al. 2009.

Heavy metals are primarily found as lithophiles (rock-loving) or chalcophiles (sulfide ore-loving). Contamination due to heavy metals occurs due to natural and anthropogenic activities. Natural sources include geological activities such as volcanic activities, weathering of minerals, and erosion. The Earth's crust is approximately 5% heavy metal, with iron constituting the major proportion, i.e. about 95% of this quantity. Human-induced heavy metal pollution in the environment from different sources includes sewage, industrialized waste, fuel combustion, mining, brick kilns, military operations, smelting, pesticides and fertilizers, small-scale industrial units (industries associated with metal smelting, cable coating, metal products, battery production, etc.), and coal burning in thermal power plants (Weyens et al. 2015). The sources of common heavy metals are shown in Table 5.1.

Table 5.1 Sources of common heavy metals.

S. no.	Name of heavy metal	Sources
1)	Arsenic (As) (Z = 33)	Pesticides, herbicides, wood preservatives, tobacco, semiconductors, mining and smelting, power plants, petroleum refining, animal feed additives, volcanoes, and wood preservatives (Gardea-Torresdey et al. 2005).
2)	Cadmium (Cd) (Z = 48)	Sewage sludge from urban stormwater, agrochemicals waste from agricultural soils (Khan et al. 2007), electroplating, plastic stabilizers, paints and pigments, incineration of wastes constituting cadmium-enriched plastics, phosphate-containing fertilizers (Rajfur et al. 2010; Seaward and Richardson 1989), incineration of waste materials and plastics, and nickel-cadmium batteries (Baird and Cann 2008).
3)	Chromium (Cr) (Z = 24)	Electroplating industry, steel industries, fly ash, cement dust, tanneries, solid waste, and sludge (Knox et al. 1999; Kuiper et al. 2004). Chlorinated copper arsenate (CCA) is widely used as a wood preservative (Baird and Cann 2008).
4)	Copper (Cu) (Z = 29)	Pesticides, fertilizers (Knox et al. 1999).
5)	Lead (Pb) (Z = 82)	Paint, municipal sewage water, metalliferous ore mining and smelting, lead-containing industrial wastes, leaded gasoline combustion, etc. (Göhre and Paszkowski 2006; Sekara et al. 2005), emissions from the burning of lead petrol, battery manufacturing processes, herbicides, and insecticides (Gardea-Torresdey et al. 2005; Zorrig et al. 2012).
6)	Mercury (Hg) (Z = 80)	Volcanic eruptions, forest fires, the burning of fossil fuel, discharges from industries producing coal, peat and wood burning, caustic soda, and gold-silver mining (Liu et al. 2005); agricultural runoff, atmospheric deposition, mining, fossil fuels combustion, and pharmaceutical waste discharge heavy metals to aquatic environments (Chekroun and Baghour 2013). Emissions from coal combustion, Au–Ag mining, and medical waste are also responsible for heavy metal contamination (Mesjasz-Przybyłowicz et al. 2004; Saha et al. 2016; Zorrig et al. 2012).

(Continued)

Table 5.1 (Continued)

S. no.	Name of heavy metal	Sources
7)	Nickel (Ni) (Z = 28)	Volcanic eruptions, industrial effluent, kitchen appliances, steel alloys, forest fires, bursting of bubbles and gaseous exchange in the ocean, weathering of rocks and geological objects, power plants, electroplating, automobile batteries (Knox et al. 1999), industrial emissions and effluents, steel alloys, kitchen devices, surgical/operational instruments, and automobile batteries (Ali et al. 2013; Thangavel and Subbhuraam 2004).
8)	Zinc (Zn) (Z = 30)	Electroplating, smelting and refining, mining, metallurgic operations involving zinc, and biosolids (Lockitch 1993).

5.3 Heavy Metals in Air

Common air polluting compounds include nitrogen oxides (NO_x), sulfur oxides (SO_x), hydrocarbons (HC), suspended particulate matter (SPM), carbon monoxide (CO), and ozone (O_3). It was observed that heavy metals form a very small proportion of air pollutants (Singh and Verma 2007).

The amount of heavy metal in the environment depends on factors like distance from the source, rate of release, extent of suspension or disturbance in nearby soil or water, seasonal variation in temperature, wind direction, wind speed, etc. However, the concentrations of heavy metals vary on a day-to-day or even an hour-to-hour basis depending on the frequent changes in the prevailing weather conditions. Heavy metals exist in the atmosphere as particulate matter and can prove to be toxic and carcinogenic for humans as a result of prolonged exposures at higher concentrations. The presence of heavy metals in the air is analyzed with the help of instruments such as the atomic absorption spectrophotometer (AAS) and inductively coupled plasma mass spectrometry (ICPMS) (Gharaibeh et al. 1998).

5.3.1 Sources of Heavy Metals in Air

Heavy metal introduction into ambient air is commonly related to particle emission processes. The sources of heavy metals in the air are listed as follows:

Arsenic (As): Weathering of rocks and sediments, volcanic emissions, geothermal activities, smoke from forest fires, and sea salt spray are common natural sources of arsenic in the environment. The anthropogenic sources of heavy metals include wood preservatives, insecticides, weedicides, and herbicides. About 80% of the arsenic released into the environment is traced from impurities in pesticides. Coal combustion also releases arsenic in the form of gases and aerosols in the air. Tobacco smoking is another major source of arsenic in ambient air (Ali et al. 2013).

Cadmium (Cd): Cadmium is released as a side product from processes like refining, smelting, and mining. Minor amounts of cadmium are added into the air from secondary sources like dust generated by recycling plants of iron and steel scrap. Zinc-containing coals also contain large amounts of associated cadmium. Naturally, cadmium enters into the water system from volcanoes, airborne and suspended soil particles, sea spray, forest fires, etc. (Ali et al. 2013).

Chromium (Cr): Chromium is released into the environment by ferrochrome production. Other sources of chromium include automobile brake lining, leather tanneries, ore refining, and cement-producing plants (Ali et al. 2013).

Copper (Cu): Primary sources include copper mining, roasting, material handling, incinerators, converters, blast furnaces, wire burning, smelting, coal-fired plants, etc.

Lead (Pb): Alkyl lead additives are in motor fuel, lead smelters, refuse incineration, manufacture and refining of lead-containing compounds and products, leaded petrol emissions, coal combustion, iron and steel industries, copper smelting plants, etc.

Mercury (Hg): Forest fires and volcano eruptions release substantial amounts of mercury into the atmosphere. In addition, the burning of wood, oil, coal, and fossil fuels also adds to its concentration in air. Incineration and poor disposal of electronic devices such as light bulbs, batteries, and thermometers also add a considerable amount of the metal to the atmosphere (Ali et al. 2013).

Nickel (Ni): Common natural sources of nickel into the air include volcanic emissions, forest fires, vegetation, wind-blown dust, and the weathering of rocks. Nickel is also added into ambient air as a result of burning coal, the ignition of diesel and oil, and the incineration of waste. Another important source is tobacco smoking, which contaminates the air, and can cause problems among passive smokers (Cempel and Nikel 2006).

Zinc (Zn): The common sources of zinc include forest fires, windborne dust, and volcanic eruptions (Sandstead 2015).

5.4 Heavy Metals: Problems and Harmful Effects

There are few heavy metals that in small amounts are nutritionally indispensable for the proper development of plants. Heavy metals can dissolve in domestic, municipal, or industrial wastewaters from where they are released into inland waters and finally can be bioaccumulated within the inhabiting organisms of the food chain. These contaminating heavy metals also have a tendency to trickle into groundwater and contaminate drinking water; hence, making the consumers vulnerable to heavy metal-related negative effects (Akpor and Muchie 2010).

Heavy metals result in oxidative stress due to their free radical formation ability. Oxidative stress is defined as the condition inside the living cell when reactive oxygen species production is enhanced up to the extent at which intrinsic antioxidant defenses of living cells can be overwhelmed, leading to the overall destruction of the cell. Heavy metals can substitute imperative elements including vital metals in enzymes or pigments that obstruct their respective function through metabolic dysfunction (Ali et al. 2013). The impact of heavy metals depends on the amount or level of exposure, duration of exposure, frequency of exposure, inhalation or consumption, the health status of the individual, etc. The general effects and problems attributed to each heavy metal are discussed in the following section.

- **Arsenic:** The problems caused by arsenic include skin lesions, obstruction of cognitive development, inhibition of growth, photosynthesis and reproduction, and effects on behavior patterns. As arsenate (As^{5+}), arsenic exists as a phosphate analog, hence it can hamper necessary cellular metabolic processes like oxidative phosphorylation, electron transport chain, and ATP production (Ali et al. 2013). It is a metalloid and its consumption might prove lethal through gastrointestinal damage, resulting in severe vomiting and diarrhea. Inorganic As (III) (Arsenite) is more toxic than As (V) because of its ability to be retained in the body for a longer period because of attachment to sulfhydryl groups in several enzymes. Due to the subsequent inactivity of the enzymes, energy production in the cell is reduced and the cell is damaged. Drinking water contaminated with arsenic may also lead to the disruption of hormonal balance and cause disorders such as diabetes and cardiovascular disease. Arsenic is carcinogenic in humans. Different kinds of cancers include lung cancer and cancers of the bladder, skin, and kidney may arise from ingestion and drinking of arsenic-contaminated food and water. Although the exact mechanism of arsenic-linked cancers is not clear, evidence indicates that Ar acts as a co-carcinogen because it inhibits DNA's self-replication process, i.e. transcription, and thus increase the tendency of other cancer-causing carcinogenic compounds (Baird and Cann 2008).
- **Cadmium(Cd):** A major incident related to cadmium toxicity was reported in Japan, where older women and children in particular suffered from a degenerative bone disorder called "itai-itai" also known as "ouch-ouch," with symptoms including brutal joint pain. Cadmium toxicity inside the body of organisms leads to the replacement of Ca^{2+} ions with Cd^{2+} ions. The bones slowly became porous and subsequently are fractured and collapsed (Baird and Cann 2008). Cadmium is a potentially toxic metal that is accumulated in animals and plants as well as humans. It decreases seed germination, lipid content, and growth in plants (Tangahu et al. 2011). Plants absorb cadmium from irrigated water fields, while humans are exposed directly through drinking and breathing cadmium-contaminated water and air, respectively. Tobacco plants grown in cadmium-contaminated soil and irrigation water can absorb and transport the heavy metal up to the leaves which directly pose a risk of contamination to smokers. Smokers burn out the tobacco leaves, which results in the liberation of heavy metal back to the environment. Seafood and organ meats have higher levels of cadmium than most other foods. The lethal dose of cadmium is about 1 g. In humans, Cd is stored in the kidneys and liver which eventually leads to a higher chance of acquiring renal disease. Although Cd is not biomagnified, it is a cumulative poison and its lifetime in the body can be several decades (Baird and Cann 2008). Cadmium being a mutagenic, teratogenic, carcinogenic, and endocrine disrupter also disturbs calcium control in biological systems, which is a reason for kidney dysfunction and chronic anemia (Ali et al. 2013). Cadmium can result in severe membrane integrity damage, subsequent inactivation of membrane-bound enzymes, and cell disintegration (Saier and Trevors 2010; Zhang et al. 2010).
- **Chromium:** Chromium normally occurs in the form of inorganic ions. It is the second most abundant inorganic pollutant of groundwater under hazardous waste sites (Baird and Cann 2008). The toxic state of chromium is hexavalent chromium (Cr^{6+}). The problems related to chromium involve a reduction in chlorosis, membrane damage, enzyme activity, and root damage in plants (Akpor and Muchie 2010), and hair loss in humans (Ali et al. 2013). It can cause headaches, vomiting, diarrhea, and nausea (Lichtfouse and Goyal 2015).

- **Copper:** Copper is a necessary micronutrient for plants at a low level. But at a higher concentration, it becomes toxic to living organisms (Kabata-Pendias and Pendias 2001). It inhibits the rate of photosynthesis, affects plant growth and reproduction, and reduces the surface area of thylakoids (Akpor and Muchie 2010). An increased concentration of copper could lead to suppressed root growth and chlorosis of leaves (Brown 1995). In humans, copper accumulation causes brain damage, renal problems, liver cirrhosis, chronic anemia, and irritation in the stomach and intestines (Ali et al. 2013).
- **Lead:** Lead is a nonessential heavy metal and toxic even at lower concentrations. In plants, high concentrations of lead result in the reduction of chlorophyll manufacturing and stunted plant growth and an increase in superoxide dismutase (Akpor and Muchie 2010). Lead is known to attack the brain and the central nervous system in humans. Lead poisoning poses severe problems in children such as the loss of short term memory, renal failure, impaired development, reduced intelligence, reduced learning ability, problems in coordination, and an enhanced risk for cardiovascular diseases (Ali et al. 2013; Iqbal 2012; Li et al. 2003; Scheper and Tsao 2003; Zhuang et al. 2007). Lead residues lead to the disruption of the nervous system and gastrointestinal and hematological dysfunction in animals. A prolonged exposure to lead can result in hypertension, chronic nephropathy, and impairment in reproductive functioning. Lead also influences enzymatic activities, alters cellular calcium metabolism, and slows down the frequency of nerve conduction (Saier and Trevors 2010; Akpor and Muchie 2010; Suzuki et al. 2001).
- **Mercury:** Mercury is a persistent environmental contaminant that possesses the capability of accumulating in the bodies of fish, animals, and human beings through bioaccumulation. Salts of mercury and compounds comprising of organo-mercury are the most lethal materials present in the environment (Tariq et al. 2006). It gets into the food chain and participates in both biomagnification and bioaccumulation phenomena. Mercury reduces photosynthetic activity, water, and nutrient uptake via roots, reduces the synthesis of antioxidant enzymes, and concentrates phenol and proline compounds in plants. Mercury exposure could also result in the reduction of autoimmunity development, in which an individual's immune system attacks its own body cells (Akpor and Muchie 2010). Mercury disturbs the electron transport chain during cellular energy production in chloroplasts and mitochondria in the form of ATPs. Mercury can lead to the inhibition of aquaporin activity and decreases water content in plants (Odjegba and Fasidi 2007). Mercury poisoning can result in the dysfunction of the nervous system and kidney disruption. It easily passes the blood–brain barrier and causes a deadly effect on the brain (Burken and Schnoor 1997). The effects of mercury contamination at varied concentrations are anxiety, depression, drowsiness, fatigue, loss of memory, insomnia, irritability, recurrent infections, restlessness, vision problems, change in temper, ulcers, tremors, dementia, brain damage, renal damage, and lung-related problems (Ali et al. 2013). A major incident of mercury poisoning was reported in Japan with the Minamata disease outbreak as a result of biomagnification and bioaccumulation of metal in fish and humans who consumed mercury poisoned fish. The disease affected bones and the central nervous system.
- **Nickel:** Nickel at a high level decreases seed germination, dry mass accumulation, and protein production, and affects chlorophylls and enzymes in plants (Akpor and Muchie 2010). It causes nausea, chronic asthma, coughing, and dermatitis (Knox et al. 1999). It also causes lung cancer, while cancers of the throat and stomach have also

been reported by its inhalation, as well as severe hair loss. It is a hematotoxic, hepatotoxic, genotoxic, immunotoxic, reproductive toxic, pulmonary toxic, neurotoxic, and nephrotoxic (Aleksandra and Urszula 2008; Das et al. 2008; Khan et al. 2007; Modaihsh et al. 2004; Scheper and Tsao 2003).

- **Zinc:** The toxicity of zinc may cause lethargy, depression, and neurologic problems (Lescourret et al. 2015). An overdose of zinc often leads to giddiness and tiredness in humans (Ali et al. 2013).

5.5 Remediation of Heavy Metals: Need and Conventional Treatments

The release of toxic heavy metal pollutants into the soil and water is of major environmental concern and the concentration of heavy metals is continuously increasing in the environment every year. Therefore, reducing or removing heavy metals from water and soil ecosystems is necessary for ecological balance and for maintaining environmental health (Govindasamy et al. 2011). All these pathways result in a diverse set of problems caused among living organisms by different heavy metals at different concentrations and levels. The major effects, common problems, and heavy metal-related diseases have led to increased research and experiments in the field of remediation of heavy metals. Various remediation methods and approaches have been formulated for the complete removal of heavy metals from living systems. To fulfill this purpose, different chemical and physical methods have been introduced and each method has its own merits and demerits. However, these old conventional techniques, in general, are expensive, destructive, labor-intensive, and are known to cause secondary problems (Peng et al. 2009; Wuana and Okieimen 2011).

The serious impact of heavy metals has motivated researchers to undertake work on remediation approaches, but according to the research, most of the work is focused on soil and water remediation approaches for the cleanup and restoration of those metals from contaminated sites. Remediation approaches of heavy metals from the air are fewer in number. The most frequently adopted techniques for the elimination of heavy metals consist of solvent extraction, oxidation/reduction, membrane technologies (especially reverse osmosis), ion exchange, and chemical precipitation. The application of each process depends on the number of metal ions, initial concentration, and other physical and chemical properties of the contaminated sample (Chekroun and Baghour 2013).

There are several conventional methods available for the treatment of heavy metals from the soil, including the following:

- **In situ vitrification:** Electrical power is used to heat up and melt the soil and sediment. The molten solidified product is then disposed of after treatment.
- **Soil incineration:** Soil is subjected to thermal treatment to get rid of the contaminants present.
- **Excavation and landfill:** Contaminated soil is removed and then disposed of in a landfill.
- **Soil washing:** Used for the removal of lower concentration of heavy metals from the soil via flowing water through it. The running water stream takes out a sufficient amount of contaminants along with it.

- **Soil flushing:** Enhances the leaching of heavy metals from the impure soil by the facilitation of chemical compounds called chelating agents, which, on reaction with metallic ions, are converted from heavy metals into stable and water-soluble complexes (Wei and Zhou 2006.)
- **Solidification:** Heavy metals are removed from soil systems by treatment with chemical agents that control the solubility of heavy metals and retain species in the soil (Yukselen and Alpaslan 2001).
- **Stabilization of electro-kinetic systems:** The procedure separates organic contaminants and heavy metals, etc. from polluted soil, sludge, and sediment. A contaminated soil site is identified and is implanted with an electrode pair at both ends. A low-intensity electric current is passed that causes the electromigration of ions, electro-osmosis, and electrophoresis. The subsurface contaminants that are present in the aqueous form are desorbed and transported to the respective electrodes depending on the charge they are carrying. The efficiency of the process is dependent on the accurate approximations of the probable geochemical interactions that may occur in the soil upon inducing electricity. It is an efficient method for removing heavy metal contaminants (Sheoran et al. 2010).

Conventional methods related to the removal of heavy metal removal from water include:

- **Reverse osmosis (RO):** RO (Figure 5.1) is a suitable technique used for heavy metal elimination both on a small and large scale. It is a membrane-based diffusion process in which the membrane acts as a molecular filter. The process is effective in removing over 99% of the total dissolved minerals. Reverse osmosis is specifically effective in removing ionic species, i.e. heavy metal ions from a soil or water solution. The eventual recovery of metals from concentrated solutions becomes more feasible. Regarding the drawbacks, the membrane processes use amplified pressure mechanisms which make the method expensive to assemble and operate. (Akpor and Muchie 2010).

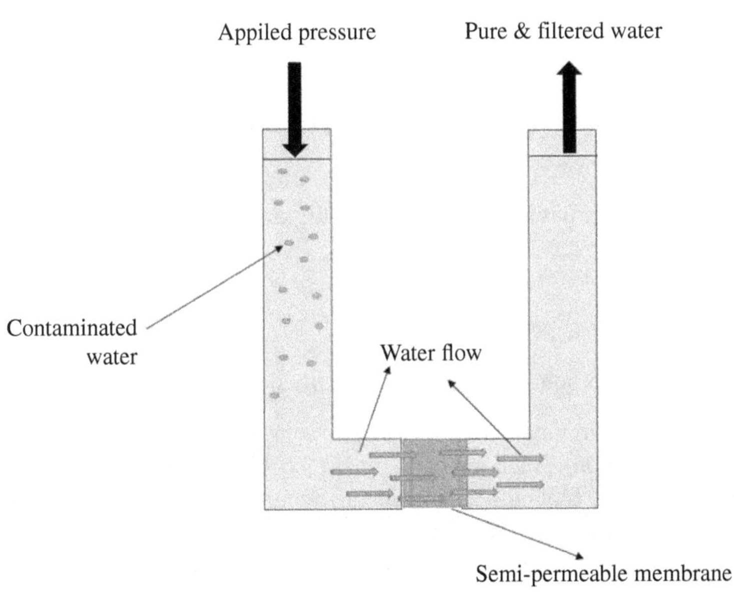

Figure 5.1 Reverse osmosis process.

- **Ultra-filtration:** This method uses low-pressure systems and removes dissolved and colloidal material along with the associated contaminants. Micellar-enhanced ultra-filtration and polymer-enhanced ultra-filtration are types of ultra-filtration. The costly set-up and installation are the main problems related to these processes (Rezaee et al. 2005).
- **Nano-filtration:** Also a membrane-based process that is applied for the removal of contaminant particle sizes ranged between ultra-filtration and reverse osmosis. Membranes used for nano-filtration at commercial scales are made up of synthetic polymers having charged groups. The set-up needs specific designing for removing a particular set of targeted charged entities (Rezaee et al. 2005).
- **Bioremediation:** An advanced process mainly uses microbes for cleaning up contaminated soils, residues, aquifers, and sludge.
- **Chemical precipitation:** Chemicals are added to contaminated water that react with heavy metals, which leads to the formulation of insoluble precipitates. Moreover, precipitates are further separated out of the treated water by means of sedimentation. The agents include sulfides, chelating compounds hydroxides, etc. (Rezaee et al. 2005).
- **Activated carbon adsorption:** Activated carbon is an excellent option for adsorbing heavy metals as it has pores or minute openings which offer a large surface area for interaction and eventual adsorption. Activated carbon is useful for the adsorption of heavy metals and is prepared from any carbon-enriched material (Rezaee et al. 2005).

5.6 Conventional Remedial Techniques: Challenges

Conventional methods of eliminating heavy metals from water and soil can prove helpful for removing a variety of contaminants but there are also many disadvantages of such methods. Most of these techniques may pose the following problems:

- Expensive treatment
- Large energy requirements
- Complex installations
- Nonselective
- Maintenance and repair costs
- Expert requirement
- Alteration of soil or water properties that may cause a disturbance to important naturally growing microbial lifeforms
- Labor expenses
- Less effectiveness
- Less public acceptance
- Release of secondary harmful products and subsequent emissions (Ali et al. 2013)

There are also technical and environmental challenges with conventional methods of heavy metals. For example, in vitrification, a high temperature (>1500 °C) is used to treat the soil area, which melts the soil, and after cooling, heavy metals are trapped in a glassy matrix (Liang et al. 2017). This treated soil is not able to grow plants and is unfertile for agricultural purposes. Similarly, chemical remediation options of heavy metals directly impact on the properties of soil/water and the floral/faunal quantity of water/soil. On the other hand, biological treatment approaches for remediation and restoration are cost-effective, eco-friendly, and a green way to sustain the surroundings from contaminated sites. Hence, phytoremediation is found to be an ecologically sustainable approach among various approaches for the bioremediation of metals. Therefore, this chapter presents an overview of possible phytoremediation options for heavy metal removal.

5.6.1 Phytoremediation: An Emerging Technique

Extensive research and investigation in the field of heavy metals and their effects have led to the development of a different technique in which living organisms are used to eliminate/remove heavy metals from the soil. Phytoremediation is a relatively new technique that has proved helpful in reducing and removing contaminants from the environment (Rodrigues et al. 2012). It has found extensive application in removing organic pollutants and heavy metals from air, water, and soil ecosystems. In comparison with the older methods, phytoremediation is economically cheaper, environmentally sound, and eco-friendly and it is an effective strategy for heavy metal contamination remediation or cleaning polluted sites (Rodrigues et al. 2012; Sakai et al. 2012; Singh 2012; UNEPA 1997).

The term "phytoremediation" originates from the words "phyto" meaning plant and "remediation" meaning recovery. It is also called "phytomining" "bioremediation," "botanical remediation," and "green remediation." Phytoremediation is related to environmental remediation technologies that use plants as the main agent. The equivalent terms of phytoremediation include agro-remediation or vegetative remediation, microbial remediation, green remediation, botano-remediation, etc. (Akpor and Muchie 2010; Poniedziałek et al. 2010). Organic and inorganic pollutants have been removed from contaminated areas by using different species of plant forms (Tariq et al. 2006). It is a remediation practice utilizing plants to eliminate selected contaminants of interest in contaminated soil, sediment, sludge, surface water, groundwater, industrial discharges, and wastewater partially or substantially.

Phytoremediation is employed for the exclusion of unwanted and toxic heavy metals and radionuclides as well as for the elimination of organic pollutants like polychlorinated biphenyls (PCBs), polyaromatic hydrocarbons (PAHs), and pesticides (Ali et al. 2013). Recently, the utilization of plants for the elimination of heavy metals enriched in soil and water has emerged as an effective and efficient alternative process. For the removal of heavy metals from contaminated water, research and a number of studies have been conducted providing several choices among aquatic macrophytes and algal species that can be used for this purpose.

Factors that may affect the process of phytoremediation include soil pH, soil structure, soil type, soil nutrients, irregular dispersal of contaminants, deviations in precipitation, temperature and moisture, herbivores, plant pathogens, etc. (Ali et al. 2013). Phytoremediation is a cost-effective affordable method that helps in the elimination of inactive heavy metals/pollutants from polluted soil or water systems. Phytoremediation facilitates selective nutrient uptake and allows plant roots to absorb nutrients (Tariq et al. 2006). Mechanisms for phytoremediation are shown in Figure 5.2.

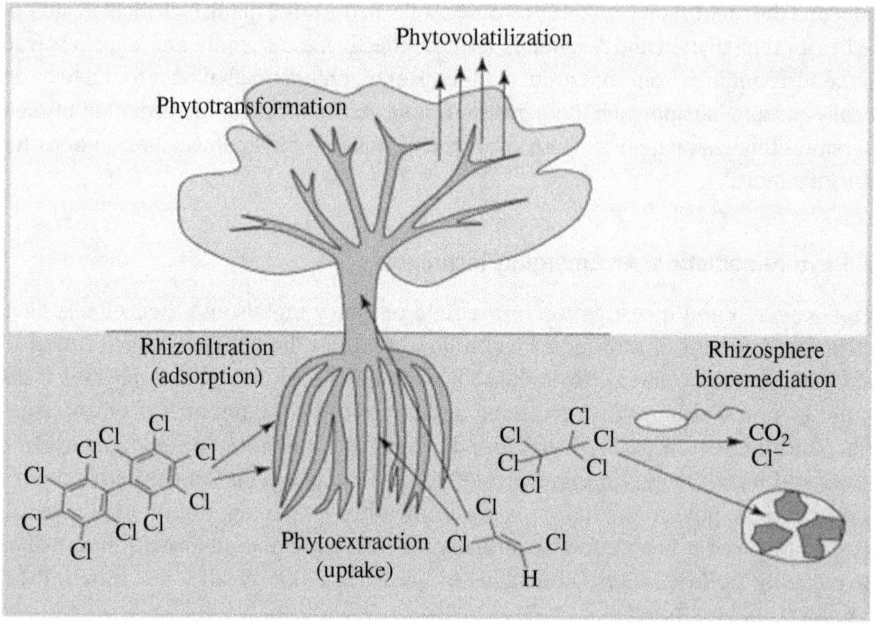

Figure 5.2 Different mechanisms of phytoremediation. (*Source:* Tariq et al. 2006.)

5.6.2 Uses of Phytoremediation

- Improve and clean degraded environments.
- Phytoremediation eliminates the pollutant from the growth media such as soil, air, or water.
- Remediate lethal chemicals found in groundwater, polluted soil, sediment, sludge, wastewater, and surface water.
- Eliminate contaminants from air, water, and soil environments or transform the toxic form of contaminants into harmless forms.
- Remove harmful environmental toxicants from the soil and water environment.
- Phytoremediation helps in degrading, extracting, containing, or immobilizing impurities from the water and soil ecosystem (Tariq et al. 2006).

5.6.3 Phytoremediation: Role of Plants

Phytoremediation is a potential technique that is being explored to mitigate contaminants present in the soil and water. Along with this, a small section of research is also focused on the scope of phytoremediation in clearing polluted air around us. Different tolerant and accumulator trees and plant species are being studied for their potential in removing heavy metals and other pollutants from the air. The pollutants in the air are usually present as particles in a suspended state and mainly interact with plants through physical processes. Plants along with their rhizobia have the potential to take up and metabolize, convert, or even lock up the pollutants from their surrounding environment. Gaseous exchange mechanisms of plants in ambient air and the role of stomata are being explored for applying remediation techniques.

Particle pollutants present in the air can be adsorbed or absorbed by plants through leaves or their stem. In phytoremediation, plants and the associated microbial lifeforms are studied and observed for the capability to degrade or detoxify pollutants from ambient air through various metabolic mechanisms. The use of plants can prove to be an effective and sustainable technique to combat air pollutants both outdoors and indoors (Weyens et al. 2015). As compared with old conventional methods, phytoremediation is cheaper, cost-effective, and environmentally friendly, and can be used for a variety of inorganic as well as organic pollutants.

Vegetation provides large biologically active surface areas to take up air pollutants by absorption or adsorption or sometimes in the form of wet or dry deposition. According to studies, organic pollutants in the air can be removed by using ornamental plants that have shown an ability to absorb and transport them.

In addition, microbial lifeforms present in the root system promote interactions that are important in phytoremediation to enhance detoxification, degradation, and sequestration of specific pollutants (Weyens et al. 2009). Plant species with photosynthetic systems phenomena like the C3 cycle, C4 cycle, and crassulacean acid metabolism (CAM) cycle are being studied under diverse environments to check if they can remove or reduce the concentration of pollutants in the air. CAM or facultative CAM plants show an additional advantage of exchanging gases during the night thus making them helpful in phytoremediation of air contaminants, specifically indoors. The role of waxy layers on leaf surfaces, trichomes, and other plant modifications are also being explored for their potential to mitigate air pollutants under real as well as simulated conditions. Leaf surfaces also provide dwelling spaces for a variety of microorganisms and can be explored whether they help in the degradation of air pollutants. The role of microorganisms present around the root system is also important in the removal of air pollutants. Soil microbes can be investigated so they can be exploited for the remediation of air pollutants.

The entry of air pollutants into plants is primarily through the exposed aerial parts having stomata and cuticles on them. After a pollutant gains entry to the plant, it may undergo various processes and modifications. It can be locked up or sequestered within the plant, detoxified, or transformed or metabolized into CO_2 and H_2O as by-products that are released back into the air. The entire process of phytoremediation includes a set of coordinated processes and chemical reactions to remove or reduce the pollutant. Such processes differ on the basis of the nature, composition, and type of pollutant and specificity of the plant.

Stomata are the air spaces or pores found on the leaf surface or the exterior of plant parts, widely considered as the main entry point for a variety of air pollutants. There can be a gaseous exchange or diffusion of the particles into the stomatal cavity that occurs in a controlled manner. The plant system can direct the channel to air contaminants based on their molecular weights and accordingly adjust the pore spaces or stomatal cavity present between two guard cells. The adjustment of these stomatal cavities is sensitive to the type of air pollutants as well as the prevailing environmental conditions.

On the other hand, stomata allow heavy metals and the associated suspended particulate matter to pass through it. The entry of such pollutants is controlled by the cuticle because they should be present in an ionic form to make their way through this waxy layer. Some of the lipophobic compounds like liquid and gaseous contaminants can gain access through the cuticle. The cuticle acts as an entrance point for contaminants, primarily at some stage in the nighttime while the stomata remain closed. In addition, the cuticle layer of young leaves is more likely to absorb pollutants as compared with older and matured leaves (Agarwal et al. 2019).

5.6.4 Phytoremediation: Agents for Remediation

Phytoremediation is the process of decontaminating soil and aquatic environments by making use of biological components including plants, fungi, or algae to facilitate heavy metal contaminants absorption by these organisms. In recent times, the usage of aquatic plants, specifically micro and macro forms of algae, has gained a lot of interest because of their potential to absorb metals along with taking up other toxicants from their respective surroundings and removing or converting the contaminants into less dangerous forms (Chekroun and Baghour 2013).

In phytoremediation, plants can be deployed for ex situ or in situ management of polluted water, soils, and sediments. Terrestrial green plant vegetation, aquatic flora, and algal forms have the potential to degrade, assimilate, metabolize, or detoxify inorganic and organic pollutants from their respective environment. Phytoextraction and phytomining are widely used practices. The harvested plant tissues are rich in accumulated contaminants and metals and can be processed by drying, converting into ash, or composing to reclaim the metals. This reduces the generation of hazardous waste and generates revenue as a result of recycling metals in a cost-effective way (Aisien et al. 2010).

Hydroponic experiments are leading to an increased thoughtfulness about metal uptake accelerated by accumulator and selective tolerant plant species. They are also used to find contaminant uptake effectiveness and metal tolerance of suggested phytoremediation plant species (Meers and Tack 2004; Padmavathiamma and Li 2007).

Research worldwide has also reported that algae can be used for the phytoremediation of wastewater (Renu et al. 2017). Plant species namely *Salsola kali*, *Prosopis* species, and *Brassica juncea* are being harvested through the hydroponic cultivation system with agar. These plant species have a significant uptake capability for heavy metals and are utilized in phytoremediation. Until now, about 400 potential plant species have been recognized for the cleaning of contaminated water and about 90% of the species have been found to be efficient in removing higher concentrations of nickel (Akpor and Muchie 2010; Macek et al. 2008).

Metallophytes is another category of plants that are specifically adapted to survive in heavy metal-rich soils. Such plants provide a variety of possibilities to be used in phytoremediation (Bothe 2011; Singh et al. 2009) of heavy metal-contaminated soils, either alone or in combination with other microorganisms in soils (Bothe 2011); they mainly belong to the Brassicaceae family. Metallophytes can function as indicators, excluders, and hyperaccumulators of metals.

Metal indicators can accumulate and store heavy metals in their above-ground biomass and tissues but are not suitable for remediation purposes. As the name suggests, the plants only indicate the type of heavy metal present and its probable concentration in the water or soil (Ali et al. 2013; Singh et al. 2009).

Metal excluders can concentrate heavy metals in their roots from growth media, however, they do not allow their transport into the shoot system (Manousaki and Kalogerakis 2010;

Singh et al. 2009). However, such plants are less suitable for metal removal but they have been proven as proficient in the phytostabilization process (Ali et al. 2013).

Advanced research involves identifying genes that are responsible for coding for the over accumulation of particular heavy metals in the plant body. Successful recognition and conversion of such genes to appropriate over- or hyperaccumulator plants make it an alternative option for developing "superbug" plants to facilitate effective and efficient phytoremediation. However, studies related to gene expression and interaction were carried out carefully to avoid any unintended accidental damage that could occur to the environment in the present or future. Regardless of having a lack of knowledge, experience, and other challenges, phytoremediation can be seen as a green or environmentally friendly remediation technology that has an immense potential to tackle heavy metal contamination in the environment (Ali et al. 2013). Systematic representation of the interdisciplinary nature of phytoremediation is shown in Figure 5.3.

5.6.5 Types of Phytoremediation

For organic contaminants, phytoremediation can be undertaken by methods such as phytostabilization, rhizodegradation, rhizofiltration, phytodegradation, and phytovolatilization. For inorganic pollutants, mechanisms that can be used are phytostabilization, rhizofiltration, phytoaccumulation, and phytovolatilization (Tariq et al. 2006). Also, different

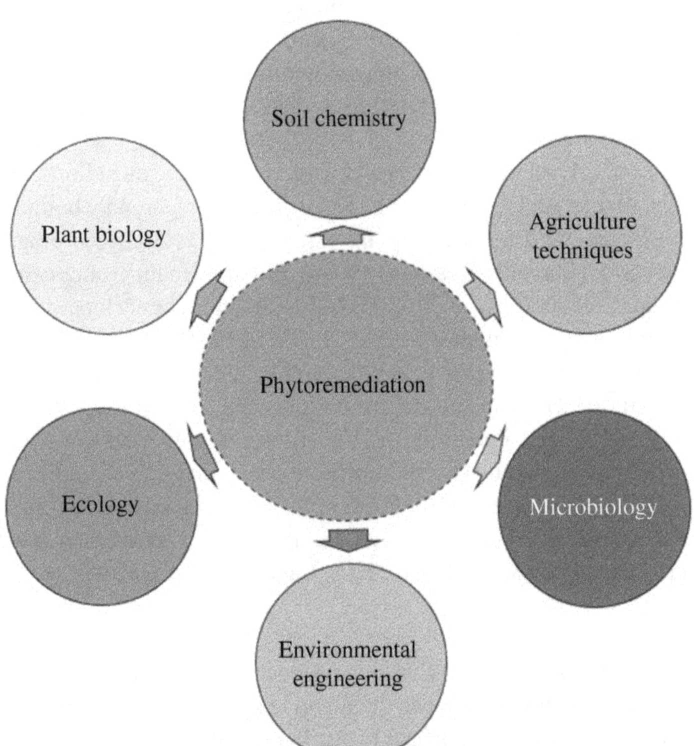

Figure 5.3 Phytoremediation as an interdisciplinary approach.

phytoremediation technologies are not mutually exclusive, which means that accumulation, stabilization, and volatilization processes can take place simultaneously (Hansen et al. 1998; Padmavathiamma and Li 2007).

5.6.5.1 Phytoextraction

Phytoextraction is the absorption and transportation of toxicants and heavy metals by plant roots into the above-ground portions or above-ground biomass of the plant shoot system, which can then be harvested. The harvested portion can also be burned to obtain energy or the metal can be retained and recycled from the left out ash (Tariq et al. 2006). Phytoextraction is often replaced with terms such as phytoabsorption, phytoaccumulation, or phytosequestration. Phytosequestration specifically means there is a permanent locking up of contaminants inside the plant body, thus removing them from the respective environment. It is the upward movement of toxicants from water or soil substratum via plant roots and their accumulation in biomass in the shoot system (Rajfur et al. 2010; Shaw and Bell 1991; Younes and Siegers 1984). Metal translocation to shoots is an important biochemical process and is desirable in the phytoextraction process because the harvest of root biomass is generally not feasible (Tariq et al. 2006; Zhang et al. 2010). It has been concluded that the phytoextraction strategy can help recover several valuable metals like palladium, silver, gold, and platinum (Akpor and Muchie 2010; Truong 2003).

Extraction can be carried out in two ways: natural and artificially induced or stimulated. The natural way of extraction, also called continuous phytoextraction, intends to remove heavy metals from plants under natural circumstances and does not include any soil alteration. Artificially induced phytoextraction uses diverse agents like elemental sulfur, citric acid, be ethylenediamine tetraacetic acid (EDTA), and ammonium sulfate that enhance the bioaccessibility or bioavailability of heavy metals within the soil. This further enhances the efficiency of the process (Ali et al. 2013).

Reduced bioavailability of nearly all heavy metals is considered to be the most important controlling aspect for phytoextraction (Ali et al. 2013). Heavy metals are strongly bound to soil particles and remain inaccessible for plant uptake (Singh et al. 2009). According to heavy metal bioavailability and metalloid accessibility in the soil, three main groups can be identified: readily bioavailable metals (Se, As, Cu, Ni, Cd, and Zn), moderately bioavailable metals (Mn, Fe, and Co), and least bioavailable metals (Cr, U, and Pb) (Prasad 2005). Also, plants have organized certain systems for heavy metals solubilization in the soil. Roots of plants exude metal-activating enzymes in the rhizosphere termed as phytosiderophores (Macek et al. 2008). The release of H^+ ions from roots acidifies the root zone along with enhancement in the dissolution of metal. H^+ ions substitute heavy metal cations that are adsorbed on the soil particles (Alford et al. 2010). Further, the microorganisms existing in the rhizosphere, including mycorrhizal fungi and bacteria, significantly boost heavy metal bioavailability in the soil (Ali et al. 2013; Singh et al. 2009; Van der Ent et al. 2013).

5.6.5.2 Phytofiltration

Phytofiltration is the removal of pollutants from wastewater with help of plants (Williams 2002). In the process, pollutants are adsorbed and absorbed, thus their concentration in wastewater is reduced. Phytofiltration can be further specified as:

- **Rhizofilteration**: The use of roots for absorbing contaminants and heavy metals and then harvesting them after the concentration in water is minimized.

- **Blastofilteration:** The use of seedlings for metal extraction from water.
- **Caulofilteration**. The utilization of excised shoots of plants to eliminate contaminants (Latin caulis = shoot) (Miller 1996).

5.6.5.3 Rhizofiltration

Rhizofiltration is the precipitation or adsorption of contaminants and heavy metals onto plant roots and subsequent storage or permanent stocking of contaminants that are present in the form of solution surrounding the root zone. The heavy metal or contaminant is stored in the roots forever, and thus, the concentration of the heavy metal in soil or water is reduced (Tariq et al. 2006). Rhizofiltration involves the decomposition of organic toxicants present in water through microbial activity enhancement in the rhizosphere of plants. The process is generally applied in polluted soil remediation. The pollutants are either taken out through adsorption onto the plant surface or by means of absorption in the roots. Plants involved for this purpose are first exposed to the pollutants *in vitro* so that they become adapted to such conditions (Lasat 2000; Mishra et al. 2010). A number of plant species such as sunflower, tobacco, spinach, rye, and Indian mustard have been observed to be effective in removing heavy metals such as Cu^{2+}, Cd^{2+}, Cr^{6+}, Ni^{2+}, Pb^{2+}, and Zn^{2+} from aqueous solutions (Akpor and Muchie 2010; EPA 1998).

5.6.5.4 Phytostabilization

Phytostabilization is the use of certain plant species to immobilize the contaminants and heavy metals in the soil and groundwater through absorption and accumulation in plant tissues, adsorption onto the root system, or precipitation within the root zone to prevent their movement in the soil, as well as their transfer by erosion and deflation (Tariq et al. 2006). The process is also called phytoimmobilization. It refers to the utilization of plants in order to facilitate contaminant stabilization in the soil (Sukumaran 2013). This method decreases pollutant mobility and bioavailability in the environment, hence limiting the subsequent movement of contaminants into groundwater and entrance into the food chain (Rew 2007). Heavy metal immobilization by plants in soils occurs through sorption by roots, precipitation, complexation, or metal valence reduction in the rhizosphere (Ghosh and Singh 2005; Poschenrieder and iColl 2003; Younes and Siegers 1984; Zhuang et al. 2007). Phytostabilization limits the accumulation of heavy metals in biotic components and minimizes their movement into underground water. However, this method may not be a permanent solution because the heavy metals continue to remain in the soil or water and only their movement is restricted. It can be categorized as a management strategy for stabilizing (inactivating) or immobilizing potentially toxic contaminants and heavy metals (Vishnoi and Srivastava 2007).

Phytostabilization can also be called as an in-place inactivation that is mainly used for the remediation of soil, sediment, and sludge (EPA 2000). The plant roots limit the mobility and bioavailability of the contaminants in the soil. Contaminants are absorbed and stored by roots, adsorbed, or precipitated in the rhizosphere. As a result of these mechanisms, the mobility of the contaminants is reduced and prevented, hence limiting their chances of getting into the food chain (Jadia and Fulekar 2009; Lasat 2000). The efficiency of phytostabilization depends upon the contaminant tolerance limit of a plant to the contaminant targeted for remediation (Akpor and Muchie 2010; Burken and Schnoor 1997).

5.6.5.5 Phytovolatilization

Phytovolatilization is the uptake and subsequent transpiration of a contaminant by a plant, followed by the release of the contaminant or its modified form into the surrounding atmosphere. Phytovolatilization takes place as growing trees and other plants absorb water along with the contaminants from the ground. Some of these contaminants can pass through the plants and may reach leaves and then volatilize into the atmosphere at comparatively low concentrations. Plant roots may also provide potential surfaces for sorption or precipitation of heavy metal contaminants. The contaminants in the soil and groundwater can be absorbed by the root and are subsequently stored and metabolized by the plant. In many cases, these contaminants are converted or locked into some volatile forms or compounds to be released into the atmosphere. Degradation of contaminants in the soil by plant enzymes released from the roots is another mechanism for phytoremediation. For many contaminants, passive uptake via micropores in the root cell walls can be a major way into the root, where degradation or conversion of contaminants into modified volatile forms can take place (Tariq et al. 2006).

Phytovolatilization is the process involving the uptake of pollutants from the soil by plants, their conversion into their respective volatile form, and then subsequent release in the atmosphere. This method can be used for the removal of organic pollutants and some heavy metals like Hg and Se. The application of phytovolatilization is limited because it does not remove the pollutant completely. On the other hand, it is just transferred from one ecosystem (soil) to another ecosystem (atmosphere) from where it can be redeposited and find its way back to the soil. Phytovolatilization is the most controversial and doubtful technique of phytoremediation methodologies (Peng et al. 2009).

This process involves a specific plant to remove pollutants from the growth medium and subsequent transforming or volatilizing them into the surrounding air through pores present in the leaves. The contaminants can undergo certain modifications along the way, while the contaminated water moves through the plant's vascular system from the roots to the leaves. Few pollutants can get volatilized into the atmosphere at relatively minor concentrations (Ghosh and Singh 2005). Phytovolatization applicability largely depends on the physical properties of the toxicant (Akpor and Muchie 2010).

5.6.5.6 Phytodegradation

Phytodegradation is the breakdown of organic pollutants by plants by releasing enzymes like dehalogenase and oxygenase without being dependent on rhizospheric microbes (Wasay et al. 1998). Plants can accumulate organic contaminants from polluted environments and detoxify them through their metabolic reactions and enzymatic activities. Because of this potential, green plants can be named as the "green liver" for the biosphere. Phytodegradation is limited for the removal of organic pollutants only because heavy metals are nonbiodegradable and hence cannot be eliminated or degraded by some kind of chemical reaction inside the plant body. Recently, scientists also diverted toward phytodegradation of various organic pollutants including synthetic herbicides and insecticides. Some studies have also reported the use of genetically modified plants (e.g. transgenic poplars) and their potential use in this technique (Doty et al. 2000).

Phytodegradation can also be defined as the degradation or disintegration of organic contaminants that are broken down by internal and external metabolic reactions occurring inside the plant body. The major processes involve the uptake, storing, metabolizing, and destroying of the contaminants within its tissue. Plants take up metal contaminants directly from the soil and water or release exudates or enzymes that help to degrade or detoxify pollutants by co-metabolism in the rhizosphere. For the environmental application of the process, it is important that the metabolites that accumulate inside plant tissues are non-toxic or at least less toxic than the primary compound (Akpor and Muchie 2010; Seaward and Richardson 1989).

5.6.5.7 Rhizodegradation

Rhizodegradation is the method in which the breakdown of organic pollutants in the soil is carried out by microorganisms present in the rhizosphere (Williams 2002). The rhizosphere extends about 1 mm around the root system and is under the influence of the plant (Pivetz 2001). The main reason for the increased degradation of contaminants in the rhizosphere is because of the increase in the numbers and metabolic activities of the microorganisms. Plants can influence and stimulate microbial activity to about 10–100 times higher in the rhizosphere by secretion of exudates that contain carbohydrates, amino acids, and flavonoids. The release of nutrient-rich exudates by the plant root system provides carbon and nitrogen supply to the soil microbes and creates a nutrient-rich environment in which the microbial activity is stimulated and enhanced. In addition to secreting organic substrates that facilitate the growth and activities of rhizospheric microorganisms, plants also release certain enzymes that are capable of degrading and detoxifying organic contaminants present in soils (Kuiper et al. 2004; Yoon et al. 2006).

5.6.5.8 Phytodesalination

Phytodesalination is a recently reported, relatively new, and emerging technique in the field of phytoremediation with substantial potential to remove contaminants and other contaminants from the environment. It is an efficient technique used for the removal of excess salts from saline soils by using halophytes growing in the natural environment. In the phytodesalination process, halophytic plants play a major role in the removal of salt from salt-enriched soils in order to allow them to maintain normal and healthy plant growth (Marchiol et al. 2004; Salem et al. 2000). Halophytic plants are naturally well adapted and effective in managing habitats with heavy metal contamination as compared with glycophytic plants (Marchiol et al. 2004).

5.6.5.9 Phytomining

Plant biomass that contains accumulated heavy metals can be burned or combusted for energy and the remaining ash is considered as "bio-ore." Phytomining offers an advantage in terms of energy production from biomass combustion (Anderson et al. 1999). Phytomining is an eco-friendly and environmentally sound alternative when a comparison is made with other old or conventional elimination techniques. However, the feasibility of phytomining is based on different aspects including phytoextraction efficiency, plant species involved, and the current market value of processed metals. Phytomining has been commercially used for nickel and it has been reported that it is less expensive than the respective conventional extraction methods. Ni phytomining has become a highly profitable agricultural technique for treating Ni-contaminated or mineralized soils (Chaney et al. 2007).

5.6.5.10 Phytostimulation

Phytostimulation is the process in which plants secrete natural substances/enzymes through their root system. Microbes are enthused which efficiently increases biological degradation. It works as a symbiotic relationship between plants and rhizospheric microorganisms, encouraging a breakdown of impurities (Lasat 2000; Mishra et al. 2010). In phytostimulation, organic substances and oxygen are released by plant roots, which stimulate microorganisms in the rhizosphere compared with bulk soils. This results in increased microbial activity and subsequent stimulation of a natural breakdown of organic contaminants (Meers and Tack 2004). Different phytoremediation techniques and their mechanisms are given in Table 5.2.

Table 5.2 Summary of different phytoremediation techniques.

Techniques	Mechanism	Useful plant species
Phytoextraction	Accumulation of pollutants in harvestable or removable above-ground biomass and tissues.	Water hyacinth, Indian mustard, *Salix* spp., *Populus* spp., *Trifolium* spp., grasses, pennycress, sunflowers, alyssum, hybrid poplars.
Phytofiltration	Heavy metals concentration reduction using plants.	Mosses, wolfia, azolla, water hyacinth.
Rhizofiltration	Adsorption of contaminants by plant roots and their sequestration, generally under observed conditions.	Sunflower, Azolla, Indian mustard, tobacco, spinach, rye, lemna.
Blastofilteration	Use of seedlings for metal extraction from water.	Seed bearing plants used for phytoextraction.
Caulofilteration	Use of harvested or excised plant shoots to remove heavy metals and other contaminants.	Water hyacinth, Indian mustard, *Salix* spp., *Populus* spp., *Trifolium* spp., grasses, pennycress, sunflowers, alyssum, hybrid poplars.
Phytostabilization	Decreasing the soil mobility and bioavailability limiting of contaminants in the soil environment.	Indian mustard, hybrid poplar, grasses.
Phytovolatilization	Conversion of contaminants into volatile and less harmful products and subsequent release into the surrounding external environment.	Alfalfa, bald cypress, black willow, poplar trees.
Phytodegradation	Deterioration of organic or carbon-rich compounds through plant metabolism and enzyme activities within the tissues.	Poplar trees, sugarcane.
Rhizodegradation	Decomposition of organic compounds by the microorganisms present in the root zone.	Grasses, rice, red mulberry, hybrid poplars.

Table 5.2 (Continued)

Techniques	**Mechanism**	**Useful plant species**
Phytodesalination	Removal of high concentration of salts by halophytes	*Suaeda maritima*, *Sesuvium portulacastrum*, and other hyperaccumulator halophytes.
Phytomining	Elimination of heavy metals through plants for contaminant accumulation.	Water hyacinth, Indian mustard, azolla, lemna.
Phytostimulation	Stimulation of rhizospheric microorganisms to boost the biological breakdown of toxicants effectively.	Maize, Indian mustard, grasses, sunflower.

5.7 Metal Hyperaccumulators: Scope of Phytoremediation

Hyperaccumulators are plants that have the potential to accumulate heavy metals in their upper body parts called the shoot system (Memon et al. 2001; Mesjasz-Przybyłowicz et al. 2004). Such plants are highly metal tolerant and can accumulate metal in the shoots. Generally, the plants taken into account as hyperaccumulators can attain about 100-fold higher shoot metal concentration (without yield reduction) as compared with crop plants or common nonaccumulator plants (Chaney et al. 2007; Lasat 2002). Mathematically, hyperaccumulators attain a shoot-to-root metal concentration ratio (called translocation factor, TF) of greater than the order of 1 (Badr et al. 2012; Tariq et al. 2006). Although it is a helpful measure in supporting the fact of hyperaccumulation, TF cannot be used alone in order to define hyperaccumulation (Vangronsveld et al. 2009). Exploring and finding more effective hyperaccumulator plant species for concentration of heavy metals can prove an important point related to the successful phytoremediation of these pollutants (Shen et al. 2002; Williams 2002).

More than 400 plant species are known as metal hyperaccumulators, and above 300 Ni-hyperaccumulators have been identified (Li et al. 2003; Pulford and Watson 2003). The Brassicaceae family consists of a large number of metal-accumulating plant species (Poniedziałek et al. 2010). Examples of hyperaccumulators are *Alyssum bertolonii,* and *Thlaspi caerulescens. T. caerulescens* (Alpine pennycress) is known as the best metal hyperaccumulator (Lasat 2002). This species is recognized as a hyperaccumulator for Ni, Cd, and Zn (Assunção et al. 2003). The most commonly anticipated theory concerning the advantage of metal hyperaccumulation in plants is the elemental defense mechanism against herbivores (having poisonous, inedible, or unpalatable leaves) and pathogens (Dipu et al. 2012; Memon and Schröder 2009; Pulford and Watson 2003).

Hyperaccumulator plant species are used for toxic heavy metals phytoremediation including Au, Pd, and Pt phytomining. Hyperaccumulators result in the formation of a bio-ore, which is commercially important for cost investment in soil remediation (Gisbert et al. 2003). The heavy metal quantity eliminated by hyperaccumulators from the soil can be calculated by multiplying the metal concentration in the tissue with the quantity of biomass produced by the plants (Malik and Biswas 2012). Several plants have the potential to accumulate specific heavy metals and are known as natural hyperaccumulators.

However, the accumulation ability of some plants for specific heavy metals can be improved unnaturally using biotechnological and bioengineering modification methods. Such modified or improved plants have shown proficient outputs for phytoremediation for various heavy metals, especially Nickel.

The plant species selected for phytoremediation should possess the following characteristics:

- An inherent potential to hyperaccumulate and tolerate heavy metals and organic compounds in above-ground biomass.
- Comparatively fast-growing biomass.
- Naturally repulsive to herbivores so that the accumulated metals and organic contaminants do not enter into the food chain.
- Preferably an extensively distributed and higher level well-developed root system.
- Harvesting and nurturing are fairly simple and easy.
- Extensive geographic allocation so that remediation can be done at different sites at the same time.
- A large accumulation of the targeted contaminant from the growth media.
- Resistance to the toxic effects of the heavy metals being accumulated.
- Well adapted to dominant outer environmental and climatic conditions.
- Hostile to pathogens and possible pest attack (Ali et al. 2013).

5.8 Advantages and Limitations of Phytoremediation

Various advantages and limitations associated with this technique in practical applications are discussed to understand its feasibility.

5.8.1 Advantages of Phytoremediation

- Phytoremediation is an inexpensive, clean, and environmentally friendly method for the removal of heavy metals from contaminated water and soils (Wei et al. 2008; Zorrig et al. 2012).
- Phytoremediation is a process that uses solar energy and does not need an external energy supply or installations.
- It has low maintenance and installation costs as compared with other treatment approaches (Ali et al. 2013).
- By-products along with their modified forms can find a wide range of other uses (Danh et al. 2009; Turan and Esringu 2007). Contaminants can also be collected in woody tissues that can be further used as a potential fuel.
- The results can be observed in low polluted sites in a relatively shorter duration on an experimental basis (Tariq et al. 2006).
- An economically viable and low-cost technique that requires small initial investments.
- Discarded places are not required any longer which reduces the threat of multiple contaminants.
- Esthetically acceptable as compared with traditional treatment methods.
- Phytoremediation shows wide public acceptance and no major social objections.
- The technique offers the potential to treat water and soil samples containing more than one type of contaminant.

- Phytoremediation is known to have fewer aftereffects than other conventional physical and chemical treatment methodologies (Akpor and Muchie 2010).

5.8.2 Limitations

- Phytoremediation is not widely used in developing countries because of a lack of knowledge. It requires additional research on site-specific conditions and plants (Danh et al. 2009).
- It is a new technique and requires prior knowledge of the mechanisms involved and expert supervision.
- Phytoremediation is generally restricted to the area that is inhabited by the plant root system, which does not help efficiency in the case of small plants with low above-ground biomass and a lower root system.
- Longer durations for remediation are required in cases of nonperennial plants as a result of slower growth and less biomass production.
- The efficiency of phytoremediation is dependent on external environmental conditions. Environmental conditions affect the survival and growth of plants, toxicity, and the general conditions of contaminated soil and water (Sas-Nowosielska et al. 2008).
- Phytoremediation techniques require different types of plants for the removal of different types of pollutants in water and soil. This adds to the intricacy of the process and the wide range of experimental actions before a recommendation for commercial application (Danh et al. 2009).
- The success of the mechanism primarily depends on the tolerance of the plant to the contaminants present in the soil or water. This tolerance may depend on climate, geology, altitude, temperature, and other physical characters of the substratum.
- Phytoremediation techniques need different agricultural gears and information to use them on a greater scale.
- There is a small risk of the release of collected contaminants from plant tissues back into the environment after they are dead or decomposed.
- It only offers the removal or reduction of contaminants from soil and water and does not remediate air pollutants.
- There is still a long way to go to specify plant species that can offer the specific elimination of contaminants with great efficiency on a large scale.
- The process of remediation mainly depends on plant growth, and when compared with other technologies, the time requirement may be long.
- The application of phytoremediation is not found to be efficient enough to treat large quantities of heavy metals and other organic contaminants. It provides better results in cases of water or soils contaminated with lower concentrations of toxic substances.
- It is appropriate in areas with very little to modest amounts of heavy metals because plant growth is not sustainable in heavily polluted sites.
- Phytoremediation techniques are mostly limited to the laboratory. Only a few studies have been done in the field using realistic world or water resources (Ali et al. 2013).
- Leaching of soluble contaminants may possibly lead to environmental damage. Furthermore, due to depth limitations, the effectiveness of the remediation that can be achieved is contained in a restricted zone (Akpor and Muchie 2010; Gardea-Torresdey et al. 2005; Negri et al. 1997; Poniedziałek et al. 2010). Leaching of soluble contaminants might pose a threat to the environment globally.

5.9 Useful Plants

Certain plant families are known to have good phytoextraction capabilities like Amaranthaceae, Asteraceae, Brassicaceae, Cucurbitaceae, Fabaceae, and Poaceae. Species like *B. juncea* (Indian mustard), *Eichhornia crassipes* (water hyacinth), *Salix* species (willow tree), *Populus deltoides* (poplar tree), *Sorghastrum nutans* (Indian grasses), and *Helianthus annuus* (sunflower) have been observed as best suited to phytoremediate soil and water contaminants. Species like *Azolla pinnata*, *Arabis paniculata*, *Ipomoea alpine*, *Eleocharis acicularis*, *Euphorbia cheiraadenia*, *Alyssum corsicum*, *Alyssum markgrafii*, *T. caerulescens*, *Potentilla griffithii*, *Pteris vittata*, *Pteris ryukyuensis*, *Phragmites australis*, and *Schima superba* have been found to be efficient for eliminating different heavy metals from soils (Ashraf et al. 2019).

Plants like *Ficus benjamina* (weeping fig), *Spathiphyllum* sp. (peace lily), *Chrysalidocarpus lutescens* (areca palm), *Dracaena fragrans* (corn plant), *Rhapis excelsa* (lady palm), *Dracaena deremensis* (warneckei), *Dieffenbachia camille* (dumb cane), *Philodendron domesticum* (elephant ear philodendron), *Epipremnum aureum* (golden pothos), *Syngonium podophyllum* (arrowhead vine), *Sansevieria trifasciata* (snake plant), *Codiaeum variegatum* (croton), and *Cyperus alternifolius* (umbrella grass) have shown potential in reducing air polluting substances present indoors (Wolverton and Wolverton 1996).

Pinus nigra Arn. and *Carpinus betulus* L have shown the potential to accumulate particulate matter and trace elements like nickel, lead, cadmium, antimony, zinc, chromium, and manganese from ambient air (Popek et al. 2018). *Tilia cordata* Mill. can be used to make the air quality better in polluted cities and is known as a bioindicator for particulate matter (PM) in ambient air (Popek et al. 2018). *Tamarix smyrnensis* can be used for the phytoremediation of cadmium, which is accumulated in its leaves and can be extracted afterward if required (Manousaki et al. 2008).

5.10 Conclusion

Phytoremediation is an evolving technology with enormous potential. It is a promising technique to tackle contamination and pollutant remediation in water, air, and soil ecosystem. It provides a variety of options that can be implemented for the removal of heavy metals with minimum damage to the natural environment. However, there is a lack of intensive research work in many disciplines and how it can be integrated with fields like environmental engineering, molecular biology, biochemistry, and microbiology for maximum benefit. Numerous plant species are standard as heavy metals hyperaccumulators, however, there is a need to develop new efficient approaches so that effective recovery of heavy metals can also be done at commercial levels.

References

Agarwal, P., Sarkar, M., Chakraborty, B., and Banerjee, T. (2019). Phytoremediation of air pollutants: prospects and challenges. *Phytomanagement of Polluted Sites*: 221–241. https://doi.org/10.1016/B978-0-12-813912-7.00007-7.

Aisien, F.A., Faleye, O., and Aisien, E.T. (2010). Phytoremediation of heavy metals in aqueous solutions. *Leonardo Journal of Sciences* 17: 7–46. http://creativecommons.org/licenses/by/3.0.

Akpor, O.B. and Muchie, M. (2010). Remediation of heavy metals in drinking water and wastewater treatment systems: processes and applications. *International Journal of Physical Sciences* 5 (12): 1807–1817. http://www.academicjournals.org/IJPS.

Aleksandra, D. and Urszula, B. (2008). The impact of nickel on human health. *Journal of Elementology* 13 (4): 685–696.

Alford, É.R., Pilon-Smits, E.A., and Paschke, M.W. (2010). Metallophytes – a view from the rhizosphere. *Plant and Soil* 337 (1–2): 33–50. https://doi.org/10.1007/s11104-010-0482-3.

Ali, H., Khan, E., and Sajad, M.A. (2013). Phytoremediation of heavy metals – concepts and a applications. *Chemosphere* 91 (7): 869–881. https://doi.org/10.1016/j.chemosphere.2013.01.075.

Anderson, C.W., Brooks, R.R., Stewart, R.B., and Simcock, R. (1999). Gold uptake by plants. *Gold Bulletin* 32 (2): 48–52.

Ashraf, S., Ali, Q., Zahir, Z.A. et al. (2019). Phytoremediation: environmentally sustainable way for reclamation of heavy metal polluted soils. *Ecotoxicology and Environmental Safety* 174: 714–727. https://doi.org/10.1016/j.ecoenv.2019.02.068.

Assunção, A.G., Schat, H., and Aarts, M.G. (2003). *Thlaspi caerulescens*, an attractive model species to study heavy metal hyperaccumulation in plants. *New Phytologist* 159 (2): 351–360. https://doi.org/10.1046/j.1469-8137.2003.00820.x.

Badr, N., Fawzy, M., and Al-Qahtani, K.M. (2012). Phytoremediation: an ecological solution to heavy-metal-polluted soil and evaluation of plant removal ability. *World Applied Sciences Journal* 16 (9): 1292–1301.

Baird, C. and Cann, M. (2008). Other toxic organic compounds of environmental concern. In: *Environmental Chemistry*, 507–516. New York, NY: WH Freeman and Company.

Bothe, H. (2011). Plants in heavy metal soils. In: *Detoxification of Heavy Metals*, 35–57. Berlin, Heidelberg: Springer https://doi.org/10.1007/978-3-642-21408-0_2.

Brooks, R.R., Chambers, M.F., Nicks, L.J., and Robinson, B.H. (1998). Phytomining. *Trends in Plant Science* 3 (9): 359–362. https://doi.org/10.1016/S1360-1385(98)01283-7.

Brown, K.S. (1995). The green clean. *Bioscience* 45 (9): 579–582. https://www.jstor.org/stable/1312760.

Burken, J.G. and Schnoor, J.L. (1997). Uptake and metabolism of atrazine by poplar trees. *Environmental Science & Technology* 31 (5): 1399–1406. https://doi.org/10.1021/es960629v.

Cempel, M. and Nikel, G.J.P.J.S. (2006). Nickel: a review of its sources and environmental toxicology. *Polish Journal of Environmental Studies* 15 (3): 375–382.

Chaney, R.L., Angle, J.S., Broadhurst, C.L. et al. (2007). Improved understanding of hyperaccumulation yields commercial phytoextraction and phytomining technologies. *Journal of Environmental Quality* 36 (5): 1429–1443. https://doi.org/10.2134/jeq2006.0514.

Chekroun, K.B. and Baghour, M. (2013). The role of algae in phytoremediation of heavy metals: a review. *Journal of Materials and Environmental Science* 4 (6): 873–880.

Danh, L.T., Truong, P., Mammucari, R. et al. (2009). Vetiver grass, *Vetiveria zizanioides*: a choice plant for phytoremediation of heavy metals and organic wastes. *International Journal of Phytoremediation* 11 (8): 664–691. https://doi.org/10.1080/15226510902787302.

Das, K.K., Das, S.N., and Dhundasi, S.A. (2008). Nickel, its adverse health effects & oxidative stress. *Indian Journal of Medical Research* 128 (4): 412.

Dipu, S., Kumar, A.A., and Thanga, S.G. (2012). Effect of chelating agents in phytoremediation of heavy metals. *Remediation Journal* 22 (2): 133–146. https://doi.org/10.1002/rem.21304.

Doty, S.L., Shang, T.Q., Wilson, A.M. et al. (2000). Enhanced metabolism of halogenated hydrocarbons in transgenic plants containing mammalian cytochrome P450-2E1. *Proceedings of the National Academy of Sciences* 97 (12): 6287–6291. https://doi.org/10.1073/pnas.97.12.6287.

EPA Environmental Protection Agency (1998). Phytoremediation of TCE in Groundwater using Populus. *EPA Report.* http://in.org/products/phytotce.htm.

EPA Environmental Protection Agency (2000). Introduction to Phytoremediation. *EPA Report* EPA/600/R-99/107.

Fulekar, M.H., Singh, A., and Bhaduri, A.M. (2009). Genetic engineering strategies for enhancing phytoremediation of heavy metals. *African Journal of Biotechnology* 8 (4): 529–535. http://www.academicjournals.org/AJB.

Gardea-Torresdey, J.L., Peralta-Videa, J.R., DeLaRosa, G., and Parsons, J.G. (2005). Phytoremediation of heavy metals and study of the metal coordination by X-ray absorption spectroscopy. *Coordination Chemistry Reviews* 249 (17–18): 1797–1810. https://doi.org/10.1016/j.ccr.2005.01.001.

Gharaibeh, S.H., Wa'il, Y., and Al-Kofahi, M.M. (1998). Removal of selected heavy metals from aqueous solutions using processed solid residue of olive mill products. *Water Research* 32 (2): 498–502. https://doi.org/10.1016/S0043-1354(97)00221-2.

Ghosh, S. (2010). Wetland macrophytes as toxic metal accumulators. *International Journal of Environmental Sciences* 1 (4): 523–528.

Ghosh, M. and Singh, S.P. (2005). A review on phytoremediation of heavy metals and utilization of it's by products. *Asian Journal of Energy and Environment* 6 (4): 18. http://www.asian-energy-journal.info.

Gisbert, C., Ros, R., De Haro, A. et al. (2003). A plant genetically modified that accumulates Pb is especially promising for phytoremediation. *Biochemical and Biophysical Research Communications* 303 (2): 440–445. https://doi.org/10.1016/S0006-291X(03)00349-8.

Göhre, V. and Paszkowski, U. (2006). Contribution of the arbuscular mycorrhizal symbiosis to heavy metal phytoremediation. *Planta* 223 (6): 1115–1122. https://sci-hub.tw/10.1007/s00425-006-0225-0.

Govindasamy, C., Arulpriya, M., Ruban, P. et al. (2011). Concentration of heavy metals in seagrasses tissue of the Palk Strait, Bay of Bengal. *International Journal of Environmental Sciences* 2 (1): 145.

Hansen, D., Duda, P.J., Zayed, A., and Terry, N. (1998). Selenium removal by constructed wetlands: role of biological volatilization. *Environmental Science & Technology* 32 (5): 591–597. https://doi.org/10.1021/es970502l.

Iqbal, M.P. (2012). Lead pollution-a risk factor for cardiovascular disease in Asian developing countries. *Pakistan Journal of Pharmaceutical Sciences* 25 (1): 289. http://ecommons.aku.edu/pakistan_fhs_mc_bbs/97.

Jadia, C.D. and Fulekar, M.H. (2009). Phytoremediation of heavy metals: recent techniques. *African Journal of Biotechnology* 8 (6): 921–928. http://www.academicjournals.org/AJB.

Kabata-Pendias, A. and Pendias, H. (2001). *Trace Elements in Soils and Plants*, 3e, 403. Boca Raton: CRC Press.

Khan, A.G. (2005). Role of soil microbes in the rhizospheres of plants growing on trace metal contaminated soils in phytoremediation. *Journal of Trace Elements in Medicine and Biology* 18 (4): 355–364. https://doi.org/10.1016/j.jtemb.2005.02.006.

Khan, M.A., Ahmad, I., and Rahman, I.U. (2007). Effect of environmental pollution on heavy metals content of *Withania somnifera*. *Journal of the Chinese Chemical Society* 54 (2): 339–343. https://doi.org/10.1002/jccs.200700049.

Knox, A.S., Gamerdinger, A.P., Adriano, D.C. et al. (1999). Sources and practices contributing to soil contamination. In: *Bioremediation of Contaminated Soils*, 37e, Chapter 4 (eds. D.C. Adriano, J.M. Bollag, W.T. Frankenberger Jr. and R.C. Sims). Madison, WI: American Society of Agronomy https://doi.org/10.2134/agronmonogr37.c4.

Kuiper, I., Lagendijk, E.L., Bloemberg, G.V., and Lugtenberg, B.J. (2004). Rhizoremediation: a beneficial plant-microbe interaction. *Molecular Plant-Microbe Interactions* 17 (1): 6–15. https://doi.org/10.1094/MPMI.2004.17.1.6.

Lasat, M.M. (2000). *The Use of Plants for the Removal of Toxic Metals from Contaminated Soils*. US Environmental Protection Agency.

Lasat, M.M. (2002). Phytoextraction of toxic metals. *Journal of Environmental Quality* 31 (1): 109–120. https://doi.org/10.2134/jeq2002.1090.

Lescourret, F., Dutoit, T., Rey, F. et al. (2015). *Agroecological Engineering* 14: 68–75.

Li, Y.M., Chaney, R., Brewer, E. et al. (2003). Development of a technology for commercial phytoextraction of nickel: economic and technical considerations. *Plant and Soil* 249 (1): 107–115. https://doi.org/10.1023/A:1022527330401.

Liang, L., Liu, W., Sun, Y. et al. (2017). Phytoremediation of heavy metal contaminated saline soils using halophytes: current progress and future perspectives. *Environmental Reviews* 25 (3): 269–281. https://doi.org/10.1139/er-2016-0063.

Lichtfouse, E. and Goyal, A. (2015). *Sustainable Agriculture Reviews*. Cham: Springer.

Lindqvist, O., Johansson, K., Bringmark, L. et al. (1991). Mercury in the Swedish environment – recent research on causes, consequences and corrective methods. *Water, Air, and Soil Pollution* 55 (1–2): xi–261. https://doi.org/10.1007/BF00542429.

Liu, W.H., Zhao, J.Z., Ouyang, Z.Y. et al. (2005). Impacts of sewage irrigation on heavy metal distribution and contamination in Beijing, China. *Environment International* 31 (6): 805–812. https://doi.org/10.1016/j.envint.2005.05.042.

Lockitch, G. (1993). Perspectives on lead toxicity. *Clinical Biochemistry* 26 (5): 371–381. https://doi.org/10.1016/0009-9120(93)90113-K.

Lone, M.I., He, Z.L., Stoffella, P.J., and Yang, X.E. (2008). Phytoremediation of heavy metal polluted soils and water: progresses and perspectives. *Journal of Zhejiang University. Science. B* 9 (3): 210–220. https://doi.org/10.1631/jzus.B0710633.

Macek, T., Kotrba, P., Svatos, A. et al. (2008). Novel roles for genetically modified plants in environmental protection. *Trends in Biotechnology* 26 (3): 146–152. https://doi.org/10.1016/j.tibtech.2007.11.009.

Malik, N. and Biswas, A.K. (2012). Role of higher plants in remediation of metal contaminated sites. *Scientific Reviews and Chemical Communications* 2 (2): 141–146.

Manousaki, E. and Kalogerakis, N. (2010). Halophytes present new opportunities in phytoremediation of heavy metals and saline soils. *Industrial & Engineering Chemistry Research* 50 (2): 656–660. https://doi.org/10.1021/ie100270x.

Manousaki, E., Kadukova, J., Papadantonakis, N., and Kalogerakis, N. (2008). Phytoextraction and phytoexcretion of Cd by the leaves of *Tamarix smyrnensis* growing on contaminated non-saline and saline soils. *Environmental Research* 106 (3): 326–332. https://doi.org/10.1016/j.envres.2007.04.004.

Marchiol, L., Assolari, S., Sacco, P., and Zerbi, G. (2004). Phytoextraction of heavy metals by canola (*Brassica napus*) and radish (*Raphanus sativus*) grown on multi-contaminated soil. *Environmental Pollution* 132 (1): 21–27. https://doi.org/10.1016/j.envpol.2004.04.001.

Meers, E. and Tack, F. (2004). The potential of foliar treatments for enhanced phytoextraction of heavy metals from contaminated soil. *Remediation Journal: The Journal of Environmental Cleanup Costs, Technologies & Techniques* 14 (4): 111–123. https://doi.org/10.1002/rem.20025.

Meharg, A.A. (2005). Mechanisms of plant resistance to metal and metalloid ions and potential biotechnological applications. *Plant and Soil* 274 (1–2): 163–174. https://doi.org/10.1007/s11104-004-0262-z.

Memon, A.R. and Schröder, P. (2009). Implications of metal accumulation mechanisms to phytoremediation. *Environmental Science and Pollution Research* 16 (2): 162–175. https://doi.org/10.1007/s11356-008-0079-z.

Memon, A.R., Aktoprakligil, D., Özdemir, A., and Vertii, A. (2001). Heavy metal accumulation and detoxification mechanisms in plants. *Turkish Journal of Botany* 25 (3): 111–121.

Mesjasz-Przybyłowicz, J., Nakonieczny, M., Migula, P. et al. (2004). Uptake of cadmium, lead nickel and zinc from soil and water solutions by the nickel hyperaccumulator Berkheyacoddii. *Acta Biologica Cracoviensia Series Botanica* 46: 75–85.

Miller, R.R. (1996). Phytoremediation. Technology Overview Report, 39:80. Pittsburgh, PA: Ground Water Remediation Technologies Analysis Center.

Mishra, S., Dwivedi, S.P., and Singh, R.B. (2010). A review on epigenetic effect of heavy metal carcinogens on human health. *Open Nutraceuticals Journal* 3: 188–193. https://doi.org/10.2174/18763960010030100188.

Modaihsh, A.S., AI-Swailem, M.S., and Mahjoub, M.O. (2004). Heavy metals content of commercial inorganic fertilizers used in the Kingdom of Saudi Arabia. *Journal of Agricultural and Marine Sciences* 9 (1): 21–25. https://doi.org/10.24200/jams.vol9iss1pp21-25.

Navarro-Aviñó, J.P., Bellés, J.M., and Serrano, R. (2003). Yeast inositol mono- and trisphosphate levels are modulated by inositol monophosphatase activity and nutrients. *Biochemical and Biophysical Research Communications* 302 (1): 41–45.

Negri, M.C., Hinchman, R.R. and Mollock, J. (1997). Biotreatment of produced waters for volume reduction and contaminant removal (No. ANL/ES/CP-93084; CONF-970990-). Argonne National Lab, Lemont, IL. https://doi.org/10.2172/554814.

Odjegba, V.J. and Fasidi, I.O. (2007). Phytoremediation of heavy metals by *Eichhornia crassipes*. *The Environmentalist* 27 (3): 349–355. https://doi.org/10.1007/s10669-007-9047-2.

Padmavathiamma, P.K. and Li, L.Y. (2007). Phytoremediation technology: hyper-accumulation metals in plants. *Water, Air, and Soil Pollution* 184 (1–4): 105–126. https://doi.org/10.1007/s11270-007-9401-5.

Peng, K.J., Luo, C.L., Chen, Y.H. et al. (2009). Cadmium and other metal uptake by *Lobelia chinensis* and *Solanum nigrum* from contaminated soils. *Bulletin of Environmental Contamination and Toxicology* 83 (2): 260–264. https://doi.org/10.1007/s00128-009-9701-0.

Pilon-Smits, E. (2005). Phytoremediation. *Annual Review of Plant Biology* 56: 15–39. https://doi.org/10.1146/annurev.arplant.56.032604.144214.

Pivetz, B.E. (2001). *Phytoremediation of Contaminated Soil and Ground Water at Hazardous Waste Sites.* US Environmental Protection Agency, Office of Research and Development, Office of Solid Waste and Emergency Response.

Poniedziałek, M., Sękara, A., Jędrszczyk, E., and Ciura, J. (2010). Phytoremediation efficiency of crop plants in removing cadmium, lead and zinc from soil. *Folia Horticulturae* 22 (2): 25–31. https://doi.org/10.2478/fhort-2013-0155.

Popek, R., Łukowski, A., Bates, C., and Oleksyn, J. (2017). Accumulation of particulate matter, heavy metals, and polycyclic aromatic hydrocarbons on the leaves of *Tilia cordata* Mill. In five polish cities with different levels of air pollution. *International Journal of Phytoremediation* 19 (12): 1134–1141. https://doi.org/10.1080/15226514.2017.1328394.

Popek, R., Przybysz, A., Gawrońska, H. et al. (2018). Impact of particulate matter accumulation on the photosynthetic apparatus of roadside woody plants growing in the urban conditions. *Ecotoxicology and Environmental Safety* 163: 56–62. https://doi.org/10.1016/j.ecoenv.2018.07.051.

Poschenrieder, C. and iColl, J.B. (2003). Phytoremediation: principles and perspectives. *Contributions to Science* 343: 333–344.

Prasad, M.N.V. (2003). Phytoremediation of metal-polluted ecosystems: hype for commercialization. *Russian Journal of Plant Physiology* 50 (5): 686–701. https://doi.org/10.1023/A:1025604627496.

Prasad, M.N.V. (2005). Nickelophilous plants and their significance in phytotechnologies. *Brazilian Journal of Plant Physiology* 17 (1): 113–128. https://doi.org/10.1590/S1677-04202005000100010.

Pulford, I.D. and Watson, C. (2003). Phytoremediation of heavy metal-contaminated land by trees – a review. *Environment International* 29 (4): 529–540. https://doi.org/10.1016/S0160-4120(02)00152-6.

Rafati, M., Khorasani, N., Moattar, F. et al. (2011). Phytoremediation potential of *Populus alba* and *Morus alba* for cadmium, chromuim and nickel absorption from polluted soil. *International Journal of Environmental Research* 5 (4): 961–970.

Rajfur, M., Kłos, A., and Wacławek, M. (2010). Sorption properties of algae *Spirogyra* sp. and their use for determination of heavy metal ions concentrations in surface water. *Bioelectrochemistry* 80 (1): 81–86. https://doi.org/10.1016/j.bioelechem.2010.03.005.

Renu, Agarwal, M., and Singh, K. (2017). Methodologies for removal of heavy metal ions from wastewater: an overview. *Interdisciplinary Environmental Review* 18 (2): 124–142. https://doi.org/10.1504/IER.2017.087915.

Revathi, K., Haribabu, T.E., and Sudha, P.N. (2011). Phytoremediation of chromium contaminated soil using sorghum plant. *International Journal of Environmental Sciences* 2 (2): 417–428.

Rew, A. (2007). Phytoremediation: an environmentally sound technology for pollution prevention, control and remediation in developing countries. *Educational Research Review* 2 (7): 151–156.

Rezaee, A., Derayat, J., Mortazavi, S.B. et al. (2005). Removal of mercury from chlor-alkali industry wastewater using *Acetobacter xylinum* cellulose. *American Journal of Environmental Sciences* 1 (2): 102–105. https://doi.org/10.3844/ajessp.2005.102.105.

Rodrigues, S.M., Henriques, B., Reis, A.T. et al. (2012). Hg transfer from contaminated soils to plants and animals. *Environmental Chemistry Letters* 10 (1): 61–67. https://doi.org/10.1007/s10311-011-0329-z.

Saha, N., Rahman, M.S., Jolly, Y.N. et al. (2016). Spatial distribution and contamination assessment of six heavy metals in soils and their transfer into mature tobacco plants in Kushtia District, Bangladesh. *Environmental Science and Pollution Research* 23 (4): 3414–3426. https://doi.org/10.1007/s11356-015-5575-3.

Saier, M.H. and Trevors, J.T. (2010). Phytoremediation. *Water, Air, and Soil Pollution* 205 (1): 61–63. https://doi.org/10.1007/s11270-008-9673-4.

Sakai, Y., Ma, Y., Xu, C. et al. (2012). Phytodesalination of a salt-affected soil with four halophytes in China. *Journal of Arid Land Studies* 22 (1): 17–20.

Salem, H.M., Eweida, E.A. and Farag, A. (2000). Heavy metals in drinking water and their environmental impact on human health. ICEHM2000, 542–556. Cairo University, Egypt.

Sandstead, H.H. (2015). Zinc. In: *Handbook on the Toxicology of Metals*, 4e, 1369–1385. Amsterdam: Elsevier.

Sangeeta, M. and Maiti, S.K. (2010). Phytoremediation of metal enriched mine waste: a review. *American-Eurasian Journal of Agricultural & Environmental Sciences* 9 (5): 560–575. http://www.idosi.org/aejaes/jaes9(5)/17.pdf.

Sas-Nowosielska, A., Galimska-Stypa, R., Kucharski, R. et al. (2008). Remediation aspect of microbial changes of plant rhizosphere in mercury contaminated soil. *Environmental Monitoring and Assessment* 137 (1–3): 101–109. https://doi.org/10.1007/s10661-007-9732-0.

Scheper, T. and Tsao, D.T. (2003). *Advances in Biochemical Engineering/Biotechnology*, vol. 78. Berlin, Germany: Phytoremediation Springer-Verlag.

Seaward, M.R.D. and Richardson, D.H.S. (1989). Atmospheric sources of metal pollution and effects on vegetation. In: *Heavy Metal Tolerance in Plants: Evolutionary Aspects* (ed. A.J. Shaw), 75–92. CRC Press.

Sekara, A., Poniedzialeek, M., Ciura, J., and Jedrszczyk, E. (2005). Cadmium and lead accumulation and distribution in the organs of nine crops: implications for phytoremediation. *Polish Journal of Environmental Studies* 14 (4): 509–516.

Shaw, G. and Bell, J.N.B. (1991). Competitive effects of potassium and ammonium on caesium uptake kinetics in wheat. *Journal of Environmental Radioactivity* 13 (4): 283–296. https://doi.org/10.1016/0265-931X(91)90002-W.

Shen, Z.G., Li, X.D., Wang, C.C. et al. (2002). Lead phytoextraction from contaminated soil with high-biomass plant species. *Journal of Environmental Quality* 31 (6): 1893–1900. https://doi.org/10.2134/jeq2002.1893.

Shenbagavalli, S. and Mahimairaja, S. (2010). Electro-kinetic remediation of contaminated habitats. *African Journal of Environmental Science and Technology* 4 (13): 930–935. http://www.academicjournals.org/AJEST.

Sheoran, V., Sheoran, A.S., and Poonia, P. (2010). Role of hyperaccumulators in phytoextraction of metals from contaminated mining sites: a review. *Critical Reviews in Environmental Science and Technology* 41 (2): 168–214. https://doi.org/10.1080/10643380902718418.

Singh, S. (2012). Phytoremediation: a sustainable alternative for environmental challenges. *International Journal of Green and Herbal Chemistry* 1: 133–139.

Singh, S.N. and Verma, A. (2007). Phytoremediation of air pollutants: a review. In: *Environmental Bioremediation Technologies* (eds. S.N. Singh and R.D. Tripathi), 293–314. Berlin, Heidelberg: Springer https://doi.org/10.1007/978-3-540-34793-4_13.

Singh, A., Eapen, S., and Fulekar, M.H. (2009). Potential of *Medicago sativa* for uptake of cadmium from contaminated environment. *Romanian Biotechnological Letters* 14 (1): 4164–4169.

Sukumaran, D. (2013). Phytoremediation of heavy metals from industrial effluent using constructed wetland technology. *Applied Ecology and Environmental Sciences* 1 (5): 92–97. http://10.12691/aees-1-5-4.

Suzuki, N., Koizumi, N., and Sano, H. (2001). Screening of cadmium-responsive genes in *Arabidopsis thaliana*. *Plant, Cell & Environment* 24 (11): 1177–1188. https://doi.org/10.1046/j.1365-3040.2001.00773.x.

Tak, H.I., Ahmad, F., and Babalola, O.O. (2013). Advances in the application of plant growth-promoting rhizobacteria in phytoremediation of heavy metals. In: *Reviews of Environmental Contamination and Toxicology*, vol. 223, 33–52. New York, NY: Springer https://doi.org/10.1007/978-1-4614-5577-6_2.

Tangahu, B.V., Abdullah, S., Rozaimah, S. et al. (2011). A review on heavy metals (As, Pb, and Hg) uptake by plants through phytoremediation. *International Journal of Chemical Engineering* https://doi.org/10.1155/2011/939161.

Tariq, M., Ali, M., and Shah, Z.J.S.E. (2006). Characteristics of industrial effluents and their possible impacts on quality of underground water. *Soil and Environment* 25 (1): 64–69.

Thangavel, P. and Subbhuraam, C.V. (2004). Phytoextraction: role of hyperaccumulators in metal contaminated soils. *Proceedings-Indian National Science Academy Part B* 70 (1): 109–130.

Truong, P. (2003). Vetiver system for water quality improvement. In Proceedings of 3rd International Vetiver Conference, 6–9, Guangzhou, China.

Turan, M. and Esringu, A. (2007). Phytoremediation based on canola (Brassica napus L.) and Indian mustard (*Brassica juncea* L.) planted on spiked soil by aliquot amount of Cd, Cu, Pb, and Zn. *Plant, Soil and Environment* 53 (1): 7.

UNEPA. (1997). Prevention, Pesticides and Toxic Substances (7508P), EPA 738-F-06-014 June 2008. Washington, DC: United States Environmental Protection Agency.

Vamerali, T., Bandiera, M., and Mosca, G. (2010). Field crops for phytoremediation of metal-contaminated land. A review. *Environmental Chemistry Letters* 8 (1): 1–17. https://doi.org/10.1007/s10311-009-0268-0.

Van der Ent, A., Baker, A.J., Reeves, R.D. et al. (2013). Hyperaccumulators of metal and metalloid trace elements: facts and fiction. *Plant and Soil* 362 (1–2): 319–334. https://doi.org/10.1007/s11104-012-1287-3.

Vangronsveld, J., Herzig, R., Weyens, N. et al. (2009). Phytoremediation of contaminated soils and groundwater: lessons from the field. *Environmental Science and Pollution Research* 16 (7): 765–794. https://doi.org/10.1007/s11356-009-0213-6.

Vishnoi, S.R. and Srivastava, P.N. (2007). Phytoremediation–green for environmental clean (1016:1021). *Proceedings of Taal 2007, 12th World Lake Conference*, Jaipur, Rajasthan, India

Wasay, S.A., Barrington, S.F., and Tokunaga, S. (1998). Remediation of soils polluted by heavy metals using salts of organic acids and chelating agents. *Environmental Technology* 19 (4): 369–379. https://doi.org/10.1080/09593331908616692.

Wei, S. and Zhou, Q.X. (2006). Phytoremediation of cadmium-contaminated soils by Rori aglobosa using two-phase planting (5pp). *Environmental Science and Pollution Research* 13 (3): 151–155. https://doi.org/10.1065/espr2005.06.269.

Wei, S., Zhou, Q., and Saha, U.K. (2008). Hyperaccumulative characteristics of weed species to heavy metals. *Water, Air, and Soil Pollution* 192 (1–4): 173–181. https://doi.org/10.1007/s11270-008-9644-9.

Weyens, N., van der Lelie, D., Taghavi, S., and Vangronsveld, J. (2009). Phytoremediation: plant–endophyte partnerships take the challenge. *Current Opinion in Biotechnology* 20 (2): 248–254. https://doi.org/10.1016/j.copbio.2009.02.012.

Weyens, N., Thijs, S., Popek, R. et al. (2015). The role of plant–microbe interactions and their exploitation for phytoremediation of air pollutants. *International Journal of Molecular Sciences* 16 (10): 25576–25604. https://doi.org/10.3390/ijms161025576.

Willey, N. (ed.) (2007). *Phytoremediation: methods and reviews*, vol. 23, 3–13. Springer Science and Business Media.

Williams, J.B. (2002). Phytoremediation in wetland ecosystems: progress, problems, and potential. *Critical Reviews in Plant Sciences* 21 (6): 607–635. https://doi.org/10.1080/0735-260291044386.

Wolverton, B.C. and Wolverton, J.D. (1996). Interior plants: their influence on airborne microbes inside energy-efficient buildings. *Journal of the Mississippi Academy of Sciences* 41 (2): 99–105.

Wright, R.T. (2007). *Environmental Science: Toward a Sustainable Future*. Burlington, MA: Jones & Bartlett Publishers.

Wu, G., Kang, H., Zhang, X. et al. (2010). A critical review on the bio-removal of hazardous heavy metals from contaminated soils: issues, progress, eco-environmental concerns and opportunities. *Journal of Hazardous Materials* 174 (1–3): 1–8. https://doi.org/10.1016/j.jhazmat.2009.09.113.

Wuana, R.A. and Okieimen, F.E. (2011). Heavy metals in contaminated soils: a review of sources, chemistry, risks and best available strategies for remediation. *Isrn Ecology* 2011: 1–20.

Yadav, R., Arora, P., Kumar, S., and Chaudhury, A. (2010). Perspectives for genetic engineering of poplars for enhanced phytoremediation abilities. *Ecotoxicology* 19 (8): 1574–1588. https://doi.org/10.1007/s10646-010-0543-7.

Yoon, J., Cao, X., Zhou, Q., and Ma, L.Q. (2006). Accumulation of Pb, Cu, and Zn in native plants growing on a contaminated Florida site. *Science of the Total Environment* 368 (2–3): 456–464. https://doi.org/10.1016/j.scitotenv.2006.01.016.

Younes, M. and Siegers, C.P. (1984). Interrelation between lipid peroxidation and other hepatotoxic events. *Biochemical Pharmacology* 33 (13): 2001–2003. https://doi.org/10.1016/0006-2952(84)90564-1.

Yukselen, M.A. and Alpaslan, B. (2001). Leaching of metals from soil contaminated by mining activities. *Journal of Hazardous Materials* 87 (1–3): 289–300. https://doi.org/10.1016/S0304-3894(01)00277-1.

Zacchini, M., Pietrini, F., Mugnozza, G.S. et al. (2009). Metal tolerance, accumulation and translocation in poplar and willow clones treated with cadmium in hydroponics. *Water, Air, and Soil Pollution* 197 (1–4): 23–34. https://doi.org/10.1007/s11270-008-9788-7.

Zhang, S., Chen, M., Li, T. et al. (2010). A newly found cadmium accumulator – *Malva sinensis* Cavan. *Journal of Hazardous Materials* 173 (1–3): 705–709. https://doi.org/10.1016/j.jhazmat.2009.08.142.

Zhuang, X., Chen, J., Shim, H., and Bai, Z. (2007). New advances in plant growth-promoting rhizobacteria for bioremediation. *Environment International* 33 (3): 406–413. https://doi.org/10.1016/j.envint.2006.12.005.

Zorrig, W., Rabhi, M., Ferchichi, S. et al. (2012). Phytodesalination: a solution for salt-affected soils in arid and semi-arid regions. *Journal of Arid Land Studies* 22 (1): 299–302. https://doi.org/10.5402/2011/402647.

6

Phytoremediation for Heavy Metal Removal

Technological Advancements

Monika Yadav[1], Gurudatta Singh[2], and R.N. Jadeja[1]

[1] *Department of Environmental Studies, Faculty of Science, The Maharaja Sayajirao University of Baroda, Vadodara, Gujarat, India*
[2] *Institute of Environment and Sustainable Development, Banaras Hindu University, Varanasi, Uttar Pradesh, India*

6.1 Introduction

The environment is a composite term consisting of air, soil, water, sunlight, and food, which are essential for carrying out the proper functioning of living beings, in general, biotic and abiotic components. The development of living organisms is favored in the environment. Pollution is defined as an undesirable change in the normal characteristics of water, air, or soil that causes an adverse impact on living conditions, human beings, cultural assets, and many other processes (Ekta and Modi 2018). The industrial revolution, agricultural techniques, and science and technology have improved lifestyles, but these developments have come at the cost of a polluted environment. The industrial era, modern lifestyles, and agricultural practices have increased the pollution load in the environment (Chua et al. 2019). A wide range of organic and inorganic pollutants such as hydrocarbons, volatile organic compounds, solvents, heavy metals, etc. have been released into the environment via industrial activities. Environmental pollution caused by heavy metals has become serious worldwide due to its toxic, carcinogenic, and mutagenic nature. Lead (Pb), cadmium (Cd), copper (Cu), chromium (Cr), mercury (Hg), zinc (Zn), and aluminum (Al), are the most common toxic heavy metals that can harm humans, for example, liver damage, kidney damage, circulatory disorders, central nervous system (CNS) damage, cancer, and skin disorders (Agwu et al. 2018).

According to the United States Environmental Protection Agency (USEPA 1996), these heavy metals are considered as "priority pollutants," thus their remediation is a requisite step to maintain a stable environment (Manoj et al. 2020). Heavy metals cannot be degraded chemically or biologically like other pollutants; they are tough to eliminate from the environment. Some of the common heavy metals like cobalt, cadmium, lead, zinc, cadmium, etc. can be phytotoxic in wastewater, whether in higher concentrations or lower concentrations (Divya et al. 2011).

Pollutants and Water Management: Resources, Strategies and Scarcity, First Edition. Edited by Pardeep Singh, Rishikesh Singh, Vipin Kumar Singh, and Rahul Bhadouria.

Heavy metals influence the biochemical and physiological procedures that can be poisonous for macro-organisms and microorganisms, reducing growth, preventing photosynthesis, and deteriorating cell organelles (Mclaughlin et al. 2000). Sources of heavy metal are discussed in Figure 6.1. The heavy metals released by the mining industry, electroplating, smelters, electronic wastes, hospital wastes, etc. increase the major portion of the pollution load in wastewater. A list of some heavy metal contaminants along with their sources are discussed in Table 6.1.

Heavy Metals

Natural
- Weathering of mineral
- Abrasion of rocks
- Forest fire
- Biogenic sources
- Particle released by vegetation
- Volcanic eruption
- Aerosol formation

Anthropogenic
- Industrial process
- Organic and inorganic fertilizer
- Fossil fuel combustion
- Mining
- Smelting
- Waste disposal
- Agriculture and forestry

Figure 6.1 Different sources of heavy metals in the environment.

Table 6.1 Heavy metal contaminants along with their source.

S. no.	Heavy metals	Source
1	Chromium (Cr)	Leather tanning, textiles, pesticides, fertilizer, dyes, mining, industrial coolants, chromium salts manufacturing
2	Manganese (Mn)	Fertilizer, municipal wastewater discharges, steel production
3	Arsenic (As)	Timber treatment, pesticides, geothermal, geogenic/natural processes, smelting operations, thermal power plants, fuel-burning
4	Cadmium (Cd)	Electroplating, batteries, fertilizers, waste batteries, e-waste, paint sludge, incinerations, and fuel combustion
5	Copper (Cu)	Fungicides, electrical, pigments, timber treatment, fertilizers, mine tailings, electroplating, smelting operations
6	Nickel (Ni)	Alloys, battery industry, smelting operations, thermal power plants
7	Iron (Fe)	Engine parts, textiles, fuel, pigments and paints, metal refining, whole meal products, vegetables
8	Cobalt (Co)	Soil contamination, food, beverages, grinding and cutting tools, paints, batteries
9	Zinc (Zn)	Dyes, paints, timber treatment, fertilizers, mine tailings, smelting operations, electroplating

(Continued)

Table 6.1 (Continued)

S. no.	Heavy metals	Source
10	Lead (Pb)	Batteries, metal products, preservatives, petrol additives, paints, e-waste, coal-based thermal power plants, ceramics
11	Mercury (Hg)	Fumigants, geothermal, thermal power plants, fluorescent lamps, hospital waste (damaged thermometers, barometers, sphygmomanometers), electrical appliances
12	Selenium (Se)	Mining smelting, coal or oil combustion, natural deposits
13	Molybdenum (Mo)	Fertilizer

To clean up metal contamination, many physiochemical techniques have been practiced but most of them are costly and produce secondary contaminants. The alternative adaptation and eco-friendly strategy for heavy metal removal using biological resources is phytoremediation (Jacob et al. 2018).

6.2 Phytoremediation

Phytoremediation is derived from the Greek prefix "phyto" relating to plants and the Latin suffix "remedium" meaning clean (Vasavi et al. 2010). It is an in situ remediation method in which the ability of plants is used to assimilate or detoxify metal and organic chemicals from water, soil, and air (Sharma et al. 2016). The roots of plants are selected for phytoremediation to access the organic nutrients and metal ions through absorption, later they will be sequestered, degraded, immobilized, or mobilized. Solar power is utilized directly in this remediation method, eliminating the need for high capital equipment (Wang et al. 2002). Fast growth with high biomass production, high tolerance, an extensive root system, adaptability, the ability to accumulate pollutants, and being native are some of the common properties of phytoremediation plants. They are grouped into three categories: excluders (metal sensitive plants), indicators (poor metal uptake and transport), and accumulators (higher uptake and accumulation) (Khan et al. 2009). Plants tolerating elevated heavy metal concentrations and that accumulate an increased level of harmful heavy metals in their tissues are hyperaccumulators (Ma et al. 2016).

According to Ghosh and Singh (2005), around 400 plant species from 45 plant families are reported for hyperaccumulating metals, the most common families among them are *Fabaceae, Brassicaceae, Asteraceae, Euphorbiaceae, Scrophulariaceae*, and *Lamiaceae* (Ghosh and Singh 2005). According to a report by Lasat (2002), the bioaccumulation potential for cadmium/zinc, copper, cobalt, selenium, and nickel are very high in plants like *Thlaspi caerulescens, Ipomoea alpine, Haumaniastrum robertii, Astragalus racemosus*, and *Sebertia acuminata*, respectively (Lasat 2002). Elevated heavy metal uptake and tolerance were reported in plants like willow, maize, Indian mustard, sunflower, etc. Many plants tend to accumulate heavy metals in their vegetative as well as reproductive parts. Accumulation of high levels of cadmium is found in the leaves rather than the roots in plants like *Brassica oleracea* L., *Lactuca sativa* L., and *Nicotiana tabacum* L. According to Lone et al. (2008), cadmium, chromium, copper, nickel, lead, and zinc can be effectively eliminated by the roots of Indian mustard, and lead, uranium, cesium, and strontium can be removed from hydroponic solutions using sunflower (Tang et al. 2003; Lone et al. 2008).

The main advantage of phytoremediation is that being an autotrophic system with huge biomass, it needs less nutrient input and is easy to manage. The plants are generally accepted by society due to their aesthetic appeal and environmental sustainability (Mahajan and Kaushal 2018).

Generally, the aerial parts of plants contribute to the accumulation, concentration, and translocation process. The initiation of metal uptake is the extraction from the soil solution by mobilizing metals toward the root surface. Selectively permeable recognition of heavy metals binds metal ions to the root cells and transports them to aerial parts, facilitating metal uptake via the roots (Mahajan and Kaushal 2018; Suman et al. 2018). The vascular system of plants enables the transport process accompanying some transporter proteins. In the aquatic ecosystem, phytoremediation takes place by direct absorption from wastewater depending on the type of hyperaccumulator plant and the level of wastewater. Passive metal transport processes are initiated when aquatic plants directly contact contaminated wastewater, which results in metal ion accumulation in aerial parts (Rezania et al. 2016). The processes of phytoremediation are discussed in Figure 6.2.

The route of metal uptake and transport in plants is differentiated in the phytoremediation process as phytostabilization (plant activities reduce the mobilization of metal), phytofiltration (plant parts adsorb and/or absorb metal ions from the aqueous environment), phytoextraction (metal uptake from the environment into the plant biomass), and phytovolatilization (some heavy metals convert into a gaseous form and vaporize into the atmosphere through leaves) (DalCorso et al. 2019). Contamination can be reduced by some mechanisms of

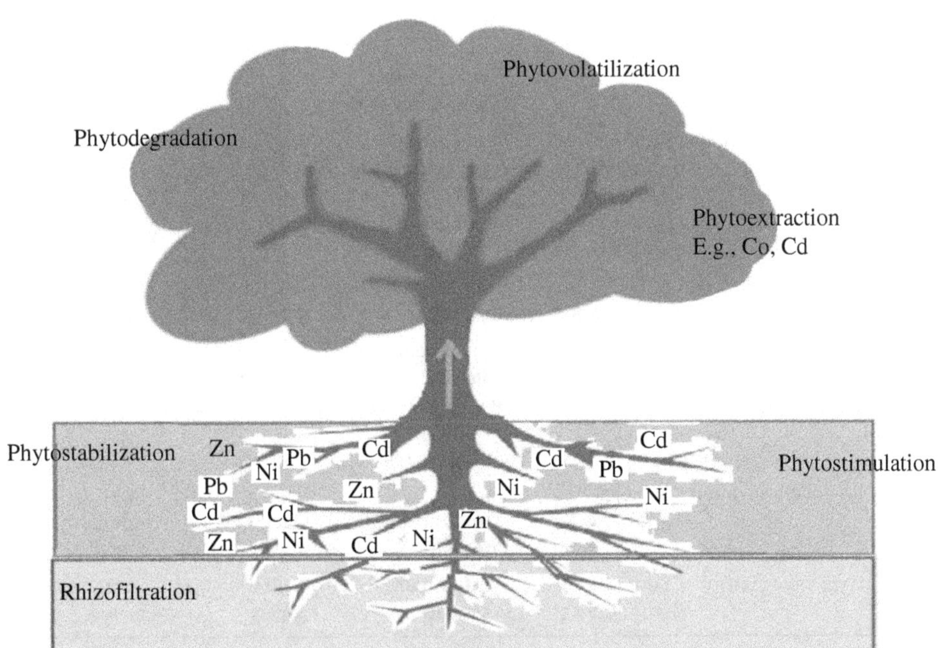

Figure 6.2 Different types of phytoremediation processes.

phytoremediation such as alteration in roots exudates and increased biodegradation with root tissues, abiotic losses (photodegradation, leachate, volatilization, irreversible sorption), and antimicrobial degradation (Wang et al. 2011). The phytoremediation process can be enhanced by the presence of microbes in the roots and the rhizosphere, metal translocation can be improved, and the mobilization of metals is reduced in the rhizosphere through the phytostabilization process (Rajkumar et al. 2012). The core of phytoremediation, i.e. hyperaccumulator efficiency, is often limited by low biomass production at a higher metal concentration, sensitivity toward multi-metals, a slow growth rate, and a poor and shallow root system. Several methods such as agricultural approaches and chemical reagents have been used to enhance biomass and stimulating uptake in plants is ultimately about increasing remediation efficiency. An overview of phytoremediation techniques is discussed in Table 6.2.

Table 6.2 Overview of phytoremediation techniques.

S. no.	Techniques	Media/Surface	Mechanism	Plants	Contaminants
1	Rhizofiltration	Surface water and groundwater	Uptake of metals into plant roots	Sunflower, water hyacinth, Indian mustard	Metals and radionuclides
2	Phytoextraction	Soil and sediment sludge	Uptake and concentration of metal via direct uptake into the plant tissue with subsequent removal of inorganic contaminants through the plant	Alyssum sunflower, pennycress, Indian mustard	Metals, Ag, Co, Cr, Cu, Hg, Mn, Mo, Pb, Ni, etc.; some radionuclides, Sr, Cs
3	Phytodegradation	Groundwater and soils, sediment sludge within the rhizosphere	Contaminant destruction and enhances microbial degradation in the rhizospheric zone	Algae, black willow, bald cypress	Some heavy metals, Ni, Cd, Zn, Mn, Mo, etc.
4	Rhizodegradation	Soil, groundwater	Contaminant destruction	Red mulberry, grasses, cattail	Metals, Ni, Cd, Zn, etc.
5	Phytostabilization	Groundwater, mine tailings, and soils	Root exudates cause the metal to precipitate and become less available	Indian mustard, hybrid poplars, grasses	As, Cd, Cr, Cu, Pb, Zn
6	Phytovolatilization	Soils and groundwater	Plant evapotranspiration rate	Poplars, alfalfa black, Indian mustard	Metals, selenium, mercury
7	Hydraulic control	Surface and groundwater	Contaminant degradation	Willow, hybrid poplar	Some heavy metals

Several plant growth regulators (PGRs) such as indole-3-acetic acid (IAA), gibberellin A3 (GA3), 6-benzylaminopurine (6-BA), and 24-epibrassinolide (EBL) are used in agriculture research by regulating several physiological and biochemical processes to increase phytoremediation efficiency (Ullah et al. 2015; Maresca et al. 2017). The cost-effective method for restoring contaminated toxic metal sites is phytoremediation at the cost of conventional technologies like solidification, soil replacement, and washing strategies (Mejáre and Bülow 2001). Plants that have been studied so far for phytoremediation purposes are discussed in detail in this chapter.

6.3 Bamboo (*Bambusa vulgaris*)

Bamboo (*Bambusa vulgaris*) belongs to the Poaceae (Gramineae) family with a unique ability to survive in the utmost marginalized soil and climatic conditions. In comparison with other grass species, bamboos are solid and hard due to the presence of sclerenchymatous cells with a joined hollow stem, large woody body (around 30 cm in diameter and 12 m in height), and a branched system with complex rhizomes. There are around 80 genera with 1000 species of bamboo growing worldwide (Yeasmin et al. 2015; Bian et al. 2019). Bamboo used for the phytoremediation of heavy metals is shown in Figure 6.3.

Vegetative growth, physiological integration, and clonal reproduction of bamboo forests enable patch expansion in native forests, which plays a crucial role in abating environmental pressure and enhancing living spaces (Zhang and Xue 2018). Quick harvesting, extensive root systems, and plant tolerance toward abiotic stress help select bamboo as a phytoremediation plant (Yang et al. 2005). Several studies have revealed that bamboo species have a high tolerance in heavy metal contaminated soils with metal absorption capacities.

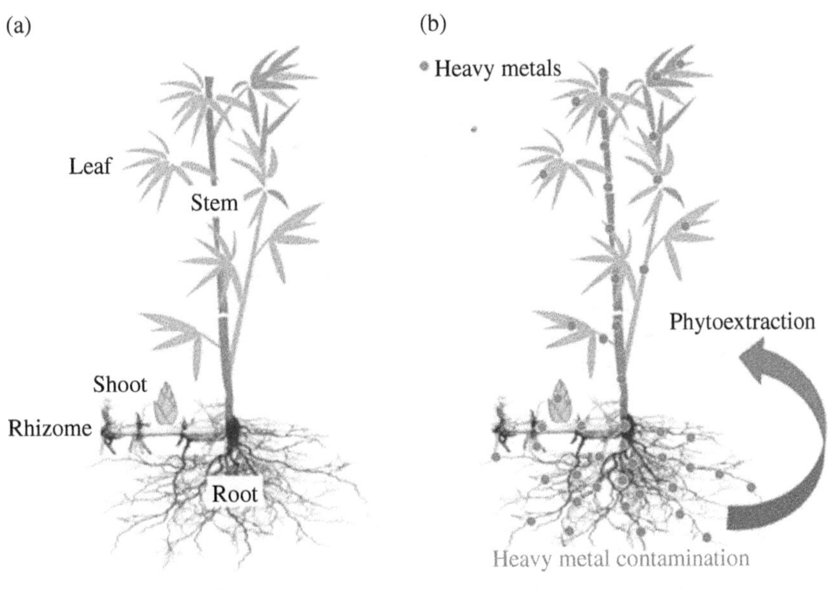

Figure 6.3 Bamboo (*Bambusa vulgaris*) used for the phytoremediation of heavy metals. (*Source:* Adapted from Bian et al. 2019).

Some species like *Arundinaria fortunei* and *Sasa auricoma* (Pb: 0–1500 mg/kg and EDTA: 0–1000 mg/kg), *Indocalamus latifolius* (Cu/Pb/Zn: 2000 mg/kg), and *Moso bamboo* (Pb/Zn: 286/210 mg/kg) show beneficial phytoremediation processes. Bamboo is a very promising plant species that could be used for phytoremediation of heavy metal contaminated sites (Emamverdian et al. 2018; Jiang et al. 2019). The biomass production of bamboo can achieve up to 82 t/hm^2 of dry weight and they have a great ability to adapt to a different environment. The high production of biomass and tolerance of *M. bamboo* in a stressed environment revealed it as an easily maintained phytoremediation plant. Studies revealed that Bamboo (*Gigantochloa* sp. "Malay dwarf") has a higher tolerance to copper concentration in hydroponic solutions. *M. bamboo* plants exposed to a higher concentration of metals, i.e. 600 mg/kg Cu, 400 mg/kg Pb, 120 mg/kg Cd, and 400 mg/kg zinc showed no visual toxicity symptoms like chlorosis and necrosis (Li et al. 2016). Accumulation of 3171 mg/kg in the roots of sympodial bamboos (*Phyllostachys fastuosa*) has been reported by Collin et al. (2014) in a four-month hydroponics experiment (Collin et al. 2014). Bamboo shoots have been used in biochar preparation for the removal of heavy metals, which could reduce global concerns by carbon sequestration; it also eliminates perrhenate from aqueous solutions and reduces the bioavailability of heavy metals in soil and water (Lu et al. 2014, 2017). The growth of *B. vulgaris* is shown in Figure 6.4.

According to the study of Were et al. (2017), the heavy metal accumulation and translocation potential of six bamboo species were estimated to restore chromium-contaminated

Figure 6.4 *Bambusa vulgaris* after two years of growth during phytoremediation of a chromium-contaminated site. (*Source:* Adapted from Were et al. 2017).

tannery soil. The bamboo species established at contaminated sites with the help of their extensive root systems prevent wind and soil erosion, minimizing the risk of chromium spreading to other areas and reducing adverse health impacts (Were et al. 2017). Cadmium can stimulate the formation of free radicals and disrupt the plant defense system in opposition to reactive oxygen species (ROS). *M. bamboo* has a tendency to eliminate cadmium from contaminated soils, thus can be used as a phytoremediation measure. It can tolerate a maximum of 120 mg/kg of cadmium and can accumulate cadmium efficiently. Cadmium toxicity was prevented in *M. bamboo* due to the presence of the antioxidant defense system (Li et al. 2016).

6.4 Mustard (*Brassica juncea*)

The Brassicaceae family has almost 80 plant species that can act as metal accumulators, among them is Indian mustard, i.e. *Brassica juncea*, a species of an annual herb, which has large biomass and a great capacity to extract, sequester, and detoxify heavy metals from contaminated soils (Figure 6.5) (Rathore et al. 2019). Metals like lead, cadmium, chromium, vanadium, etc. have been eliminated by using mustard in the phytoremediation process. Reclamation of smelter polluted soil was done by chemically enhancing phytoremediation using potherb mustard. The bioavailability of cadmium and zinc can be significantly increased in smelter sites by the application of ethylenediaminetetraacetic acid (EDTA), citric acid, and oxalic acid, but the effect is time limited. Phytoextraction of potentially toxic metals by potherb mustard can be enhanced by chelates. A higher accumulation and transfer capacity of cadmium in comparison with zinc has been observed in potherb mustard (Guo et al. 2019).

Studies have shown that mustard can easily accumulate certain heavy metals including cadmium and uranium from contaminated sites, thus it can be considered as a good aspirant for phytoremediation. Application of PGRs with mustard improves the stress tolerance and removal efficiency of cadmium (Cd) and uranium (U). Cadmium and uranium stress tend to inhibit chlorophyll synthesis and plant growth, whereas PGRs hamper the negative effects of cadmium and uranium. Oxidative stress of cadmium and uranium can be reduced in soil with the help of PGRs, as it decreases the malondialdehyde (MDA) and H_2O_2 content of

Figure 6.5 Indian mustard plant use for the phytoremediation of heavy metals.

mustard and enhances antioxidant enzyme (superoxide dismutase [SOD], peroxidases [POD], catalase [CAT], and aseorbateperoxidase [APX]) activities. The maximum translocation factor and removal efficiency of cadmium and uranium were studied at 500 mg/l IAA treatment (Chen et al. 2020). PGRs enhance the remediation, translocation, and uptake of cadmium and uranium in mustard. The evaluations of phytoremediation rely on the total concentration of cadmium and lead in aboveground parts. According to the report of Yang et al. (2018), results showed a significant difference in shoot cadmium and lead content in 80 *B. juncea* L. after 90 days of cultivation grown on bimetal contaminated soils. A noteworthy positive correlation was found between cadmium and lead concentrations (Gurajala et al. 2019). The phytoremediation of lead and other metals has been studied using Indian mustard (*B. juncea*). One of the broadly studied chelates is EDTA, which has been extensively used due to its relatively low cost and ability to enhance the solubility of several metals. According to several reports, the effectiveness of chelating agents enhances the accumulation of lead in Indian mustard shoot. Using electric potential along with EDTA enhances the phytoremediation efficiency by two to four times in comparison with the use of EDTA alone (Lim et al. 2004). The most promising advantage of phytoremediation is restoring lead-contaminated soil using Indian mustard along with EDTA. The copper and lead concentration in the shoot of Indian mustard has been enhanced by the addition of EDTA (Wu et al. 2004).

6.5 Rhizobacteria

Rhizosphere bacteria or rhizobacteria are also known as plant growth-promoting rhizobacteria (PGPR). They play a crucial role in phytoremediation with plants. They are capable of extensively promoting plant growth and colonizing plant roots. Studies have revealed that interactions of rhizobacteria and plants effectively eliminate arsenic (Yang et al. 2012). Fayiga and Ma (2006). The study showed that phosphate rocks can be helpful when applied as a successful alteration for arsenic phytoremediation since they enhance the arsenic uptake in *Pteris vittata* (Fayiga and Ma 2006). The interaction of native rhizosphere soil bacteria and maize plays an important role in clarifying the mechanism and accelerates the efficiency of phytoremediation (Vigliotta et al. 2016). Phytoremediation of chromium by using rhizobacteria with sunflower (*Helianthus annuus*) is shown in Figure 6.6.

The efficiency of phytoremediation depends on the growth of plants. *Scirpus triqueter* is inoculated with plant-associated bacteria to enhance remediation efficiency by encouraging plant growth (Adediran et al. 2015). Rhizobacteria promote plant growth by accelerating phytohormones synthesis to increase the bioavailability of nutritional elements. The winter rapeseed co-inoculated with barium and vanadium strains proved the most promising tested combination of the inoculate for phytoextraction of cadmium and zinc from contaminated soils by this plant species (Dąbrowska et al. 2017). Plants stimulate the secretion of chemicals like amino acids, proteins, and antibiotics due to PGPR that helps in stabilizing heavy metals to lessen the adverse effects of heavy metals. Siderophores enhance the bioaccessible portion of heavy metals that are produced by the plant in the presence of PGPR, enhancing the absorption of heavy metals by plants. The study reveals that PGPR inoculation enhances heavy metals (Zn, Pb, Ni, etc.) in plants (Wu et al. 2006). The rhizofiltration process has been

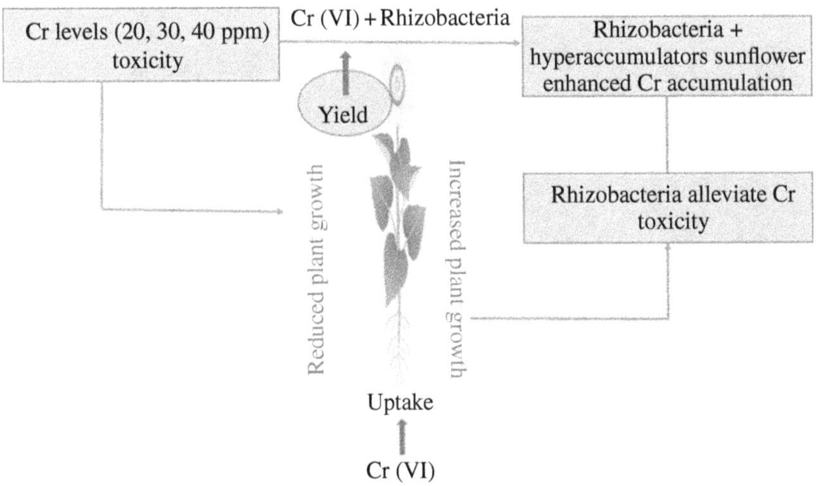

Figure 6.6 Use of rhizobacteria with sunflower (*Helianthus annuus*) for the phytoremediation of chromium. (*Source:* Adapted from Bahadur et al. 2017.)

adopted by plants to eliminate heavy metals as well as radioactive metals such as cadmium, copper, nickel, lead, chromium, cesium, arsenic, uranium, and strontium from aqueous media. Plant roots absorb metal pollutants from wastewater, and their suitability depends on the root system because they are responsible for eliminating metals. Plants showing this mechanism have hairy root systems and large surface areas (Tomé et al. 2008). The interaction between microorganisms and plants present in soil is the prominent mechanism for the increased depletion of polyaromatic hydrocarbons (PAHs). Study shows that *S. triqueter* inoculated with PGPR increases phytoremediation efficiency of pyrene and nickel contaminated soils. The resistance of *S. triqueter* increases against different environmental stresses when inoculated with PGPR and it improves the efficiency of phytoremediation (Zhang et al. 2020).

6.6 Seagrass

Seagrasses are submerged aquatic angiosperm found throughout the world in shallow coastal waters, which provide enormous ecological services and functions. Beds of seagrass are considered a highly productive ecosystem that provide shelter and nursery areas for different invertebrates, fish, and mammals. Phytoremediation by seagrass is shown in Figure 6.7.

It also enhances the water quality by stabilizing sediments and eliminating nutrients. They also act as biomonitors/bioindicators of environmental conditions because of their ability to concentrate and retain anthropogenic chemicals in their tissues that are not essential (Huesemann et al. 2009; Lewis and Devereux 2009). Several studies reveal that these plants are tolerant of many anthropogenic chemicals in comparison with other marine flora (macroalgae, kelp, salt marsh plants, and mangroves) and macro-invertebrates. They are useful in the abatement of contamination in coastal areas, thus can act as a potent tool for phytoremediation (Malea et al. 2018). The sensitivity of aquatic ecosystems due to heavy metals is greater in comparison with terrestrial ecosystems. The main source of

Figure 6.7 Seagrass used for the phytoremediation of heavy metals.

metal contamination in the aquatic ecosystem is due to atmospheric deposition in domestic wastewater, industrial discharge, fertilizers, pesticides, and municipal sewage sludge (Haplern et al. 2008; Bazzano et al. 2014). They can eliminate a huge amount of inorganic contaminants from the surrounding environment (sediments and water column) via root tissues and leaves (Lee and Dunton 1999). Several investigations have revealed the mechanism of heavy metal accumulation in seagrass tissues and the contribution to the heavy metal cycle. The concentration of heavy metals in seagrass tissues shows the level of heavy metals in the ambient environment (Bonanno and Borg 2018). The transplantation of seagrass has been investigated to assist the recovery of degraded coastal environments (Pereda-briones et al. 2018). The phytoremediation potential of seagrass has been investigated by transplanting them as a biological agent to eliminate heavy metal contamination from polluted coastal and estuarine environments (Lee et al. 2019).

The majority of seagrasses absorb cadmium at a different extent, whereas species like *Zostera marina* can accumulate up to 10 mg/g concentration without showing any toxic effects. Some other species like *Posidonia oceanica* reveal altered leaf physiology and metabolism when exposed to a lower concentration of cadmium (Lewis and Devereux 2009). According to the study of Zhang et al. (2016), *Zostera japonica* can reduce the toxic effects of higher concentrations of copper by both primary, i.e. chelation, and secondary, i.e. antioxidant, pathways. *Z. japonica* can be considered as a suitable candidate for the decontamination of marine bodies moderately contaminated with copper due to their high accumulation potential (Lin et al. 2016). Metal uptake studies of seagrass species like *Hetero Zostera tasmanica*, *P. oceanica*, and *Halophila stipulacea*, and most commonly, *Zostera* species, have been done. Accumulation was mostly found in leaves or roots/rhizomes. Cadmium, lead, and zinc uptake are widely studied in seagrass experiments (Malea et al. 1995; Malea et al. 2018).

6.7 Sunflower (*Helianthus annuus*)

H. annuus L. commonly known as sunflower belongs to the Asteraceae family, and is one of the most important and profitable oilseed crops grown globally (Bahadur et al. 2017). It has a large flowering head (inflorescence) and the stem can grow up to 3 m along with a flower head of up to 30 cm in diameter with large seeds (Alaboudi et al. 2018). A *H. annuus* plant is shown in Figure 6.8 for the remediation of heavy metals.

Figure 6.8 Sunflower (*Helianthus annuus* L.) plant used in the phytoremediation of heavy metals.

It has high biomass and it ranks in second position among all oilseeds at a global level after soybean as a source of vegetable oil. Sunflower oil composition includes monounsaturated fat like oleic acid, polyunsaturated fat like linoleic acid, and vitamins such as vitamins A, B, and E (Quiles et al. 2002). They contain high protein value with several essential amino acids such as valine, isoleucine, and leucine, which facilitate muscle growth and production. The germplasm of sunflower can be used in phytoremediation to remove lead-contaminated soils (Zehra et al. 2020a). Sunflower is a documented metal accumulator used in the phytoremediation process. Since phytoremediation is an effective technology to decrease heavy metal concentration from soil and water, it can be evaluated as a potential procedure contributing to sustainable soil utilization and increase agro-production (Oh et al. 2014). Plants possess a natural ability for trace elements uptake through their roots and investigation reveals that sunflowers can eliminate toxic substances and heavy metals from contaminated farmlands (Yanqun et al. 2005). A study showed that *H. annuus* is a fast-growing, high biomass crop, easy to harvest, and can accumulate elevated concentrations of cadmium in their aboveground parts (Motesharezadeh et al. 2016). Investigations have shown the phytoextraction ability of sunflower under controlled environmental conditions generally in pots, soil, and hydroponic trials. According to studies, sunflower is grown in oxisol and inceptisol and has been proved as a prospective candidate for phytoremediation in different cadmium contaminated soil types and for safe agro-economic production (Zehra et al. 2020b). It has the ability to eliminate lead and cadmium from degraded soil and accumulate them in their shoots and roots. The accumulation of cadmium in plant shoots has been more favorable than lead (Alaboudi et al. 2018).

6.8 Water Hyacinth (*Eichhornia crassipes* Mart)

Eichhornia crassipes Mart. also known as water hyacinth belongs to the Pontederiaceae family and originated in the rainforests of the Amazon basin (Figure 6.9). It is the fastest-growing (doubling in number and biomass every two weeks) submerged or free-floating aquatic weed (Jafari 2010). It has thick, glossy, and broad ovate leaves of 10–20 cm floating

Figure 6.9 Water hyacinth (*Eichhornia crassipes* Mart.) plant used in the phytoremediation of heavy metals.

above the water surface on long bulbous stalks. It consists of six to ten leaves arranged in a rosette attached to rhizome with a developed fibrous root system (unbranched with conspicuous root cap) (Greenwood et al. 2009). In response to climatic and environmental conditions, the morphology and physiological function of foliage show foliar plasticity. It can flourish in both saline and freshwater as well as reproduce via the sexual and asexual processes (Barret 1980). They have shown a great possibility for decontamination of wastewater via physical, chemical, and biological mechanisms (Jones et al. 2018). There are various uses of water hyacinth, such as in fish feed/animal fodder, biogas and bioethanol production, biosorbents for toxic metal removal, paper manufacturing, compost, and as phytoremediation agents (Rezania et al. 2015).

Water hyacinth used in phytoremediation involves the reduction of nitrogen, suspended solids, fecal bacteria, organic matter, dyes, and heavy metals, demonstrating it as suitable, inexpensive, effective, and eco-friendly technology for treating wastewater (Priya and Selvan 2017). In the natural environment, water hyacinth creates serious challenges in the field of irrigation, navigation, and power generation. To overcome these problems, the phytoremediation process is carried out by controlling water hyacinth overgrowth. Their roots are generally suspended in water and the root structure provides a suitable environment for the uptake of pollutants (Mahamadi 2011). Water hyacinth obtained an extensive interest in environmental phytoremediation application due to their higher heavy metal sorption and accumulation abilities in comparison with other aquatic macrophytes (Malik 2007). The sorption performances differ depending upon metal species, concentration, and solution chemistry. Studies have successfully demonstrated high sorption of heavy metals like cadmium (Cd), zinc (Zn), chromium (Cr), lead (Pb), and copper (Cu) using water hyacinth (Mitra et al. 2014).

According to the study of Fazal et al. (2015), water hyacinth efficiently reduced 92.5% of turbidity, 83.7% of chemical oxygen demands (COD), 91.8% of total dissolved solids (TDS), 80% of total suspended solids (TSS), 67.5% of nitrates, 71.6% of ammonia, 90.2% of phosphates, 97.5% of cadmium, 95.1% of nickel, 99.9% of mercury, and 83.4% of lead from industrial wastewater (Ting et al. 2018). The study of Hasan et al. (2007), showed that the sorption of the metal mixture using water hyacinth favors the higher accumulation of zinc over cadmium. Another study conducted by Mahamadi and Nharingo (2010) revealed that lead suppresses the sorption of cadmium and zinc (Li et al. 2016).

6.9 Willow (*Salix alba* L.)

Salix alba L. also known as willow belongs to the Salicaceae family. They have watery bark sap, tough wood, and slender branches. Some species of willow-like *Salix caprea*, *S. alba*, and others are considered as useful because of their application in phytoremediation, living snow fences, and riparian buffers (Volk et al. 2006; Harada et al. 2011). The phytoremediation of heavy metals through *S. alba* is presented in Figure 6.10.

Investigations have revealed that willows demonstrate phytoextraction, i.e. uptake of heavy metals and organics from soils and assists the breakdown of organics into nontoxic compounds (rhizodegradation) (Ebbs et al. 2003). According to the study, willows are commonly used for phytoremediation purposes in the northern hemisphere due to their fast growth rate, coppicing ability, higher biomass production, lower nutrient demands, effective photosynthesis, and high plantation density tolerance (Kuzovkina and Volk 2009).

According to the report of Marmiroli et al. (2011) and Yang et al. (2018), willows have been considered as the most important phytoremediators as *Salix* sp. have genetic variability in the accumulation and tolerance of cadmium (Marmiroli et al. 2011; Yang et al. 2018). Fast growth and productivity in willows, even at juvenile stages, as well as a substantial and continuous fibrous root system, allow them to evaporate huge amounts of water in growing seasons (Yıldırım and Kasım 2018). According to Pulford et al. (2001), *Salix viminalis* accumulates metals like lead, nickel, and chromium, mostly in the below-ground tissues of trees, showing phytostabilization was the phytoremediation mechanism dominating in this study. Phytostabilization has more potential than phytoextraction according to the results of the studies in the case of willow (Pulford et al. 2001).

Along with heavy metal-specific uptake, other factors also influence phytoremediation potentials such as metabolic pathways and soil chemistry. Bioavailability of heavy metals depends on a particular fraction of heavy metals in the water and soil matrix, the presence of calcium carbonate that immobilizes metal, and the pH value of the substrate (Doni et al. 2015;

Figure 6.10 *Salix* sp. used for the phytoremediation of heavy metals (adapted from Harada et al. (2011)).

Pilipovic et al. 2019). A report by Pulford and Watson (2003), revealed that *S. viminalis* has the potential for the phytoremediation of soils having heavy metals (Ni, Cu, Cd, and Zn). Eleven species have shown phytoremediation potential along with better survival, enhanced biomass production, and higher metal uptake out of 20 species of *Salix* studied. The study of Zhivotovsky et al. (2011) demonstrated that *Salix* has the potential for heavy metal (lead) uptake and translocation to aboveground tissues with satisfactory growth rates in contaminated soil (Salam et al. 2016).

6.10 Description of Heavy Metal Removal from Water

A hybrid, bioenergy, perennial crop, *Miscanthus* sp. Goedae-Uksae 1 developed in South Korea, has the potential of phytoremediation of heavy metals. Goede-Uksae 1 absorbed all the metals from polluted water except arsenic. The metals used in the study conducted by Bang et al. (2015) were arsenic, copper, lead, nickel, cadmium, and zinc. Metals like cadmium, lead, and zinc were removed from contaminated water. A higher concentration of metals was observed in roots compared with shoots. The phytoremediation rate was highly influenced by the initial concentration of metals. According to this study, the fast removal of lead and zinc was observed in the first ten days, whereas complete removal was seen on the 15th day. Cadmium, nickel, and copper also exhibited fast removal during the first five days and attained maximum removal on the 15th and 40th day. Arsenic showed a minimum removal of 8–12%. It was concluded that on the basis of the bioconcentration factor, translocation factor, and enrichment coefficient, Goedae-Uksae 1 could be utilized in phytoremediation.

A common green algae, *Sc. abundans* was studied for the removal of cadmium and copper from contaminated water via biosorption. The study was also conducted in the multicomponent system to detect the competitive effects between the metals. The measurement of chlorophyll was also conducted at different cadmium and copper concentrations to discover the effect of metal on cell viability. The study revealed that contaminated water moved into the treatment pond would have to be diluted below 15 mg/l cadmium or copper so that the viability of algae should be maintained (Terry and Stone 2002).

In the case of free-floating macrophytes, roots remain submerged in water while the whole plant body is above water. Phytoremediation potential has been shown by various plants species such as water hyacinths, i.e. *Eichhornia* sp. (*E. crassipes, E. azurea, E. diversifolia, E. paniculata*), *Salvinia* sp. (*Sal. molesta, Sal. auriculata, Sal. minima*), duckweed (*Lemna minor*), water lettuce (*Pistia stratiotes*), and *Azolla*. Several submerged species such as *Cyperus* sp., *Spartina* sp., *Potamogeton* sp., *Myriophyllum* sp., and *Limnocharis flava* have shown good potential for the removal of metals from contaminated water (Rezania et al. 2016). *Pistia stratiotes* (water lettuce) is an aquatic plant that grows rapidly with high biomass along with an extensive root system enhancing the elimination of heavy metals. Water lettuce belongs to the Araceae family. Water lettuce has been studied by Gupta et al. (2012) to eliminate iron, magnesium, manganese, calcium, cadmium, cobalt, and potassium through the attachment of an external root surface (Gupta et al. 2012).

According to the study of Hua et al. (2012), *P. stratiotes* can effectively remove Mn from contaminated water showing its phytoremediation ability. The effectiveness of water lettuce depends on the growth rate in wetlands; it is easy harvestable and has high coverage of the water surface (Hua et al. 2012). According to the study of Baharudin (2008), 15 plants grown in constructed wetlands showed the highest removal of lead (99.28%) and cadmium (65.58%) at the neutral condition in comparison with basic and acidic conditions (Baharudin 2008). *Sal. minima* shows good phytoremediation potential and is considered as a hyperaccumulator of nickel. The study of Prado et al. (2010) revealed that the accumulation of chromium was found more in the submerged leaves of *Sal. minima* than in the floating leaves (Prado et al. 2010). Dhir et al. (2011) demonstrated that *Sal. natans* shows the phytoremediation capacity of removing several heavy metals as leaves of this species eliminate metal ions more than other parts (Dhir et al. 2011).

Duckweed is a free-floating aquatic plant on the surface of the water, it adapts easily in different aquatic conditions and shows fast growth. It is a flowering plant belonging to the family Lemnaceae. *Lemna* sp. duckweeds are commonly studied for the removal of heavy metal from contaminated water and they play an important role in the accumulation and extraction of metals from wastewater (Cayuela et al. 2007). According to the report of Divya et al. 2011, *L. minor* shows 90% removal of soluble lead from polluted water (Divya et al. 2011). *Hy. verticillata*, a fast-growing submerged aquatic weed, forms dense mats on the surface of the water body. Its whole plant body is beneficial in terms of metal removal. Studies have shown 98% removal of lead using *Hydrilla* after exposing the plant for one week (Singh et al. 2012). The study of de Souza and Silva (2019) demonstrated the removal of copper using *P. stratiotes* (de Souza and Silva 2019).

6.11 Conclusion and Future Research Recommendations

The contamination of heavy metals to the environment, i.e. soil, water, plant, and air, is of great concern due to its potential impact on human and animal health. Cheaper and effective technologies are needed to protect precious natural resources and biological lives. Today, phytoremediation is known for its economic viability and easily operated technology for the removal of heavy metals from a contaminated environment. Substantial efforts have been made in identifying plant species and their mechanisms of uptake and hyperaccumulation of heavy metals in the last decade such as metal uptake, accumulation, exclusion, translocation, osmoregulation, and compartmentation that can vary with each plant species. The use of cleaning technologies is site-specific due to spatial and climatic variations and is not economically feasible everywhere. Therefore, cheaper technologies are being sought for practical use. Nevertheless, recent advances in plant biotechnology have created new hope for the development of hyperaccumulating species.

Since the adverse effects of heavy metals are well known, it becomes important to remove them naturally from contaminated sites (surface, groundwater, and soil). Various hyperaccumulator plants such as bamboo, Indian mustard, sunflower, water hyacinth, *Miscanthus* sp., common green algae, water lettuce, duckweed, *Hy. verticillata*, *Sal. Minima*, and rhizobacteria are relatively inexpensive and practically used in comparison with other technologies. In-depth water and soil microbial studies are required to identify the microorganisms highly associated with metal solubility or precipitations. To date, the available methods for the recovery of heavy metals from the plant biomass of hyperaccumulators are still limited. Traditional disposal approaches such as burning and ashing are not applicable to volatile metals, therefore, investigations are needed to develop new methods for the effective recovery of metals:

- Using hyperaccumulator plants
- Using biomass

The accumulation of heavy metals has different sites during phytoremediation, among them, roots, shoots, and leaves are the most common. Based on different studies, we can conclude that roots are extensively used in the elimination process, enhancing the phytoremediation capacity of plants.

References

Adediran, G.A., Ngwenya, B.T., Mosselmans, J.F.W. et al. (2015). Mechanisms behind bacteria induced plant growth promotion and Zn accumulation in *Brassica juncea*. *Journal of Hazardous Materials* 283: 490–499.

Agwu, K.K., Okoye, C.M.I., Okeji, M.C., and Clifford, E.O. (2018). Potential health impacts of heavy metal concentrations in fresh and marine water fishes consumed in southeast, Nigeria. *Pakistan Journal of Nutrition* 17 (12): 647–653.

Alaboudi, K.A., Ahmed, B., and Brodie, G. (2018). Phytoremediation of Pb and cd contaminated soils by using sunflower (*Helianthus annuus*) plant. *Annals of Agricultural Science* 63 (1): 123–127.

Bahadur, A., Ahmad, R., Afzal, A. et al. (2017). The influences of Cr-tolerant rhizobacteria in phytoremediation and attenuation of Cr (VI) stress in agronomic sunflower (*Helianthus annuus* L.). *Chemosphere* 179: 112–119.

Baharudin, M.S. (2008). Lead and cadmium removal in synthetic wastewater using constructed wetland. Doctoral dissertation. UMP.

Bang, J., Kamala-Kannan, S., Lee, K.J. et al. (2015). Phytoremediation of heavy metals in contaminated water and soil using Miscanthus sp. Goedae-Uksae 1. *International Journal of Phytoremediation* 17 (6): 515–520.

Barrett, S.C. (1980). Sexual reproduction in *Eichhornia crassipes* (water hyacinth). II. Seed production in natural populations. *Journal of Applied Ecology* 17: 113–124.

Bazzano, A., Rivaro, P., Soggia, F. et al. (2014). Anthropogenic and natural sources of particulate trace elements in the coastal marine environment of Kongsfjorden, Svalbard. *Marine Chemistry* 163: 28–35.

Bian, F., Zhong, Z., Zhang, X. et al. (2019). Bamboo – an untapped plant resource for the phytoremediation of heavy metal contaminated soils. *Chemosphere* 246: 125750.

Bonanno, G. and Borg, J.A. (2018). Comparative analysis of trace element accumulation in seagrasses *Posidonia oceanica* and *Cymodocea nodosa*: biomonitoring applications and legislative issues. *Marine Pollution Bulletin* 128: 24–31.

Cayuela, M.L., Millner, P., Slovin, J., and Roig, A. (2007). Duckweed (*Lemna gibba*) growth inhibition bioassay for evaluating the toxicity of olive mill wastes before and during composting. *Chemosphere* 68 (10): 1985–1991.

Chen, L., Long, C., Wang, D., and Yang, J. (2020). Phytoremediation of cadmium (Cd) and uranium (U) contaminated soils by *Brassica juncea* L. enhanced with exogenous application of plant growth regulators. *Chemosphere* 242: 125112.

Chua, J., Banua, J.M., Arcilla, I. et al. (2019). Phytoremediation potential and copper uptake kinetics of Philippine bamboo species in copper contaminated substrate. *Heliyon* 5 (9): e02440.

Collin, B., Doelsch, E., Keller, C. et al. (2014). Evidence of sulfur-bound reduced copper in bamboo exposed to high silicon and copper concentrations. *Environmental Pollution* 187: 22–30.

Dąbrowska, G., Hrynkiewicz, K., Trejgell, A., and Baum, C. (2017). The effect of plant growth-promoting rhizobacteria on the phytoextraction of Cd and Zn by *Brassica napus* L. *International Journal of Phytoremediation* 19 (7): 597–604.

DalCorso, G., Fasani, E., Manara, A. et al. (2019). Heavy metal pollutions: state of the art and innovation in phytoremediation. *International Journal of Molecular Sciences* 20 (14): 3412.

De Souza, C.B. and Silva, G.R. (2019). Phytoremediation of effluents contaminated with heavy metals by floating aquatic macrophytes species. In: *Biotechnology and Bioengineering* (eds. E. Jacob-Lopes and L.Q. Zepka), 1–18. IntechOpen.

Dhir, B., Sharmila, P., Saradhi, P.P. et al. (2011). Heavy metal induced physiological alterations in *Salvinia natans*. *Ecotoxicology and Environmental Safety* 74 (6): 1678–1684.

Divya, S., Richa, G., and Archana, T. (2011). Phytoremediation of lead from wastewater using aquatic plants. *International Journal of Biomedical Research* 2 (7): 411–421.

Doni, S., Macci, C., Peruzzi, E. et al. (2015). Heavy metal distribution in a sediment phytoremediation system at pilot scale. *Ecological Engineering* 81: 146–157.

Ebbs, S., Bushey, J., Poston, S. et al. (2003). Transport and metabolism of free cyanide and iron cyande complexes by willow. *Plant, Cell & Environment* 26 (9): 1467–1478.

Ekta, P. and Modi, N.R. (2018). A review of phytoremediation. *Journal of Pharmacognosy and Phytochemistry* 7 (4): 1485–1489.

Emamverdian, A., Ding, Y., and Xiea, Y. (2018). Phytoremediation potential of bamboo plant in China. *Ecology, Environment and Conservation* 24 (1): 530–539.

Fayiga, A.O. and Ma, L.Q. (2006). Using phosphate rock to immobilize metals in soil and increase arsenic uptake by hyperaccumulator *Pteris vittata*. *Science of the Total Environment* 359 (1–3): 17–25.

Fazal, S., Zhang, B., and Mehmood, Q. (2015). Biological treatment of combined industrial wastewater. *Ecological Engineering* 84: 551–558.

Ghosh, M. and Singh, S.P. (2005). A review on phytoremediation of heavy metals and utilization of it's by products. *Asian Journal on Energy & Environment* 6 (4): 1–18.

Greenwood, M.S., Day, M.E., and Berlyn, G.P. (2009). Regulation of foliar plasticity in conifers: developmental and environmental factors. *Journal of Sustainable Forestry* 28 (1–2): 48–62.

Guo, D., Ali, A., Ren, C. et al. (2019). EDTA and organic acids assisted phytoextraction of Cd and Zn from a smelter contaminated soil by potherb mustard (*Brassica juncea*, Coss) and evaluation of its bioindicators. *Ecotoxicology and Environmental Safety* 167: 396–403.

Gupta, P., Roy, S., and Mahindrakar, A.B. (2012). Treatment of water using water hyacinth, water lettuce and vetiver grass – a review. *Resources and Environment* 2 (5): 202–215.

Gurajala, H.K., Cao, X., Tang, L. et al. (2019). Comparative assessment of Indian mustard (*Brassica juncea* L.) genotypes for phytoremediation of Cd and Pb contaminated soils. *Environmental Pollution* 254: 113085.

Harada, E., Hokura, A., Nakai, I. et al. (2011). Assessment of willow (Salix sp.) as a woody heavy metal accumulator: field survey and in vivo X-ray analyses. *Metallomics* 3 (12): 1340–1346.

Hasan, S.H., Talat, M., and Rai, S. (2007). Sorption of cadmium and zinc from aqueous solutions by water hyacinth (Eichchorniacrassipes). *Bioresource Technology* 98 (4): 918–928.

Hua, J., Zhang, C., Yin, Y. et al. (2012). Phytoremediation potential of three aquatic macrophytes in manganese contaminated water. *Water Environment Journal* 26 (3): 335–342.

Huesemann, M.H., Hausmann, T.S., Fortman, T.J. et al. (2009). In situ phytoremediation of PAH-and PCB-contaminated marine sediments with eelgrass (*Zostera marina*). *Ecological Engineering* 35 (10): 1395–1404.

Jacob, J.M., Sivan, S., and Aji, M. (2018). Selenium and lead tolerance in fungi isolated from sea water. *Journal of Environmental Management* 217: 56–70.

Jafari, N. (2010). Ecological and socio-economic utilization of water hyacinth (*Eichhornia crassipes* Mart Solms). *Journal of Applied Sciences and Environmental Management* 14 (2): 43–49.

Jiang, M., Liu, S., Li, Y. et al. (2019). EDTA-facilitated toxic tolerance, absorption and translocation and phytoremediation of lead by dwarf bamboos. *Ecotoxicology and Environmental Safety* 170: 502–512.

Jones, J.L., Jenkins, R.O., and Haris, P.I. (2018). Extending the geographic reach of the water hyacinth plant in removal of heavy metals from a temperate Northern Hemisphere river. *Scientific Reports* 8 (1): 1–15.

Khan, M.S., Zaidi, A., Wani, P.A., and Oves, M. (2009). Role of plant growth promoting rhizobacteria in the remediation of metal contaminated soils. *Environmental Chemistry Letters* 7 (1): 1–19.

Kuzovkina, Y.A. and Volk, T.A. (2009). The characterization of willow (Salix L.) varieties for use in ecological engineering applications: co-ordination of structure, function and autecology. *Ecological Engineering* 35 (8): 1178–1189.

Lasat, M.M. (2002). Phytoextraction of toxic metals. *Journal of Environmental Quality* 31 (1): 109–120.

Lee, K.S. and Dunton, K.H. (1999). Inorganic nitrogen acquisition in the seagrass Thalassia testudinum: development of a whole-plant nitrogen budget. *Limnology and Oceanography* 44 (5): 1204–1215.

Lee, G., Suonan, Z., Kim, S.H. et al. (2019). Heavy metal accumulation and phytoremediation potential by transplants of the seagrass *Zostera marina* in the polluted bay systems. *Marine Pollution Bulletin* 149: 110509.

Lewis, M.A. and Devereux, R. (2009). Nonnutrient anthropogenic chemicals in seagrass ecosystems: fate and effects. *Environmental Toxicology and Chemistry: An International Journal* 28 (3): 644–661.

Li, Q., Chen, B., Lin, P. et al. (2016). Adsorption of heavy metal from aqueous solution by dehydrated root powder of long-root *Eichhornia crassipes*. *International Journal of Phytoremediation* 18 (2): 103–109.

Li, S., Chen, J., Islam, E. et al. (2016). Cadmium-induced oxidative stress, response of antioxidants and detection of intracellular cadmium in organs of *Moso bamboo* (*Phyllostachys pubescens*) seedlings. *Chemosphere* 153: 107–114.

Lim, J.M., Salido, A.L., and Butcher, D.J. (2004). Phytoremediation of lead using Indian mustard (*Brassica juncea*) with EDTA and electrodics. *Microchemical Journal* 76 (1–2): 3–9.

Lin, H., Sun, T., Zhou, Y., and Zhang, X. (2016). Anti-oxidative feedback and biomarkers in the intertidal seagrass *Zostera japonica* induced by exposure to copper, lead and cadmium. *Marine Pollution Bulletin* 109 (1): 325–333.

Lone, M.I., He, Z.L., Stoffella, P.J., and Yang, X.E. (2008). Phytoremediation of heavy metal polluted soils and water: progresses and perspectives. *Journal of Zhejiang University Science B* 9 (3): 210–220.

Lu, K., Yang, X., Shen, J. et al. (2014). Effect of bamboo and rice straw biochars on the bioavailability of Cd, Cu, Pb and Zn to *Sedum plumbizincicola*. *Agriculture, Ecosystems & Environment* 191: 124–132.

Lu, K., Yang, X., Gielen, G. et al. (2017). Effect of bamboo and rice straw biochars on the mobility and redistribution of heavy metals (Cd, Cu, Pb and Zn) in contaminated soil. *Journal of Environmental Management* 186: 285–292.

Ma, Y., Oliveira, R.S., Freitas, H., and Zhang, C. (2016). Biochemical and molecular mechanisms of plant-microbe-metal interactions: relevance for phytoremediation. *Frontiers in Plant Science* 7: 918.

Mahajan, P. and Kaushal, J. (2018). Role of phytoremediation in reducing cadmium toxicity in soil and water. *Journal of Toxicology* 2018 https://doi.org/10.1155/2018/4864365.

Mahamadi, C. (2011). Water hyacinth as a biosorbent: a review. *African Journal of Environmental Science and Technology* 5 (13): 1137–1145.

Mahamadi, C. and Nharingo, T. (2010). Competitive adsorption of Pb^{2+}, Cd^{2+} and Zn^{2+} ions onto Eichhorniacrassipes in binary and ternary systems. *Bioresource Technology* 101 (3): 859–864.

Malea, P., Kevrekidis, T., and Haritonidis, S. (1995). The short-term uptake of zinc and cell mortality of the seagrass *Halophila stipulacea* (Forsk.) Aschers. *Israel Journal of Plant Sciences* 43 (1): 21–30.

Malea, P., Kevrekidis, T., Chatzipanagiotou, K.R., and Mogias, A. (2018). Cadmium uptake kinetics in parts of the seagrass *Cymodocea nodosa* at high exposure concentrations. *Journal of Biological Research (Thessaloniki)* 25 (1): 1–11.

Malik, A. (2007). Environmental challenge Vis a Vis opportunity: the case of water hyacinth. *Environment International* 33 (1): 122–138.

Manoj, S.R., Karthik, C., Kadirvelu, K. et al. (2020). Understanding the molecular mechanisms for the enhanced phytoremediation of heavy metals through plant growth promoting rhizobacteria: a review. *Journal of Environmental Management* 254: 109779.

Maresca, V., Sorbo, S., Keramat, B., and Basile, A. (2017). Effects of triacontanol on ascorbate-glutathione cycle in *Brassica napus* L. exposed to cadmium-induced oxidative stress. *Ecotoxicology and Environmental Safety* 144: 268–274.

Marmiroli, M., Pietrini, F., Maestri, E. et al. (2011). Growth, physiological and molecular traits in Salicaceae trees investigated for phytoremediation of heavy metals and organics. *Tree Physiology* 31 (12): 1319–1334.

McLaughlin, M.J., Zarcinas, B.A., Stevens, D.P., and Cook, N. (2000). Soil testing for heavy metals. *Communications in Soil Science and Plant Analysis* 31 (11–14): 1661–1700.

Mejáre, M. and Bülow, L. (2001). Metal-binding proteins and peptides in bioremediation and phytoremediation of heavy metals. *Trends in Biotechnology* 19 (2): 67–73.

Mitra, T., Singha, B., Bar, N., and Das, S.K. (2014). Removal of Pb (II) ions from aqueous solution using water hyacinth root by fixed-bed column and ANN modeling. *Journal of Hazardous Materials* 273: 94–103.

Motesharezadeh, B., Navabzadeh, M., and Liyaghat, A. (2016). Modeling phytoremediation of cadmium contaminated soil with sunflower (*Helianthus annus*) under salinity stress. *International Journal of Environmental Research* 10 (1): 109–118.

Oh, K., Cao, T., Li, T., and Cheng, H. (2014). Study on application of phytoremediation technology in management and remediation of contaminated soils. *Journal of Clean Energy Technologies* 2 (3): 216–220.

Pereda-Briones, L., Tomas, F., and Terrados, J. (2018). Field transplantation of seagrass (*Posidonia oceanica*) seedlings: effects of invasive algae and nutrients. *Marine Pollution Bulletin* 134: 160–165.

Pilipovic, A., Zalesny, R.S. Jr., Rončević, S. et al. (2019). Growth, physiology, and phytoextraction potential of poplar and willow established in soils amended with heavy-metal contaminated, dredged river sediments. *Journal of Environmental Management* 239: 352–365.

Prado, C., Rodríguez-Montelongo, L., González, J.A. et al. (2010). Uptake of chromium by Salvinia minima: effect on plant growth, leaf respiration and carbohydrate metabolism. *Journal of Hazardous Materials* 177 (1–3): 546–553.

Priya, E.S. and Selvan, P.S. (2017). Water hyacinth (*Eichhornia crassipes*) – an efficient and economic adsorbent for textile effluent treatment – a review. *Arabian Journal of Chemistry* 10: S3548–S3558.

Pulford, I.D. and Watson, C. (2003). Phytoremediation of heavy metal-contaminated land by trees – a review. *Environment International* 29 (4): 529–540.

Pulford, I.D., Watson, C., and McGregor, S.D. (2001). Uptake of chromium by trees: prospects for phytoremediation. *Environmental Geochemistry and Health* 23 (3): 307–311.

Quiles, J.L., Ramírez-Tortosa, M.C., Gomez, J.A. et al. (2002). Role of vitamin E and phenolic compounds in the antioxidant capacity, measured by ESR, of virgin olive, olive and sunflower oils after frying. *Food Chemistry* 76 (4): 461–468.

Rajkumar, M., Sandhya, S., Prasad, M.N.V., and Freitas, H. (2012). Perspectives of plant-associated microbes in heavy metal phytoremediation. *Biotechnology Advances* 30 (6): 1562–1574.

Rathore, S.S., Shekhawat, K., Dass, A. et al. (2019). Phytoremediation mechanism in Indian mustard (*Brassica juncea*) and its enhancement through agronomic interventions. *Proceedings of the National Academy of Sciences, India Section B: Biological Sciences* 89 (2): 419–427.

Rezania, S., Ponraj, M., Talaiekhozani, A. et al. (2015). Perspectives of phytoremediation using water hyacinth for removal of heavy metals, organic and inorganic pollutants in wastewater. *Journal of Environmental Management* 163: 125–133.

Rezania, S., Taib, S.M., Din, M.F.M. et al. (2016). Comprehensive review on phytotechnology: heavy metals removal by diverse aquatic plants species from wastewater. *Journal of Hazardous Materials* 318: 587–599.

Salam, M.M.A., Kaipiainen, E., Mohsin, M. et al. (2016). Effects of contaminated soil on the growth performance of young Salix (*Salix schwerinii* EL wolf) and the potential for phytoremediation of heavy metals. *Journal of Environmental Management* 183: 467–477.

Sharma, S., Rana, S., Thakkar, A. et al. (2016). Physical, chemical and phytoremediation technique for removal of heavy metals. *Journal of Heavy Metal Toxicity and Diseases* 1 (2): 1–15.

Singh, D., Tiwari, A., and Gupta, R. (2012). Phytoremediation of lead from wastewater using aquatic plants. *Journal of Agricultural Technology* 8 (1): 1–11.

Suman, J., Uhlik, O., Viktorova, J., and Macek, T. (2018). Phytoextraction of heavy metals: a promising tool for clean-up of polluted environment? *Frontiers in Plant Science* 9: 1476.

Tang, S., Xi, L., Zheng, J., and Li, H. (2003). Response to elevated CO_2 of Indian mustard and sunflower growing on copper contaminated soil. *Bulletin of Environmental Contamination and Toxicology* 71 (5): 988–997.

Terry, P.A. and Stone, W. (2002). Biosorption of cadmium and copper contaminated water by *Scenedesmus abundans*. *Chemosphere* 47 (3): 249–255.

Ting, W.H.T., Tan, I.A.W., Salleh, S.F., and Wahab, N.A. (2018). Application of water hyacinth (*Eichhornia crassipes*) for phytoremediation of ammoniacal nitrogen: A review. *Journal of Water Process Engineering* 22: 239–249.

Tomé, F.V., Rodríguez, P.B., and Lozano, J.C. (2008). Elimination of natural uranium and 226Ra from contaminated waters by rhizofiltration using *Helianthus annuus* L. *Science of the Total Environment* 393 (2–3): 351–357.

USEPA (1996). The Metal Translator: Guidance for Converting a Total Recoverable Permit Limit from a Dissolved Criterion, vol. 823, (96-007). Washington, DC: Office of Water, EPA.

Ullah, A., Heng, S., Munis, M.F.H. et al. (2015). Phytoremediation of heavy metals assisted by plant growth promoting (PGP) bacteria: a review. *Environmental and Experimental Botany* 117: 28–40.

Vasavi, A., Usha, R., and Swamy, P.M. (2010). Phytoremediation – an overview review. *Journal of Industrial Pollution Control* 26 (1): 83–88.

Vigliotta, G., Matrella, S., Cicatelli, A. et al. (2016). Effects of heavy metals and chelants on phytoremediation capacity and on rhizobacterial communities of maize. *Journal of Environmental Management* 179: 93–102.

Volk, T.A., Abrahamson, L.P., Nowak, C.A. et al. (2006). The development of short-rotation willow in the northeastern United States for bioenergy and bioproducts, agroforestry and phytoremediation. *Biomass and Bioenergy* 30 (8–9): 715–727.

Wang, Q., Cui, Y., and Dong, Y. (2002). Phytoremediation of polluted waters potentials and prospects of wetland plants. *Acta Biotechnologica* 22 (1–2): 199–208.

Wang, H., Zhang, H., and Cai, G. (2011). An application of phytoremediation to river pollution remediation. *Procedia Environmental Sciences* 10: 1904–1907.

Were, F.H., Wafula, G.A., and Wairungu, S. (2017). Phytoremediation using bamboo to reduce the risk of chromium exposure from a contaminated tannery site in Kenya. *Journal of Health and Pollution* 7 (16): 12–25.

Wu, L.H., Luo, Y.M., Xing, X.R., and Christie, P. (2004). EDTA-enhanced phytoremediation of heavy metal contaminated soil with Indian mustard and associated potential leaching risk. *Agriculture, Ecosystems & Environment* 102 (3): 307–318.

Wu, S.C., Chung Cheung, K., Luo, Y., and Wong, M. (2006). Effects of inoculation of plant growth-promoting rhizobacteria on metal uptake by *Brassica juncea*. *Environmental Pollution* 140 (1): 124–135.

Yang, X., Feng, Y., He, Z., and Stoffella, P.J. (2005). Molecular mechanisms of heavy metal hyperaccumulation and phytoremediation. *Journal of Trace Elements in Medicine and Biology* 18 (4): 339–353.

Yang, Q., Tu, S., Wang, G. et al. (2012). Effectiveness of applying arsenate reducing bacteria to enhance arsenic removal from polluted soils by *Pteris vittata* L. *International Journal of Phytoremediation* 14 (1): 89–99.

Yang, W., Wu, F., Ding, Z. et al. (2018). Cadmium accumulation and tolerance in seven ornamental willow genotypes. *Bulletin of Environmental Contamination and Toxicology* 101 (5): 644–650.

Yanqun, Z., Yuan, L., Jianjun, C. et al. (2005). Hyperaccumulation of Pb, Zn and Cd in herbaceous grown on lead–zinc mining area in Yunnan, China. *Environment International* 31 (5): 755–762.

Yeasmin, L., Ali, M.N., Gantait, S., and Chakraborty, S. (2015). Bamboo: an overview on its genetic diversity and characterization. *3 Biotech* 5 (1): 1–11.

Yıldırım, K. and Kasım, G.Ç. (2018). Phytoremediation potential of poplar and willow species in small scale constructed wetland for boron removal. *Chemosphere* 194: 722–736.

Zehra, A., Sahito, Z.A., Tong, W. et al. (2020a). Assessment of sunflower germplasm for phytoremediation of lead-polluted soil and production of seed oil and seed meal for human and animal consumption. *Journal of Environmental Sciences* 87: 24–38.

Zehra, A., Sahito, Z.A., Tong, W. et al. (2020b). Identification of high cadmium-accumulating oilseed sunflower (*Helianthus annuus*) cultivars for phytoremediation of an oxisol and an inceptisol. *Ecotoxicology and Environmental Safety* 187: 109857.

Zhang, H. and Xue, J. (2018). Spatial pattern and competitive relationships of *Moso bamboo* in a native subtropical rainforest community. *Forests* 9 (12): 774.

Zhang, X., Su, C., Liu, X. et al. (2020). Effect of plant-growth-promoting rhizobacteria on phytoremediation efficiency of *Scirpus triqueter* in pyrene-Ni co-contaminated soils. *Chemosphere* 241: 125027.

Zhang, X., Zhou, Y., Xue, D.X., and Liu, J.X. (2016). Genetic divergence of the endangered seagrass *Zostera japonica* Ascherson & Graebner between temperate and subtropical coasts of China based on partial sequences of *matK* and ITS. *Biochemical Systematics and Ecology* 68: 51–57.

Zhivotovsky, O.P., Kuzovkina, Y.A., Schulthess, C.P. et al. (2011). Lead uptake and translocation by willows in pot and field experiments. *International Journal of Phytoremediation* 13 (8): 731–749.

Part III

Microbial Remediation of Water Pollution

7

Advances in Biological Techniques for Remediation of Heavy Metals Leached from a Fly Ash Contaminated Ecosphere

Krishna Rawat and Amit Kumar Yadav

School of Environment and Sustainable Development, Central University of Gujarat, Gandhinagar, Gujarat, India

7.1 Introduction

In India, low sulfur and high ash content coal such as bituminous and subbituminous coal, as well as a lignite type, are used in thermal power plants (TPPs) for electricity generation. By 2040, India's energy demand will be 1908MT. This growth of energy demand is the highest in the world and India will account for a 25% rise in global energy use. Also, India will account for the largest absolute growth in coal consumption (CEA 2017).

Coal-based electricity generation in India will remain dominant for the present and into the near future (Figure 7.1). Coal in India has a high ash content and is of low grade. However, its projected demand is high as it is the major primary fuel for energy generation. The percentage share of various energy sources is shown in Figure 7.1. As per a report of the Central Electricity Authority of India (CEA 2017), in 2016–2017, for 509.46 MT of coal consumed in TPPs, fly ash generation was 169.25 MT. This figure is expected to increase in the coming years.

Thus, the practice of using coal as a dominant fuel for energy generation is causing concern in terms of pollution where the by-products generated during coal combustion are a cause of major concern owing to their physicochemical properties and minute nature. Metal contamination by fly ash is of major concern as the eventual release of heavy metals in the environment causes pollution in the long term with acute and chronic effects. These toxic metals cause problems in the normal functioning of living beings, which includes plants and animals from large to small (Singh et al. 2011).

7.1.1 Status of Fly Ash Generation, Utilization, and Disposal in India

Around 70% of the total energy requirement in India is met by pulverized coal combustion in TPPs. This practice will likely continue for the foreseeable future due to the excessive coal reserves in India, estimated at 319.02 billion tonnes in geographical area (Ministry of Coal, India 2020). Because 1 MW thermal power results in the production of 1800 t of ash,

Pollutants and Water Management: Resources, Strategies and Scarcity, First Edition. Edited by Pardeep Singh, Rishikesh Singh, Vipin Kumar Singh, and Rahul Bhadouria.

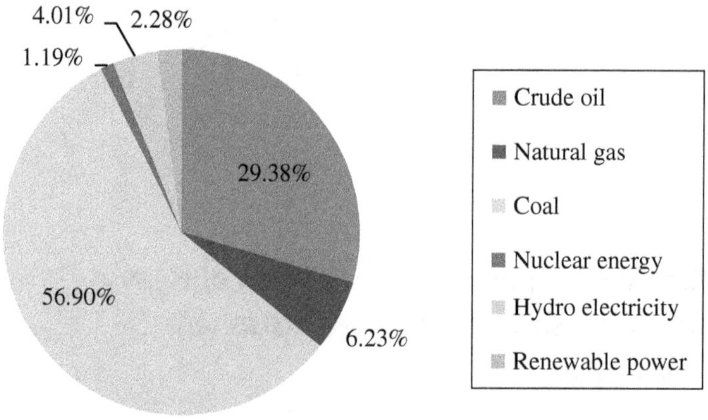

Figure 7.1 Percentage share of various energy sources in India. (*Source:* Modified from CEA 2017.)

Table 7.1 Fly ash generation and utilization in 2016–2017.

Description	2016–2017
No. of thermal power stations	155
Installed capacity (MW)	157 377
Coal consumed (MT)	509.46
Fly ash generation (MT)	169.25
Fly ash utilization (MT)	107.10
Percentage utilization	63.28
Average percentage ash content (%)	33.22

Source: CEA (2017).

fly ash generation is likely to surpass 300 MT by the year 2017 (end of 12th plan); 155 coal or lignite based thermal power station data from different power utilities in India were analyzed for fly ash generation and utilization for 2016–2017 (Table 7.1).

Seventy-six power utilities were assessed for fly ash generation and utilization – 100% or more fly ash utilization was achieved by 21 power utilities, while 75–100% was found in 23 power utilities. Table 7.2 shows the status of fly ash generation and utilization for the year 2016–2017 for various power utilities throughout the country.

More than 10 MT of ash was generated by Andhra Pradesh, Chhattisgarh, Madhya Pradesh, Maharashtra, Odisha, Uttar Pradesh, and West Bengal. More than 100% was utilized by the territory of Delhi and the states of Gujarat, Haryana, Jharkhand, and Punjab, while less than 50% was utilized by Chhattisgarh, Madhya Pradesh, Telangana, and Uttar Pradesh.

Initially, fly ash was considered as a polluting industrial waste, but after concerted efforts by various organizations under the fly ash mission, different avenues have been marked for effective management and utilization of this residue (Kumar et al. 2005).

Table 7.2 Fly ash generation and its utilization for 2016–2017.

S. no.	State	No. of TPPs	Installed capacity (MW)	Fly ash generation (MT)	Fly ash utilization (MT)	Percentage utilization %
1	Andhra Pradesh	10	10 972.5	17.8322	12.6832	71.13
2	Assam	1	250	0.2070	0	0
3	Bihar	4	4100	7.3916	2.3929	32.37
4	Chhattisgarh	20	17 740	25.1658	10.8393	43.07
5	Delhi	2	840	0.3650	0.5720	156.71
6	Gujarat	11	15 472	3.4842	3.5465	101.79
7	Haryana	5	5550	4.1413	5.2922	127.79
8	Jharkhand	7	5612.50	6.7453	7.2876	108.04
9	Karnataka	4	5480	3.0739	1.7080	55.56
10	Madhya Pradesh	7	10 640	11.4186	4.1064	35.96
11	Maharashtra	20	21 556	16.4984	11.7649	71.31
12	Odisha	5	5188	11.4560	6.3975	55.84
13	Punjab	3	2640	0.8799	1.4296	162.48
14	Rajasthan	8	7840	6.7657	6.4489	95.32
15	Tamil Nadu	10	8430	6.1763	4.6297	74.96
16	Telangana	3	1820	2.6751	0.9652	36.08
17	Uttar Pradesh	18	19 104	28.2732	13.2480	46.86
18	West Bengal	17	14 142	16.7039	13.7837	82.52
	Grand Total	155	157 377	169.2533	107.0956	63.2754

Source: CEA (2017).

7.1.2 Fly Ash Utilization

To replace fly ash disposal, its gainful utilization is practiced and such research is promoted under the Fly Ash Mission (CEA 2017). The first notification of fly ash utilization was issued in September 1999 by the Ministry of Environment and Forest, Government of India (GOI), and was subsequently amended in 2003, 2009, and 2016, stipulating targets for thermal power stations for fly ash utilization and the use of fly ash by construction agencies within a prescribed circumference of any thermal power station. GOI initiatives and policy decisions have increased the utilization of fly ash in different construction activities such as in the manufacturing of Portland pozzolana cement, the development of wasteland, manufacturing fly ash-based products, the construction of dams and highways, and the reclamation and backfilling of low-lying areas and mines. The utilization target achieved during the year 2016–2017 was 63.28% (107.10 MT) as compared with the figures of the year 1996–1997, i.e. less than 10% (6.64 MT) more effort is required for achieving the target of 100% utilization of fly ash (CEA 2017).

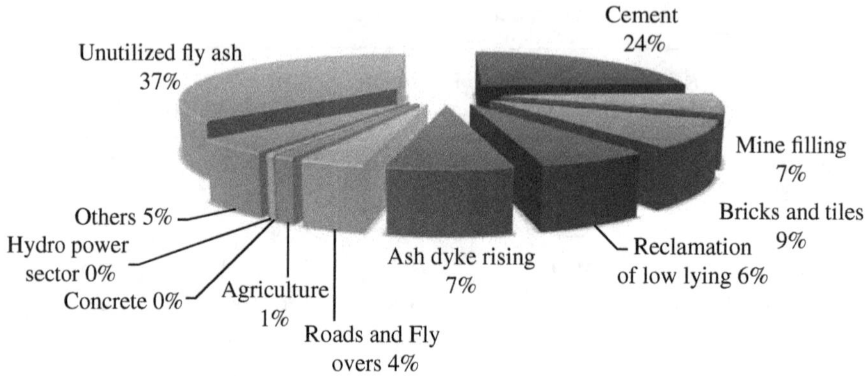

Figure 7.2 Major modes of fly ash utilization during the year 2016–2017. (*Source:* Modified from CEA 2017.)

TPPs dispose of most of the fly ash generated in energy production in the vicinity of the plant, which covers many hectares of land in the waste material of coal combustion. Ash dyke construction, the filling of low-lying areas, and mine filling are possible avenues for bulk utilization of this waste material of thermal stations. Coal ash has been used successfully in structural fills in many developed countries. However, this particular method of bulk utilization is yet to be implemented in India – this can be an economic alternative to natural soil (CEA 2017). Figure 7.2 shows the different modes of fly ash utilization.

7.1.3 Disposal of Fly Ash in India

By 2022, coal-based power generation is expected to reach 260 GW from 186 GW as of March 2016 (CEA 2017), with this pace, pollution is expected to worsen if emission regulations are left unchanged. This is due to the gap between the generation and utilization of fly ash as a result of coal combustion in TPPs. Fly ash that is produced during coal combustion needs proper ash handling. Primarily, fly ash is disposed of by two methods: one is dry disposal and the other is wet disposal. Dry disposal of fly ash is accomplished by transporting ash by trucks, chutes, or conveyors from the TPP to the disposal site by constructing a dry embankment (dyke). In this process, a vacuum system or a pressure system is used for removing dry ash collected in the electrostatic precipitator (ESP) hoppers in dry form, this ash is conveyed to a buffer hopper located adjacent to the ESP. For further use, a part of the ash in dry form can be transported by pneumatic conveying to the storage site from this buffer hopper. Transportation ash is conditioned using a minimum amount of water, and in this form, it is transported through conveyors, where at the disposal site, it is compacted. In wet disposal, fly ash is transported through pipelines and is disposed of in an impoundment called an "ash pond" as slurry. Two methods are practiced in this disposal: lean concentration slurry disposal (LCSD) and high concentration slurry disposal (HCSD).

7.1.4 Introduction to Fly Ash Heavy Metals

Fly ash is the by-product generated as a result of coal combustion in TPPs. It rises with hot air and is composed of fine particles. Several heavy metals are present and their concentration depends on raw material particle size, the mineralogy of the substrate being burnt, and the temperature at which it is burnt. These toxic metals are enriched in fly ash as compared with its parent substrate, this may be due to the vaporization of many other trace metals in the parent coal during combustion in TPPs. Silicon dioxide (SiO_2) and calcium oxide (CaO) are preliminary content found in all kinds of ashes. The quantity of heavy metals released from fly ash depends primarily on pH, elemental bonding, and the physicochemical properties of the solute found in nearby areas of fly ash (Fulekar and Dave 1991).

Various studies have revealed that the ecosphere of fly ash contains heavy metals in higher concentrations than the permissible limit, which may be fatal for living organisms residing in nearby areas. Verma et al. (2016) showed a high concentration of heavy metals in the groundwater released from fly ash.

7.2 Status of Fly Ash Heavy Metals: Composition and Remediation

Fly ash is composed of almost all types of earth metals, as it is formed from coal that is a composite reserve of all kinds of metals. Table 7.3 shows metal concentration and the expected availability in their surroundings.

The release of trace metals from ash can be determined by leaching tests, which may be used as an indicator of the potential release of metals in order to assess their impact on the environment in the process of rendering or deposition of fly ash under laboratory conditions (Sarode et al. 2010). In addition, they allow the classification of trace elements of ash based on their behavior during leaching under different conditions and to identify the mechanisms of leaching (Fytianos and Tsaniklidi 1998; Praharaj et al. 2002; Sitarz-Palczak and Kalembkiewicz 2012).

7.3 Treatment of Fly Ash Heavy Metal Contaminated Ecospheres

Several methods (physical, chemical, and thermal methods) such as in situ vitrification, soil incineration, excavation and landfill, soil washing, soil flushing, solidification, and stabilization of electrokinetic systems are already used to clean up the environment from these kinds of contaminants, but most of them are costly, require intensive labor, and may cause secondary pollution and irreversible changes in soil properties. Various forms of conventional treatments are either practiced individually or in combinations with each other, this makes them quicker for treatment, but may cause secondary problems such as an increase in cost and secondary pollutants may be released into the environment (Sharma et al. 2018).

Table 7.3 Elemental composition of fly ash.

	Fly ash	
Metals(mg/kg)	**Total**	**Available**
Fe	68	10–15
Al	4.8–312	0.1–822
Ca	0.029–34	460–4400
Mg	0.017–1.4	0.8–179
K	10.8	32–8900
B	17–38	0.5–3
Mn	500–739	0.09–19
P	10.8	6.2
Co	21.1–58	0.05–0.15
Cu	40–80	0.5–11
Zn	52–203	0.4–4.6
Mo	4–33.3	0.1–0.6
Ni	50–204.8	0.15–3
Se	0.6–2.6	0.1–0.4
As	1–4	0.1–16
Cd	5–43	0.03–0.07
Cr	38.2–330	0.3–1.3
Hg	BD	BD
Pb	20–70	<0.1

Source: Modified from Kumar et al. (2017).

7.3.1 Conventional Practices for Heavy Metal Remediation

Conventional practices of remediation involve leaving land unused, excavation, or physically removing the contaminated soil from the site of contamination. They prove to be a rapid solution to the problem of contamination, however, the disadvantage is that they are not in a treatment plan, rather, they are simply moved from one place to another, causing contamination in a new site. Besides, they need to be monitored during transportation, which is also a very costly process. Another conventional process involved in metal treatment is the stabilization of metals. In this treatment process, metals are made unavailable at the place of contamination, i.e. metals are fixed in situ to avoid metal intake by living organisms. This process is not very economical, and it may result in secondary metal formation (Wuana and Okieimen 2011).

7.3.2 Advanced Practices for Heavy Metal Remediation

Different advanced techniques involve thermal, biological, and chemical treatments. In thermal treatment, detoxification of metals is undertaken by the high temperature treatment of fly ash. Bioleaching of heavy metals present in fly ash can be useful in metal recovery, i.e.

organisms convert insoluble metals into a soluble form, making them leach out and the metals can then be recovered from the substrate. It may be a uni- or a multi-step procedure (Meer and Nazir 2017). Although the process is eco-friendly and economical, in this case, the consumption process time will be high. Additionally, the metal leaching end product may be quantitatively not satisfactory. Another approach that can be used is chemical leaching, but the process is very costly and may not be economical. Also, it may produce secondary metabolites that could be a problem to resolve. Their combination studies have shown satisfactory results while others need to be modified to improve their economy or performance. Thus, we need to develop an approach that fits to remediate heavy metals and is economical and eco-friendly. In this direction, synergistic approaches seem to be a new ray of hope (Selvi et al. 2019).

7.3.2.1 Synergistic Approach Using Plants and Microbes Together for Metal Removal

Bioaugmentation (one of the aspects of bioremediation) is the addition of microbes to enhance the decontamination process (Stephenson and Stephenson 1992). Lebeau et al. (2008) have reviewed bioaugmentation-assisted phytoextraction implementation in metal contaminated soils. The combination of phytoremediation and bioaugmentation techniques could certainly be helpful for the decontamination of contaminated sites. A sizable number of statistics are available for published reports regarding metal removal using plants. Plants by physical, chemical, and biological processes extract, sequester, and detoxify toxic metals. Some work has also been done on the metal removal approach using microbial combinations with plants on contaminated soils. However, few studies are available regarding the use of indigenous microbes for metal remediation. Thus, green technology in the true sense can be developed. Its success depends on various factors: (i) plants should have sufficient biomass for the accumulation of heavy metals from the contaminated site. Increased biomass helps to remove more heavy metals from the contaminated site with less concentration of metal within the respective plant. (ii) Metal accumulation must in the shoots as compared with the roots as this improves the efficiency of the phytoremediation technology. (iii) Metal-tolerant plant–microbe associations have been the objective of particular attention due to the potential of microorganisms for bioaccumulating metals from polluted environments or the effects on metal mobilization/immobilization and consequently enhancing metal uptake and plant growth. The synergistic use of plants and microbes has been profitable in the cleanup of metalliferous soils (Glick 2003; Jing et al. 2007). Genus *Prosopis* L. is characteristic of arid and semiarid zones, and its widespread distribution includes ecosystems in Asia, Africa, and the Americas (Perez-Garcia et al. 2012). The species of this genus are characterized by high individual genetic variability at the population level, which influences the responsiveness of these species to extrinsic variables such as climate, drought, alkalinity, and salinity (Shiferaw et al. 2004).

7.4 Case Study of Gandhinagar Thermal Power Plant, Gandhinagar, India

Heavy metals are natural ingredients that have a great impact on humans. By default, their concentration is very low in nature, but in higher concentrations, they are dangerous to living beings when consumed or inhaled (Ali et al. 2019). A major source of anthropogenic pollution includes the use of coal fossil fuels as an energy source for electricity generation.

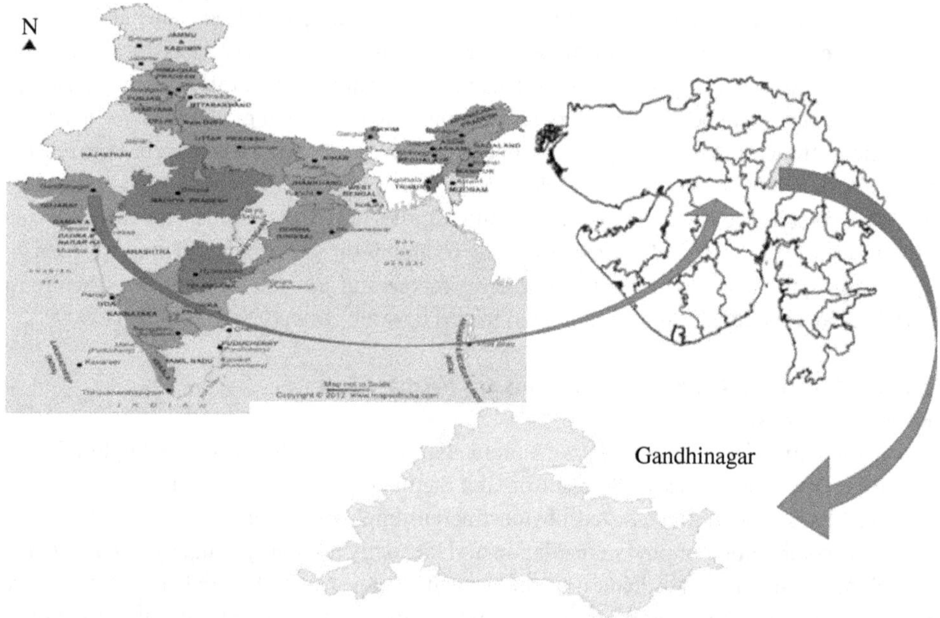

Figure 7.3 Gandhinagar thermal power plant.

Heavy metals can be defined under various criteria such as their position in the periodic table, density, atomic weight, etc. (Heavy Metals 2020). In this study, fly ash generated as coal combustion for electricity generation was studied for its heavy metal constitution and its remediation approach using plants and microbes.

Gandhinagar Thermal power station was selected as a study site situated in Gandhinagar, Gujarat, India (Figure 7.3). Samples of fly ash were collected from the dumpsite of the TPP. These sites were also marked for a potential floral diversity study. The climate of Gandhinagar is tropical wet and dry with summer, monsoon, and winter as the prominent three seasons. The months from March to June are hot to severely hot with a maximum temperature range of 36 to 42 °C (97–108 °F) and the minimum recorded temperature is 19–27 °C (66–81 °F). The winter days of December to February are pleasant but nights are quite cold. The climate of the area is extremely dry with an average maximum temperature of approximately 29 °C (84 °F) and a minimum of 14 °C (57 °F). Annually, average rainfall is about 803.4 mm (31.63 in); from mid-June to mid-September, monsoons from the southwest makes the climate humid (Gandhinagar 2020).

Physicochemical analysis of ash was carried out using standard methods to find its constituents. The analysis included color, pH, EC, TDS, alkalinity, moisture content (MC), cation exchange capacity (CEC), organic carbon (OC), sulfate, nitrate, phosphate, etc. pH value ranged between 8.59 and 7.07. It showed an alkaline nature due to low sulfur content in parent coal and the presence of hydroxides, carbonates of calcium, and magnesium. Alkalinity value was found in the range of 35–46. Values of electrical conductivity (EC), total dissolved solids (TDS), and CEC ranged between 73 and 361, 160 and 184, and 12.99 and 120.6, respectively. MC and organic carbon showed a lower percentage with values

Table 7.4 Physicochemical characterization of ash.

Parameter	ESP	Ash dyke	Ash pond
Color	Light gray	gray	gray
pH	8.53 ± 0.23	7.07 ± 0.30	8.59 ± 0.29
EC	322 ± 0.05	73.97 ± 0.23	157.76 ± 0.22
TDS (mg/l)	176 ± 0.02	44.31 ± 0.05	78.76 ± 0.02
Alkalinity (mg/l)	35 ± 0.07	31.5 ± 0.02	46 ± 0.05
Moisture content (%)	0.169 ± 0.05	6.778 ± 0.11	7.9 ± 0.06
Organic carbon (%)	4.90 ± 0.25	6.754 ± 0.01	5.881 ± 0.08
Porosity (%)	62 ± 0.05	65.72 ± 0.05	56 ± 0.05
WHC (%)	45 ± 0.10	38.506 ± 0.01	39 ± 0.11
Bulk density	1.01 ± 0.12	1.03 ± 0.01	1.02 ± 0.03
CEC	51.35 ± 0.05	13.91 ± 0.20	57.19 ± 0.01
Sulfate (mg/l)	0.669 ± 0.01	0.063 ± 0.24	0.362 ± 0.08
Nitrate (mg/l)	0.006 ± 0.07	6.789 ± 0.06	0.027 ± 0.2
Phosphate (mg/l)	0.297 ± 0.05	0.492 ± 0.06	0.241 ± 0.05

Values in triplicate ± SD, $p < 0.05$.

ranging between 0.033 and 0.262 and 4.33 and 7.05, respectively. MC value was found to be less than 1%. Other nutrient parameters such as sulfate, nitrate, and phosphate showed values between 0.667 and 0.749, 0.0012 and 0.009, and 0.25 and 0.455, respectively. Table 7.4 shows the results of the physicochemical analysis. The average values were for samples collected from four silos of ESP from TPPs and five sites selected for sample collection from ash dykes and ponds. One-way ANOVA analysis was done which indicated significant variation among the values for different parameters studied for different ash samples.

Chemical analysis of fly ash showed the presence of different elements in varying proportions. Figure 7.4 shows the presence of different elements in varying concentrations that occurred in oxide form, mainly in samples. Relative distributions of the oxides were observed with dominant elements, which were Fe and Ca followed by other heavy metals and earth metals. X-ray fluorescence (XRF) spectra showed the presence of Si, Al, Fe, Na, Ti, K, Ca, Mn, S, V, Cu, Zn, Sr, Co, As, Ni, and Br as the bulk oxide of major elements in the dry and wet disposal fly ash samples.

7.4.1 Heavy Metal Analysis of Fly Ash

Heavy metal analysis of fly ash (Table 7.5) showed a maximum concentration of heavy metals in ESP fly ash followed by ash dyke and ash pond. Lower values of ash pond may be attributed to the leaching of metals with water when dumped as slurry. Here, values found were high for ESP followed by ash dyke and ash pond. Relative enrichment factor (RE) was also found to be high for ESP followed by dyke ash and ash pond.

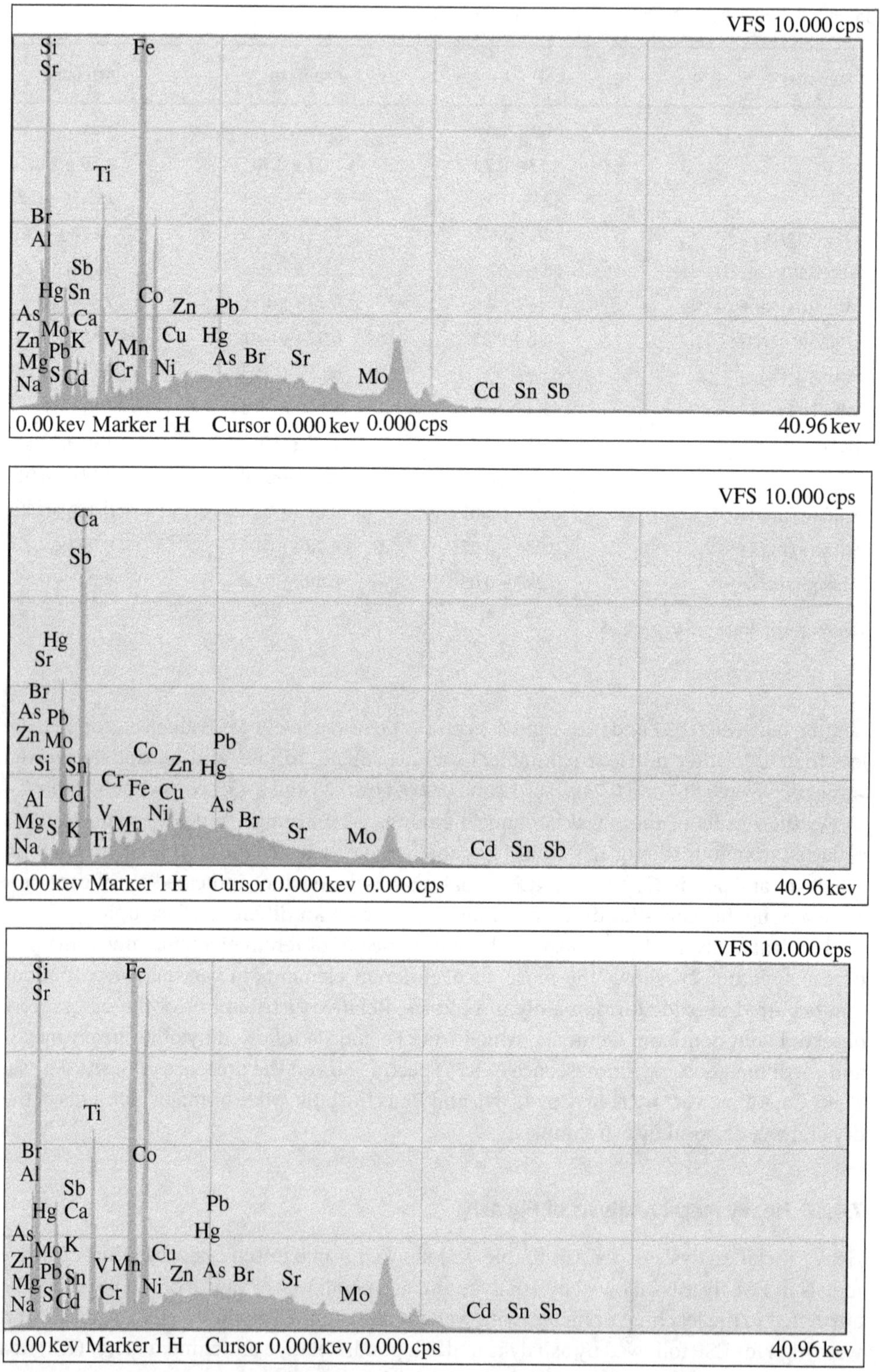

Figure 7.4 X-ray fluorescence peaks of fly ash.

Table 7.5 Heavy metal analysis (μg/g) and relative enrichment factor.

Metals	Ash dyke	RE	Ash pond	RE	ESP	RE
Cr	9.39	0.3158	7.45	0.2506	10.82	0.3639
Fe	2681	0.1713	2533	0.1619	3714	0.2374
Ni	6.85	0.3393	5.41	0.2679	8.13	0.4027
Cu	7.48	0.3410	5.349	0.2438	8.61	0.3925
Zn	11.62	0.3948	9.29	0.3156	17.13	0.5820
As	2.47	0.2612	2.32	0.2453	2.91	0.3077
Cd	0.09	0.3773	0.06	0.2515	0.19	0.7966
Pb(206)	4.61	0.5011	4.12	0.4478	7.31	0.7946
Pb(207)	4.33	0.5083	3.83	0.4496	6.92	0.8124
Pb(208)	4.52	0.5053	3.98	0.445	7.12	0.7960

This objective concluded that the fly ash samples were heterogeneous in nature with irregular surface morphology. Elemental components varied in different percentages in fly ash samples with C as the dominant element in coal while Si, Al, and Fe dominated the elemental fraction with other elements in fly ash. The presence of almost all elements in fly ash was confirmed by XRF in varying fractions. Lastly, an enhanced concentration of heavy metals was found in fly ash. That makes fly ash a matter of concern when disposed of in the environment, as heavy metals present in fly ash may be released into the environment and will contaminate water bodies in the surrounding area as well as groundwater once they seep down from the surface. Thus, their remediation approach is of prime concern when using native plants in the studied areas.

7.4.2 Heavy Metal Remediation Using Plants

For the treatment of fly ash heavy metal problems, lab plant bioremediation apparatus was developed using pot culture and a wild species of the plant *Prosopis juliflora*. At the end of two months, plant metal accumulation was analyzed in the roots and shoots (Table 7.6). The results revealed that the plant is a potential survivor in a harsh fly ash environment and is capable of accumulating considerable metals in its roots and shoot. The roots showed more metal concentration compared with the shoots.

7.4.3 Metal Remediation Using Microbes

Another analysis of metal accumulation was carried out using indigenous microbes of fly ash. For this analysis, microbes were isolated from fly ash samples and were tested to analyze the metal remediation potential of microbes. Potential microbes isolated from fly ash samples mostly included *Bacillus* species with *Pseudomonas aeruginosa* and *Agromyces sp.* (Figure 7.5). These species were used for a metal accumulation study which showed that a consortium was a better option than individual microbes for metal remediation (Table 7.7).

Table 7.6 Metal accumulation in a potential plant.

Plant		Conc. (μg/g)									
		Cr	Fe	Ni	Cu	Zn	As	Cd	Pb	Pb	Pb
0%	Root	0.87	7.538	0.9	0.043	2.32	0.122	0.009	1.3	0.063	1.04
	Shoot	0.095	7.412	0.64	0.032	0.728	0.063	0.002	1.21	0.048	1.02
25%	Root	0.103	11.6	1.54	0.045	3.591	0.123	0.036	1.97	0.118	1.35
	Shoot	0.07	8.163	0.96	0.038	1.159	0.116	0.002	1.19	0.114	1.15
50%	Root	0.107	12.02	1.81	0.046	3.937	0.125	0.038	2.41	0.123	1.44
	Shoot	0.044	10.96	1.02	0.04	1.587	0.06	0.007	1.56	0.116	1.2
75%	Root	0.142	12.85	2.08	0.047	4.77	0.21	0.041	3.47	1.04	2.31
	Shoot	0.11	11.51	1.05	0.038	1. 923	0.048	0.026	1.64	0.118	1.23
100%	Root	1.9	90.16	3.47	7.38	4.843	0.39	0.047	3.55	1.96	3.25
	Shoot	0.116	61.22	1.17	8.82	2.031	0.118	0.031	1.9	0.122	1.24

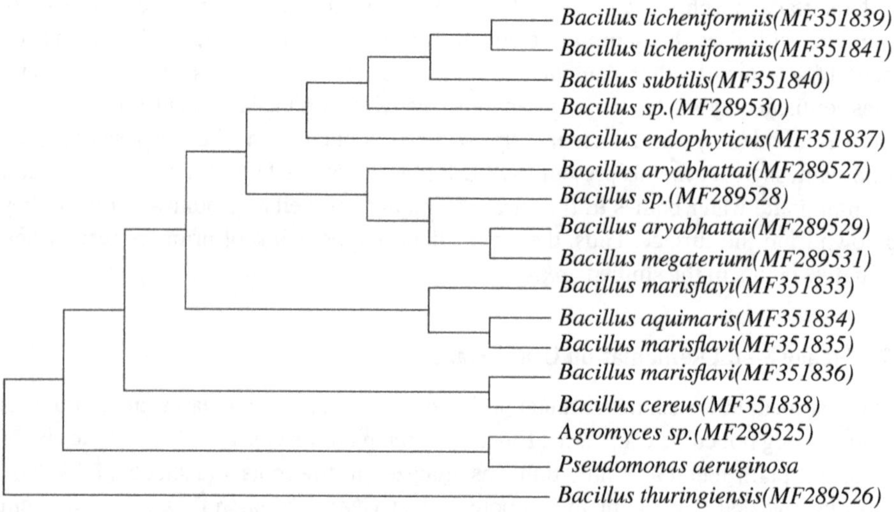

Figure 7.5 Phylogenetic tree for isolated organisms.

7.4.4 Bioaugmentation of Potential Indigenous Plants with Potential Indigenous Microbes

Potential plants were tested with potential microbes for metal remediation efficiency. The best results were obtained using a combination of microbes and plants (Table 7.8).

Consortium 1 comprised *Bacillus endophyticus*, *Bacillus aquimaris*, *Bacillus subtilis*, *Bacillus thuringiensis*, *Bacillus cereus*, *Bacillus licheniformis*, *Bacillus marisflavi*, and *P. aeruginosa*. Consortium 2 comprised *B. endophyticus*, *B. Aquimaris*, *B. subtilis*, and *B. thuringiensis*, while consortium 3 comprised *B. cereus*, *B. licheniformis*, *B. marisflavi*, and *P. aeruginosa*.

Table 7.7 Comparative analysis between the bioremediation of heavy metals by potential microorganism and microbial consortium.

Fe (% removal)	25%	50%	75%	100%
Bacillus endophyticus	60.16	54.42	52.43	50.91
Bacillus aquimaris	78.58	78.64	78.31	75.45
Bacillus subtilis	95.78	91.87	83.54	81.54
Bacillus thuringiensis	67.74	67.50	65.00	64.71
Bacillus cereus	92.00	88.00	85.48	83.63
Bacillus licheniformis	79.75	78.92	78.31	77.12
Bacillus marisflavi	91.08	89.52	88.53	87.36
Pseudomonas aeruginosa	66.13	64	62.5	59.98
Consortium 1	98.72	97.05	96.52	94.81
Consortium 2	98.24	97.68	97.16	94.91
Consortium 3	98.34	97.93	96.87	94.12
Cu (% removal)	25%	50%	75%	100%
Bacillus endophyticus	90.57	83.12	81.25	66.97
Bacillus aquimaris	59.77	56.25	53.43	45.6
Bacillus subtilis	79.85	78.43	77.94	77.32
Bacillus thuringiensis	94.97	87.46	87.23	86.35
Bacillus cereus	79.85	77.47	76.83	73.12
Bacillus licheniformis	69.53	68.92	65.86	63.4
Bacillus marisflavi	67.03	63.38	63.05	62.27
Pseudomonas aeruginosa	71.15	69.35	68.51	60.36
Consortium 1	99.98	99.98	99.96	99.93
Consortium 2	99.97	99.96	99.90	99.95
Consortium 3	99.98	99.97	99.93	99.91
As (% removal)	25%	50%	75%	100%
Bacillus endophyticus	31.2	24.00	20.00	16.00
Bacillus aquimaris	34.11	30.86	26.57	24.35
Bacillus subtilis	51.72	48.31	47.97	38.20
Bacillus thuringiensis	47.00	40.95	39.16	38.45
Bacillus cereus	56.25	51.75	50.95	49.37
Bacillus licheniformis	43.98	38.84	33.98	30.88
Bacillus marisflavi	49.62	46.36	44.87	43.68
Pseudomonas aeruginosa	42.89	42.03	41.98	40.68
Consortium 1	93.69	93.63	89.55	80.82

(Continued)

Table 7.7 (Continued)

Fe (% removal)	**25%**	**50%**	**75%**	**100%**
Consortium 2	95.25	94.32	92.41	87.45
Consortium 3	93.37	92.21	88.46	80.57
Cd (% removal)	25%	50%	75%	100%
Bacillus endophyticus	40.16	35.61	31.50	23.73
Bacillus aquimaris	41.64	31.52	29.90	27.37
Bacillus subtilis	56.93	55.05	54.89	52.6
Bacillus thuringiensis	51.34	55.23	53.21	50.19
Bacillus cereus	63.42	62.74	62.34	61.89
Bacillus licheniformis	48.10	44.26	44.06	36.16
Bacillus marisflavi	56.25	59.61	58.91	51.00
Pseudomonas aeruginosa	82.68	81.25	80.99	77.5
Consortium 1	99.98	99.96	99.95	99.93
Consortium 2	99.93	99.92	99.88	99.77
Consortium 3	99.92	99.9	99.86	99.76
Zn (% removal)	25%	50%	75%	100%
Bacillus endophyticus	78.21	67.83	65.10	62.17
Bacillus aquimaris	84.86	84	80	75.58
Bacillus subtilis	84.11	83.75	83.12	81.27
Bacillus thuringiensis	78.18	76.52	75.97	72.74
Bacillus cereus	86.10	84.63	83.46	80.87
Bacillus licheniformis	91.38	91.34	89.71	80.97
Bacillus marisflavi	72.11	70.68	71.38	68.92
Pseudomonas aeruginosa	49.85	47.86	45.88	45.12
Consortium 1	99.88	99.85	99.81	99.60
Consortium 2	99.92	99.91	99.89	99.53
Consortium 3	99.94	99.90	99.87	99.66
Cr (% removal)	25%	50%	75%	100%
Bacillus endophyticus	56	36.33	34.75	26.57
Bacillus aquimaris	71.25	67.74	59.21	57.12
Bacillus subtilis	75.6	74.97	73.68	72.48
Bacillus thuringiensis	67.24	64	63.10	62.82
Bacillus cereus	61.9	57.24	56.73	55.9
Bacillus licheniformis	62.5	62.12	61.87	60.73
Bacillus marisflavi	68.42	67.74	67.23	67.07
Pseudomonas aeruginosa	77.14	74.82	71.35	70.90

Table 7.7 (Continued)

Fe (% removal)	25%	50%	75%	100%
Consortium 1	98.93	98.36	97.30	95.22
Consortium 2	98.45	98.11	97.41	96.52
Consortium 3	99.00	98.28	97.29	94.94
Pb (% removal)	25%	50%	75%	100%
Bacillus endophyticus	50.64	32.58	31.43	24.42
Bacillus aquimaris	62.74	51.71	53.82	29.52
Bacillus subtilis	64.86	63.84	59.76	58.35
Bacillus thuringiensis	59.68	59.45	58.37	55.38
Bacillus cereus	67.48	66.35	65.98	65.41
Bacillus licheniformis	60.12	59.25	54.83	54.03
Bacillus marisflavi	60.86	59.2	57.14	55.38
Pseudomonas aeruginosa	79.36	75.50	74.29	72.46
Consortium 1	99.93	99.90	99.88	99.81
Consortium 2	99.94	99.91	99.90	99.82
Consortium 3	99.93	99.92	99.89	99.83
Ni (% removal)	25%	50%	75%	100%
Bacillus endophyticus	80.85	77.28	75.21	65.04
Bacillus aquimaris	55.38	48.12	44.96	27.61
Bacillus subtilis	71.56	67.08	64.93	62.74
Bacillus thuringiensis	87.58	86	89.6	81.25
Bacillus cereus	72.31	69.87	69.24	68.74
Bacillus licheniformis	78.8	78.12	72.72	71.2
Bacillus marisflavi	78.1	77.59	73.48	72.98
Pseudomonas aeruginosa	56.5	55.2	53.21	52.38
Consortium 1	99.93	99.91	99.88	99.81
Consortium 2	99.94	99.91	99.90	99.82
Consortium 3	99.98	99.97	99.95	99.832

1/S: *B. endophyticus*; 2R/S: *B. Aquimaris*; 3R/S: *B. subtilis*; 4R/S: *B. thuringiensis*; 5R/S: *B. cereus*; 6R/S: *B. licheniformis*; 7R/S: *B. marisflavi*; 8R/S: *P. aeruginosa*.

Results concluded that coal-based TPPs generated enormous amounts of fly ash, which is enriched with all kinds of earth metals and several percentage points of heavy metals, which are potential degraders of the environment and can be hazardous in both the short and long term. Heavy metals that are leached out in the environment are persistent in the long run, making the area unfit for survival (Tchounwou et al. 2012). In such harsh environmental conditions, naturally grown flora has proven to be a boon due to their easy and luxuriant growth in such nutrient-deficient barren lands. These naturally growing weeds provide us an opportunity to utilize their

Table 7.8 Metal accumulation in roots and shoots of plants (μg/g).

Root	Ni	Cd	Cu	Zn	Fe	As	Pb	Cr
1R	16.85	0.054	38.02	36	66.98	0.55	2.76	2.84
2R	19.63	0.08	17.5	38.2	53.35	0.109	1.59	1.26
3R	12.43	0.086	15.6	25.2	96.27	0.35	1.64	2.68
4R	12.95	0.073	13.6	33.5	95.67	0.74	1.36	1.51
5R	18.75	0.072	37.5	19.5	91.66	0.59	1.75	5.72
6R	15.41	0.064	27.7	17.65	72.02	0.1	1.6	1.45
7R	16.23	0.051	32.9	12.13	74.22	0.53	3.99	6.52
8R	19.73	0.046	34.5	11.64	56.64	0.26	1.67	2.27
C1	19.82	0.182	41.27	**42.12**	116.8	0.87	4.65	7.38
C2	21.94	0.173	39.46	41.52	**224.67**	0.943	**8.37**	8.42
C3	**26.71**	**0.241**	**45.82**	39.81	101.11	**0.982**	7.42	**9.38**
Shoot	**Ni**	**Cd**	**Cu**	**Zn**	**Fe**	**As**	**Pb**	**Cr**
1S	11.31	0.046	19	10.58	23.81	0.048	0.1	0.4
2S	14.95	0.045	8	17.95	50.94	0.05	0.17	0.7
3S	10.06	0.051	3.6	10.53	51.92	0.114	0.14	0.66
4S	11.35	0.043	2.7	12.14	50.29	0.087	0.18	0.82
5S	9.48	0.053	**23.2**	11.67	50.51	0.068	0.68	0.14
6S	10.56	0.02	10.2	11.42	15.61	0.019	0.19	0.24
7S	11.4	0.031	11.5	16.66	26.44	0.051	0.51	0.3
8S	9.57	0.044	12.5	10.28	33.47	0.034	0.34	0.34
C1	11.45	0.58	16.52	14.82	35.71	0.274	**2.73**	2.024
C2	**11.669**	0.56	14.06	17.23	97.140	0.626	2.43	1.7
C3	17.4	**0.69**	18.91	**18.31**	**98.370**	**0.314**	0.634	**2.53**

Note: Bold values indicate the highest metal accumulated in the respective setup for different metals.

potential for heavy metal remediation from contaminated barren wastelands, which have been lying unused for years. These barren lands were also found to have a low microbial count, thus, indigenous microbes can be seen as potential organisms for metal remediation. Finally, the combination of native weeds (such as *P. juliflora*, *Saccharum spontaneum*, *Saccharum munja*, etc.) and indigenous microbes (such as *Bacillus* species, etc.) can be an easily managed economic and eco-friendly technique for the remediation of acres of fly ash polluted wastes lands.

7.5 Conclusion and Future Prospectives

Heavy metals are a global pollutant with a significant threat to living beings. Although they occur naturally, anthropogenic addition makes their condition worse. Coal-based fly ash serves as an enriched reservoir of many heavy metals present in nature. In slurry form, they serve as major pollutants to water resources such as groundwater, ponds, etc. Due to the leaching of these metals, water quality greatly differs compared with its natural composition. These conditions greatly affect the population dependent on such water resources. Deficiencies of essential nutrients, high alkalinity, low soil microbial activity, and high-soluble salt concentrations of trace elements are some of the major concerns for the reclamation of fly ash ponds. Remediation of contaminated sites may be facilitated by the selection of tolerant plant species as well as microorganisms that play a critical role in establishing early ecosystem development as a result of functional abilities such as nitrogen fixation, organic matter turnover, mycorrhizal symbiosis, and potential facilitation of plant establishment (Hodkinson et al. 2002). In this direction, indigenous plants and microbes can be the helping hands for the eco-restoration of such heavily contaminated sites. Weed species naturally growing in the area do not require much maintenance and can withstand these harsh conditions. Therefore, these species, without interfering with food crops, can be helpful for the reclamation of these metal polluted sites. Indigenous microbes assist metal mobility in these plants, thus enhancing their ability to remove metals from sites. Later, the metal of interest can be recovered from plants by biomining, etc.

Bioremediation using plants and microbes, specifically native species, needs to be studied in detail. Still, there is a need to find indigenous plants and microbes for the potential rehabilitation of the contaminated site. The screening of such biological agents is required to analyze the diversity of such plants and microbes for their metal removal ability. Such analysis is helpful to impart information regarding various populations and structures of communities of plants and microbes. It also determines the role of microbes in the biochemical cycle. More work is needed to explore indigenous bacterial colonies inhabiting such extreme conditions. These indigenous colonies can be used as inoculums to rehabilitate such dumpsites for sustainable development. We need to find out the diversity of plants and microbes growing at fly ash dumpsites. They should be studied for determining their synergistic effects and potential in the remediation of fly ash heavy metals.

References

Ali, H., Khan, E., and Ilahi, I. (2019). Review article environmental chemistry and ecotoxicology of hazardous heavy metals: environmental persistence, toxicity, and bioaccumulation. *Journal of Chemistry*: 1–14. https://doi.org/10.1155/2019/6730305.

CEA (2017). June Report. www.cea.nic.in/monthlyinstalledcapacity.html (accessed 11 February 2020).

Fulekar, M.H. and Dave, J.M. (1991). Release and behaviour of Cr, Mn, Ni and Pb in a fly ash/soil//water environment: column experiment. *International Journal of Environmental Sciences* 38 (4): 281–296. https://doi.org/10.1080/00207239108710673.

Fytianos, K. and Tsaniklidi, B. (1998). Leachability of heavy metals in greek fly ash from coal combustion. *Environment International* 24 (4): 477–486. https://doi.org/10.1016/S0160-4120(98)00027-0.

Gandhinagar (2020). Gandhinagar. https://en.wikipedia.org/wiki/Gandhinagar (accessed 11 February 2020).

Glick, B.R. (2003). Phytoremediation: synergistic use of plants and bacteria to clean up the environment. *Biotechnology Advances* 21 (5): 383–393. https://doi.org/10.1016/s0734-9750(03)00055-7.

Heavy Metals (2020). Heavy metals. https://en.wikipedia.org/wiki/Heavy_metals (accessed 21 February 2020).

Hodkinson, I.D., Webb, N.R., and Coulson, S.J. (2002). Primary community assembly on land- the missing stages: why are the heterotrophic organisms always there first? *Journal of Ecology* 90: 569–577. doi: 10.1046/j.1365-2745.2002.00696.x.

Jing, Y.D., He, Z.L., and Yang, X.E. (2007). Role of soil rhizobacteria in phytoremediation of heavy metal contaminated soils. *Journal of Zhejiang University. Science. B* 8 (3): 192–207. https://doi.org/10.1631/jzus.2007.B0192.

Kumar, V., Mathur, M. and Sinha, S. S. (2005). A case study: Manifold increase in fly ash utilization in India. *Fly Ash Utilization Programme* (FAUP), TIFAC, DST, New Delhi.

Kumar, T., Tedia, K., Samadhiya, V., and Kumar, R. (2017). Review on effect of fly ash on heavy metals status of soil and plants. *International Journal of Chemical Studies* 5 (4): 11–18.

Lebeau, T., Braud, A., and Jezequel, K. (2008). Performance of bioaugmentation – assisted phytoextraction applied to metal contaminated soil: a review. *Environmental Pollution* 153 (3): 497–522. https://doi.org/10.1016/j.envpol.2007.09.015.

Meer, I. and Nazir, R. (2017). Removal techniques for heavy metals from fly ash. *Journal of Material Cycles and Waste Management* 20 (10) https://doi.org/10.1007/s10163-017-0651-z.

Ministry of Coal, India (2020). Coal reserves. New Delhi: Ministry of Coal. https://coal.nic.in/content/coal-reserves (accessed 30 September 2020).

Perez-Garcia, E.A., Meave, J.A., and Cevallos-Ferriz, S.R.S. (2012). Flora and vegetation of the seasonally dry tropics in Mexico: origin and biogeographical implications. *Acta Botanica Mexicana* 100: 149–193. https://doi.org/10.21829/abm100.2012.35.

Praharaj, T., Powell, M.A., Hart, B.R., and Tripathy, S. (2002). Leachability of elements from sub bituminous coal fly ash from India. *Environmental International* 27 (8): 609–615. https://doi.org/10.1016/s0160-4120(01)00118-0.

Sarode, D.B., Jadhaw, R.N., Khatik, V.A. et al. (2010). Extraction and leaching of heavy metals from thermal power plant fly ash and its admixtures. *Polish Journal of Environmental Studies* 19 (6): 1325–1330.

Selvi, A., Rajasekar, A., Theerthagiri, J. et al. (2019). Integrated remediation processes toward heavy metal removal/recovery from various environments – a review. *Frontiers in Environmental Science* 7 https://doi.org/10.3389/fenvs.2019.00066.

Sharma, S., Tiwari, S., Hasan, A. et al. (2018). Recent advances in conventional and contemporary methods for remediation of heavy metal-contaminated soils. *3Biotech* 8 https://doi.org/10.1007/s13205-018-1237-8.

Shiferaw, H., Teketay, D., Nemomissa, S., and Assefa, F. (2004). Some biological characteristics that foster the invasion of *Prosopis juliflora* (Sw.) DC. at Middle Awash Rift Valley area, North-Eastern Ethiopia. *Journal of Arid Environmental* 58 (2): 135–154. https://doi.org/10.1016/j.jaridenv.2003.08.011.

Singh, R., Gautam, N., Mishra, A., and Gupta, R. (2011). Heavy metals and living systems: an overview. *Indian Journal of Pharmacol* 43 (3): 246–253. https://doi.org/10.4103/0253-7613.81505.

Sitarz-Palczak, E. and Kalembkiewicz, J. (2012). Study of remediation of soil contamined with heavy metals by coal fly ash. *Journal of Environmental Protection* 3: 1373–1383. https://doi.org/10.13140/2.1.3952.0963.

Stephenson, D. and Stephenson, T. (1992). Bioaugmentation for enhancing biological waste water treatment. *Biotechnology Advances* 10 (4): 549–559. https://doi.org/10.1016/0734-9750(92)91452-K.

Tchounwou, P.B., Yedjou, C.G., Patlolla, A.K., and Sutton, D.J. (2012). Heavy metals toxicity and the environment. *Molecular, Clinical and Environmental Toxicology* 101: 133–164. https://doi.org/10.1007/978-3-7643-8340-4_6.

Verma, C., Madan, S., Hussain, A., and Dubey, S. (2016). Heavy metals contamination of groundwater due to fly ash disposal of coal fired thermal power plant, Parichha, Jhansi, India. *Cogent Engineering* 3 (1): 1–8. https://doi.org/10.1080/23311916.2016.1179243.

Wuana, R.A. and Okieimen, F.E. (2011). Heavy metals in contaminated soils: a review of sources, chemistry, risks and best available strategies for remediation. *International Scholarly Research Network ISRN Ecology* https://doi.org/10.5402/2011/402647.

8

Microbial Degradation of Organic Contaminants in Water Bodies

Technological Advancements

Deepak Yadav[1], *Sukhendra Singh*[2], *and Rupika Sinha*[3]

[1] *Chemical Engineering Department, Harcourt Butler Technical University, Kanpur, Uttar Pradesh, India*
[2] *School of Biochemical Engineering, IIT (BHU) Varanasi, Varanasi, Uttar Pradesh, India*
[3] *Department of Biotechnology, MNNIT, Prayagraj, India*

8.1 Organic Contaminants in Water

Clean freshwater is a vibrant resource with limited reserves, which is persistently burdened under pollution growth and rapid urbanization. These events are the reason for the over-exploitation of freshwater. For example, 17% of European territory and 11% of its population have been severely affected by water paucity (Stein et al. 2016). The situation became threatening when developed countries discharged high volumes of untreated wastewater directly into main rivers, causing detrimental effects on human health and the environment (Wwap 2017). According to the Water Framework Instruction (European Commission 2012a), water quality norms should have been achieved by 2015, and, to date, only half of European water reserves have met their standard directives (European Commission 2012b). Industrial and municipal wastes have been accepted as the fundamental cause of environmental contaminants, viz. pesticides (Pintado-Herrera et al. 2014), personal care products (PCPs) (Lara-Martín et al. 2014), pharmaceutically active compounds (PhACs) (Lara-Martín et al. 2014), polycyclic aromatic hydrocarbons (PAHs) (Qiao et al. 2018), polychlorinated biphenyls (PCBs), antibiotics (Thai et al. 2018), metals (Fu and Wang 2011), and new embryonic pollutants.

PAHs and PCBs can be classified under hydrophobic organic compounds firmly bound to sediments; therefore, they function as lasting contaminants in water bodies. Developments in modern analytical techniques assist us to measure in parts per million (ppm) level the anthropogenic impurities in water. The characteristic heterogeneity of sediment demands physicochemical restrictions, for instance, surface area and overall organic carbon content, besides particle size distribution of organic contaminant management and control. The exposure of organic pollutants are persistent and toxic in water, and mitigation effects reduce our health

Pollutants and Water Management: Resources, Strategies and Scarcity, First Edition. Edited by Pardeep Singh, Rishikesh Singh, Vipin Kumar Singh, and Rahul Bhadouria.

risks by the addition of carbonaceous material to assist the contaminants' binding planned for PCBs, for example, in the Hudson River (National Research Council 2004).

Several landfill sites (a hotspot of pollution) cover distinctive organic and inorganic contaminants with xenobiotic compounds (Christensen et al. 2001). Pesticides are communal xenobiotic compounds in old landfills as their usage increases through population growth and the succeeding agricultural practices (Baun et al. 2004). In recent food creation studies, diffuse contamination has attained the limelight from point-sources of pesticide pollution (Bjerg et al. 2011). The xenobiotic compounds of environmental concern include aromatic hydrocarbons (AHs) and chlorinated solvents in landfills (Kjeldsen et al. 2002). There are supplementary methods or sampling policies appropriate for the analysis of organic contaminants. The most challenging zone for the study is the hyporheic zone where the dynamics of microbial structure, chemistry, activity, and hydrology are studied. The challenge of seasonal disparities of these constraints and interruptions of the results can be managed by the practice of passive samplers. For example, during diffusive gradients in thin films (DGT) study, soil pollutant adsorption takes place on sediments and colloids. The core benefit of passive samplers includes the study of chemical availability and accessibility of substrates to microorganisms that promote the chemical activity of targeted compounds; these methods cannot always be optimized for intricate pollutants (Milosevic 2012). This chapter reviews some of the current strategies being adopted for the treatment of these organic compounds to improve water quality. The characteristics and xenobiotic nature of organic pollutants are discussed first followed by the role of microorganisms for the degradation of these compounds. Bioreactors that have been used for the removal and degradation of organic compounds from wastewater are discussed in detail. Future strategies such as nanobioremediation and microbial fuel cells (MFCs), which can be helpful in wastewater pretreatment, are discussed.

8.1.1 Characterization of Organic Contaminants

Currently, environmental contamination is cumulative because of several human activities. Organic pollutants are the main difficulties confronted by human society. Many trials have been conducted on various activities, these pose a warning to ecologists, scientists, and researchers (Assessment 1996). Impurities cause countless acute and chronic respiratory health effects, neurological toxicity, fatigue, headaches, depression, dizziness, nausea, lung cancer, lethargy, and throat and eye irritation (Fowles et al. 2016). Out of various PAHs, benzene, ethyl-benzene, toluene, and phenolic compounds are classified as pollutants that are a severe danger to the atmosphere (Castillo et al. 2017).

With the aim of eliminating the harmful consequences of organic contaminants, in recent times, a number of new skills have been recognized. Several approaches, for example, biological (Gao et al. 2016), physical, chemical (Bhanot et al. 2020), and advanced oxidation processes (AOPs) (Huang et al. 2017) are applied to reduce these organic impurities to treat industrialized wastewater. However, the most frequently used approaches are bacterial deprivation and AOPs (Singh and Borthakur 2018).

A complete depiction of an organic pollutant comprises the measurement of sample purity, determination of physical properties, functional groups, elemental structure, empirical formula, clarification of spatial relationships, and structural formulation. The characterization of organic compounds is found by several methods and techniques. The optimal standards and practical steps are frequently reliant on the availability of distinct apparatus.

High-pressure liquid chromatography (HPLC) along with fluorescence detector and gas chromatography (GC) are ideal techniques for analytical quality control and research in the detection of organic contaminants. The leading restrictions of the liquid chromatography–mass spectrophotometer (LC–MS)-based procedures are identical such as direct-infusion MS, i.e. environmental properties, with an overall decrease in electro-spray fluctuation. The recent advances in online MS equipment methods, but deprived of some perspiration, have several limitations, comprising a narrow in-spectrum ion suppression and dynamic array. GC–MS is the traditional method for the analysis of aerosols and has been used for quantitative analysis of different organic compounds, including PAHs, alkanes, ketones, aldehydes, carboxylic acids, monoterpenes, and sugars. Solvent-based extraction is still the most common sample preparation method for GC analysis. GC–MS has been revealed to be a reliable technique for targeted analysis; it has limitations in terms of separation efficiency and sensitivity (Parshintsev and Hyötyläinen 2015). X-ray diffraction (XRD) and Infrared absorption spectra (IR) and patterns are appropriate to classify many identified constituents and recognize chemical procedures, but they occasionally take a longer time. A comprehensive discourse on detailed characterization procedures and practices may be available in the future. The latest methods have concerns about formative sample purity, starting physical methods, and chemical properties of classification (Later et al. 1981).

8.1.2 Xenobiotic Nature of Organic Contaminants

Xenobiotic compounds are human-made compounds distinct to a specific environmental system which has a biological action. Xenobiotics comprise engineering chemicals, drugs, naturally arising poisons, and ecological impurities. Specific microbes have the capacity of cracking xenobiotic compounds partly or completely. However, certain xenobiotics are inflexible as they cannot be used as a substrate by microbes, nor can they be transported because of the absence of appropriate transporting enzymes. Some of them have a massive structure and insolubility and are unavailable to microorganisms, while others are distributed as per their chemical composition into different groups. The inexpensive biological remediation surveys include biodegradation used in landfilling by the bacterial route and composting, while nonbiological xenobiotic degradation techniques are assembled under thermal and nonthermal practices.

Xenobiotic compounds are chemicals that are distinct to the biosphere. Depending on the nature of water and soil, xenobiotic compounds are prone to microorganisms. Different xenobiotic structures include aliphatic and aromatic halogenated compounds, cyclic biphenyls, nitroaromatic compounds, organic sulfonic acid, polyaromatic hydrocarbons, triazines, azo dyes, etc. Xenobiotics have adverse effects on human health by disrupting multi-cellular communication routes, which are very toxic. In addition to that, these chemicals are persistent with long half-life periods in the environment prominent to biomagnification. Xenobiotics are found in several food chains in high concentrations and persist for generations (Godheja et al. 2016; Vanek et al. 2010).

Xenobiotics in the atmosphere embrace an extensive variability of compounds, e.g. PCPs, drugs, pesticides, textile dyes, stabilizers, and numerous others. These xenobiotics compounds move in running water by rain washing of cured zones, through leaching via the soil from places of use, and by wastewaters of industrial services or municipal wastewaters (evacuation of undigested drugs, clearance of unexploited drugs). In physical, biological, and chemical routes of natural waters, subsidization occurs to reduce the concentration of xenobiotics. Materials that are sturdy to chemical routes (hydrolysis, etc.) and biological degradation routes may denote significant deprivation paths.

Xenobiotic degradation by photocatalysis can be classified into two major assemblies: those occurring in natural waters and individuals tested in the treatment of wastewater. Set number one is aimed at homogeneous photocatalysis, whereas the following set includes AOPs, based on heterogeneous photocatalysis (semiconductors), which is the best-explored practice. Both practices bring aquatic vulnerability estimation with the consequence of a xenobiotic effect in water populations, with interacting mechanisms neither freely measured nor understood. In this respect, environmental risk calculation becomes more complicated than social hazard evaluation. Although, as society is worried about social health, the level of acceptable hazard is not as frequently on human risk valuation as compared with water risk assessment. Subsequently, the unit of high-to-low dose decision is usually dominant in audits considered to identify anthropoid hazards (Stara et al. 1980).

An outline of micropollutants, sources, their effects, and incidence in various grades of water comprising analytical procedures and concentration ranges was delivered in an unresolved review by Tijani et al. (2016). By means of published reviews, we understood from the lists of the constituents enclosed as evolving xenobiotic compounds and were grouped as antibiotics, analgesics, and antidepressants as pharmaceuticals, whereas preservatives and disinfectants as PCPs were in the finest rich congresses (Tijani et al. 2016).

Most of the xenobiotics observed in natural waters are persistent compounds that are obstinate to microbial deterioration and repel chemical degradation over hydrolysis or other chemical reactions. Most of them contain heteroatoms, aromatic rings, and functional groups in natural waters, absorb solar energy, or react with photogenerated transitory forms. Such compounds carry structures and functional groups, for example, nitro, carboxyl, and phenol, which undertake photodegradation, and chlorine derivatives incline them to dechlorination. Each reaction under radiation-absorption can be classified as a photochemical reaction, conferring the photo-initiated reaction mechanism allied to xenobiotic degradation (Klementova 2018).

8.1.3 Environmental Relevance

Commonly, three categories of photochemical degradation of xenobiotic compounds can be recognized:

a) Environmental (photochemical) investigations, viz. ground or laboratory study beneath natural environments (sunlight irradiation with solar spectrum, concentrations of natural water schemes).
b) Environmental studies (solar radiation, organic substrate concentrations above natural waters).
c) Additional photochemical investigations with small wavelengths not pertinent to ecological circumstances, adding of O_3 and H_2O_2, and other constituents, e.g. semiconductors, act as photocatalysts.

In environmental (especially photochemical) investigations, and those pertinent to the eco-friendly state, the solar radiation range must be measured. In a UV region, only a sketch of the solar spectrum is assessed: the region from 290 to 380 nm forms about 5% of the overall (ultraviolet plus visible) range (Klementova 2018). While UV solar radiation falling to the surface of the Earth signifies a minor portion of solar spectra, it shows a significant part in photochemical changes as it comprises the radiation of higher energy ranges. However, while considering photochemical reactions of xenobiotics, it is essential to consider that these compounds frequently show insignificant absorption in the region of wavelengths extended to 290 nm that fall under the lower wavelength range, consequently, other compounds cannot respond below solar radiation. The photodegradation mechanisms such as photolysis, photocatalysis and photoinitiation can aid breakdown of some of the resistant xenobiotic compounds (Klementova 2018).

The environmental effect of xenobiotics belonging to the families of PCPs and pharmaceuticals has not been adequately agreed since the ecological diligence of these compounds has, so far, not been examined. There are numerous suggestions that photochemical degradation may be a dominant factor in deciding the eco-friendly outcome of these compounds. A subsidiary proof depicts the structure of their constituents; they frequently cover physical structures that have been photodegradable in added groups of compounds like pesticides (Burrows et al. 2002). An important group of organic xenobiotics in water are parabens, extensively used as food preservatives, cosmetics, and pharmaceutical products.

Parabens were principally cast-off as antimicrobial preservers in pharmaceuticals in the 1920s and persisted as preserving choices for nearly a century, subsequently, they were selected as an ideal preservative: they display many antimicrobial actions, they have been measured safe to use, and they are steady over a wide pH range. They are adequately water-soluble to yield active concentrations. Currently, apprehension has risen about their security as some parabens have been stated to have estrogenic activity in tentative cell schemes and animal reproductions (Fenech et al. 2013).

As previously highlighted, xenobiotics (organic) present in water signify tremendously variegated, complex chemical compounds with the aforesaid pesticides, drugs and their additives, chemicals recycled in various process industries, their derivatives, solvents, etc. The current survey will help scientists and policymakers to recognize and calculate concentrations of organic contaminants and xenobiotics with some natural pollutants in wastewater (Bester et al. 2008). The focus on these organic water pollutants has considered these issues, or indirectly in the context of water directives (Eu 2008).

8.1.4 Hazards Associated with Xenobiotics

The total volume of water on Earth is assessed at ca. 1500 million cubic kilometers, with 97.1% in oceans, 2.23% in icecaps, and 0.6% in groundwater. Hazards associated with organic contaminants in water bodies should be properly characterized. The hazards of these compounds to human health have been accurately documented, but their capacity to pass through the aquatic environment has only been explored recently and ecological diligence is still mysterious. As a result of their sources, physical and chemical properties, behavior, and allied hazards, subsequent forms of micro-contaminants may be measured as incipient in groundwater periodically (Jońca et al. 2013).

The status of judging health hazards linked with chemical contamination of the aqueous situation frequently rises. There is an essential need to continue checking water quality in the closed system due to global finite resources. The aim of health effect evaluations is a basis for water quality measures for the estimation of ambient water ingredient concentrations to protect community health. Preferably, the norms should characterize levels for chemical compounds in ambient water reservoirs and would not pose any threat to social health. However, in certain accurate calculations of human health risks as a result of eco-friendly chemicals, a necessary dissimilarity must be prepared between total protection and adequate hazard. Standards for total protection based on comprehensive awareness of quantity/reaction associations in individuals include all causes of chemical disclosure from numerous green media, the types of toxic effects provoked because of precise contaminants or their mixtures, the presence or existence of "thresholds" for stated toxic properties, the implication of synergistic/antagonistic connections, and sympathy and contact levels within inhabitants (Stara et al. 1980).

A toxicity assessment of environmental pollutants is created on the quantification and criteria of the detected health hazards in humans and trial organisms. In a challenge to establish the scale of various toxic indexes, following classification can be adopted:

1) Acute, subchronic, and chronic toxicity
2) Mutagenicity
3) Teratogenicity
4) Carcinogenicity

The critical variance between acute and chronic toxicity and the further three classes is that the last three, i.e. mutagenic, teratogenic, and carcinogenic properties, may comprise the inherited scheme. The effects of all four groups are "toxic."

8.1.5 Chemical Versus Biological Treatment of Organic Contaminants

Earlier chemical treatment techniques dominated biological methods due to their costs, ease of application on a large scale, and improved control over the process. These chemical processes include adsorption, precipitation, neutralization, solvent extraction, ion exchange, and oxidative degradation. Typical chemical processes include the dehalogenation of halogenated benzenes and chemical oxidation. In the dehalogenation process, halogens, like chlorine or bromine, are removed step by step to yield less halogenated benzene and finally transformed into benzene. In chemical oxidation, different chemical oxidants like H_2O_2, O_3, ClO_2, K_2MnO_4, K_2FeO_4, and others are used to oxidize organic pollutants to be less toxic and transform them into a manageable form (Fennell et al. 2019). Nowadays, advanced oxidation technologies (Fenton, ozonation, wet oxidation, and photocatalysis) have exploited the high reactivity of hydroxyl radicals for the oxidation of organic pollutants and they have emerged as a promising technology for wastewater treatment (Brillas 2014).

However, the segregation of physical and chemical processes is a complicated process, therefore, most of the processes are termed as physicochemical processes. These processes may include a photo and electro-catalytic degradation by using TiO_2, graphene oxide, Cu-Co-Al layered double oxides, manganese oxides, silver modified Zn-GeO nanorods, and nanocomposites hydrogels (Afzal et al. 2018). Photocatalytic or catalytic oxidation is believed to be a sustainable and efficient method for organic pollutant degradation. There are many catalysts used for the photocatalytic degradation of pollutants. TiO_2 is preferred among the different catalysts because of its low cost, nontoxicity, and chemical stability. In many studies, TiO_2 nanoparticles are also employed as a photocatalyst due to its high reactivity and physical and chemical stability (Deng and Zhao 2015).

The main reason associated with the continued use of these processes along with biological processes for organic contaminant removal is to remove particulate matter of sizes greater than 0.45 μm and dissolved matter of sizes less than 0.045 μm. The method of flocculation is effective in more than 50% removal of compounds such as benzo[a]anthracene, benzo[a]pyrene, and mirex. The process of adsorption has been reported to withdraw more than 90% of contaminants such as meprobamate, trimethoprim, and sulfamethoxazole. The chlorination process has shown a positive eradication of more than 90% of pollutants such as 17β-estradiol, triclosan, and oxybenzone, whereas ozonation was a promising method for the removal of compounds such as carbamazepine, trimethoprim, and progesterone. Flocculation singularly was not effective in the removal of endocrine-disrupting chemicals (EDC) and p (PPCP), although activated carbon was efficient in adsorbing many of them (Shon et al. 2009).

Biological wastewater depends on the exploration of the ability of natural systems of bacteria, fungi, yeasts, algae, and plants. These systems can be used either in their de novo form or engineered for improved efficiency; for instance, items recycled in our garden/kitchen that are composed of carbohydrates, fats, starches, sugars, and proteins (approximating lipids/plant-based oils). We can also enhance hydrocarbons (viz. ethane, hexane, octane, pentane, and nonane), along with lower aromatic compounds (namely, benzene, phenols, xylene, toluene, etc.), which are simple compounds easy to break down and digest by bacteria if pollutants are more complex like certain polymers incline to be more stringent for the bacteria to reduce. In these circumstances, chemistry can be used in the primary wastewater management stages to break nondegradable complexes by partially oxidizing and transforming polymer (complex) to monomers (simple), so the bacteria can conquer and treat the rest of the simpler molecules. Coupling of bioremediation with advanced oxidation processes such as use of O_3 and H_2O_2 have paved the way to effective degradation of organic contaminants. The coupling of biological and chemical wastewater action might be mandatory where industrial streams mainly constitute hydrocarbons. These challenging streams are segregated for chemical pretreatment – wherever a majority of the toxicity is removed – then links the streams to the biological treatment (Mandal et al. 2016). Biological systems can be both aerobic and anaerobic depending on the specific requirement of the system. A well-established strategy is the use of an activated sludge process where biodegradation is facilitated under aerobic conditions for biological conversion of organic pollutants into oxidized end products containing carbonate, sulfate, nitrate, and phosphate groups. Fungal systems are also high oxygen consumers and are characterized by pollutant degrading enzymes such as laccases. Whereas anaerobic degradation can also be coupled with energy generation such as biogas production. Algal biorefineries, coupled with organic waste degradation, are also promising alternatives for wastewater treatment. These biological systems will be discussed later in the chapter.

8.2 Treatment of Organic Contaminants Using Microbes

Biodegradation is a process of degrading or converting pollutants into a less harmful state from polluted sites by employing microorganisms like fungi, bacteria, and algae, and their enzymes. The organic contaminants of wastewater such as phenolic compounds, persistent organic pollutants (POPs), PAHs, PCBs, and benzene-containing compounds must be removed or degraded; otherwise, they cause environmental pollution. Several microorganisms have been reported for organic pollutants degrading abilities. Microorganisms like *Pseudomonas aeruginosa* and *Achromobacter xylosoxidans* were identified to degrade phenolic compounds into less toxic and harmful short-chain hydrocarbon compounds. Chen et al. (2019) isolated two strains and stated that the organic pollutants like phenolic compounds present in water used for gas washing could be efficaciously degraded by isolated bacteria. They also proposed that the efficiency of degradation of total organic pollutants was exceeded by 79% and the degradation efficacy of total phenolic compounds was 98.8 and 99.0% by *P. aeruginosa* and *A. xylosoxidans*, respectively (Chen et al. 2019). Liu et al. (2016) isolated and selected a phenol degrading bacteria *Acinetobacter calcoaceticus* PA from petrochemical wastewater, and reported that the selected strain had a high tolerance of phenol concentration at 1700 mg/l, and could efficiently remove 91.6% of the initial phenol concentration (Liu et al. 2016).

8.2.1 Methods of Selection of Microorganisms

Microbes that have to be employed for the degradation of organic pollutants must have some basic properties: (i) a high pollutant degrading ability; (ii) the ability to survive in a high concentration of organic pollutants; and (iii) the ability to produce less toxic by-products. These microorganisms are mostly identified by their physiological, morphological, and biochemical characteristics and are further confirmed by 16S rDNA sequence analysis. Methods through which microorganisms selected for the biodegradation of organic pollutants are discussed here in detail.

8.2.1.1 Random Screening

Random screening is performed to isolate microorganisms of interest that have desired characteristics from a large population of organisms. This screening method is further divided into primary and secondary screening methods. Primary screening is a qualitative screening method where microorganisms are selected either directly or indirectly for a specific type of activity from a large population. In secondary screening, microorganisms are chosen based on both qualitative and quantitative properties. The objective of secondary screening is to determine and verify the precise activity of organisms such as production or degradation of products, and to evaluate the production or degradation potential of microorganisms identified in primary screening. Secondary screening also provides further information such as whether the product has a simple, complex, or even a macromolecular structure.

8.2.1.2 Metagenomic Screening

Most of the microorganisms taken from environmental samples cannot be cultured in laboratories, and thus, are unable to be screened by using conventional technologies. Metagenomic screening has the capability for screening uncultured microorganisms by utilizing the abundant genetic resources of microorganisms. Metagenomic libraries can be constructed by isolating the genomic DNA directly from potential pollutant degraders. These libraries can be explored for screening the uncultured microorganisms in two ways: (i) sequence-based and (ii) function-based screening.

8.2.1.2.1 Sequence-Based Screening This screening method includes the design of primers and DNA probes for the identification of novel genes or proteins. The primers or probes used in this method are part of conserved sequences of the known genes or protein families. A sequence-based screening strategy helps in finding the genes encoding novel proteins or enzymes that take part in pollutant degradation such as monooxygenase, dioxygenase, dehalogenases, laccases, nitrite reductase, oxidoreductase, chitinases, and dehydrogenases (Simon and Daniel 2011). For example, a PCR-based screening approach combined with the pyrosequencing of small fragments of DNA was used to isolate genes coded for aromatic dioxygenases from PCB-polluted soil samples (Iwai et al. 2010). A quantitative kinetic PCR-based analysis was performed to study the variation and abundance of dioxygenases, herbicide-degrading enzymes, encoded by *tfdA*-like genes in a soil-dwelling bacterium. About 437 sequences were obtained by using five sets of primers. The phylogenetic analysis suggested that soil *tfdA*-like genes were associated with *Bradyrhizobium* sp., *Sphingomonas* sp., and uncultured soil-based bacteria (Zaprasis et al. 2010). Wang et al. (2012) discovered a functional operon *nag2* for the degradation of naphthalene by using the stable isotope probing (SIP) with metagenomic sequencing of ^{13}C labeled DNA of naphthalene degrading microbes.

The sequence-driven screening is more commonly used as compared with function-based approaches because of the ease of accessibility of the metagenomic sequencing data and the availability of software for data analysis. On the other hand, sequence-based screening has some inherent limitations, with their efficacy due to the reduced accuracy of genome annotation and data availability.

8.2.1.2.2 Function-Based Screening Screening based on function is employed mainly for discovering genes encoded for a specific function or identifying the sequence diversity of protein families with novel functions. This screening approach is not only helpful in efficiently identifying the microbiome with a novel function but also efficient in collecting the biochemical data related to substrate specificity. The most important procedure of this screening method includes the construction of metagenomic libraries for the identification of novel genes with desired functions using different assay techniques (Ufarté et al. 2015). Three different function-based strategies have been applied to identify novel genes or enzymes: (i) phenotype detection; (ii) heterologous complementation; and (iii) induced gene expression. In phenotype detection, chemical dyes or chromophore-bearing derivatives are added into the growth medium, which is helpful in the identification of individual clones having specific metabolic capabilities. The second approach of function-based screening depends on heterologous complementation of wild strains or mutant strains, where the desired gene of interest is required for the growth under selective pressure. The third and last approach, which is also known as product-induced gene expression, is designed based on induced gene expression, where the enzymatic activities are determined by the expression of green fluorescent protein, which is induced by the product formation (Simon and Daniel 2011).

Several investigations have been done in the last decade for identifying the novel functions of unknown genes. A functional metagenomic strategy was used for the identification of clones having β-D-glucuronidase activity and 19 positive metagenomic clones identified using chromogenic glucuronides as probes from intestinal metagenomic libraries. However, only one clone exhibited an intense β-D-glucuronidase activity when cloned into a vector. The genomes identified were isolated from *Clostridiaceae*, *Lachnospiraceae*, and *Ruminococcaceae* strains (Gloux et al. 2011). In another study, functional metagenomics was used for the identification of resistant genes by inserting a large fragment of foreign DNA into *Escherichia coli* for evaluating the expression functions of clones. Two genes encode for bi-functional proteins such that one of the bi-functional proteins provides resistance to ceftazidime and a beta-lactamase, whereas the second bi-functional protein containing the aminoglycoside acetyltransferase domain is resistant to kanamycin (Donato et al. 2010).

8.2.1.3 High-Throughput Screening

High-throughput screening is a time-saving technique for rapid isolation of the desired microorganism using various methods including ultraviolet–visible spectroscopy (UV–Vis), fluorescence, and mass spectrophotometry. These methods present an advantage over traditional random isolation techniques in terms of the quality of screening. These rapid measurement methods can screen many culture samples at a time; however, they should be coupled with reliable data analysis software to yield reproducible results. A recent study has utilized this method to screen wild-type bacteria that can degrade crude petroleum oil. This study was based on the robust and straightforward method of using Nile red dye as the reagent, which fluoresces with crude oil. The test was carried out in a 96 well plate containing minimal salt medium, microbial culture sample, crude oil, and the dye. The reduction in oil concentration due to degradation was correlated to the decrease in fluorescence intensity (French and Terry 2019).

8.2.1.4 Microfluidics-Based Screening

A microfluidics-based approach is one of the platforms to enable high-throughput screening where the sample size is reduced to microscale. This system miniaturization has caused a reduction in system footprints and increases in analytical performance. This droplet-based technique can be used to study the cell–cell variability in a population and thus is a potent tool to isolate a promising candidate for the degradation of organic pollutants (Dressler et al. 2017). The response of various microorganisms, including bacteria, yeast, protozoa, and microalgae, to the presence of organic contaminants in varying concentrations can facilitate their screening. Another study has shown the application of microfluidics in aquatic ecotoxicology (Campana and Wlodkowic 2018).

8.2.1.5 Reporter-Based Screening

Another method of selection of microorganisms is based on their ability to incorporate reporter genes such as lux operon for expression in response to the concentration of the contaminant. The culture isolated from the site of contamination can be engineered to produce bioluminescence as a function of the contaminant concentration, which leads to an accurate quantitative analysis. In a study, four strains of *Pseudomonas* isolated from an industrial wastewater effluent were developed to carry a copy of lux operon. Further, it was observed that the decay in the luminescence signal could be correlated to the phenolic concentration in the sample. Thus, it is an ecologically relevant method to monitor the toxicity in aquatic systems (Wiles et al. 2003).

8.2.1.6 Use of Computational Techniques

The use of computational methods can be a refined or directed approach to the selection of microorganisms for the degradation of organic contaminants present in water bodies. A study was conducted using the National Center for Biotechnology Information (NCBI) database to detect phylogenetic similarity between aromatic compounds degrading bacteria and quorum sensing bacteria. It was found that greater than 11% of the search population contained the aromatic ring-hydroxylating-dioxygenase (RHD) gene and the acyl-homoserine-lactone (AHL) or auto-inducer synthase gene. This study reported the isolation of six *Sphingomonadale*s and four *Rhizobiales* based on this relationship between quorum sensing and aromatics degradation, which showed pyrene or phenanthrene degrading efficiency along with AHL production (Huang et al. 2013).

System biology can be used to characterize the effect of an organic contaminant on the degrading microorganism at the cellular, population, and ecological levels. A critical review detailing the advanced application of system biology and metabolic engineering for solving the bottlenecks of traditional bioremediation processes has been published (Dangi et al. 2019).

8.2.1.7 Procurement from Repositories

Based on characterization results obtained from computational analysis or literature review, microorganisms can be procured directly from different repositories or microbial culture banks. The World Federation of Culture Collections (WFCC) has registered about 708 microbial banks that aid in bioprospecting (Valsala and Sugathan 2017).

8.2.2 Mechanism of Degradation

The interconnected metabolic pathways of biological systems are responsible for the degradation of the complex structured organic pollutants found in wastewater. The degradation mechanisms are diverse and also depend on environmental conditions. The study of natural degradation pathways is essential to design the strategy for metabolic engineering for enhanced removal of contaminants. Natural degradation methods using biological systems are discussed in the following section.

8.2.2.1 Biological Processes

Among different methods of organic pollutant removal, a biological approach has been considered as a promising method owing to its economic and eco-friendly nature. In the biological process, the organic pollutants are decomposed from water and soil by microorganisms or plants (phytoremediation) (Megharaj et al. 2011). Microorganisms utilize organic contaminants as a source of carbon or energy, and they assimilate to support the growth of cells. The degradation of organic contaminants depends on the approachability and metabolic efficiency of microorganisms to detoxify the organic material into smaller and less toxic substances (Perelo 2010). The biodegradation process involves several important chemical reactions such as hydrogenation, dehalogenation, dealkylation, hydrolysis, O_2 double-bond oxidation, sulfur conversion to O_2, amino-groups oxidation, nitro-groups reduction, ring cleavage, conjugation, methylation, and addition of hydroxyl groups to aromatic compounds (Joutey et al. 2013). These reactions are catalyzed by several enzymes such as dioxygenase, laccase, monooxygenase, dehydrogenase, cytochrome P_{450}, esterase, lignin peroxidase, and dehalogenase, which are encoded by biodegradation genes of microorganisms (Fang et al. 2013).

8.2.2.1.1 Microbial Degradation Microbial degradation techniques not only have the ability to remove organic materials but also immobilize unwanted toxic substances such as POPs, hydrocarbons, acids, and their derivatives for degradation (Perelo 2010). Several investigations have been carried out for the removal of organic pollutants using microorganisms such as bacteria, fungi, and algae (Kadri et al. 2017; Srivastava et al. 2019). Some species of *Bacillus, Pseudomonas, Rhodococcus, Sphingomonas*, and *Mycobacterium* have been found as potential microorganisms for the transformation of toxic pollutants into a harmless end product (Table 8.1).

Biodegradation techniques such as bioattenuation, bioaugmentation, biostimulation, and bioventing have been employed for the degradation of organic contaminants from the environment (Megharaj et al. 2011). A natural attenuation method is also known as bioattenuation, a transformation technique where pollutants are biodegraded or immobilized by naturally occurring microorganisms into less harmful forms. Biostimulation is a process of microbial degradation where the growth of microbes is enhanced by modifying the culture conditions, and the biodegradation rate of pollutants depends on the availability of nutrients such as carbon, nitrogen, phosphorus, oxygen, and type of organic pollutant. Bioaugmentation is the process of the addition of an active culture of microorganisms that have acclimatized into organic pollutants to enhance the rate of degradation of contaminants. Bioventing is a method of improving microbial biodegradation by aerating the polluted water or soil (Ossai et al. 2019). Microorganisms can degrade organic materials, either aerobically or anaerobically. In aerobic conditions, molecular oxygen like mono and dioxygenase act as an electron acceptor whereas in anaerobic conditions, molecules such as nitrate, sulfate, or sulfur acts as an electron acceptor. After transformation, the pollutants are degraded into CO_2, H_2O, and inorganic salts. The organic pollutants, which are recalcitrant such as chlorinated compounds are degraded anaerobically (Megharaj et al. 2011). The enzymes secreted by bacteria or fungi are also used instead of whole-cell microbial degradation.

8.2.2.1.1.1 Bacterial Degradation Bacteria play a crucial and critical role in the sustainability of the environment by degrading organic contaminants. Hydrocarbon-degrading bacteria mainly degrade these pollutants. Several bacterial species such as *Pseudomonas putida, Rhodococcus erythropolis, Bacillus subtilis, Arthrobacter citreus, Bacillus brevis, Alcaligenes faecalis, Sphingomonas, Serratia marcescens*, and *Acinetobacter* have been reported for organic pollutant degradation (Sandoval et al. 2005; Tsoi et al. 1991; Zhao et al. 2017a). Many bacterial species, such as *Mycobacterium, Arthrobacter, Sphingomonas, Pseudomonas*, and *Rhodococcus* are reported for alkyl-aromatic degradation (Srivastava et al. 2019). The facultative bacteria convert the toxic pollutants into harmless by-products by using the available nitrate, sulfate, metal ions, and carbon dioxide as electron donors under anaerobic conditions.

Several bacterial species have been isolated from polluted sites of petroleum industries which could degrade crude oil. Many scientists reported that complex pollutants are degraded

Table 8.1 A list of some of the microorganisms reported for the biodegradation of organic contaminants.

Organic pollutant	Microorganism	Source of microorganisms	Degradation product	Reference
Nonylphenol, 4-*tert*-octylphenol, and 2,4-dichlorophenol	*Thielavia* sp. HJ22	Soil samples collected from Tunisian arid regions, Tunisia	2-(4-hydroxyphenyl) acetic acid, 2-hydroxy-2-(4-hydroxyphenyl) acetic acid, 4-hydroxybenzoic acid, 1-(4-hydroxyphenyl) ethan-1-one, 4-hydroxybenzaldehyde, 3,4-dihydroxybenzoic acid, 3-hydroxy-3-(3-hydroxyphenyl) propanoic acid, hydroquinone, 2,4-dichlorophenol, 3,5-dichlorocatechol, and 2,4-dichloro-6-(2,4 dichlorophenoxy) phenol	Mtibaà et al. (2020)
Hexabromocyclododecane (HBCD)	*Rhodopseudomonaspalustris*YSC3	Paddy soil in Taiwan	Pentabromocyclododecanols and pentabromocyclododecenes	Chang et al. (2020)
Naphthalene, phenanthrene, and fluoranthene	*Rhodococcus opacus*	Microbial type cell culture (MTCC, Chandigarh, India)	Salicylate, salicylaldehyde, salicylic acid, and 1-hydroxymethyl-2-hydroxy-acenaphthylene	Goswami et al. (2018)
Creosote	*Pseudomonas mendocina*, *Brevundimonas olei*, *Sphingobacterium multivorum*, and *Serratia marcescens*	Soil sample	NA	Smułek et al. (2020)
PAHs	*Ochrobactrum halosaudia* AJH1, *Ochrobactrum halosaudia* AJH2, and *Pseudomonas* aeruginosa AJH3	Brine water samples collected from Saline Water Conversion Company, Jeddah, Saudi Arabia	NA	Pugazhendi et al. (2017)

Phenanthrene, pyrene, and benzo[a]pyrene (BaP) (PAHs)	*Rhodococcus wratislaviensis* strain 9	Polluted sites	NA	Subashchandrabose et al. (2019)
PAHs	*Aeribacillus pallidus* (UCPS2), *Bacillus axarquiensis* (UCPD1), *Bacillus siamensis* (GHP76), and *Bacillus subtilis* subsp. *Inaquosorum* (U277)	Unkeshwar hot spring (India)	NA	Mehetre et al. (2019)
Lindane	*Scenedesmus* sp. ISTGA1 and *Paenibacillus* sp. ISTP10	Marble mining sites Alwar, Rajasthan, India	Pentachlorocyclohexenes, tetrachlorocyclohexenes, trichlorocyclohexane, dichlorocyclohexane aliphatic chlorinated compounds, and phenols	Kumari et al. (2020)
Phenol and PAHs	*Pleurotus dryinus* and *Trametes hirsuta*	Institute of Biochemistry and Biotechnology in Georgia, and Microbial Type Culture Collection Centre in India	NA	Ariste et al. (2020)
Pyrene and phenanthrene	*Proteus mirabilis*	A site chronically polluted with animal charcoal, Nigeria	NA	Obayori et al. (2017)
Phenanthrene and anthracene	*Pleurotus ostreatus* D1 and *Agaricus bisporus* F-8	Microbiology and Mycology Laboratory of the Institute of Biochemistry and Physiology of Plants and Microorganisms, Russian Academy of Sciences, Saratov	9, 10 –anthraquinone, phenanthrene-9,10-quinone, phthalic acid, and 2,2′-diphenic acid	Pozdnyakova et al. (2018)
Petroleum and PAHs	*Streptomyces* sp. isolate ERI-CPDA-1	Oil contaminated soil in Chennai, India	Phenylacetic acid, cinnamic acid, catechol, and Protocatechuic acid	Balachandran et al. (2012)
PAHs	*Candida parapsilosis*, *Candida tropicalis*, and *Bjerkandera* sp	Activated sludge from Qinghe WWTP in Beijing, China	NA	Qiao et al. (2016)

NA, not available.

by a mixed population of microorganisms with large enzymatic capacities. Lal et al. (2010) reported that the biodegradation of Lindane (pesticide) or γ- hexachlorocyclohexane (γ-HCH) occurred under both aerobic as well as anaerobic conditions (Lal et al. 2010). However, the mineralization of γ-HCH was generally found under aerobic conditions. Cyanobacteria have been reported for the degradation of naturally present organic contaminants, including AHs and xenobiotics (Cerniglia et al. 1979). Two cyanobacterial species including *Anabaena* sp. and *Nostoc ellipsosporum* were reported for degradation of Lindane and their ability to degrade another chlorinated pollutant, 4-chlorobenzoate, could be improved by genetic engineering (Kuritz and Wolk 1995). *Sphingomonas paucimobilis* has been reported for exceptionally efficient biodegradation by dechlorinating and transforming polyaromatic hydrocarbons such as Lindane or γ-HCH under aerobic and anaerobic degradation pathways (Srivastava et al. 2019).

8.2.2.1.1.2 Fungal Degradation Fungi are eukaryotic microorganisms and classified as molds and yeast. Molds are filamentous microorganisms; however, yeasts are unicellular microorganisms. The biodegradation by fungal species is known as myco-degradation technology where various fungi are used to degrade the organic compounds because they can grow on various substrates and can survive for a longer duration under harsh conditions. Mold genera, such as *Talaromyces*, *Amorphoteca*, and *Graphium neosartorya*, and yeast genera, such as *Candida*, *Pichia*, *Trichosporon*, and *Yarrowia,* were isolated from petroleum-contaminated soil and worked as potential organisms for the degradation of AHs (Gargouri et al. 2015; Prenafeta-Boldu et al. 2006). Several scientists reported that many terrestrial fungal species (*Aspergillus* sp., *Cephalosporium* sp., *Penicillium* sp., *Candida lipolytica*, *Geotrichum* sp., *Rhodotorula mucilaginosa*, and *Trichosporon mucoides*) were found to be potential organisms for the degradation of POPs (Aydin et al. 2017; Kadri et al. 2017; Srivastava et al. 2019). These microorganisms exhibit more significant potential for the degradation of industrial contaminants by their coordinated and enzyme assisted biodegradation mechanism. Fungal species can degrade PAHs, POPs, and PCBs into less harmful products. For example, *Irpex lacteus* and *Pleurotus ostreatus* have been reported to degrade PAHs. The biodegradation of these compounds depends on several operating conditions, such as nutrient availability, oxygen concentration, temperature, and agitation rate (Novotný et al. 2001).

Some fungal species such as *Phanerochaete chrysosporium*, *Bjerkandera adusta*, and *P. ostreatus* use various ligninolytic enzymes like lignin peroxidases, laccase, and manganese peroxidase for the degradation of PAHs (Kadri et al. 2017). The lignin degradation capability of these fungal species helps to remove the persistent organic-pollutants from the environment (Bumpus and Aust 1987). Several *Trichoderma* species (*Trichoderma reesei*, *Trichoderma hamatum*, *Trichoderma viride*, *Trichoderma harzianum*, *Trichoderma virens*, *Trichoderma koningii*, and *Trichoderma asperellum*) have been reported for having the capability of degradation of both low and high molecular weight of PAHs such as phenanthrene, naphthalene, pyrene, chrysene, and benzo[a]pyrene (Zhao et al. 2017b). These species use other enzyme systems such as multicopper laccase, ring cleavage dioxygenases, and peroxidases for the degradation of PAHs (Zafra and Cortés-Espinosa 2015).

8.2.3 The Development of Recombinant Strains

Wild-type strains of microorganisms are not sufficiently capable in the biodegradation of organic pollutants. Hence there is a need to develop new recombinant strains to improve biocatalytic ability to degrade organic pollutants. Biomolecular engineering exploits the engineered biomolecules or strategies for the development of genetically modified enzymes or microorganisms for the biodegradation of organic pollutants (Liu et al. 2019). A hydrocarbon-degrading bacterial strain, *Acinetobacter baumannii* S30, was isolated and inserted with the lux gene for enhancing the degradation ability for total petroleum hydrocarbon (TPH) of crude oil. The bioaugmentation with *A. baumannii* S30 pJES drastically reduced the TPH level in crude oil contaminated soil (Mishra et al. 2004). The large and small subunits of oxygenase enzymes are encoded by six different pairs of genes, which had earlier been detected in a gene cluster, and have a role in the degradation of monocyclic as well as PAHs (Kim et al. 2006). Chao et al. performed the knockout study for salicylate oxygenase activity and observed the accumulation of salicylate from naphthalene and 1-hydroxy-2-naphthoic acid from phenanthrene. They demonstrated that the insertion of *bphA3-bphA2c-bphA1c* genes in *E.coli* enhanced the capability of salicylate oxygenase enzyme to convert salicylate to catechol and 3-, 4-, and 5-methyl salicylate to methyl catechols (Cho et al. 2005).

Most biphenyl-degrading bacteria can metabolize biphenyl via an upper pathway, which is composed of four enzymes, encoded *BphA*, *BphB*, *BphC*, and *BphD*. These enzyme are biphenyl 2,3-dioxygenase (*BphA*), dihydrodiol dehydrogenase (*BphB*), 2,3-dihydroxybiphenyl dioxygenase (*BphC*), and 2-hydroxyl-6-oxo-6-phenylhexa-2,4-dienoic acid hydrolase (*BphD*) (Li et al. 2009). Biphenyl 2, 3-dioxygenase (BDO) is a crucial enzyme that helps in the biodegradation of PCBs (Shumkova et al. 2015). *Sphingobium yanoikuyae* B1 can effectively use biphenyl as a carbon and energy source by decomposing a number of monocyclic and heterocyclic PAHs (naphthalene, toluene, carbazole, dibenzo-p-dioxin, benz(a)anthracene, m- and p-xylene, and phenanthrene) (Cho et al. 2005; Liu et al. 2019; Zhao et al. 2017a). The dioxygenase enzyme is a three-component type IV dioxygenase that performs the dihydroxylation and hydrogenation of the benzene ring. Dihydrodiol dehydrogenase catalyzes the dehydrogenation of biphenyl cis-dihydrodiol, which is a rate-limiting step in the process of biphenyl decomposition and degradation. The 2,3-dihydroxybiphenyl dioxygenase acts as a catalyst in the meta-cleavage reaction of di-hydroxylated aromatic rings for biphenyl degradation (Li et al. 2009). A strain of *Pseudomonas pseudoalcaligenes* KF707-D34 was prepared by biphenyl dioxygenase gene recombination of *Burkholderia cepacia* LB400, a PCB-degrader. The homologous recombination led to a small variation in amino acid sequence resulting in enhanced PCB degradation ability in biphenyl-utilizing *P. pseudoalcaligenes* KF707-D34 (Suenaga et al. 1999). Chlorobenzoic acid is a common toxic organic contaminant that is produced as an intermediate in the process of PCB aerobic degradation (Adebusoye 2017) Dehalogenase or 4-chlorobenzoate-4-hydroxylase enzyme catalyzes the biodegradation of chlorobenzoic acid. Overexpression of the *fcb* gene encoding dehalogenase from *Arthrobacter globiformis* KZT1 in *E. coli* led to the development of a recombinant strain possessing enhanced degradation capability for chlorobenzoic acid as compared with that of the original strain (Tsoi et al. 1991). Table 8.2 shows a list of genetically modified microorganisms developed for the biodegradation of organic pollutants.

Table 8.2 Some genetically modified microorganisms for the degradation of organic pollutants.

Organic pollutant	Engineered microorganism	Gene	Enzyme	Gene isolated from microorganism	Reference
Chlorobenzoic acid	*E. coli*	*fcbA*	Dehalogenase (4-chlorobenzoate-4-hydroxylase)	*Arthrobacter globiformis* KZT1	Tsoi et al. (1991)
PCBs	*Pseudomonas pseudoalcaligenes* KF707-D34	*bph*	Biphenyl dioxygenase	*Pseudomonas pseudoalcaligenes* KF707 and *Burkholderia cepacia* LB400	Suenaga et al. (1999)
PAHs	*Saccharomyces cerevisiae*	*MtL*	Laccase enzyme	*Myceliophthora thermophila*	Bulter et al. (2003)
TPH	*Acinetobacter baumannii* S30 pJES	*lux*	Luciferase	*Vibrio fischeri*	Mishra et al. (2004)
PAHs	*Escherichia coli*	*bphA3-bphA2c-bphA1c* genes	Salicylate oxygenase	*Sphingomonas yanoikuyae* B1	Cho et al. (2005)
n-Phenyl-alkanoic acids	*P. putida* U ΔfadBA-phaZ	*phaZ*	Poly-3-hydroxy-n-phenyl alkanoate (PHPhA) depolymerase	*Pseudomonas putida* U	Sandoval et al. (2005)
4-Chlorobenzoic acid	*Pseudomonas putida* PaW340/pDH5	*fcbA*	Dehalogenase	*Arthrobacter* sp. FG1	Massa et al. (2009)
2,3-dihydroxybiphenyl	*Escherichia coli*	*bphC*	2,3-dihydroxybiphenyl 1,2-dioxygenase	*Dyella ginsengisoli* LA-4	Li et al. (2009)
Pyrene and phenanthrene	*Pichia pastoris*	L324F and W129F/L324F	P450 CYP5136A3 enzyme	*Phanerochaete chrysosporium*	Syed et al. (2013)

8.2.4 Determination of the Rate of Degradation

The biodegradation rate of the organic pollutants mainly depends on the characteristic of the substance and the adaptation of microorganisms on it. Polar groups containing organic pollutants such as organophosphorus pesticides are more water soluble and have low acute toxicity. Thus, they have more bioavailability and are easily degradable. However, materials that have properties like higher bioaccumulation, biomagnification, and biotoxicity are hard to degrade by microorganisms in natural conditions. Examples of these materials are POPs and xenobiotic compounds such as PCBs, PAHs, and heterocyclic compounds.

Most of the organic pollutants follow the first-order degradation kinetics and follow the given kinetic model.

$$\frac{C_t}{C_i} = e^{-kt} \tag{8.1}$$

where C_i is the initial concentration of organic pollutant, C_t is the concentration of an organic pollutant at time t, k and t are the rate constant of degradation and time of degradation, respectively.

Chang et al. (2020) investigated the biodegradation of hexabromocyclododecane (HBCD), a brominated flame retardant, by *Rhodopseudomonas palustris* YSC3 isolated from soil. They reported that the degradation of HBCD follows the first-order degradation kinetics with an R^2 value that varies from 0.59 to 0.94 and degrades 81% of HBCD (1 ppm) within 0.72 days (Chang et al. 2020). Greskowiak et al. (2017) reported the biodegradation rate constants of first-order degradation kinetics of 82 emerging organic compounds in soil and groundwater (Greskowiak et al. 2017). However, a modified Luedeking–Piret equation was used for describing the biodegradation kinetics of phenanthrene by *Achromobacter* sp. LH-1 (a biodemulsifier producing strain), which utilized phenanthrene as a carbon and energy source and produced a biodemulsifier that enhanced the bioavailability of PAHs (Hou et al. 2018). A similar study, done by Włóka et al. (2017), studied the kinetics of PAH biodegradation in soil fertilized with organic materials by measuring the cyclic CO_2 and PAHs in soil samples (Włóka et al. 2017).

8.3 The Process Design for Microbial Treatment

8.3.1 Parameters Affecting the Degradation Process

Several parameters affect the microbial degradation of organic contaminants in wastewater. These parameters are discussed in the following sections. These parameters need special consideration while designing the process for the microbial treatment of contaminants. Various studies have been conducted and described in the literature to emphasize the impact of these factors, some of which have been mentioned in Table 8.3.

8.3.1.1 Impact of Organic Contaminants on Microorganisms

Studies have been done to analyze the toxic impact of organic pollutants on the growth of microorganisms. The toxic effect of petroleum products on the growth of microbial biomass has been analyzed (Xu et al. 2018). Growth inhibitory effects of wastewater organic contaminants such as phenol, organophosphorus, and organochlorine pesticides have been studied. The effect of phenol on decreasing the membrane fluidity of microorganisms such as *E. coli*, *Vibrio* sp., and *P. putida* has been reviewed elsewhere (Wasi et al. 2013). The choice of microbes

Table 8.3 A list of various studies conducted to determine the effect of different parameters affecting microbial biodegradation.

S. no.	Name of organic contaminant	Microorganisms	Parameters evaluated	Reference
1	Crude oil	*Arthrobacter* sp.	Exogenous source of nitrogen and phosphorus; pH; concentration of crude oil; K_2HPO_4 and $(NH_4)_2SO_4$	Reisfeld et al. (1972)
2	Active pharmaceutical ingredients and endocrine-disrupting compounds	Mixed culture of wastewater sample	Acclimatization of the microbes to the trace organic contaminants	Blunt et al. (2018)
3	Crude oil	*Pseudomonas aeruginosa* NCIM 5514	Temperature; agitation rate; NaCl concentration; pH; carbon and nitrogen source concentration and inoculum ratio	Varjani and Upasani (2017)
4	Polyaromatic aromatic hydrocarbons	Use of activated sludge and augmentation with microorganisms such as *Proteiniphilum acetatigenes* and *Acidithiobacillus ferrooxidans*	Solubilization based on *Kow*[a] values; use of additives; use of anaerobic treatment; membrane bioreactor; ozonation; biofiltration; wetlands; electrochemical oxidation; photocatalysis	Zhang et al. (2019a)
5	Insecticides as DDT	*Proteus vulgaris, E. coli, Pseudomonas aeruginosa, Bacillus* sp., *Cyanobacteria, Ralstonia eutropha*, and many more	Bioavailability; anaerobic and aerobic pathways of degradation	Gohil and Ogram (2020)
6	Nitrated polycyclic aromatic hydrocarbons and PAH	Activated sludge	Use of different reactors (batch and moving-bed biofilm reactors); use of anaerobic-anoxic-aerobic processes	Zhao et al. (2019)
7	PCBs	Municipal sludge and industrial sludge	Solvent extraction	Macherzyński et al. (2016)
8	Polybrominated diphenyl ethers	*Dehalococcoides*	Different nutrient levels in wastewater	Chen et al. (2018)
9	Different organic micropollutants	Microorganisms present in wastewater sludge	Different factors affecting biosorption and biodegradation efficiencies	Alvarino et al. (2018)
10	Organic micropollutants	Wastewater sludge consortium	Substrate concentration; redox conditions; a combination of different aerobic and anaerobic processes; SRT[b]	Falås et al. (2016)
11	Phenanthrene	*Burkholderia fungorum* FM-2	pH; temperature; PHE concentration and salinity	Huang et al. (2019)

[a] Octanol–water partition coefficient.
[b] Solid retention time.

for specific degradation of a particular organic contaminant is a crucial factor. If the concentration of the organic pollutant is greater than the toxic threshold, the biocatalyst is not able to degrade the contaminant. Also, if during the degradation another secondary metabolite is produced with increased toxicity, there are alterations in the growth and metabolic processes of the biocatalysts. This is evident in the case of anaerobic biodegradation of trichloroethylene which produces vinyl chloride, a compound with increased toxicity (Bressler and Gray 2003). Thus, the cytotoxic effect of pollutants on the growth of degrading microorganisms should be quantified. The minimum inhibitory concentration (MIC) of the pollutant should be quite high as compared with the concentration in the polluted water for efficient remediation.

8.3.1.2 Choice of Microorganisms

The efficacy of biological remediation depends on the adaptability of microorganisms. Degradation studies have been conducted using microorganisms under aerobic and anaerobic conditions by oxidative and reductive pathways, respectively (Al-Jawhari et al. 2015; Das and Adholeya 2015; Lei et al. 2002; Xiong et al. 2018). The degradation of efficacies was analyzed for pure and mixed culture. It has been found that many microorganisms act in a coordinated manner where the end product obtained after degradation of the organic contaminant may act as a substrate for another microbial community. Thus, bioaugmentation using mixed microbial consortia is a well-known approach for bioremediation (Nogueira 2018). Studies have revealed that during granular sludge formation, extracellular polymeric substances play a critical role in the tolerance and adaptability of microorganisms toward organic pollutants (Rusanowska et al. 2019). Many researchers have compared the enzymatic xenobiotic compound removal efficiency of bacterial and fungal species (Deshmukh et al. 2016; Gangola et al. 2019; Zou 2015). This is governed by the production of enzymes required for the degradation process. The synthesis of enzymes is regulated by the availability of suitable substrate and metabolic constraints. Many recent studies have been done on the factors affecting the efficiency of the activated sludge on the removal of organic contaminants (Boonnorat et al. 2018; Martinez-Alcala et al. 2017; Nguyen et al. 2018; Tobajas et al. 2016; Vuković Domanovac et al. 2019; Wang et al. 2017). The mechanism of degradation can be aerobic or anaerobic (Han et al. 2020; He et al. 2018; Nolte et al. 2018; Yoshikawa et al. 2017). It includes a variety of microbial community behavior under various reactor operational strategies for the effective removal of micropollutants (Cydzik-Kwiatkowska and Zielińska 2016). A recent report suggested that the aerobic degradation of high total organic load in industrial wastewater in a process using sequential batch reactors (SBR) and ozonation. This has led to a significant reduction in high pollutant concentration in wastewater streams (Chávez et al. 2019). A study was carried out to determine the mechanism of removal of hydrophilic and hydrophobic organic wastes such as phytoestrogens, pesticides, steroid hormones, and pharmaceuticals in a membrane-based reactor (MBR) which was inoculated with activated sludge (Wijekoon et al. 2013). Research has also been conducted for the identification of the pathways of microbial degradation of pesticides (Huang et al. 2018). The control and kinetics of biodegradation depend on the interactive effects of enzymes, microbes, and microbial communities, which are highly specific (Sannino et al. 2016).

8.3.1.3 Mass Transfer Limitation of Pollutants to the Site of Degradation

The mass transfer limitation for biodegradation is more evident in aerobic processes as compared with anaerobic processes because the maximum rate of biotransformation is usually higher in the former case. The mass transfer limitation of wastewater organic pollutants to microbial cells is governed by three different parameters. The *first factor* is the spatial distribution and availability of organic contaminants to microbial cells. This factor is determined by aqueous solubility. To quantify the solubility in biological membranes and lipids, the log K_{ow} value can be estimated. The compounds with comparatively low log K_{ow} values pose a difficulty in biodegradation as they tend to be adsorbed on solid suspended particles. The accessibility of the pollutants also affects the rate of microbial degradation due to nonhomogeneous spatial distribution. The *second factor* is the multi-phase distribution of the contaminant in water. In case more than one phase is present, the partitioning of the contaminant in various phases also governs the mass transfer limitation for biodegradation. The compound with a reduced aqueous solubility has a greater concentration in the hydrophobic phase if present. This partitioning varies with the mixing, interfacial properties, solute equilibrium, and viscosity. The chemical structure of the pollutant is the key player in this phasic distribution. For example, the addition of an alkyl group to compounds such as benzene will reduce the solubility and hence biodegradability of the organic compound (Bressler and Gray 2003). Most of the fungal enzymes are extracellular, however, the majority of the bacterial biodegradative enzymes are located within the cells. The *third factor* responsible for mass transfer limitation is the barrier for transport of pollutants across the bacterial cell membrane made up of a lipid bilayer, where enzymes transform them into less toxic substances. Earlier reports suggested that contaminants that can be taken up by the cell membrane are those that have an appropriate combination of functional group, polarity, and size (Bressler and Gray 2003). According to Bressler and Gray (2003), the relation to determine the pollutant flux across the cell membrane is given by the following equation:

$$F = P(X_e - X_i) \tag{8.2}$$

Where, $P = \dfrac{Permeability\ of\ the\ membrane}{Area\ of\ the\ membrane}$

Furthermore, X_e and X_i are the external and internal concentrations of the pollutant, respectively.

In the above equation, when the rate of biocatalysis is significantly high, $X_i \cong 0$. Equation (8.3) can be used to determine the value of P for cell membranes in terms of K_{ow} and molecular weight of the pollutants.

$$P = 0.03\left[K_{oil-water}{}^{*} \Big/ \sqrt{M}\right] \tag{8.3}$$

The value of $K_{oil-water}$ is one-tenth of the K_{ow} value. Thus, the value of the maximum pollutant flux (F^m) can be given by:

$$F^m = 0.003\left[\frac{\left(K_{oil-water} {}^{*} S_{aq}\right)}{\sqrt{M}}\right] \tag{8.4}$$

where S_{aq} is the aqueous solubility of the pollutant at saturation.

These issues of mass transfer limitations are more evident in soil contaminants than those found in wastewater. Some of the methods suggested for improving the rate of degradation are mechanical mixing and the addition of surfactants (Harms and Bosma 1997).

8.3.1.4 Operational Parameters

The effect of operational parameters should be understood before designing the processing facility for the microbial biodegradation of wastewater. Factors such as biomass concentration, pH, temperature, use of surfactant, chemical oxygen demand (COD), dissolved oxygen, mode of feeding, sludge retention time (SRT), and hydraulic retention time (HRT) of wastewater to be treated also affect the rate of biodegradation. The effect of these parameters can be estimated, and their significance can be determined using many statistical tools such as response surface methodology (RSM), artificial neural networks (ANNs), and genetic algorithms (GA). Previous studies were conducted to evaluate the effect of adding rhamnolipid surfactant on the removal of different types of COD (degradable, inert, and metabolic products) and PAHs from petrochemical wastewater in a continuous stirred tank reactor (Sponza and Gök 2010). Similarly, in another recent study, the parameters for the removal of p-Cresol containing wastewater were conducted. In this study, the role of factors such as HRT, initial concentration, the ratio of COD to nitrate, and nitrate-free influent and salinity, were determined in a denitrification baffled reactor (Mahdavianpour et al. 2018). However, another study discussed the crucial factors influencing the performance of microbial biodegradation in constructed wetlands. These included redox condition, presence or absence of plants, availability of organic carbon, temperature, pH, and media composition. Mode of operation is another factor that affects the rate of biodegradation. It was proposed that systems with low redox potential are less effective with continuous mode than with batch or fed-batch modes (Meng et al. 2014).

8.3.2 The Design of a Large-Scale Facility for a Microbial Degradation System

Many bioreactors have been evaluated for biodegradation efficacies. The design of the reactor, size, and mode of operation are some of the crucial parameters affecting the biotransformation rate. Some of the bioreactors used for bioremediation of the organic waste in wastewater are shown in Figure 8.1. The choice of these bioreactors depends on the microbial system and the nature of wastewater. Many studies report the use of hybrid systems for the removal of organic contaminants from wastewater. A study was performed to evaluate the combinatorial effect of membrane bioreactors with UV oxidation or high-pressure filtration systems such as reverse osmosis and nanofiltration. Membrane bioreactors were found to effectively degrade hydrophobic organic contaminants and some easily degradable hydrophilic contaminants. However, other hydrophilic and some trace organic contaminants were removed by the coupling of UV oxidation with the membrane bioreactor (Nguyen et al. 2013). In another study, a hybrid design containing a series of four different reactors were used for the remediation of hospital wastewater. About 26 compounds were assessed for degradability, and the study was conducted at pilot scale after adaptation for ten months. The series of bioreactors included an activated sludge reactor followed by two-hybrid reactors (activated sludge system with biofilm) and one biofilm reactor. This study determined the kinetics of the degradation process and also compared the degradation efficiency in batch and continuous processes (Casas et al. 2015). A report also suggested that another hybrid system (adsorption based activated carbon granules and submerged membrane bioreactor) was efficient in reducing the concentration of 19 pharmaceutical compounds up to a significant level. It was suggested that hydrophobicity and positive and

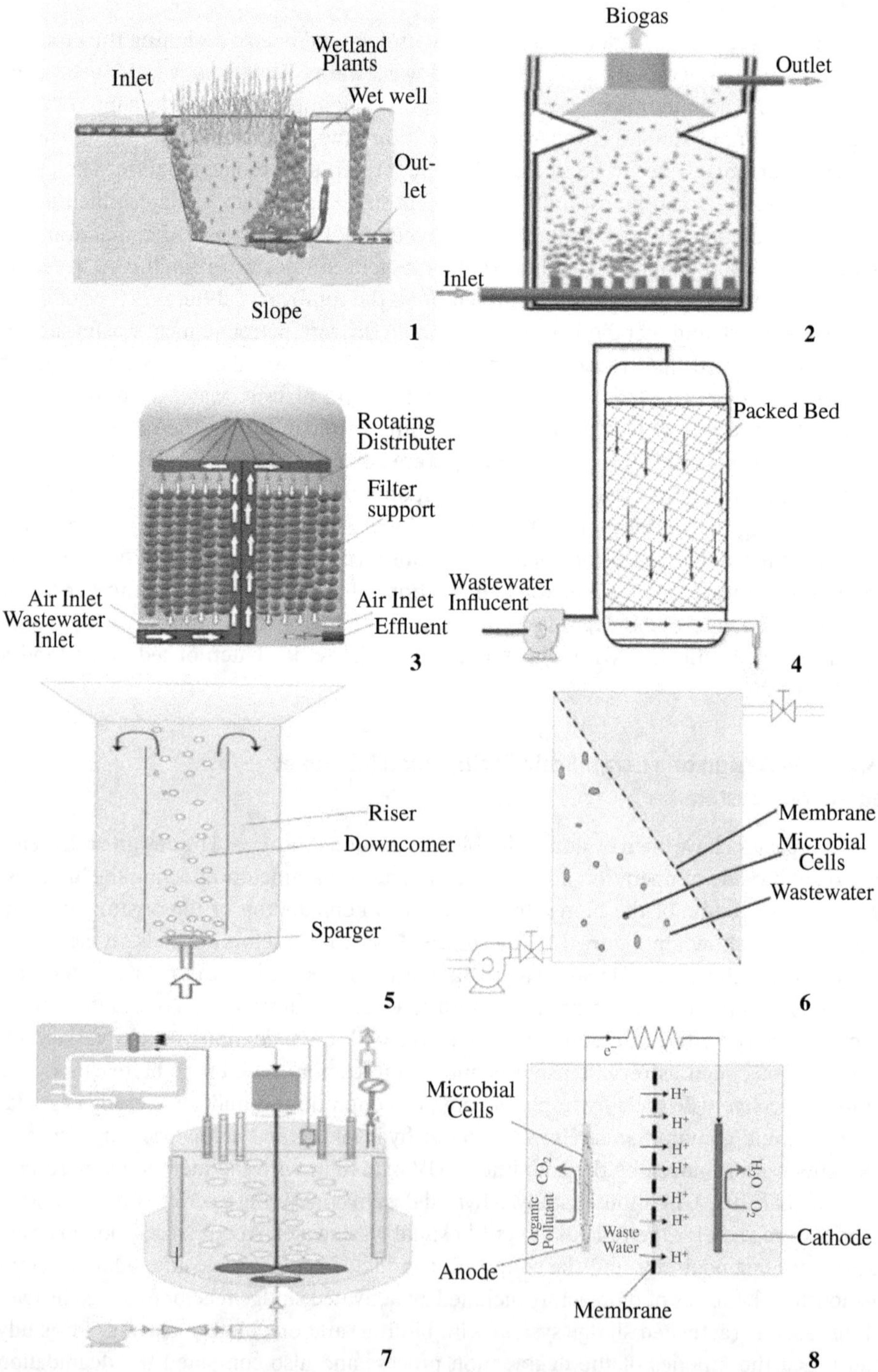

Figure 8.1 Different bioreactors used for the microbial degradation of organic contaminants. 1. Constructed wetland; 2. up-flow anaerobic sludge blanket bioreactor (UASB); 3. biotrickling filter; 4. packed bed reactor; 5. airlift reactor; 6. membrane bioreactor; 7. stirred tank reactor; 8. microbial fuel cell.

neutral charges favored the degradation of the removal of the organic compounds (Shanmuganathan et al. 2017).

To determine the removal efficiency under different aeration conditions, a hybrid system was developed and studied at the lab scale. In this work, an anaerobic system was in series with an anoxic and an oxic reactor each having working volume 6L, 12L, and 18L, respectively. The reduction of polyaromatic hydrocarbons was evaluated and was found to reduce from 254 to 488 μg/l to 4.1–4.5 μg/l after optimization (Zhao et al. 2018). In order to reduce a load of biological nutrient loading (BNR) from the wastewater, research was conducted that supported the use of a hybrid system comprising a five-stage Bardenpho process with a moving bed biofilm bioreactor. This process was optimized using RSM (Ashrafi et al. 2019).

8.3.2.1 Coupling of Energy Generation with the Contaminant Removal Process

To improve the efficiency of the process, microbial organic pollutant degradation can be coupled with bioenergy generation methods. In this direction, a summary was presented by some authors to evaluate the efficiency of wastewater pollutant removal in high rate algal ponds (HRAP). The algal system can be explored for the dual benefits of energy generation and degradation of organic contaminants. Based on the removal efficiency, the emerging organic contaminants have been divided into four classes, which are: high removal (>90%; such as ibuprofen and methyl dihydrojasmonate); moderate-high deduction (60–90%; such as oxybenzone and triclosan); moderate-low elimination (40–60%; such as atrazine and diazinon); and low-no elimination (<30%; 2, 4-D and methylparaben).

HRT is one of the significant factors in the elimination of the organic contaminant. These organic contaminants are utilized by the algal system as substrate and result in the biosynthesis of metabolites such as starch and lipids, which can be utilized for bioenergy production. Although the success of this process depends on the amount of algal biomass produced and specific accumulation of the starch or lipid, the large-scale production of microalgae-based on wastewater nutrients can be carried out in reactors such as turf scrubbers, photobioreactors, raceway ponds, and hybrid systems (Salama et al. 2017). Another reactor well known to assist in energy generation is the MFC, where the anodic chamber contains electrochemically active microbes that oxidize the organic contaminants. The electrons so generated are transferred to the anode which is captured as electrical energy while the protons are transferred through a membrane to the cathodic chamber. The techno-economic feasibility of such systems has been analyzed (Cao et al. 2019; Xie et al. 2018; Zhang et al. 2019b).

8.4 Future Prospects

Undoubtedly, these organic pollutants pose a massive threat to all ecosystems. Diseases including cancer, cardiovascular issues, obesity, and hormonal disturbances can be caused by the presence of these contaminants, even in trace amounts. Thus, the existing state of the art for microbial degradation should be continuously improved by understanding and meeting the current challenges of this process. Some of the major issues are as follows: inability to remove many organic pollutants such as POPs; accurate determination of toxicity caused by them; development of online detection methods; development of recombinant strains for effective removal which are more tolerant to pollutants, temperature, and pH; reduction in

HRT; improvement in the design of reactors in terms of efficiency scale and economics; improved mass transfer for immobilized biological systems; and more efficient coupling of energy generation technologies. Nanotechnology has found a large role in the development of efficient and economical systems. Bioremediation is no exception to this statement.

8.4.1 Nanobioremediation

"Nanobioremediation" is a relatively novel concept that explains the application of nanoparticles in bioremediation for the removal of contaminants (Basak et al. 2020; Cecchin et al. 2017; Tripathi et al. 2018). Similarly, some researchers have compared the application of nanotechnology with stand-alone bioremediation processes for the effective removal of PCBs. They integrated the application of nano-sized zero-valent iron with the process of bioaugmentation using microbes that significantly increased the percentage removal of persistent organic contaminants by up to 99% (Horváthová et al. 2019). These nano-sized zero-valent irons have reduced toxicity and have a low cost of production. However, the other side of the coin is that beyond the concentration of 10 mg/l, iron nanoparticles have shown developmental toxicity in zebrafish embryos (Boruah et al. 2019). However, research suggests a promising role in wastewater remediation. A study suggests that indeno(1, 2, 3-cd)pyrene can be remediated in the presence of a biosurfactant and iron nanoparticles. The factors for biodegradation were optimized using the RSM tool (Ojha et al. 2019). Another study demonstrated the application of nanobioremediation for the treatment of textile wastewater containing azo dyes using peroxidase enzyme immobilized on iron-magnetic nanoparticles in a bioreactor system (Darwesh et al. 2019). Thus, the development of such strategies that involve the development of cost-effective nanoparticles with reduced toxicity can have a future in the bioremediation process.

8.4.2 Microbial Fuel Cells

The MFC is an encouraging eco-friendly biotechnology method anticipated chiefly for waste treatment in addition to power production. However, a small power yield obliges its request for straight effective electrical strategies. Excessive growth in its electrochemical, chemical, and microbiological features has extended the desires of MFCs into other areas, for example, chemicals generation, bioremediation, and biosensors. In recent decades, MFC-based biosensors have gained growing use as a consequence of ease and sustainability. Applications of this technology extend from water quality observation to air quality recognition (e.g. CO, HCHO) (Seelam et al. 2018).

There is a necessity to discuss low power densities in the MFC process by current optimization of the design to diminish the losses affected by activation, ohmic, and concentration overpotentials (Nastro et al. 2015). Additionally, losses due to side reactions, for example, the direct oxidation of fuel by O_2 diffusion, do not assist the process and must be battered. The action is crucial to increase the microbial population density and seems to curb the causes eliminating the availability of additional locations on the surface of electrodes, which will possibly be regulated by bioaugmentation. Several constant exertions among the electrodes are prepared to create better electron transfer mechanisms. The amendment in electrode surfaces and the biocatalyst is doped over the active catalyst layer on the electrode surface (Chaturvedi and Verma 2016).

The MFC is an advanced technology for electricity production and the metabolism of microorganisms. Allocated with key xenobiotics and wastes, for example, agro-wastes, azo dyes, hexavalent chromium, and nitrates, they are very toxic to the environment. In MFC, possible practices of pollution control and waste management are used to transform them into less toxic metabolites for electricity production. Previously, a wide range of substrates (comprising waste plus xenobiotics) and microorganisms produced a meager power output and efforts are ongoing to make this technology more competent, relevant, and broadly conventional (Seelam et al. 2018).

At present, we are observing increasing universal energy demand and an ongoing crisis due to depleting nonrenewable energy sources and less awareness of renewable energy. An urgent requirement is a substitute for energy generation. MFC is a promising alternative practice to convert organic compounds (chemical energy) using microorganisms into electricity. Comprehensive training has validated new avenues in the field of MFC, which demonstrate the usage of a wide range of organic pollutants as carbon sources for the growth of electricity-producing microorganisms. Therefore, organic waste conversion using unique bioremediation technology such as MFC for energy generation is a practical and eco-friendly approach.

The different modules of xenobiotics and wastes can be engaged for bioenergy yield, microorganism production, and biodegradation (Chaturvedi and Verma 2016). The commercialization of MFC technology can be influenced by commercial viability when it is advertised in vast quantities and used across the globe. Subsequently, MFCs deal with power production using waste materials. Their socialization proposes numerous benefits, which are as follows:

1) Low-cost electricity production from waste.
2) Readily available waste and xenobiotics can ensure an uninterrupted power supply.
3) Easily viable homemade solutions can produce electricity.
4) Helpful for citizens of developing countries where extensive infrastructure is required.
5) An alternate method for bioremediation as MFCs can clean up wastes and xenobiotics.

MFC undergoes several challenges that hamper its commercialization. Some of them are as follows:

1) Power density gained from wastes and xenobiotics is very low compared with carbon sources, waste management, and electricity generation.
2) Pure carbon resources cannot usually be active for electricity generation as they are costly, as related to xenobiotics.
3) The scaling up of MFCs, cathode/anode material, and membranes is expensive.

MFCs have a prodigious future as MFC technology is an innovative method for bioelectricity production by the oxidation of organic wastes using microbes. Effective MFC under augmented anodic situations will permit the anodic biocatalyst to procure electrogenic biofilm fuel cell performance in long-term processes. Besides, open-air cathode MFCs or single chambers do not require any external energy input (Chaturvedi and Verma 2016). In the future, it could be beneficial for widespread use in site deficient electrical facilities. Although current applications and research for MCF-based bioremediation have significantly increased, however, certain limitations are still the roadblocks such as choice of electrodes, efficient utilization of mixed culture and scale-up of the system. The current growth in nanotechnology with material science may offer exclusive utensils to proficiently yield, transport, and operate the electrical energy gathered from MFCs for valuable uses in the future.

The MFC-based biosensor has created new interest worldwide as it is a self-powered method with whole cells. As shown previously, substantial development has ended in the field, particularly in BOD and toxicity recognition. However, some problems need to be discussed for a mature detecting capability recognized by technical societies. First, more surveys on the constancy of MFC-based biosensors are crucial, which are frequently ignored. MFC uses catalysts (bacteria), which are self-renewable and progress fast in response to environmental alterations for long-term operations (Cui et al. 2019).

Consequently, the selectivity, sensitivity, and reproducibility of biosensors will be poorly affected. Additionally, the detection limit of MFC-based biosensors, particularly for toxicity detection, is usually considered as inferior per World Health Organization (WHO) water quality standards. In conclusion, water quality standards have a significant impact on the electrical signal output of MFC-based biosensors (Cui et al. 2019). Therefore, recent research has applied modern techniques to obtain MFC-based biosensors with an improvement in materials, especially MFC-based biosensors electrogenic bacteria, which may ultimately develop appropriate commercial MFC systems.

8.5 Conclusion

With the advent of high-end technologies, substantial advances have been witnessed in both aerobic and anaerobic biological treatments of organic pollutants. Anaerobic digestion has proved to be an economical and energy-efficient process. The design improvements of conventional up-flow anaerobic sludge blanket bioreactors and anaerobic membrane bioreactors have been reviewed by many researchers. On the other hand, the conventional activated sludge process has been coupled with high retention membrane bioreactors (HRMB) to improve the biodegradation of trace organic contaminants. Wastewater treatment coupled with the bioelectrochemical system is also a novel strategy for remediation of organic pollutants. It has been found to be efficient in the removal of persistent organic compounds. These studies indicate that the technologies employed for bioremediation are being continuously improvised. However, their application is still controlled by the nature of the organic pollutants being treated. This research has significantly increased the pollutant removal percentage and optimized process economics.

Abbreviations Used

AHL	Acyl-homoserine-lactone
AHs	Aromatic hydrocarbons
ANN	Artificial neural network
AOPs	Advanced oxidation processes
BDO	Biphenyl 2, 3-dioxygenase
BNR	Biological nutrient loading

DGT	Diffusive gradients in thin films
EDC	Endocrine-disrupting chemicals
GA	Genetic algorithm
GC	Gas chromatography
HBCD	Hexabromocyclododecane
HPLC	High-pressure liquid chromatography
HRAPs	High rate algal ponds
HRMB	High retention membrane bioreactors
HRT	Hydraulic retention time
IR	Infrared absorption spectra
LC–MS	Liquid chromatography–mass spectrophotometer
MBR	Membrane-based reactor
MFC	Microbial fuel cell
MIC	Minimum inhibitory concentration
NCBI	National Center for Biotechnology Information
PAHs	Polycyclic aromatic hydrocarbons
PCBs	Polychlorinated biphenyls
PCPs	Personal Care products
PhACs	Pharmaceutically active compounds
PHPA	Poly-3-hydroxy-n-phenyl alkanoate
POPs	Persistent organic pollutants
PPCP	Pharmaceutical and personal care products
RHD	Ring-hydroxylating-dioxygenase
RSM	Response surface methodology
SBR	Sequential batch reactors
SIP	Stable isotope probing
SRT	Sludge retention time
TPH	Total petroleum hydrocarbon
WFCC	World Federation of Culture Collections
WHO	World Health Organization
XRD	X-ray diffraction (XRD)
γ- HCH	γ- Hexachlorocyclohexane

References

Adebusoye, S.A. (2017). Biological degradation of 4-chlorobenzoic acid by a PCB-metabolizing bacterium through a pathway not involving (chloro) catechol. *Biodegradation* 28 (1): 37–51.

Afzal, T., Isa, M.H., and ul Mustafa, M.R. (2018). Removal of organic pollutants from produced water using Fenton oxidation. *Paper presented at the E3S Web of Conferences* 34 (2): 02035. https://doi.org/10.1051/e3sconf/20183402035.

AI-Jawhari, I.F.H., Mhail, N.J., Saher Abed, A. et al. (2015). Efficiency of some filamentous fungi to treatment of effluent petroleum wastewaters from refinery. *International Journal of Current Microbiology and Applied Sciences* 4 (4): 625–641.

Alvarino, T., Suarez, S., Lema, J. et al. (2018). Understanding the sorption and biotransformation of organic micropollutants in innovative biological wastewater treatment technologies. *Science of the Total Environment* 615: 297–306.

Ariste, A.F., Batista-García, R.A., Vaidyanathan, V.K. et al. (2020). Mycoremediation of phenols and polycyclic aromatic hydrocarbons from a biorefinery wastewater and concomitant production of lignin modifying enzymes. *Journal of Cleaner Production* 253: 119810.

Ashrafi, E., Zeinabad, A.M., Borghei, S.M. et al. (2019). Optimising nutrient removal of a hybrid five-stage Bardenpho and moving bed biofilm reactor process using response surface methodology. *Journal of Environmental Chemical Engineering* 7 (1): 102861.

Assessment, C. R (1996). Proposed guidelines for carcinogen risk assessment. *Federal Register* 61 (79): 17960–18011.

Aydin, S., Karaçay, H.A., Shahi, A. et al. (2017). Aerobic and anaerobic fungal metabolism and Omics insights for increasing polycyclic aromatic hydrocarbons biodegradation. *Fungal Biology Reviews* 31 (2): 61–72.

Balachandran, C., Duraipandiyan, V., Balakrishna, K. et al. (2012). Petroleum and polycyclic aromatic hydrocarbons (PAHs) degradation and naphthalene metabolism in Streptomyces sp.(ERI-CPDA-1) isolated from oil contaminated soil. *Bioresource Technology* 112: 83–90.

Basak, G., Hazra, C., and Sen, R. (2020). Biofunctionalized nanomaterials for in situ clean-up of hydrocarbon contamination: a quantum jump in global bioremediation research. *Journal of Environmental Management* 256: 109913.

Baun, A., Ledin, A., Reitzel, L. et al. (2004). Xenobiotic organic compounds in leachates from ten Danish MSW landfills – chemical analysis and toxicity tests. *Water Research* 38 (18): 3845–3858.

Bester, K., Scholes, L., Wahlberg, C. et al. (2008). Sources and mass flows of xenobiotics in urban water cycles – an overview on current knowledge and data gaps. *Water, Air, & Soil Pollution: Focus* 8 (5–6): 407–423.

Bhanot, P., Celin, S.M., Sreekrishnan, T. et al. (2020). Application of integrated treatment strategies for explosive industry wastewater – a critical review. *Journal of Water Process Engineering* 35: 101232.

Bjerg, P.L., Tuxen, N., Reitzel, L.A. et al. (2011). Natural attenuation processes in landfill leachate plumes at three Danish sites. *Groundwater* 49 (5): 688–705.

Blunt, S.M., Sackett, J.D., Rosen, M.R. et al. (2018). Association between degradation of pharmaceuticals and endocrine-disrupting compounds and microbial communities along a treated wastewater effluent gradient in Lake Mead. *Science of the Total Environment* 622: 1640–1648.

Boonnorat, J., Techkarnjanaruk, S., Honda, R. et al. (2018). Use of aged sludge bioaugmentation in two-stage activated sludge system to enhance the biodegradation of toxic organic compounds in high strength wastewater. *Chemosphere* 202: 208–217.

Boruah, T., Dowarah, B., and Mili, C. (2019). Nanobioremediation: Potential and Future Prospects. In: *Advances in Biotechnology and Bioscience* (ed. S. Kumar), 91–104. Delhi: AkiNik Publications.

Bressler, D.C. and Gray, M.R. (2003). Transport and reaction processes in bioremediation of organic contaminants. 1. Review of bacterial degradation and transport. *International Journal of Chemical Reactor Engineering* 1 (1): 1–16.

Brillas, E. (2014). A review on the degradation of organic pollutants in waters by UV photoelectro-Fenton and solar photoelectro-Fenton. *Journal of the Brazilian Chemical Society* 25 (3): 393–417.

Bulter, T., Alcalde, M., Sieber, V. et al. (2003). Functional expression of a fungal laccase in Saccharomyces cerevisiae by directed evolution. *Applied and Environmental Microbiology* 69 (2): 987–995.

Bumpus, J.A. and Aust, S.D. (1987). Biodegradation of environmental pollutants by the white rot fungus Phanerochaete chrysosporium: involvement of the lignin degrading system. *BioEssays* 6 (4): 166–170.

Burrows, H.D., Santaballa, J., and Steenken, S. (2002). Reaction pathways and mechanisms of photodegradation of pesticides. *Journal of Photochemistry and Photobiology B: Biology* 67 (2): 71–108.

Campana, O. and Wlodkowic, D. (2018). The undiscovered country: ecotoxicology meets microfluidics. *Sensors and Actuators B: Chemical* 257: 692–704.

Cao, X., Zhang, S., Wang, H. et al. (2019). Azo dye as part of co-substrate in a biofilm electrode reactor–microbial fuel cell coupled system and an analysis of the relevant microorganisms. *Chemosphere* 216: 742–748.

Casas, M.E., Chhetri, R.K., Ooi, G. et al. (2015). Biodegradation of pharmaceuticals in hospital wastewater by a hybrid biofilm and activated sludge system (Hybas). *Science of the Total Environment* 530: 383–392.

Castillo, A.S.R., Guiheneuf, S., Biard, P.-F. et al. (2017). Toluene degradation in a two-phase partitioning bioreactor involving a hydrophobic ionic liquid as a non-aqueous phase liquid. *International Biodeterioration & Biodegradation* 117: 31–38.

Cecchin, I., Reddy, K.R., Thomé, A. et al. (2017). Nanobioremediation: integration of nanoparticles and bioremediation for sustainable remediation of chlorinated organic contaminants in soils. *International Biodeterioration & Biodegradation* 119: 419–428.

Cerniglia, C.E., Gibson, D.T., and Van Baalen, C. (1979). Algal oxidation of aromatic hydrocarbons: formation of 1-naphthol from naphthalene by Agmenellum quadruplicatum, strain PR-6. *Biochemical and Biophysical Research Communications* 88 (1): 50–58.

Chang, T.-H., Wang, R., Peng, Y.-H. et al. (2020). Biodegradation of hexabromocyclododecane by Rhodopseudomonas palustris YSC3 strain: a free-living nitrogen-fixing bacterium isolated in Taiwan. *Chemosphere* 246: 125621.

Chaturvedi, V. and Verma, P. (2016). Microbial fuel cell: a green approach for the utilization of waste for the generation of bioelectricity. *Bioresources and Bioprocessing* 3 (1): 38.

Chávez, A., Gimeno, O., Rey, A. et al. (2019). Treatment of highly polluted industrial wastewater by means of sequential aerobic biological oxidation-ozone based AOPs. *Chemical Engineering Journal* 361: 89–98.

Chen, J., Wang, P.-F., Wang, C. et al. (2018). How wastewater with different nutrient levels influences microbial degradation of 2, 2′, 4, 4′-tetrabromodiphenyl ether (BDE-47) in anaerobic sediments. *Chemosphere* 211: 128–138.

Chen, L., Li, C., Xu, B. et al. (2019). Microbial degradation of organic pollutants in groundwater related to underground coal gasification. *Energy Science & Engineering* 7 (5): 2098–2111.

Cho, O., Choi, K.Y., Zylstra, G.J. et al. (2005). Catabolic role of a three-component salicylate oxygenase from Sphingomonas yanoikuyae B1 in polycyclic aromatic hydrocarbon degradation. *Biochemical and Biophysical Research Communications* 327 (3): 656–662.

Christensen, T.H., Kjeldsen, P., Bjerg, P.L. et al. (2001). Biogeochemistry of landfill leachate plumes. *Applied Geochemistry* 16 (7–8): 659–718.

European Commission (2012a). A blueprint to safeguard Europe's water resources. Communication from the Commission to the European Parliament, the Council, the European Economic and Social Committee and the Committee of the Regions. Brussels: European Commission.

European Commission (2012b). River Basin Management Plans. Report from the Commission to the European Parliament and the Council on the Implementation of the Water Framework Directive (2000/60/EC). Brussels: EU Publications.

Cui, Y., Lai, B., and Tang, X. (2019). Microbial fuel cell-based biosensors. *Biosensors* 9 (3): 92.

Cydzik-Kwiatkowska, A. and Zielińska, M. (2016). Bacterial communities in full-scale wastewater treatment systems. *World Journal of Microbiology and Biotechnology* 32 (4): 66.

Dangi, A.K., Sharma, B., Hill, R.T. et al. (2019). Bioremediation through microbes: systems biology and metabolic engineering approach. *Critical Reviews in Biotechnology* 39 (1): 79–98.

Darwesh, O.M., Matter, I.A., and Eida, M.F. (2019). Development of peroxidase enzyme immobilized magnetic nanoparticles for bioremediation of textile wastewater dye. *Journal of Environmental Chemical Engineering* 7 (1): 102805.

Das, M. and Adholeya, A. (2015). Potential uses of immobilized bacteria, fungi, algae, and their aggregates for treatment of organic and inorganic pollutants in wastewater. In: *Water Challenges and Solutions on a Global Scale* (eds. S. Ahuja, J.B. de Andrade, D.D. Dionysiou, et al.), 319–337. ACS Publications.

Deng, Y. and Zhao, R. (2015). Advanced oxidation processes (AOPs) in wastewater treatment. *Current Pollution Reports* 1 (3): 167–176.

Deshmukh, R., Khardenavis, A.A., and Purohit, H.J. (2016). Diverse metabolic capacities of fungi for bioremediation. *Indian Journal of Microbiology* 56 (3): 247–264.

Donato, J.J., Moe, L.A., Converse, B.J. et al. (2010). Metagenomic analysis of apple orchard soil reveals antibiotic resistance genes encoding predicted bifunctional proteins. *Applied and Environmental Microbiology* 76 (13): 4396–4401.

Dressler, O.J., Casadevall i Solvas, X., and DeMello, A.J. (2017). Chemical and biological dynamics using droplet-based microfluidics. *Annual Review of Analytical Chemistry* 10: 1–24.

EU. (2008). European Union Risk Assessment Report, 2nd priority list, Volume 80, bis (2-ethylhexyl) phthalate (DEHP). Brussels: EU Risk Assessment Report.

Falås, P., Wick, A., Castronovo, S. et al. (2016). Tracing the limits of organic micropollutant removal in biological wastewater treatment. *Water Research* 95: 240–249.

Fang, H., Cai, L., Yu, Y. et al. (2013). Metagenomic analysis reveals the prevalence of biodegradation genes for organic pollutants in activated sludge. *Bioresource Technology* 129: 209–218.

Fenech, C., Nolan, K., Rock, L. et al. (2013). An SPE LC–MS/MS method for the analysis of human and veterinary chemical markers within surface waters: an environmental forensics application. *Environmental Pollution* 181: 250–256.

Fennell, D., Du, S., Liu, F. et al. (2019). Deehalogenation of polychlorinated dibenzo-p-dioxins and dibenzofurans, polychlorinated biphenyls, and brominated flame retardants, and potential as a bioremediation strategy. In: *Comprehensive Biotechnology*, 135–149. Oxford: Elsevier Inc.

Fowles, J.R., Banton, M.I., Boogaard, P.J. et al. (2016). Assessment of petroleum streams for thyroid toxicity. *Toxicology Letters* 254: 52–62.

French, K.E. and Terry, N. (2019). A high-throughput fluorescence-based assay for rapid identification of petroleum degrading bacteria. *Frontiers in Microbiology* 10: 1318.

Fu, F. and Wang, Q. (2011). Removal of heavy metal ions from wastewaters: a review. *Journal of Environmental Management* 92 (3): 407–418.

Gangola, S., Joshi, S., Kumar, S. et al. (2019). Comparative analysis of fungal and bacterial enzymes in biodegradation of xenobiotic compounds. In: *Smart Bioremediation Technologies* (ed. P. Bhatt), 169–189. Oxford: Elsevier.

Gao, B., Liu, H., and Gu, Z. (2016). An exothermic chip for point-of-care testing using a forehead thermometer as a readout. *Lab on a Chip* 16 (3): 525–531.

Gargouri, B., Mhiri, N., Karray, F. et al. (2015). Isolation and characterization of hydrocarbon-degrading yeast strains from petroleum contaminated industrial wastewater. *BioMed Research International* 2015: 1–11.

Gloux, K., Berteau, O., Béguet, F. et al. (2011). A metagenomic β-glucuronidase uncovers a core adaptive function of the human intestinal microbiome. *Proceedings of the National Academy of Sciences of the United States of America* 108 (Supplement 1): 4539–4546.

Godheja, J., Shekhar, S., Siddiqui, S. et al. (2016). Xenobiotic compounds present in soil and water: a review on remediation strategies. *Journal of Environmental & Analytical Toxicology* 6 (392) https://doi.org/10.4172/2161-0525.1000392.

Gohil, H. and Ogram, A. (2020). The microbial degradation of DDT and potential remediation strategies. In: *Bioremediation Technology: Hazardous Waste Management*, 1–24.

Goswami, L., Manikandan, N.A., Dolman, B. et al. (2018). Biological treatment of wastewater containing a mixture of polycyclic aromatic hydrocarbons using the oleaginous bacterium Rhodococcus opacus. *Journal of Cleaner Production* 196: 1282–1291.

Greskowiak, J., Hamann, E., Burke, V. et al. (2017). The uncertainty of biodegradation rate constants of emerging organic compounds in soil and groundwater–a compilation of literature values for 82 substances. *Water Research* 126: 122–133.

Han, Y., Yang, L., Chen, X. et al. (2020). Removal of veterinary antibiotics from swine wastewater using anaerobic and aerobic biodegradation. *Science of the Total Environment* 709: 136094.

Harms, H. and Bosma, T. (1997). Mass transfer limitation of microbial growth and pollutant degradation. *Journal of Industrial Microbiology and Biotechnology* 18 (2–3): 97–105.

He, Y., Langenhoff, A.A., Comans, R.N. et al. (2018). Effects of dissolved organic matter and nitrification on biodegradation of pharmaceuticals in aerobic enrichment cultures. *Science of the Total Environment* 630: 1335–1342.

Horváthová, H., Lászlová, K., and Dercová, K. (2019). Bioremediation vs. Nanoremediation: degradation of Polychlorinated Biphenyls (PCBS) using integrated remediation approaches. *Water, Air, & Soil Pollution* 230 (8): 204.

Hou, N., Zhang, N., Jia, T. et al. (2018). Biodegradation of phenanthrene by biodemulsifier-producing strain Achromobacter sp. LH-1 and the study on its metabolisms and fermentation kinetics. *Ecotoxicology and Environmental Safety* 163: 205–214.

Huang, Y., Zeng, Y., Yu, Z. et al. (2013). In silico and experimental methods revealed highly diverse bacteria with quorum sensing and aromatics biodegradation systems–a potential broad application on bioremediation. *Bioresource Technology* 148: 311–316.

Huang, D., Hu, C., Zeng, G. et al. (2017). Combination of Fenton processes and biotreatment for wastewater treatment and soil remediation. *Science of the Total Environment* 574: 1599–1610.

Huang, Y., Xiao, L., Li, F. et al. (2018). Microbial degradation of pesticide residues and an emphasis on the degradation of cypermethrin and 3-phenoxy benzoic acid: a review. *Molecules* 23 (9): 2313.

Huang, L., Liu, X., Hu, X. et al. (2019). Biodegradation of phenanthrene and heavy metal removal by acid-tolerant Burkholderia fungorum FM-2. *Frontiers in Microbiology* 10: 408.

Iwai, S., Chai, B., Sul, W.J. et al. (2010). Gene-targeted-metagenomics reveals extensive diversity of aromatic dioxygenase genes in the environment. *The ISME Journal* 4 (2): 279–285.

Jońca, J., Comtat, M., and Garçon, V. (2013). In situ phosphate monitoring in seawater: today and tomorrow. In: *Smart Sensors for Real-Time Water Quality Monitoring* (eds. S.C. Mukhopadhyay and A. Mason), 25–44. Springer.

Joutey, N.T., Bahafid, W., Sayel, H. et al. (2013). Biodegradation: involved microorganisms and genetically engineered microorganisms. *Biodegradation-Life of Science* 1: 289–320.

Kadri, T., Rouissi, T., Brar, S.K. et al. (2017). Biodegradation of polycyclic aromatic hydrocarbons (PAHs) by fungal enzymes: a review. *Journal of Environmental Sciences* 51: 52–74.

Kim, S.-J., Kweon, O., Freeman, J.P. et al. (2006). Molecular cloning and expression of genes encoding a novel dioxygenase involved in low-and high-molecular-weight polycyclic aromatic hydrocarbon degradation in mycobacterium vanbaalenii PYR-1. *Applied and Environmental Microbiology* 72 (2): 1045–1054.

Kjeldsen, P., Barlaz, M.A., Rooker, A.P. et al. (2002). Present and long-term composition of MSW landfill leachate: a review. *Critical Reviews in Environmental Science and Technology* 32 (4): 297–336.

Klementova, S. (2018). Photochemical degradation of organic xenobiotics in natural waters. *Photochemistry and Photophysics: Fundamentals to Applications* 67–68.

Kumari, M., Ghosh, P., and Thakur, I.S. (2020). Development of artificial consortia of microalgae and bacteria for efficient biodegradation and detoxification of lindane. *Bioresource Technology Reports* 10: 100415.

Kuritz, T. and Wolk, C.P. (1995). Use of filamentous cyanobacteria for biodegradation of organic pollutants. *Applied and Environmental Microbiology* 61 (1): 234–238.

Lal, R., Pandey, G., Sharma, P. et al. (2010). Biochemistry of microbial degradation of hexachlorocyclohexane and prospects for bioremediation. *Microbiology and Molecular Biology Reviews* 74 (1): 58–80.

Lara-Martín, P.A., González-Mazo, E., Petrovic, M. et al. (2014). Occurrence, distribution and partitioning of nonionic surfactants and pharmaceuticals in the urbanized Long Island Sound Estuary (NY). *Marine Pollution Bulletin* 85 (2): 710–719.

Later, D.W., Lee, M.L., Bartle, K.D. et al. (1981). Chemical class separation and characterization of organic compounds in synthetic fuels. *Analytical Chemistry* 53 (11): 1612–1620.

Lei, A., Wong, Y., and Tam, N. (2002). Removal of pyrene by different microalgal species. *Water Science and Technology* 46 (11–12): 195–201.

Li, A., Qu, Y., Zhou, J. et al. (2009). Enzyme–substrate interaction and characterization of a 2, 3-dihydroxybiphenyl 1, 2-dioxygenase from Dyella ginsengisoli LA-4. *FEMS Microbiology Letters* 292 (2): 231–239.

Liu, Z., Xie, W., Li, D. et al. (2016). Biodegradation of phenol by bacteria strain Acinetobacter calcoaceticus PA isolated from phenolic wastewater. *International Journal of Environmental Research and Public Health* 13 (3): 300.

Liu, L., Bilal, M., Duan, X. et al. (2019). Mitigation of environmental pollution by genetically engineered bacteria – current challenges and future perspectives. *Science of The Total Environment* 667: 444–454.

Macherzyński, B., Włodarczyk-Makuła, M., Skowron-Grabowska, B. et al. (2016). Degradation of PCBs in sewage sludge during methane fermentation process concerning environmental management. *Desalination and Water Treatment* 57 (3): 1163–1175.

Mahdavianpour, M., Moussavi, G., and Farrokhi, M. (2018). Biodegradation and COD removal of p-cresol in a denitrification baffled reactor: performance evaluation and microbial community. *Process Biochemistry* 69: 153–160.

Mandal, S., Kunhikrishnan, A., Bolan, N. et al. (2016). Application of biochar produced from biowaste materials for environmental protection and sustainable agriculture production. In: *Environmental Materials and Waste* (eds. M.N.V. Prasad and K. Shih), 73–89. Elsevier.

Martinez-Alcala, I., Guillén-Navarro, J.M., and Fernandez-Lopez, C. (2017). Pharmaceutical biological degradation, sorption and mass balance determination in a conventional activated-sludge wastewater treatment plant from Murcia, Spain. *Chemical Engineering Journal* 316: 332–340.

Massa, V., Infantino, A., Radice, F. et al. (2009). Efficiency of natural and engineered bacterial strains in the degradation of 4-chlorobenzoic acid in soil slurry. *International Biodeterioration & Biodegradation* 63 (1): 112–115.

Megharaj, M., Ramakrishnan, B., Venkateswarlu, K. et al. (2011). Bioremediation approaches for organic pollutants: a critical perspective. *Environment International* 37 (8): 1362–1375.

Mehetre, G.T., Dastager, S.G., and Dharne, M.S. (2019). Biodegradation of mixed polycyclic aromatic hydrocarbons by pure and mixed cultures of biosurfactant producing thermophilic and thermo-tolerant bacteria. *Science of the Total Environment* 679: 52–60.

Meng, P., Pei, H., Hu, W. et al. (2014). How to increase microbial degradation in constructed wetlands: influencing factors and improvement measures. *Bioresource Technology* 157: 316–326.

Milosevic, N. (2012). Attenuation of xenobiotic organic leachate compounds from a landfill to surface water: transition in clay till settings. Kongens Lyngby: DTU Environment.

Mishra, S., Sarma, P.M., and Lal, B. (2004). Crude oil degradation efficiency of a recombinant Acinetobacter baumannii strain and its survival in crude oil-contaminated soil microcosm. *FEMS Microbiology Letters* 235 (2): 323–331.

Mtibaà, R., Ezzanad, A., Aranda, E. et al. (2020). Biodegradation and toxicity reduction of nonylphenol, 4-tert-octylphenol and 2, 4-dichlorophenol by the ascomycetous fungus Thielavia sp HJ22: identification of fungal metabolites and proposal of a putative pathway. *Science of the Total Environment* 708: 135129.

Nastro, R., Falcucci, G., Toscanesi, M. et al. (2015). Performances and microbiology of a microbial fuel cell (MFC) fed with the organic fraction of municipal solid waste (OFMSW). *Paper presented at the Proceedings of EFC15 Conference* (16–18 December), Naples, Italy.

National Research Council (2004). Water and Sustainable Development: Opportunities for the Chemical Sciences: A Workshop Report to the Chemical Sciences Roundtable. Washington, DC: National Academies Press.

Nguyen, L.N., Hai, F.I., Kang, J. et al. (2013). Removal of emerging trace organic contaminants by MBR-based hybrid treatment processes. *International Biodeterioration & Biodegradation* 85: 474–482.

Nguyen, L.N., Nghiem, L.D., and Oh, S. (2018). Aerobic biotransformation of the antibiotic ciprofloxacin by Bradyrhizobium sp. isolated from activated sludge. *Chemosphere* 211: 600–607.

Nolte, T.M., Pinto-Gil, K., Hendriks, A.J. et al. (2018). Quantitative structure–activity relationships for primary aerobic biodegradation of organic chemicals in pristine surface waters: starting points for predicting biodegradation under acclimatization. *Environmental Science: Processes & Impacts* 20 (1): 157–170.

Nogueira, A.R.L. (2018). Building novel bioaugmentation consortia for degradation of polycyclic aromatic hydrocarbons: From selection to metabolic and genetic characterization of adaptive evolved microbial strains. Doctoral Thesis.

Novotný, Č., Rawal, B., Bhatt, M. et al. (2001). Capacity of Irpex lacteus and Pleurotus ostreatus for decolorization of chemically different dyes. *Journal of Biotechnology* 89 (2–3): 113–122.

Obayori, O.S., Salam, L.B., Oyetibo, G.O. et al. (2017). Biodegradation potentials of polyaromatic hydrocarbon (pyrene and phenanthrene) by Proteus mirabilis isolated from an animal charcoal polluted site. *Biocatalysis and Agricultural Biotechnology* 12: 78–84.

Ojha, N., Mandal, S.K., and Das, N. (2019). Enhanced degradation of indeno (1, 2, 3-cd) pyrene using Candida tropicalis NN4 in presence of iron nanoparticles and produced biosurfactant: a statistical approach. *3 Biotech* 9 (3): 86.

Ossai, I.C., Ahmed, A., Hassan, A. et al. (2019). Remediation of soil and water contaminated with petroleum hydrocarbon: A review. *Environmental Technology & Innovation* https://doi.org/10.1016/j.eti.2019.100526.

Parshintsev, J. and Hyötyläinen, T. (2015). Methods for characterization of organic compounds in atmospheric aerosol particles. *Analytical and Bioanalytical Chemistry* 407 (20): 5877–5897.

Perelo, L.W. (2010). In situ and bioremediation of organic pollutants in aquatic sediments. *Journal of Hazardous Materials* 177 (1–3): 81–89.

Pintado-Herrera, M.G., González-Mazo, E., and Lara-Martín, P.A. (2014). Atmospheric pressure gas chromatography–time-of-flight-mass spectrometry (APGC–ToF-MS) for the determination of regulated and emerging contaminants in aqueous samples after stir bar sorptive extraction (SBSE). *Analytica Chimica Acta* 851: 1–13.

Pozdnyakova, N., Dubrovskaya, E., Chernyshova, M. et al. (2018). The degradation of three-ringed polycyclic aromatic hydrocarbons by wood-inhabiting fungus Pleurotus ostreatus and soil-inhabiting fungus Agaricus bisporus. *Fungal Biology* 122 (5): 363–372.

Prenafeta-Boldu, F.X., Summerbell, R., and Sybren de Hoog, G. (2006). Fungi growing on aromatic hydrocarbons: biotechnology's unexpected encounter with biohazard? *FEMS Microbiology Reviews* 30 (1): 109–130.

Pugazhendi, A., Qari, H., Basahi, J.M.A.-B. et al. (2017). Role of a halothermophilic bacterial consortium for the biodegradation of PAHs and the treatment of petroleum wastewater at extreme conditions. *International Biodeterioration & Biodegradation* 121: 44–54.

Qiao, M., Qi, W., Liu, H. et al. (2016). Formation of oxygenated polycyclic aromatic hydrocarbons from polycyclic aromatic hydrocarbons during aerobic activated sludge treatment and their removal process. *Chemical Engineering Journal* 302: 50–57.

Qiao, M., Bai, Y., Cao, W. et al. (2018). Impact of secondary effluent from wastewater treatment plants on urban rivers: polycyclic aromatic hydrocarbons and derivatives. *Chemosphere* 211: 185–191.

Reisfeld, A., Rosenberg, E., and Gutnick, D. (1972). Microbial degradation of crude oil: factors affecting the dispersion in sea water by mixed and pure cultures. *Applied and Environmental Microbiology* 24 (3): 363–368.

Rusanowska, P., Cydzik-Kwiatkowska, A., Świątczak, P. et al. (2019). Changes in extracellular polymeric substances (EPS) content and composition in aerobic granule size-fractions during reactor cycles at different organic loads. *Bioresource Technology* 272: 188–193.

Salama, E.-S., Kurade, M.B., Abou-Shanab, R.A. et al. (2017). Recent progress in microalgal biomass production coupled with wastewater treatment for biofuel generation. *Renewable and Sustainable Energy Reviews* 79: 1189–1211.

Sandoval, Á., Arias-Barrau, E., Bermejo, F. et al. (2005). Production of 3-hydroxy-n-phenylalkanoic acids by a genetically engineered strain of Pseudomonas putida. *Applied Microbiology and Biotechnology* 67 (1): 97–105.

Sannino, F., Nuzzo, A., Ventorino, V. et al. (2016). Effective degradation of organic pollutants in aqueous media by microbial strains isolated from soil of a contaminated industrial site. *Chemical and Biological Technologies in Agriculture* 3 (1): 2.

Seelam, J.S., Rundel, C.T., Boghani, H.C. et al. (2018). Scaling up of MFCs: challenges and case studies. In: *Microbial Fuel Cell* (ed. D. Das), 459–481. Cham: Springer.

Shanmuganathan, S., Loganathan, P., Kazner, C. et al. (2017). Submerged membrane filtration adsorption hybrid system for the removal of organic micropollutants from a water reclamation plant reverse osmosis concentrate. *Desalination* 401: 134–141.

Shon, H., Phuntsho, S., Vigneswaran, S. et al. (2009). Physico-chemical processes for organic removal from wastewater effluent. *Waste Water Treatment Technologies* I: 205.

Shumkova, E., Egorova, D., Boronnikova, S. et al. (2015). Polymorphism of the bphA genes in bacteria destructing biphenyl/chlorinated biphenils. *Molecular Biology* 49 (4): 569–580.

Simon, C. and Daniel, R. (2011). Metagenomic analyses: past and future trends. *Applied and Environmental Microbiology* 77 (4): 1153–1161.

Singh, P. and Borthakur, A. (2018). A review on biodegradation and photocatalytic degradation of organic pollutants: a bibliometric and comparative analysis. *Journal of Cleaner Production* 196: 1669–1680.

Smułek, W., Sydow, M., Zabielska-Matejuk, J. et al. (2020). Bacteria involved in biodegradation of creosote PAH–A case study of long-term contaminated industrial area. *Ecotoxicology and Environmental Safety* 187: 109843.

Sponza, D.T. and Gök, O. (2010). Effect of rhamnolipid on the aerobic removal of polyaromatic hydrocarbons (PAHs) and COD components from petrochemical wastewater. *Bioresource Technology* 101 (3): 914–924.

Srivastava, V., Srivastava, T., and Kumar, M.S. (2019). Fate of the persistent organic pollutant (POP) Hexachlorocyclohexane (HCH) and remediation challenges. *International Biodeterioration & Biodegradation* 140: 43–56.

Stara, J.F., Kello, D., and Durkin, P. (1980). Human health hazards associated with chemical contamination of aquatic environment. *Environmental Health Perspectives* 34: 145–158.

Stein, U., Özerol, G., Tröltzsch, J. et al. (2016). European drought and water scarcity policies. In: *Governance for Drought Resilience* (eds. H. Bressers, N. Bressers and C. Larrue), 17–43. Cham: Springer.

Subashchandrabose, S.R., Venkateswarlu, K., Naidu, R. et al. (2019). Biodegradation of high-molecular weight PAHs by Rhodococcus wratislaviensis strain 9: overexpression of amidohydrolase induced by pyrene and BaP. *Science of the Total Environment* 651: 813–821.

Suenaga, H., Nishi, A., Watanabe, T. et al. (1999). Engineering a hybrid pseudomonad to acquire 3, 4-dioxygenase activity for polychlorinated biphenyls. *Journal of Bioscience and Bioengineering* 87 (4): 430–435.

Syed, K., Porollo, A., Miller, D. et al. (2013). Rational engineering of the fungal P450 monooxygenase CYP5136A3 to improve its oxidizing activity toward polycyclic aromatic hydrocarbons. *Protein Engineering, Design & Selection* 26 (9): 553–557.

Thai, P.K., Binh, V.N., Nhung, P.H. et al. (2018). Occurrence of antibiotic residues and antibiotic-resistant bacteria in effluents of pharmaceutical manufacturers and other sources around Hanoi, Vietnam. *Science of the Total Environment* 645: 393–400.

Tijani, J.O., Fatoba, O.O., Babajide, O.O. et al. (2016). Pharmaceuticals, endocrine disruptors, personal care products, nanomaterials and perfluorinated pollutants: a review. *Environmental Chemistry Letters* 14 (1): 27–49.

Tobajas, M., Verdugo, V., Polo, A.M. et al. (2016). Assessment of toxicity and biodegradability on activated sludge of priority and emerging pollutants. *Environmental Technology* 37 (6): 713–721.

Tripathi, S., Sanjeevi, R., Anuradha, J. et al. (2018). Nano-bioremediation: nanotechnology and bioremediation. In: *Biostimulation Remediation Technologies for Groundwater Contaminants* (ed. A.K. Rathoure), 202–219. IGI Global.

Tsoi, T.V., Zaitsev, G.M., Plotnikova, E.G. et al. (1991). Cloning and expression of the Arthrobacter globiformis KZT1 fcbA gene encoding dehalogenase (4-chlorobenzoate-4-hydroxylase) in Escherichia coli. *FEMS Microbiology Letters* 81 (2): 165–169.

Ufarté, L., Laville, É., Duquesne, S. et al. (2015). Metagenomics for the discovery of pollutant degrading enzymes. *Biotechnology Advances* 33 (8): 1845–1854.

Valsala, G. and Sugathan, S. (2017). Microbial repositories in bioprospecting. In: *Bioresources and Bioprocess in Biotechnology* (eds. S. Abdulhameed, N.S. Pradeep and S. Sugathan), 397–420. Springer.

Vanek, T., Podlipna, R., Fialova, Z. et al. (2010). Uptake of xenobiotics from polluted waters by plants. In: *Xenobiotics in the Urban Water Cycle* (eds. D. Fatta-Kassinos, K. Bester and K. Kümmerer), 431–444. Springer.

Varjani, S. J., and Upasani, V. N. (2017). Crude oil degradation by Pseudomonas aeruginosa NCIM 5514: Influence of process parameters. New Delhi: NISCAIR Publications.

Vuković Domanovac, M., Šabić Runjavec, M., and Meštrović, E. (2019). Bioaugmentation effect of Aeromonas hydrophila and pseudomonas putida on kinetics of activated sludge process in treating pharmaceutical industrial wastewater. *Journal of Chemical Technology & Biotechnology* 94 (8): 2721–2728.

Wang, Y., Chen, Y., Zhou, Q. et al. (2012). A culture-independent approach to unravel uncultured bacteria and functional genes in a complex microbial community. *PLoS One* 7 (10): 263–271.

Wang, L., Qiang, Z., Li, Y. et al. (2017). An insight into the removal of fluoroquinolones in activated sludge process: sorption and biodegradation characteristics. *Journal of Environmental Sciences* 56: 263–271.

Wasi, S., Tabrez, S., and Ahmad, M. (2013). Toxicological effects of major environmental pollutants: an overview. *Environmental Monitoring and Assessment* 185 (3): 2585–2593.

Wijekoon, K.C., Hai, F.I., Kang, J. et al. (2013). The fate of pharmaceuticals, steroid hormones, phytoestrogens, UV-filters and pesticides during MBR treatment. *Bioresource Technology* 144: 247–254.

Wiles, S., Whiteley, A.S., Philp, J.C. et al. (2003). Development of bespoke bioluminescent reporters with the potential for in situ deployment within a phenolic-remediating wastewater treatment system. *Journal of Microbiological Methods* 55 (3): 667–677.

Włóka, D., Placek, A., Rorat, A. et al. (2017). The evaluation of polycyclic aromatic hydrocarbons (PAHs) biodegradation kinetics in soil amended with organic fertilizers and bulking agents. *Ecotoxicology and Environmental Safety* 145: 161–168.

WWAP. (2017). The United Nations World Water Development Report 2017. Wastewater: The Untapped Resource. Paris: UNESCO.

Xie, T., Jing, Z., Hu, J. et al. (2018). Degradation of nitrobenzene-containing wastewater by a microbial-fuel-cell-coupled constructed wetland. *Ecological Engineering* 112: 65–71.

Xiong, J.-Q., Kurade, M.B., and Jeon, B.-H. (2018). Can microalgae remove pharmaceutical contaminants from water? *Trends in Biotechnology* 36 (1): 30–44.

Xu, X., Liu, W., Tian, S. et al. (2018). Petroleum hydrocarbon-degrading bacteria for the remediation of oil pollution under aerobic conditions: a perspective analysis. *Frontiers in Microbiology* 9: 2885.

Yoshikawa, M., Zhang, M., and Toyota, K. (2017). Integrated anaerobic-aerobic biodegradation of multiple contaminants including chlorinated ethylenes, benzene, toluene, and dichloromethane. *Water, Air, & Soil Pollution* 228 (1): 25.

Zafra, G. and Cortés-Espinosa, D.V. (2015). Biodegradation of polycyclic aromatic hydrocarbons by Trichoderma species: a mini review. *Environmental Science and Pollution Research* 22 (24): 19426–19433.

Zaprasis, A., Liu, Y.-J., Liu, S.-J. et al. (2010). Abundance of novel and diverse tfdA-like genes, encoding putative phenoxyalkanoic acid herbicide-degrading dioxygenases, in soil. *Applied and Environmental Microbiology* 76 (1): 119–128.

Zhang, X., Yu, T., Li, X. et al. (2019a). The fate and enhanced removal of polycyclic aromatic hydrocarbons in wastewater and sludge treatment system: a review. *Critical Reviews in Environmental Science and Technology* 49 (16): 1425–1475.

Zhang, Y., Liu, M., Zhou, M. et al. (2019b). Microbial fuel cell hybrid systems for wastewater treatment and bioenergy production: synergistic effects, mechanisms and challenges. *Renewable and Sustainable Energy Reviews* 103: 13–29.

Zhao, Q., Bilal, M., Yue, S. et al. (2017a). Identification of biphenyl 2, 3-dioxygenase and its catabolic role for phenazine degradation in Sphingobium yanoikuyae B1. *Journal of Environmental Management* 204: 494–501.

Zhao, Q., Yue, S., Bilal, M. et al. (2017b). Comparative genomic analysis of 26 Sphingomonas and Sphingobium strains: dissemination of bioremediation capabilities, biodegradation potential and horizontal gene transfer. *Science of the Total Environment* 609: 1238–1247.

Zhao, W., Sui, Q., and Huang, X. (2018). Removal and fate of polycyclic aromatic hydrocarbons in a hybrid anaerobic–anoxic–oxic process for highly toxic coke wastewater treatment. *Science of the Total Environment* 635: 716–724.

Zhao, J., Tian, W., Liu, S. et al. (2019). Existence, removal and transformation of parent and nitrated polycyclic aromatic hydrocarbons in two biological wastewater treatment processes. *Chemosphere* 224: 527–537.

Zou, X.-l. (2015). Treatment of heavy oil wastewater by UASB–BAFs using the combination of yeast and bacteria. *Environmental Technology* 36 (18): 2381–2389.

9

The Fate of Organic Pollutants and Their Microbial Degradation in Water Bodies

Gurudatta Singh, Anubhuti Singh, Priyanka Singh, Reetika Shukla, Shashank Tripathi, and Virendra Kumar Mishra

Institute of Environment and Sustainable Development, Banaras Hindu University, Varanasi, Uttar Pradesh, India

9.1 Introduction

The past century has witnessed a dramatic increase in the production of a range of artificial organic chemicals (xenobiotic), for example, pesticides, plastics of various grades, hydrocarbons, soaps, and detergents. These compounds trigger a series of biochemical reactions in the environment that ultimately affects all biota. Most of these chemicals are of great concern from a human health perspective due to their persistence, bioaccumulation capacity, and high toxicity to wildlife and human (Xing et al. 2005). A wide spectrum of hazardous organic pollutants with diverse structures originates from anthropogenic sources and continuously pollutes the environment (Alaee et al. 2003; Covaci et al. 2011; Sverko et al. 2008). These organic pollutants have a hazardous impact on various components of environments, which are mostly aquatic ecosystems, i.e. water bodies. Important organic contaminants include polycyclic aromatic hydrocarbons (PAHs), polychlorinated biphenyls (PCBs), hydrocarbons (HCs), pharmaceutical compounds, dioxins, organophosphates, biochemical oxygen demand (BOD), and organometallic compounds (Connell et al. 2009).

The major sources of organic pollutants of water bodies can be identified, viz. industries, municipal sewage, and agricultural activities (Figure 9.1). Further, these sources can be classified as point sources such as industrial effluent outlets, effluent discharges from domestic sewage treatment plants, and nonpoint sources, which are characterized by discharges over huge areas due to diffusive sources. For example, pesticides used in agriculture can be widespread in low but potentially hazardous concentrations, causing chronic effects (Schwarzenbach et al. 2006). Biodegradation has been defined as "the biologically catalyzed reduction in complexity of chemicals." This can lead to the conversion of carbon, nitrogen, phosphorous, sulfur, or other elements bound within the original organic chemical to inorganic products, this is called mineralization (Mattes et al. 2010). If a compound is biologically transformed into products that are either not metabolized or metabolized very slowly,

Pollutants and Water Management: Resources, Strategies and Scarcity, First Edition. Edited by Pardeep Singh, Rishikesh Singh, Vipin Kumar Singh, and Rahul Bhadouria.

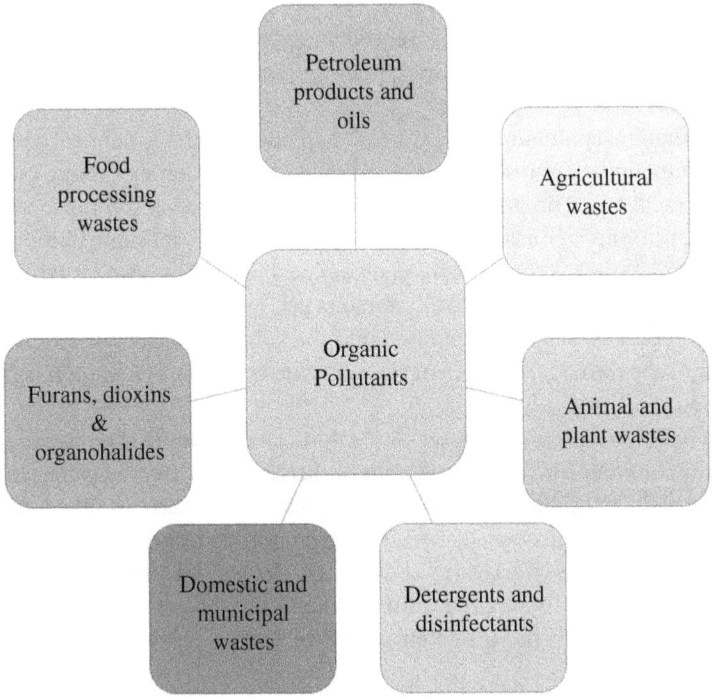

Figure 9.1 Sources of organic pollutants.

a transformation takes place. The transformation of hazardous chemicals into harmless chemicals is certainly a desirable biodegradation process, but sometimes the opposite is observed, a toxic compound is transformed into an even more toxic one, which accumulates in the environment. For example, the anaerobic dechlorination of higher chlorinated ethenes can lead to the accumulation of vinyl chloride, which is more toxic and volatile than its precursors (Mattes et al. 2010). Also, pesticides can be transformed into compounds that are more toxic than the parent molecule; the transformation of aldrin to dieldrin is one example (Jorgenson 2001).

9.2 Classification of Organic Pollutants in the Environment

Developing countries have been producing, utilizing, importing, and releasing numerous varieties of carbon-based contaminants in our surroundings for the last few years. Carbon-based pollutants have been key issues in the environment, making abundant technical as well as community considerations because of their bioaccumulation possibility, tenacity, and noxiousness. There is a lack of knowledge regarding the eco-toxicological effects of organic contaminants on the environment and humans (Zhang et al. 2005). For this reason, it is essential to evaluate the level of carbon-based contaminants similar to insecticides, endocrine-disrupting contaminants (EDCs), and additional pollutants on the well-being of humans and ecosystems (Kabir et al. 2015). Carbon-based pollutants are tremendously resilient in natural putrefaction in their surroundings as well as within the riverine ecosystem, soils, the food cycle, and within humans bodies for many years. These compounds have the capability to disturb endocrine, reproductory, immune, and nervous systems, and further cause developmental difficulties, diabetes, thyroids complications, and sarcoma (Islam et al. 2018).

The classification of organic contaminants is generally characterized by chemical reaction as well as through functional groups. Cataloging through the functioning chemicals is most logical due to chemical-based mixtures, although it has related well-being impacts as well as being acquiescent to the identical logical technique. Berner (1981) suggested that the redox potential might be used as the basis for geo-chemical cataloging systems within the deposits. The cataloging of carbon-based contaminants was grounded upon likenesses in the functions of chemicals as well as fragmented into the sufficient subclasses so that one or two logical approaches would be sufficient for a specific analysis of all members of each class (Garrison and Keith 2015).

The classification of pollutants present in water can be done on the basis of their application or their nature (Geissen et al. 2015). It is important to classify their potential persistent compounds, which is strictly associated with the prioritizing as well as selecting the tenacious, bioaccumulative, as well as noxious, compounds that are used as well as produced and are spread on global regulations (Karlaganis et al. 2001). The half-life of carbon-based contaminants in several sections is among the utilized criteria for analyzing their persistence, as well as being considered in the valuation of environmental and well-being hazards (Muir and Howard 2006).

The forthcoming sections deal with the major groups of organic compounds contaminating water bodies and are also presented in Figure 9.2.

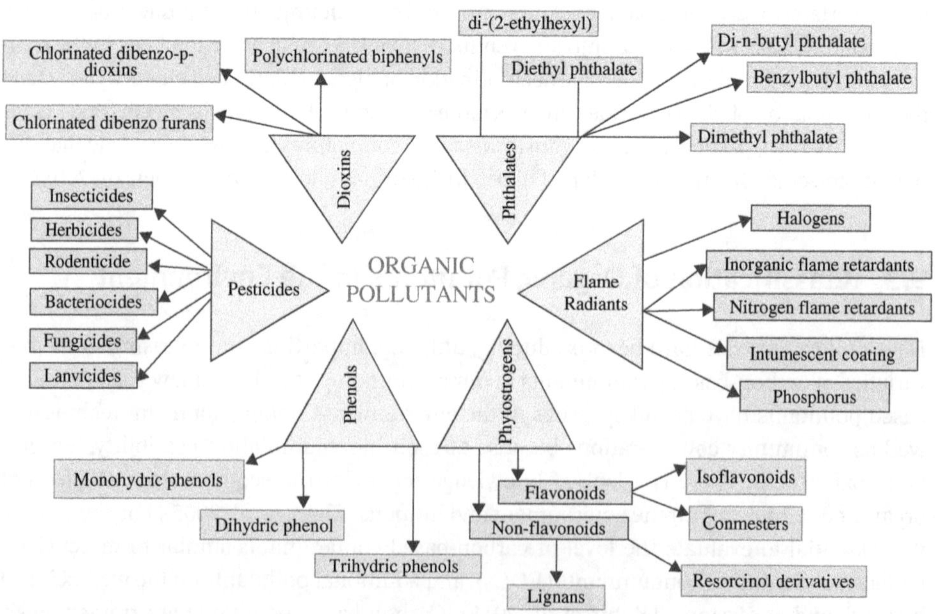

Figure 9.2 Types of organic pollutants in aquatic environments.

Table 9.1 Types of pesticides used for various purposes.

S. no.	Type of pesticide	Action	Reference
1	Algicides	Controls algae in different water bodies	Schrader (2003)
2	Antimicrobials	Kills microbes like bacteria and viruses	Laxminarayan et al. (2016)
3	Disinfectants	Kills disease-producing microbes	Bridier et al. (2011)
4	Herbicides	Kills weeds, removes unwanted plants	Saari and Mauvais (2018)
5	Insecticides	Kills insects	O'Brien et al. (2014)
6	Miticides	Kills mites	Mullin et al. (2010)
7	Molluscicides	Kills mollusks like snails	Luna et al. (2006)
8	Pheromones	Biochemicals used to disrupt the mating behavior of insects	Shorey (2013)
9	Rodenticides	Control rodents	Watt et al. (2005)
10	Biopesticides	Pesticides derived from natural materials	Pavela and Benelli (2016)

9.2.1 Pesticides

Pesticides are chemicals that are used for killing pests. They include insecticides, herbicides, rodenticides, bactericides, fungicides, larvicides, and others (Grube et al. 2011). These pesticides are generally used in the farming sector as well as in the industrial sector. Research has revealed that exposure to pesticides causes toxicity in humans. Children are generally more susceptible to pesticides, which are carcinogenic in nature, and exposure to pesticides can be through various sources (Zahm and Ward 1998). Pesticides are an important toxicant that causes pollution of water bodies and may affect human beings in various ways. There are a large number of pesticides used for various purposes and their residues remain in the environment, which causes pollution. When these compounds come into contact with rainwater, they are discharged into the surrounding water body along with surface run-off. These pesticides are generally classified depending on the type of pests. Some of the pesticide groups are displayed in Table 9.1.

9.2.2 Dioxins

Dioxins, as well as dioxin-like compounds (DLCs), are groups of chemical compounds that are persistent contaminants in the environment. Dioxins pose toxicity to living beings that depend on the numbers as well as positions of the chlorine molecule. Since dioxins refer to such a wide class of compounds, and their toxicities vary, the concept of the toxic equivalency factor (TEF) has been established to ease hazard assessments as well as supervisory controls (Wikoff et al. 2020). Dioxins are nearly not soluble in water, however, they have a comparatively higher solubility in lipids. Consequently, they tend to subordinate through carbon-based matters like planktons and plant leaves, as well as fats present in animals. Additionally, dioxins tend to be adsorbed to the inorganic constituent part, like ashes and soils (Weber et al. 2019). Dioxins are tremendously constant and tend to be accumulated in the food chain. Most of the dioxins and their congeners are mentioned in Table 9.2.

Table 9.2 Categorization of dioxins in the environment.

S. no.	Class of dioxins	Congeners	Reference
1	Polychlorinated dioxins	2,3,7,8-Tetrachlorodibenzodioxin	Fennell (2019)
		1,2,3,7,8-Pentachlorodibenzo-P-dioxin	
		1,2,3,4,7,8-Hexachlorodibenzo-P-dioxin	
		Octachlorodibenzodioxin	
2	Polychlorinated dibenzofurans	2,3,7,8-Tetrachlorodibenzofuran	Hou et al. (2019)
		1,2,3,7,8-Pentachlorodibenzofuran	
		1,2,3,4,7,8-Hexachlorodibenzofuran	
		Octachlorodibenzofuran	
3	Non-ortho-substituted PCBs	3,3′,4,4′-Tetrachlorobiphenyl	Habibullah-Al-Mamun et al. (2019)
		3,4,4′,5-Tetrachlorobiphenyl	
		3,3,4,4,5-Pentachlorobiphenyl	
		3,3′,4,4′,5,5′-Hexachlorobiphenyl	
4	Mono-ortho-substituted PCBs	2,3,3′,4,4′-Pentachlorobiphenyl	Wang et al. (2019)
		2,3,3′,4,4′,5-Hexachlorobiphenyl	
		2,3′,4,4′,5,5′-Hexachlorobyphenyl	
		2,3,3′,4,4′,5,5′-Heptachlorobiphenyl	

9.2.3 Phthalates

Phthalates are the esters of phthalic anhydride and they are generally used as plasticizers, i.e. materials that are mixed with plastics to improve the plasticity, pellucidity, and resilience, as well as permanency. Phthalates have been utilized in a huge variety of materials through the enteric coating of therapeutic tablets as well as dietary supplements to viscosity regulator agents, crystallizing agents, stabilizers, dispersants, lubricants, binders, emulsifying agents, and dangling agents (Luo et al. 2018). End applications include epoxy resin as well as gums, agricultural adjuvants, building constituents, beauty products, therapeutic equipment, and laundry detergent, as well as surfactants, packaging, toys, modeling clay, candlewax, paints, the ink used in printing, drugs, foodstuffs, and fabric materials. Beauty products contain phthalates, these include body sprays, eye shadow, lotions, nail polish, liquefied soaps, and hairspray (Lorz et al. 2007). Within the technical communities, phthalates are largely categorized as endocrine disruptors, although technical research has specified the probability that phthalates act like endocrine disruptors in humans (Table 9.3) (Bansal et al. 2018; Noutsopoulos et al. 2019; Zamkowska et al. 2018).

9.2.4 Flame Retardants

The ecological behavior of fire retardants has been discussed since the 1990s. Brominated flame retardants originated in most ecological sections as well as organisms, including humans, and in certain individual materials in which some of the toxic properties were

Table 9.3 Different types of phthalates in the environment.

S. no.	Name of phthalate	Chemical formula	Reference
1	Dimethyl phthalate	$C_6H_4\ (COOCH_3)_2$	Salazar-Beltrán et al. (2019)
2	Di-n-propyl phthalate	$C_6H_4\ [COO(CH_2)_2CH_3]_2$	Assens et al. (2019)
3	Di-n-butyl phthalate	$C_6H_4\ [COO(CH_2)_3CH_3]_2$	Bello et al. (2019)
4	Butyl cyclohexyl phthalate	$CH_3\ (CH_2)_3OOCC_6H_4COOC_6H_{11}$	Schwedler et al. (2020)
5	Butyl benzyl phthalate	$CH_3\ (CH_2)_3OOCC_6H_4COOCH_2C_6H_5$	Zhang et al. (2019)
6	Diisoheptyl phthalate	$C_6H_4\ [COO(CH_2)_4CH(CH_3)_2]_2$	Silva et al. (2019)
7	Di (n-octyl) phthalate	$C_6H_4\ [COO(CH_2)_7CH_3]_2$	Wang et al. (2020)

Table 9.4 Different flame retardants in the environment.

S. no.	Flame retardant	Reference
1	Polybrominated diphenyl ethers (PBDEs)	Noziglia et al. (2018)
2	Tetrabromobenzene (TBB)	Koenig et al. (2013)
3	Hexabromocyclododecane (HBCD)	Remberger et al. (2004)
4	Tetrabromobisphenol A (TBBPA)	Alaee et al. (2003)
5	Pentabromotoluene	Arp et al. (2011)

present. Because of the adverse impact on the surroundings, flame retardants (polychlorinated biphenyls) have been banned in many countries (Betts 2008). Roze et al. (2009) conducted a study in the Netherlands on mothers and children for the assessment of associated flame retardants among 12 organohalogen compounds (OHCs), including polychlorinated biphenyls (PCBs) and brominated diphenyl ether (PBDE). The research revealed that the transplacental transfer of polybrominated flame retardants was associated with the growth of children of school age. Some of the important flame retardants are discussed in Table 9.4.

9.2.5 Phenols

Phenols are manufactured in industry as well as being produced by flora and microorganisms within different organisms. Phenolic complexes are categorized into simple phenols or polyphenols, which are centered on the number of phenol units in the molecules (Hättenschwiler and Vitousek 2000). These complexes are the solvable molecule, but the minor particles can be impulsive. Phenols chemically interrelate through numerous other materials. Chemical properties of particles with aromatization, which occurs among phenolic particles, can be observed. Phenols have dissimilar assets and are usually different from former alcohol. These complexes are identified as lethal and may impose long-lasting impacts on different organisms. These phenols may act as carcinogens as well as cause the destruction of red blood cells and the liver, even at a very low concentration. Interactions

Table 9.5 Different types of phenols and their functions.

S. no.	Type of phenol	Function	Reference
1	Bisphenol-A	Produced from ketones and phenol/cresol	Fukahori et al. (2003)
2	Butylated hydroxytoluene (BHT)	Fat-soluble antioxidant and food additive	Ortiz-Vazquez et al. (2011)
3	4-Nonylphenol	Breakdown product of detergents and nonoxynol-9	Fries and Püttmann (2003)
4	Orthophenyl phenol	Fungicide used for waxing citrus fruits	Bakhtiar et al. (2019)
5	Trinitrophenol	An explosive material	Qu et al. (2020)
6	Xylenol	Used in antiseptics and disinfectants	Jaroenram et al. (2019)

of the mixtures through microorganisms, inorganics, and other biological mixtures in water can produce substitute mixtures that can be more toxic than the original phenol complexes (Qu et al. 2020). The classification of phenols is done on the basis of the number of molecules present. Some phenol types are shown in Table 9.5.

9.3 Organic Contaminants in Water and Their Sources

Every natural, as well as artificial carbon-based compound, could be estimated to be found in water samples if exposure limits were dropped appropriately (Donaldson 1977). Carbon-based contamination takes place once an extra carbon-based material, like manure or domestic waste, enters the aquatic system. Once carbon-based material rises in the water body, the number of microbes is increased. Those microbes grow frequently as they use the excess of oxygen throughout their development, which leads to a decrease in the oxygen level due to the process that takes place during the decaying of matter. The absence of fresh air could be responsible for the survival of riverine creatures (Zhang et al. 2013). The dead bodies of riverine creatures are decomposed by microbes and this causes a decrease in oxygen level. Carbon-based contamination can take place as the inorganic contaminants like nitrates and phosphates are stored in riverine environments. Increasing nutrients in riverine water can lead to the overgrowth of flora.

The sources of various carbon-based contaminants are agricultural practices, commercial sectors, and sewage from houses. These carbon-based contaminants are nonbiodegradable due to their high stability on a large level (Dong et al. 2015). They can be accumulated in the environment and they have the ability to travel a long distance. These contaminants cause complications as they are accumulated inside the fatty tissues or body parts of organisms as they are toxic and affect the immunity of living creatures. They also have the ability to disturb the endocrine system, causing cancers or mutations. Numerous studies have detected the effects of organic contaminants on the skeletal system (La Merrill et al. 2013). Different organic pollutants mixed with freshwater are presented in Figure 9.3.

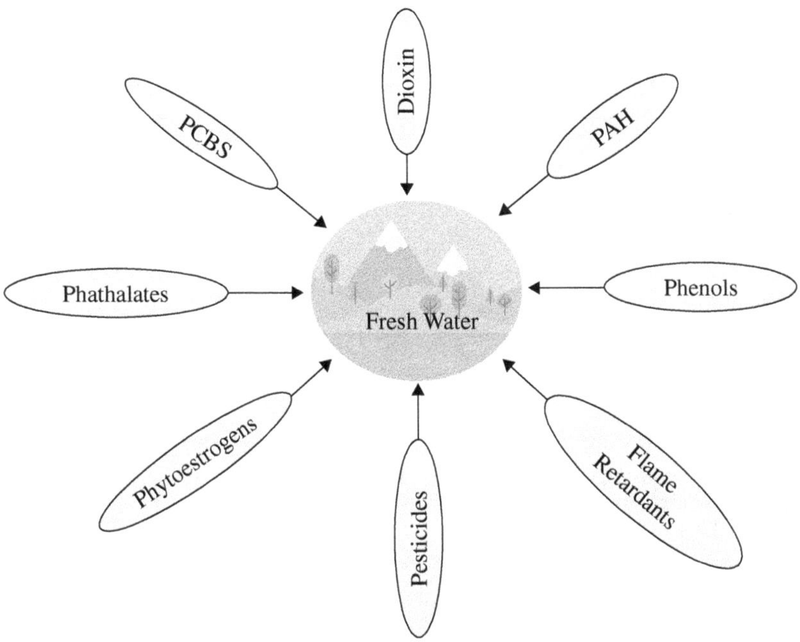

Figure 9.3 Different organic pollutants mixed with fresh water. (PCB, polychlorinated biphenyls; PAH, polycyclic aromatic hydrocarbons.)

Studies and research in these quickly progressing areas of environmental chemistry are being done in various sections, i.e. chemistry of contaminants, techniques and methods of investigation, and identification of organic contaminants. Concentrations, separation, documentation, and quantities of definite carbon-based complexes originate in all kinds of water and constitute the main concern of this volume. One study demonstrated that youngsters between 10 and 17 years were found to have a grouping of concentrated thyroxin intensities as well as amplified thyroid-stimulating hormone levels within the level of persistence PCBs, which reflects previous exposure, but not in relation to PCBs more reflective of existing exposures. These studies indicated that prenatal or newborn exposure is more influential (Schell et al. 2008). Within similar groups, there are modifications in performances on certain intellectual assessments as well as the specific effect on reminiscence (Newman et al. 2006). It has been investigated that the impact of phthalates in children has been exaggerated, and that in recent times, these materials have been examined in other products as well as in most of the population. The impact of phthalates on the thyroid has been reported, as well as on growth and reproductive developments, which could also be exaggerated (Boas et al. 2010). Most of the carbon-based contaminants basically resemble sexual steroids and thyroxine hormone.

It has been noted that most of the carbon-based contaminants, such as chlorophenoxy pesticide, which could be persistent within the environment, polluting shallow water as well as freshwater, have resistance to biodegradation as well as reasonable aqueous solubilities because of its prevalent use. The herbicide 4-chlorophenoxyacetic acid (4CPA) has greater aqueous solubilities compared with the comparable

Table 9.6 Different types of organic pollutants and concentrations in the river system.

S. no.	Organic pollutant	River	Concentration	Reference
1	Methyl orange	River water, France	150 mg/l	Wang et al. (2020)
2	Polychlorinated biphenyl	Gironde estuary	515 mg/l	Bodin et al. (2014)
3	Polycyclic aromatic hydrocarbons	France	572–4235 μg/kg	Kanzari et al. (2014)
4	Polychlorinated biphenyl	Huveaune, France	2.8–435 μg/kg	Kanzari et al. (2014)
5	Pesticides	France	0.07–1.25 μg/kg	Kanzari et al. (2014)
6	Polychlorinated biphenyl	Rhone River	11.50–417.1 μg/kg	Mourier et al. (2014)
7	Polychlorinated biphenyl, polybrominated diphenyl ethers, phthalates	Orge River	120–170 ng/g 12–18 ng/g 2250–5125 ng/g	Teil et al. (2012)

components in this class (Derylo-Marczewska et al. 2017). Phenylethylamine (PhEA) is an anti-depressant as well as psycho-active stimulating, which causes restlessness, impulsiveness, and tachycardia (Halberstadt et al. 2019). Moreover, 4-aminobenzoic acid (4ABA) has been utilized for diazotization reactions within synthesizing dyes and in associated applications like the denaturing of dyes; this could be an emerging contaminant (EC) (Naik et al. 2018). Wang et al. (2020) have reported on methyl orange in river water with a 150 mg/l concentration. Iglesias-González et al. (2020) conducted a study in France to check the impact of carbon-based contaminants (organochlorines, organophosphate, pyrethroids, azoles, dinitroanilines, oxadiazines, phenylpyrazoles, and carboxamides) on the hair of youngsters. Epidemiological studies have concentrated on children, and a correlation has been described among them to the contaminants as well as breathing-related syndromes, neurotoxicity, behavioral disorders, and immune health (Raherison et al. 2019). Different organic pollutants and their concentrations are given in Table 9.6.

9.4 Effects of Organic Pollutants on Aquatic Ecosystems

Ecological influences and various anthropogenic actions are the main reasons behind the stress of aquatic ecosystems. Out of various sources of contamination for surface water, agricultural run-off, domestic sewage, and industrial effluents have been recognized as the main contributors for ecological disturbance in water bodies that affect basic ecological mechanisms (Figure 9.4). Aquatic ecosystems are affected by reversible damage caused by contaminants if it is within the self-cleaning capacity of the ecosystems. Hence, the level of contamination depends on the amount released into the system. These sources contribute toward the discharge of a large number of organic pollutants into water bodies. A list of some organic pollutants in water and their effects on related organisms are shown in Table 9.7.

Organic load from sewage, pesticides, industrial chemical nutrients, and organic augmentation through sediments are the main stressors of the aquatic ecosystem (CEAP, USDA 2008). To conserve aquatic life, protecting their niche is an effective solution. The potential of newer technologies may provide solutions to numerous contaminations associated with problems in aqueous ecosystems. Discharge of organic contaminants into water

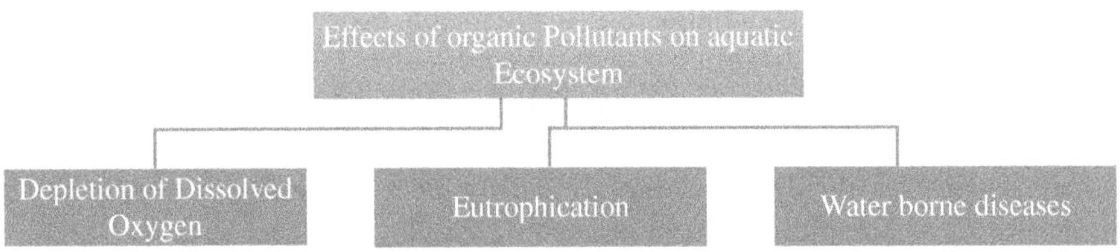

Figure 9.4 The effect of organic pollutants on the aquatic environment.

Table 9.7 List of several organic pollutants, their limit in water bodies specified by the World Health Organization, and health risks.

S. no.	Name of organic pollutant	Maximum limit (μg/l)	Effects on human health
1	Chlorinated alkanes	–	May cause skin and eye irritation; possible carcinogen (IARC)
2	Carbon tetrachloride	2	Liver and kidney damage (EPA)
3	Dichloromethane	20	Damages liver, nervous system; carcinogenic (EPA)
4	1,2-dichloroethane	30	Nervous system disorders, liver and kidney damage (ATSDR)
5	1,1,1-trichloroethane	2000 (p)	Depression of CNS, cardiac dysrhythmia
6	Aromatic hydrocarbons		Eye irritation, nausea, diarrhea, kidney and liver damage, jaundice; carcinogenic (ATSDR)
7	Benzene	10	At a high level can cause vomiting, irritation of the stomach, dizziness, sleepiness, convulsions, rapid heart rate, coma, and death. Effects at a low level are not yet known (ATSDR).
8	Toluene	700	CNS effects (ATSDR)
9	Xylenes	500	High-level contact can cause annoyance on the skin, eyes, nose, and throat, problems breathing, weakened functioning of the lungs, late response to a visual stimulus, reduced memory, stomach distress, and probable variations in the liver and kidneys (ATSDR).
10	Benzo (a) pyrene	0.7	Can cause skin and bladder cancer (IARC).

IARC, International Agency for Research on Cancer; EPA, Environmental Protection Agency (USA); ATSDR, Agency for Toxic Substances and Disease Registry.

bodies tends to alter various physicochemical properties such as temperature, light, substrate, dissolved oxygen (DO), total dissolved solids (TDS), total suspended solids (TSS), cations and anions, and so on. Changing parameters can affect various ecological functions and the quality of the ecosystem might become harmful to its inhabitants.

Substrates make up a major constituent of aquatic ecosystems (Cooper 1993). The range in habitat in natural ecosystems represents an opportunity for the adaptable organism to flourish. In diverse natural ecosystems with dissimilar habitat and niches, groups of animals and plants have established a self-balanced system. This information is vital when deliberating on the effects of contamination on aquatic life because contamination in an aqueous system occasionally disturbs a sole organism (Cooper 1993). If contamination of an ecosystem deliberately disturbs the functioning of a solitary species, it is on the way to affecting the survival of other species by altering the predator–prey balance. Moreover, contamination can also modify habitation or environment limitations, thus, upsetting types of organisms or communities and altering their stability (Cooper 1993).

Carbon-based constituents are the main freshwater contaminants, arising from household sewage releases, even after treatment in treatment plants, and from certain industries such as food processing. Carbon-based materials are normally biodegradable or they can habitually degrade with the aid of microbes that have been adapted to exploiting them. There are several effects of organic pollutants on the aquatic ecosystem. Some of them are illustrated in Figure 9.4.

9.4.1 Depletion of Dissolved Oxygen

The chief contaminating consequences of recyclable organic resources is the drop in the quantity of DO in the water. With a low level of DO, stress is created in aquatic animals in which they react differently and stressfully to survive. For example, a low oxygen level in water significantly increases the breathing rate of fish, as they can be seen pumping more water over their gills than normal (saturated DO level). There are several microorganisms like bacteria and fungi that decompose the organic matter present in their environment to fulfill their nutrient requirements. They require oxygen in this process, and in the aquatic ecosystem, this requirement can be achieved by oxygen dissolved in the water body called dissolved oxygen (DO).

Contaminated effluents undergo several changes after various discharges pass through river bodies. There is a remarked diversity in the community of upstream river flow and downstream river flow. After traveling a certain distance, there is an encroachment of nature's degradation process of conversion of pollutants toward its original makeup. The fate of the species in the ecosystem depends on the number of pollutants and carrying capacity of the water body.

If the amount of organic pollutants is minor, there will be a slight alteration in the type of flora and fauna existing in the water body and little difference in the natural sequences. Initially, the DO will be near the saturation level. Organic pollutants result due to any source of contamination discharges downstream before and reaching its destination will transform or be removed by the self-purification system. If the amount of organic pollutants rises, the DO level drops significantly as the BOD increases and the water body will be polluted for a considerable distance from the location of the discharging water. This leads

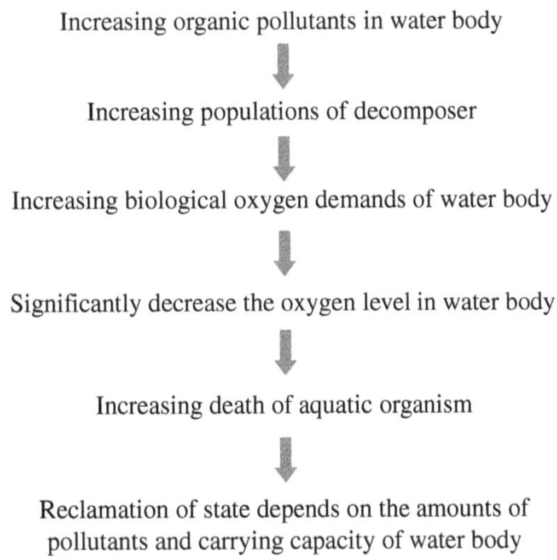

Figure 9.5 Fate of organic pollutants in an aquatic ecosystem.

to the rapid growth of some species, but at the same place, some species might be eliminated. In the nonappearance of more pollution, the water bodies will possibly recover in the course of flow (downstream), but if the area around the discharge remains polluted, then this can act as a barrier to the passage of migratory fish.

If the polluting amount increases significantly or beyond a limit, the natural ecosystem of the water body is crudely distorted. In this case, the ability of the water body to deal with its effect is significantly reduced. The decomposers require a large amount of oxygen that significantly increases in the BOD in the water body and a fall in the DO of the water. This is a situation where only a few organisms can survive. This includes some sewage fungus, certain worms, and fly larvae, representing the contaminated property of the river. Similarly, anaerobic microbes may flourish and give an unpleasant odor to the water by the breakdown of organic contaminants and forming CH_4, H_2S, and ammonia. Few algae can thrive under unadorned organic contamination, therefore, re-oxygenation by the process of photosynthesis will be prohibited. The whole process is summarized in Figure 9.5.

9.4.2 Eutrophication

In the same manner, another phenomenon is the eutrophication of water bodies that have now become a global problem. It is triggered by the excessive rise in nutritious compounds in water bodies, especially rich in nitrogen, phosphates, etc., due to these nutrients, there is an upsurge in the degradation activities of the aqueous environment by altering the types of species or the formation of noxious algal blooms causing an anoxic environment (Maier et al. 2009). Eutrophication causes vigorous growth of algae in the water body due to which the surface of the water body is covered by algae, the phenomenon is called algal bloom. Although all algal blooms are not destructive, they cause discoloration of the water body, forming a stinking odor or a buildup of bad taste. Algal bloom impairs the mixing of atmospheric O_2 in the water body. That is the reason aquatic animals like fish die – due to the nonavailability of O_2 for respiration. Algal bloom also adds several toxins. Consumption of these toxins also leads to the death of fish.

Several diseases are caused by ingesting unclean water. These diseases include typhoid, amoebiasis, giardiasis, ascariasis, hookworm, etc. Waterborne sicknesses produced by contaminated water at beaches are rashes, earache, pink eye, breathing infections, hepatitis, encephalitis, gastroenteritis, diarrhea, vomiting, stomach aches, etc. When a river is polluted with substances such as pesticides, hydrocarbons, persistent organic pollutants, or heavy metals, it can lead to cancers, such as prostate cancer and non-Hodgkin lymphoma, hormonal complications that can upset multiplicative and growing processes, injury to the nervous system, liver and kidney harm, and damage to DNA. Specifically, mercury in water bodies can cause abnormal behavior, slower growth and development, reduction in reproduction, and death (Murali and Valli 2017). A list of some organic pollutants, their specified limits in water by the World Health Organization (WHO), and their effects on human health are given in Table 9.7.

9.5 The Fate of Organic Pollutants in Natural Water Bodies

There are some natural processes through which organic pollutants are converted into less harmful and destructive forms. These processes tend to degrade the pollutants into smaller compounds. They can be used for the treatment of these compounds in several forms and can be classified as photocatalytic degradation, advanced oxidation technology, photochemical reactors, etc.

9.5.1 Photocatalytic Degradation

Although there are many conventional physical, chemical, and biological techniques that have been developed for the treatment of polluted water, this technique eliminates carbon-based contaminants by photocatalytic degradation (Hoffmann et al. 1995). This procedure has the capability to eradicate a wide range of carbon-based contaminants like pesticides, herbicides, and micropollutants like endocrine-disrupting compounds (EDCs). It has recently been gaining the attention of researchers. Although important research had been undertaken on titanium dioxide (TiO_2) photocatalysis in research laboratories, its application on manufacturing needs restrictions. A photocatalyst is well-defined like substances that are stimulated through adsorbing photons, as it has the capability for fast-tracking reactions without being consumable (Fox 1988). The application of this technique has been done to degrade bisphenol-A (BPA). It may reach water bodies through sewage discharged by production units from several industries involved within the productions of phenoplasts, phenol, concentrated poly-ester mastics, polyvinyl chloride, and thermal paper. The photolysis of bisphenol-A could be an essential but unnoticed process in freshwater.

In recent years, alkyl-phenols have been extensively utilized in complexes for examining dissolved organic matter (DOM) photoreactivity (Canonica et al. 1995, 2000; Canonica and Freiburghaus 2001). Although various applications of the treatments were inhibited through features like wide band gap (3.2 eV), shortage, and the incapability of the well-organized and profitable catalysts for the higher efficiencies of photons to operate in the broader solar spectrum (Umar and Aziz 2013). Additional research to examine the degradations of water constituents is essential for the improved comprehension of the application process.

9.5.2 Advanced Oxidation Technology

Semi-conducting oxides, as photocatalysts, have been progressively concentrated in recent years in the prospective application of solar energy conversion as well as ecosystem decontamination. Semi-conductor varied photocatalysis has a huge potential to treat carbon-based pollutants in water (Castellote and Bengtsson 2011). This type of process is known as an advanced oxidation process (AOP), and it is appropriate for the oxidation of extensive ranges of carbon-based complexes. It has been verified to be of interest because of its proficiency in obstinate carbon-based complexes. Various research has been approved on the application of assorted photocatalytic oxidation processes by decay and mineralization refractory carbon-based complexes. The photocatalytic reactor can be used either by ultraviolet (UV) or solar radiation. The solar photocatalytic reactor has an interest in the photo-oxidation of carbon-based pollutants in water. Such kinds of receptacles could be distributed into concentrating or nonconcentrating receptacles (de Lasa et al. 2005). Advanced oxidation technology has been used in advanced nations, however, this technology is not generally practiced in developing countries because of the energy consumed as well as the operating costs (Shannon et al. 2010).

9.5.3 Photochemical Reactors

The important character of this process is the probability of accomplishing the whole oxidation of the recalcitrant organic compound in water as well as carbon dioxide at room temperature and atmospheric pressure. The photo-activated chemical reaction is characterized through the free radical mechanisms introduced through the interactions of the photon of the proper energy levels by the molecules of the chemical species existing within the solutions or through the catalysts. The radical could be simply formed with the help of UV radiation through the standardized photochemical degrading the oxidizing complexes such as hydrogen peroxide as well as ozone (Benitez et al. 1996; Stefan et al. 1996). The photocatalytic processes could be approved through the slurry of the elements of the solid semi-conductors discrete in the liquid phases. In this case, the performance of the reactors may be rigorous through the low irradiation proficiency because of the opaqueness of the slurry. After the oxidation treatments, the solid catalysts should be detached from the liquid (Mazzarino et al. 1999). Apart from these natural reactions, there are certain other processes (discussed in the forthcoming sections) that facilitate the degradation of organic pollutants.

9.6 Microbial Enzymes in the Degradation of Organic Pollutants

Organic contaminants (OCs), such as pharmaceuticals, personal care products, flame retardants, and plasticizers, are societally ubiquitous, environmentally hazardous, and structurally diverse compounds that show resistance to treatment by the most commonly used conventional wastewater treatment plants. Microbial enzymes are useful in the bioremediation of environmental pollutants from various sources due to their high specificity to a broad range of substrates (pollutants). The enzymes are biological catalysts that facilitate the conversion of substrates into products by providing favorable conditions that lower the activation energy of the reaction (Karigar and Rao 2011). An enzyme may be a protein or a glycoprotein and consists of at least one polypeptide moiety. The following are some important enzymes working for bioremediation.

9.6.1 Laccases

Laccases [EC 1.10.3.2, para-benzenediol: dioxygen oxidoreductases] are multi-copper proteins that use molecular oxygen to oxidize various aromatic and nonaromatic compounds by a radical-catalyzed reaction mechanism (Strong and Claus 2011). Laccase alone, or with redox mediators, has been reported to oxidize numerous dangerous compounds, which include phenols, halogenated phenols, aromatic amines, nitro-substituted compounds such as trinitrotoluene, hydroxy-indoles, the herbicide dymron, organophosphorus compounds, PAHs, chlorinated hydroxybiphenyls, bisphenol-A and nonylphenol, hydroxyphenylureas, endocrine-disrupting chemicals, pesticides, and insecticides (Duran and Esposito 2000; Strong and Claus 2011).

9.6.2 Monooxygenase

Monooxygenases are classified into two subclasses based on the presence of cofactor: flavin-dependent monooxygenases and P450 monooxygenases (Karigar and Rao 2011). Methane monooxygenase (MMO) enzyme is the best characterized one among monooxygenases. This enzyme is involved in the degradation of hydrocarbons such as substituted methane, alkanes, cycloalkanes, alkenes, haloalkenes, ethers, and aromatic and heterocyclic hydrocarbons (Karigar and Rao 2011).

9.6.3 Cellulase

Cellulases now promise the potential of converting waste cellulosic material into foods to meet the burgeoning population and have been the subject of intense research (Bennet et al. 2002; Karigar and Rao 2011). Cellulase enzymes are capable of degrading crystalline cellulose to glucose. Alkaline cellulases are produced by *Bacillus* strains and neutral and acidic cellulases by *Trichoderma* and *Humicola* fungi (Karigar and Rao 2011). In the paper and pulp industry, cellulases have been employed for the removal of ink during the recycling of paper (Karigar and Rao 2011).

9.7 Approaches for Organic Pollutant Bioremediation

As much as the diversity in sources and chemical complexities in organic pollutants exist, there is probably more diversity in microbial members and their capabilities to synthesize or degrade organic compounds (Ramakrishnan et al. 2010; Watanabe 2001). Bioremediation chiefly depends on the metabolic capability of microbes to degrade/detoxify or transform the pollutants, which is also affected by the accessibility of pollutants and their bioavailability (Antizar-Ladislao 2010). It can be applied as both in situ (remediation at the site) and ex situ (remediation elsewhere) remediation technology. Bioremediation involves (Maszenan et al. 2011) bioattenuation (the natural process of degradation and can be monitored by a decrease in pollutant concentration with increasing time), biostimulation (intentional stimulation of pollutant degradation by the addition of water, nutrients, and electron donors or acceptors), and bioaugmentation (addition of laboratory-grown microbes with potential for degradation).

9.7.1 Bioattenuation

Bioattenuation includes a variety of chemical, physical, and biological processes that reduce the mass, toxicity, volume, or concentration of contaminants (Pathak et al. 2019). These processes include aerobic and anaerobic biodegradation, sorption, volatilization, chemical or biological stabilization, and the transformation of contaminants. Time is not a limiting factor and it is usually applied on sites with a low concentration of contaminants where no other remedial techniques are applicable (Vasquez-Murrieta et al. 2016). During bioattenuation (natural attenuation), the pollutants are transformed into less harmful forms or immobilized. Such transformation and immobilization processes are largely due to biodegradation by microorganisms (Smets and Pritchard 2003), and to some extent by the reactions with naturally occurring chemicals and sorption on geologic media. Natural attenuation processes are contaminant-specific, accepted as methods for treating fuel components such as benzene, toluene, ethylbenzene, and xylenes (BTEX) (Atteia and Guillot 2007), but not for many other classes. The time required for natural attenuation varies considerably with site conditions. Many polluted sites may not require an aggressive approach to remediation, and bioattenuation is efficient and cost-effective (Davis et al. 1994; Mulligan and Yong 2004).

9.7.2 Biostimulation

Biostimulation is a process in which beneficial microbes modify the environment to enhance the process of bioremediation. This is done in several ways including the addition of various forms of rate-limiting nutrients and electron acceptors, such as phosphorus, nitrogen, and oxygen, or carbon, modification of regime, or changing the environment by a change in pH. Biostimulation can be enhanced by bioaugmentation. This process, overall, is referred to as bioremediation and is an EPA-approved method for reversing the presence of oil or gas spills. While biostimulation is usually associated with the remediation of hydrocarbon or high production volume chemical spills, it is also potentially useful for the treatment of less frequently encountered contaminant spills, such as pesticides, particularly herbicides (Carberry and Wik 2001). To stimulate microbial degradation, nutrients in the form of fertilizers (water-soluble KNO_3, $NaNO_3$, NH_3NO_3, K_2HPO_4, and $MgNH_4PO_4$), slow-release Customblen, and oleophilic (Inipol EAP22) are added (Nikolopoulou and Kalogerakis 2008).

9.7.3 Bioaugmentation

Often, the biological response lags behind, up to weeks or months, in polluted sites with no exposure history. "Soil activation" is a concept based on the cultivation of biomass from a fraction of contaminated soil and the subsequent use as an inoculum for bioaugmentation for the same soil. This was attempted by Otte et al. (1994) for the degradation of pentachlorophenol (PCP) and PAHs. Soils with microbiota, adapted by prior exposure to degradation of organic pollutants such as hydrocarbons, can be a source of microorganisms for remediating soils freshly contaminated with hydrocarbons.

9.7.4 Mechanism of Degradation of Organic Pollutants

Stimulation of microbial growth in situ offers the prospect of converting a wide variety of dissolved and absorb contaminants into less harmful products. This approach entails perfusion of chemicals (nutrients, electron donors, and/or electron acceptors) through the contaminated subsurface to encourage microbial growth. For organisms to grow, the presence of electron donors and acceptors, a carbon source, and nutrients are essential. In addition to compounds that are the natural substrates, many contaminants can also satisfy these growth requirements. Most organic contaminants can be classed as either aliphatic or aromatic compounds that contain different functional groups, such as —OH, —Cl, $—NH_2$, $—NO_2$, and $—SO_3$. As electron donors, these contaminants undergo oxidation as a result of microbial metabolism; in the best case, they are mineralized. Some of the breakdown intermediates may be assimilated as a carbon source for microbial growth. Functional groups ($—NH_2$, $—NO_2$, and $—SO_3$) may either be used as nutrients or cleaved from the carbon skeleton when the compound is reduced or oxidized. Oxidation can occur aerobically or anaerobically, and the fate of aliphatic and aromatic compounds that undergo biological oxidation is governed primarily by the presence or absence of oxygen. Oxygen serves two distinct functions. It can act as the terminal acceptor of electrons that are released during the oxidation of electron donors or it can react directly with the organic molecule. As an electron acceptor, oxygen can be replaced by other oxidized inorganic compounds such as nitrate, metal ions, sulfate, or carbon dioxide, although the energy gains are then smaller.

The general mechanism of degradation of organic pollutants (Figure 9.6a and b) based on different scavengers including ascorbic acid, t-butanol, sodium azide, etc. have been tested to identify the active species which may be involved in the degradation of organic pollutants and it was found that singlet oxygen and carbonate radicals may play a role in degradation processes (Zhou et al. 2013).

9.7.5 Degradation of Aliphatic Compounds

The general strategy is to convert the alkane chain into fatty acid. In the presence of oxygen, the alkane chain is attacked by a monooxygenase, which introduces one atom of oxygen from molecular oxygen into the molecule, thus forming fatty alcohol. This alcohol is oxidized to an aldehyde, and finally to a carboxylic acid. The carboxylic acid is channeled into central metabolism for further oxidation by β-oxidation. Alkenes and alkynes are degraded aerobically by mechanisms similar to those by which saturated compounds

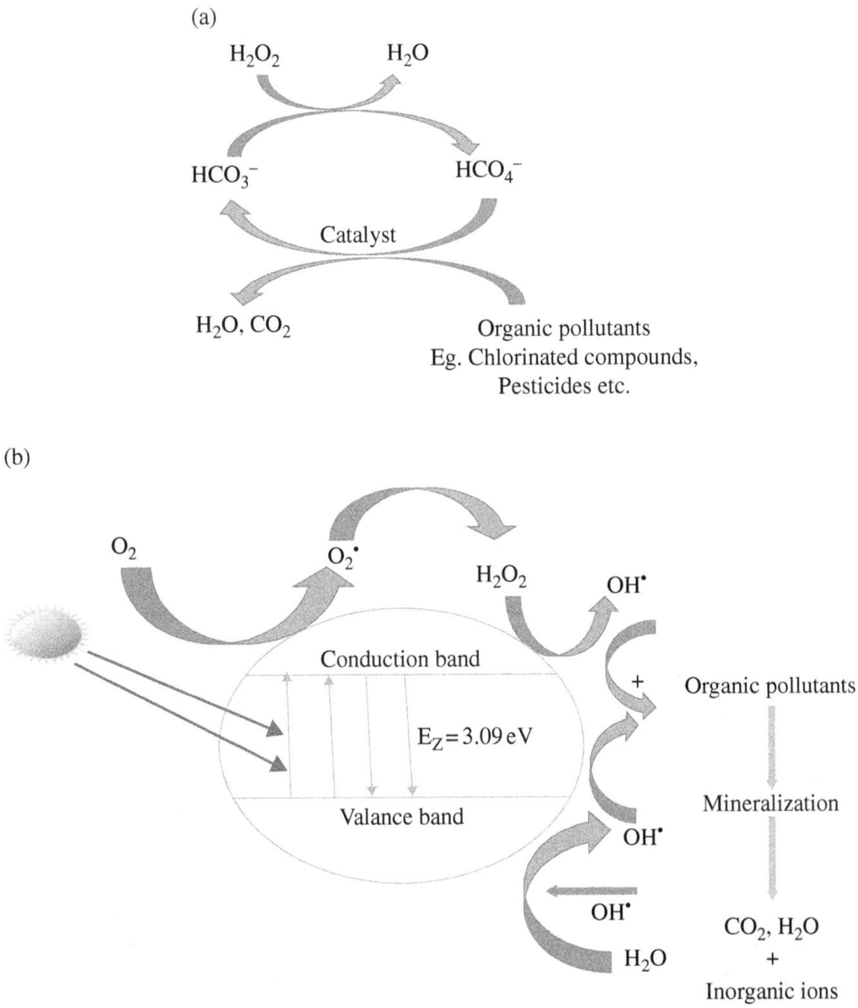

Figure 9.6 (a) General mechanism of degradation of organic pollutants. (b) Presentation of degradation mechanism of organic pollutants based on the semi-conductor.

(alkanes) are degraded. However, the double and triple bonds are more chemically reactive and can undergo additional reactions, such as epoxidation and hydration. There are some indications that the double bond of alkenes can be hydrated anaerobically to form an alcohol, which is then further converted as in the aerobic pathway. Alkynes, such as acetylene, are used as both carbon and energy sources by aerobes and anaerobes: both types of organisms initially attack the molecule by hydration.

9.7.6 Degradation of Aromatic Compounds

Aerobic biodegradation of many classes of the aromatic compound is common and proceeds through the key intermediate, catechol. Eukaryotic organisms produce catechols from single-ring aromatic compounds via an epoxide and a trans-diol using a monooxygenase.

Prokaryotes introduce the entire oxygen molecule by a dioxygenase reaction forming first a cis-diol. In both types of organisms, the aromatic ring of the catechol is opened by a further dioxygenase reaction by either ortho- or meta-fission. Halogenated aromatic compounds are most often degraded in this manner. Anaerobically, the aromatic ring is not oxidized but reduced, as illustrated for benzoate, phenol, and catechol. The key intermediate in this pathway is cyclohexanone. The ring is opened through the hydration of the cyclohexanone.

9.8 Microbial Bioaccumulation of Organic Pollutant in Water Bodies

Bioaccumulation is defined as the accumulation of pollutants by actively growing cells. A number of workers have investigated the feasibility of using growing cells for the bioaccumulation of heavy metal ions. In metal bioaccumulation processes, there is an initial rapid accumulation step, i.e. metabolism- and temperature-independent, and is thought to involve ion binding at the surface. This step is followed by a second process, i.e. metabolism-dependent, it is much slower and can accumulate larger quantities of ions than the first process. Bioaccumulation could also be achieved for the removal of various kinds of textile dyes if the growing cells find a sufficient quantity of easily used carbon and nitrogen sources instead of dye in growth media. Decolorization of the dye solution could also be due to either adsorption on the biomass or bioaccumulation. Using growing cultures in bioremoval could avoid the need for a separate biomass production process (e.g. cultivation, harvesting, drying, processing, and storage prior to use). There are significant practical limitations to biouptake, which employs living cell systems. Perhaps the most significant limitation is that cell growth is inhibited when dye concentrations are too high. Other components or constraints of the wastewater may be toxic to living cells, e.g. extremes of pH and high salt concentration. Active uptake processes also require externally provided metabolic energy. However, the fact that many conventional sewage treatment processes are based on living microbes suggests that such limitation would not preclude their applications in treatment schemes involving the bioremoval of dyes. If the problem of dye toxicity to the growing cell is overcome by the use of dye-resistant organisms, the continually self-replenishing system can be left to run continuously for extended periods. The essential characteristics of living biomass used in a dye removal process are tolerance and uptake capacities. Bioaccumulation of aquatic pollutants by higher organisms has been an area of intensive environmental research. Thus, it is surprising to find that so little attention has been given to the segment of biota that forms the base of the food chain and comprises most of the biomass. The importance of bioaccumulation for environmental purposes can hardly be overstated. This stems from the need to predict three factors precisely and reproducibly: (i) extent of bioconcentration of a specified chemical by specified organisms; (ii) extent of bioaccumulation of a specified chemical by unstudied organisms; and (iii) extent of accumulation of unstudied chemicals by specific organisms. These factors can be obtained in most cases by direct measurement.

Bacteria often improve their chances of reproductive success and survival by the production of extracellular polymeric substances (EPSs). The functional roles of EPSs and their involvement in attachment, microcolony formation, floc formation, protection against heavy metals, protection against predation and environmental fluctuations, increased resistance against antimicrobial agents, and localization of extracellular enzymes are well described in the literature. One functional aspect of microbial behavior that is not clearly understood is the involvement of EPS in the accumulation and subsequent utilization of nutrients.

Many natural environments contain low nutrient concentrations, and to survive under such conditions, it is necessary for microbial communities to adapt. Accumulation of growth-limiting nutrients by EPS may be an adaptive strategy for survival in environments where these nutrients are available at levels below threshold concentrations required by microorganisms to remain viable. It has been suggested that nutrients are accumulated in EPS by the same mechanisms responsible for the accumulation of metals. Metal-binding is a common phenomenon in wastewater treatment, involving complexation of EPS with metals, and in ion exchange, mostly via carboxyl groups on uronic acids. Most EPS have cation exchange properties that may enable bacteria to use EPS as a nutrient trap under oligotrophic conditions; it is unlikely that they can function effectively as a trap for soluble nutrients in environments where divalent cations are abundant. Bacteria may adhere to surfaces and grow as sessile biofilm communities in response to oligotrophic (low-nutrient) conditions (Decho and Gutierrez 2017). Vandevivere and Kirchman (1993) demonstrated a fivefold-greater production of EPS by attached cells than free-living cells of the same bacterium. It is possible that the increased production of EPS by attached cells improves the efficiency of biofilms. EPS may play an additional role in the utilization of trapped nutrients by the localization of extracellular enzymes that hydrolyze high-molecular-weight organic matter into a more utilizable form before it enters the cell.

Sugiura and Futagawa (1984), have suggested that benzene hexachloride (BHC) accumulation by bacteria occurs via adsorption followed by internal diffusion. Adsorption of colloidal aldrin by flocculant bacteria as a removal mechanism was proposed by Baughman et al. (1981), where DDT and *Chlorella* led them to the conclusion that absorption was the dominant process. Kennedy and Finnerty (1975) have shown that an *Acinetobacter* species grown on pentadecane, hexadecane, heptadecane, and hexadec-1-ene could accumulate these compounds against a concentration gradient to form internal "pools" or inclusion bodies.

Ludvik and co-workers (1968), have suggested that when the yeast *Candida lipolytica* is grown on hydrocarbons, the hydrocarbons penetrate the cell wall and accumulate at the cytoplasmic membrane. Based on electron microscopy, these authors further suggest pinocytosis as a possible mechanism for transport across the membrane. Yeasts are an inexpensive, readily available source of biomass. Furthermore, yeast cells retain their ability to accumulate a broad range of textile dyes to varying degrees under a wide range of external conditions. Yeast has long been known to be capable of rapid bioaccumulation of metal ions from solution, but little work has been carried out investigating the ability of yeast to act as a bioaccumulator for textile dyes in wastewater effluents.

The organisms were actually grown either on the compounds studied or in their presence under conditions not likely to be found in the environment. Also, the organisms were

Table 9.8 Different microorganisms used for the treatment of organic pollutants.

S. no.	Name of microbe	Types of organic pollutant	Reference
1	*Saccharomyces cerevisiae* Yeast	Dye (Remazol Blue, Remazol Black B, and Remazol Red RB)	Aksu and Dönmez (2005)
2	*Pseudomonas fluorescens*	Herbicide Diclofop	Wolfaardt et al. (1995)
3	*Aerobacter aerogenes*	Dichloro-diphenyl-trichloroethane (DDT)	Mansouri et al. (2017)
4	*Acinetobacter* sp.	Pentadecane, hexadecane, and heptadecane	Wilkinson et al. (2002)
5	*Bacillus megaterium*	Polycyclic aromatic hydrocarbon benzo[a]pyrene (BAP)	Lily et al. (2010)
6	*Candida lipolytica*	Hydrocarbon	Osumi (2012)
7	*Streptomyces venezuelae*	Dichloro-diphenyl-trichloroethane (DDT)	Pinero-Bonilla (2019)
8	*Trichoderma viride*	Dieldrin	Matsumoto et al. (2009)
9	*Chlorella* sp.	Dieldrin	Katagi (2010)
10	*Aspergillus* sp.	Photodieldrin	Tripathi et al. (2013)
11	*Agrobacterium tumefaciens*	Polychlorinated biphenyl (PCB)	Wang et al. (2015)
12	*Bacillus subtilis*	Methoxychlor	Johnson and Kennedy (1973)
13	*Acinetobacter species*	Pentadecane	Salam et al. (2016)
14	*Anabaena cylindrica*	Dieldrin	Salam and Das (2012)

capable of metabolizing the study compounds. Xenobiotics not adsorbed by the soil can become mobile and when the molecule is persistent it may be dispersed by water or air through the biosphere, polluting large geographical areas and affecting the biodiversity of the flora and fauna (Table 9.8) (Langenbach 2013).

9.9 Factors Affecting Microbial Degradation

Microorganisms can degrade numerous organic pollutants owing to their metabolic machinery and to their capacity to adapt to inhospitable environments. Thus, microorganisms are major players in the site remediation. However, their efficiency depends on many factors, including the chemical nature and concentration of pollutants, their availability to microorganisms, and the physicochemical characteristics of the environment (El Fantroussi and Agathos 2005). Thus, factors that influence the rate of pollutant degradation by microorganisms are either related to the microorganisms and their nutritional requirements (biological factors) or associated with the environment (environmental factors).

9.9.1 Biological Factors

A biological factor is the metabolic ability of microorganisms. The biological factors that affect the microbial degradation of organic compounds include direct inhibition of enzymatic activities and the proliferation processes of degrading microorganisms (Riser-Roberts 1998). This inhibition can occur, for example, if there is competition between microorganisms for limited carbon sources, antagonistic interactions between microorganisms, or the predation of microorganisms by protozoa and bacteriophages (Environmental Response Division 1998). The rate of contaminant degradation is often dependent on the concentration of the contaminant and the amount of "catalyst" present. In this context, the amount of "catalyst" represents the number of organisms able to metabolize the contaminant as well as the number of enzymes(s) produced by each cell. Furthermore, the extent to which contaminants are metabolized is largely a function of the specific enzymes involved and their "affinity" for the contaminant and the availability of the contaminant. In addition, sufficient amounts of nutrients and oxygen must be available in a usable form and in proper proportions for unrestricted microbial growth to occur (Environmental Response Division 1998). Other factors that influence the rate of biodegradation by controlling the rates of enzyme-catalyzed reactions are temperature, pH, and moisture. Biological enzymes involved in the degradation pathway have an optimum temperature and will not have the same metabolic turnover for every temperature (Van der Heul 2011). Indeed, the rate of biodegradation is decreased by roughly one-half for each 10 °C decrease in temperature. Biodegradation can occur under a wide range of pH; however, a pH of 6.5–8.5 is generally optimal for biodegradation in most aquatic and terrestrial systems (Cases and Lorenzo 2005). Moisture influences the rate of contaminant metabolism because it influences the kind and amount of soluble materials that are available as well as the osmotic pressure and pH of terrestrial and aquatic systems (Cases and Lorenzo 2005).

9.9.2 Environmental Factors

Soil type and soil organic matter content affect the potential for adsorption of an organic compound to the surface of solids. Absorption is an analogous process wherein a contaminant penetrates the bulk mass of the soil matrix. Both adsorption and absorption reduce the availability of the contaminant to most microorganisms and the rate at which the chemical is metabolized is proportionately reduced (Environmental Response Division 1998). Variations in the porosity of the unsaturated and saturated zones of the aquifer matrix may influence the movement of fluids and contaminant migration in groundwater. The ability of the matrix to transmit gases, such as oxygen, methane, and carbon dioxide, is reduced in fine-grained sediments and also when soils become more saturated with water. This can affect the rate and type of biodegradation taking place (Environmental Response Division 1998). The oxidation–reduction potential of soil provides a measurement of the electron density of the system. Biological energy is obtained from the oxidation of compounds in which electrons are transferred to various more oxidized compounds referred to as electron acceptors. A low electron density (Eh greater than 50 mV) indicates oxidizing, aerobic conditions, whereas high electron density (Eh less than 50 mV) indicates reducing, anaerobic conditions (Environmental Response Division 1998).

9.10 Conclusion and Future Recommendations

The ecofriendly and cost-effective bioremediation technique is well thought out as one of the new methods for the remediation of polluted areas. In summary, there are numerous approaches utilized for organic contaminant removal and degradation such as physical, chemical, and biological processes, which have been reported in the literature. In this chapter, a critical, up-to-date, and comprehensive assessment has been described for the biological, microbial enzymatic, approach for organic pollutant degradation. The literature reviewed showed that microbial peroxidases are an adaptable group of enzymes that have much probability for organic pollutant degradation and wastewater remediation. Organic pollutants can be biodegraded by indigenous microorganisms. The action of microorganisms against different organic pollutants can be achieved by bioaccumulation, bioconcentration, bioaugmentation, and biodegradation, and depends on various influencing factors such as microorganism type and culture conditions, nutrient conditions, agitation time, pH, temperature, oxygen supply and transfer, and the presence of different organic contaminants. In this chapter, we discussed different microbial processes and their mechanisms and the degradation of organic contaminants in the aquatic system. Further research efforts are required to use microorganisms in environmental remediation in a more sustainable and economically feasible manner. Some future recommendations regarding enhanced remediation are described below:

- Researchers are familiar with the fact that the frequency of degradation for a composite mixture or recalcitrant pollutants by a single species of microorganism is very time-consuming. This can be made time effective and enhanced by using the varied mixture of local microorganisms present in the natural ecosystem.
- A novel method like the "biofilm system" has also been established to remediate the recalcitrant pollutants from wastewater. The application of "biofilm systems" for bioremediation purposes can be a future scope to work with on this topic.
- Application of nanotechnology for remediation of pollutants can be taken into consideration due to various advantages like (i) giving rise to sustainable products; (ii) remediation of hazardous substances; and (iii) development of biosensors to monitor environmental pollution. Dioxin can be absorbed by carbon nanotube more easily than activated carbon.
- Biostimulation of indigenous microorganisms and bioaugmentation through introducing desired microorganisms can be used as an alternative tool to satisfy bioremediation goals
- Literature studies have specified that biosorption studies of organics are very limited and only sorption of selected toxic organics onto a few types of bacterial, fungal, and yeast biomass have been investigated. There is a need to study other organic pollutants and to develop new strains that can be provided easily as waste and/or abundant biomass, or can grow in simple, inexpensive medium and have a high production rate, while possessing a high biosorption capacity.

References

Aksu, Z. and Dönmez, G. (2005). Combined effects of molasses sucrose and reactive dye on the growth and dye bioaccumulation properties of Candida tropicalis. *Process Biochemistry* 40 (7): 2443–2454.

Alaee, M., Arias, P., Sjödin, A., and Bergman, Å. (2003). An overview of commercially used brominated flame retardants, their applications, their use patterns in different countries/regions and possible modes of release. *Environment International* 29 (6): 683–689.

Antizar-Ladislao, B. (2010). Bioremediation: working with bacteria. *Elements* 6 (6): 389–394.

Arp, H.P.H., Møskeland, T., Andersson, P.L., and Nyholm, J.R. (2011). Presence and partitioning properties of the flame retardants pentabromotoluene, pentabromoethylbenzene and hexabromobenzene near suspected source zones in Norway. *Journal of Environmental Monitoring* 13 (3): 505–513.

Assens, M., Frederiksen, H., Petersen, J.H. et al. (2019). Variations in repeated serum concentrations of UV filters, phthalates, phenols and parabens during pregnancy. *Environment International* 123: 318–324.

Atteia, O. and Guillot, C. (2007). Factors controlling BTEX and chlorinated solvents plume length under natural attenuation conditions. *Journal of Contaminant Hydrology* 90 (1–2): 81–104.

Bakhtiar, S., Bhawani, S.A., and Shafqat, S.R. (2019). Synthesis and characterization of molecular imprinting polymer for the removal of 2-phenylphenol from spiked blood serum and river water. *Chemical and Biological Technologies in Agriculture* 6 (1): 1–10.

Bansal, A., Henao-Mejia, J., and Simmons, R.A. (2018). Immune system: an emerging player in mediating effects of endocrine disruptors on metabolic health. *Endocrinology* 159 (1): 32–45.

Baughman, G.L., Paris, D.F., and Hodson, R.E. (1981). Microbial bioconcentration of organic pollutants from aquatic systems—a critical review. *CRC Critical Reviews in Microbiology* 8 (3): 205–228.

Bello, U.M., Aire, T.A., Imam, J. et al. (2019). Dose-specific morphological changes in the Sertoli cell of the adult male Japanese quails testes exposed to di (n-butyl) phthalate DBP prepubertally. *The FASEB Journal* 33 (1_supplement): 802–878.

Benitez, F.J., Beltran-Heredia, J., Acero, J.L., and Gonzalez, T. (1996). Degradation of protocatechuic acid by two advanced oxidation processes: ozone/UV radiation and H_2O2UV radiation. *Water Research* 30 (7): 1597–1604.

Bennet, J.W., Wunch, K.G., and Faison, B.D. (2002). Use of fungi biodegradation. In: *Manual of Environmental Microbiology* (ed. C.J. Hurst), 960–997. Washington, DC: ASM Press.

Berner, R.A. (1981). A new geochemical classification of sedimentary environments. *Journal of Sedimentary Research* 51 (2): 359–365.

Betts, K.S. (2008). New thinking on flame retardants. *Environmental Health Perspectives* 116 (5): A210–A213.

Boas, M., Frederiksen, H., Feldt-Rasmussen, U. et al. (2010). Childhood exposure to phthalates: associations with thyroid function, insulin-like growth factor I, and growth. *Environmental Health Perspectives* 118 (10): 1458–1464.

Bodin, N., Tapie, N., Le Ménach, K. et al. (2014). PCB contamination in fish community from the Gironde Estuary (France): blast from the past. *Chemosphere* 98: 66–72.

Bridier, A., Briandet, R., Thomas, V., and Dubois-Brissonnet, F. (2011). Comparative biocidal activity of peracetic acid, benzalkonium chloride and ortho-phthalaldehyde on 77 bacterial strains. *Journal of Hospital Infection* 78 (3): 208–213.

Canonica, S. and Freiburghaus, M. (2001). Electron-rich phenols for probing the photochemical reactivity of freshwaters. *Environmental Science & Technology* 35 (4): 690–695.

Canonica, S., Jans, U.R.S., Stemmler, K., and Hoigne, J. (1995). Transformation kinetics of phenols in water: photosensitization by dissolved natural organic material and aromatic ketones. *Environmental Science & Technology* 29 (7): 1822–1831.

Canonica, S., Hellrung, B., and Wirz, J. (2000). Oxidation of phenols by triplet aromatic ketones in aqueous solution. *The Journal of Physical Chemistry A* 104 (6): 1226–1232.

Carberry, J.B. and Wik, J. (2001). Comparison of ex situ and in situ bioremediation of unsaturated soils contaminated by petroleum. *Journal of Environmental Science and Health, Part A* 36 (8): 1491–1503.

Cases, I. and Lorenzo, V.D. (2005). Genetically modified organisms for the environment: stories of success and failure and what we have learned from them. *International Microbiology* 8 (3): 213–222.

Castellote, M. and Bengtsson, N. (2011). Principles of TiO_2 photocatalysis. In: *Applications of Titanium Dioxide Photocatalysis to Construction Materials*, 5–10. Dordrecht: Springer.

Connell, D.W., Wu, R.S., Richardson, B.J., and Lam, P.K. (2009). Chemistry of organic pollutants, including agrochemicals. *Environmental and Ecological Chemistry-Volume III* 6: 181.

Cooper, C.M. (1993). Biological effects of agriculturally derived surface water pollutants on aquatic systems—a review. *Journal of Environmental Quality* 22 (3): 402–408.

Covaci, A., Harrad, S., Abdallah, M.A.E. et al. (2011). Novel brominated flame retardants: a review of their analysis, environmental fate and behaviour. *Environment International* 37 (2): 532–556.

Davis, J.W., Klier, N.J., and Carpenter, C.L. (1994). Natural biological attenuation of benzene in ground water beneath a manufacturing facility. *Groundwater* 32 (2): 215–226.

de Lasa, H., Serrano, B., and Salaices, M. (2005). Novel photocatalytic reactors for water and air treatment. In: *Photocatalytic Reaction Engineering*, 17–47. Boston, MA: Springer.

Decho, A.W. and Gutierrez, T. (2017). Microbial extracellular polymeric substances (EPSs) in ocean systems. *Frontiers in Microbiology* 8: 922.

Derylo-Marczewska, A., Blachnio, M., Marczewski, A.W. et al. (2017). Adsorption of chlorophenoxy pesticides on activated carbon with gradually removed external particle layers. *Chemical Engineering Journal* 308: 408–418.

Donaldson, W.T. (1977). Trace organics in water. *Environmental Science & Technology* 11 (4): 348–351.

Dong, H., Zeng, G., Tang, L. et al. (2015). An overview on limitations of TiO2-based particles for photocatalytic degradation of organic pollutants and the corresponding countermeasures. *Water Research* 79: 128–146.

Duran, N. and Esposito, E. (2000). Potential applications of oxidative enzymes and phenoloxidase-like compounds in wastewater and soil treatment: a review. *Applied Catalysis B: Environmental* 28 (2): 83–99.

CEAP, USDA. (2008). Effects of agricultural conservation practices on fish and wildlife: A Conservation Effects Assessment Project (CEAP) bibliography. Vol. 7. Washington, DC: Water Quality Information Center, National Agricultural Library, Agricultural Research Service, US Department of Agriculture.

El Fantroussi, S. and Agathos, S.N. (2005). Is bioaugmentation a feasible strategy for pollutant removal and site remediation? *Current Opinion in Microbiology* 8 (3): 268–275.

Environmental Response Division (1998). *Fundamental principles of bioremediation*. Seattle, WA: Environmental Response Division.

Fennell, D.E. (2019). Anaerobic Dechlorination of polychlorinated Dibenzo-p-dioxins in Passaic River sediments. For environmental chemists. *Environmental Science & Technology* 40 (23): 7157–7166.

Fox, M.A. (1988). Photocatalytic oxidation of organic substrates. In: *Photocatalysis and Environment*, 445–467. Dordrecht: Springer.

Fries, E. and Püttmann, W. (2003). Occurrence and behaviour of 4-nonylphenol in river water of Germany. *Journal of Environmental Monitoring* 5 (4): 598–603.

Fukahori, S., Ichiura, H., Kitaoka, T., and Tanaka, H. (2003). Photocatalytic decomposition of bisphenol A in water using composite TiO_2-zeolite sheets prepared by a papermaking technique. *Environmental Science & Technology* 37 (5): 1048–1051.

Garrison, A.W. and Keith, L.H. (2015). Occurrence, Registry, and Classification of Organic Pollutants in Water, with Development of a Master Scheme for their. *Aquatic Pollutants: Transformation and Biological Effects* 1: 39.

Geissen, V., Mol, H., Klumpp, E. et al. (2015). Emerging pollutants in the environment: a challenge for water resource management. *International Soil and Water Conservation Research* 3 (1): 57–65.

Grube, A., Donaldson, D., Kiely, T., and Wu, L. (2011). *Pesticides Industry Sales and Usage*. Washington, DC: US EPA.

Habibullah-Al-Mamun, M., Ahmed, M.K., Islam, M.S. et al. (2019). Occurrence, distribution and possible sources of polychlorinated biphenyls (PCBs) in the surface water from the Bay of Bengal coast of Bangladesh. *Ecotoxicology and Environmental Safety* 167: 450–458.

Halberstadt, A.L., van der Zee, J.V., Chatha, M. et al. (2019). Chronic treatment with a metabotropic mGlu2/3 receptor agonist diminishes behavioral response to a phenethylamine hallucinogen. *Psychopharmacology* 236 (2): 821–830.

Hättenschwiler, S. and Vitousek, P.M. (2000). The role of polyphenols in terrestrial ecosystem nutrient cycling. *Trends in Ecology & Evolution* 15 (6): 238–243.

Hoffmann, M.R., Martin, S.T., Choi, W., and Bahnemann, D.W. (1995). Environmental applications of semiconductor photocatalysis. *Chemical Reviews* 95 (1): 69–96.

Hou, S., Altarawneh, M., Kennedy, E.M. et al. (2019). Formation of polychlorinated dibenzo-p-dioxins and dibenzofurans (PCDD/F) from oxidation of 4, 4′-dichlorobiphenyl (4, 4′-DCB). *Proceedings of the Combustion Institute* 37 (1): 1075–1082.

Iglesias-González, A., Hardy, E.M., and Appenzeller, B.M. (2020). Cumulative exposure to organic pollutants of French children assessed by hair analysis. *Environment International* 134: 105332.

Islam, R., Kumar, S., Karmoker, J. et al. (2018). Pathways of persistent organic pollutants (POPs) bioaccumulation and its adverse effects on ecosystems and human exposure: A review study on Bangladesh perspectives. *Environmental Technology & Innovation* 12: 1–16.

Jaroenram, W., Cecere, P., and Pompa, P.P. (2019). Xylenol orange-based loop-mediated DNA isothermal amplification for sensitive naked-eye detection of Escherichia coli. *Journal of Microbiological Methods* 156: 9–14.

Johnson, B.T. and Kennedy, J.O. (1973). Biomagnification of p, p'-DDT and methoxychlor by bacteria. *Applied and Environmental Microbiology* 26 (1): 66–71.

Jorgenson, J.L. (2001). Aldrin and dieldrin: a review of research on their production, environmental deposition and fate, bioaccumulation, toxicology, and epidemiology in the United States. *Environmental Health Perspectives* 109 (suppl 1): 113–139.

Kabir, E.R., Rahman, M.S., and Rahman, I. (2015). A review on endocrine disruptors and their possible impacts on human health. *Environmental Toxicology and Pharmacology* 40 (1): 241–258.

Kanzari, F., Syakti, A.D., Asia, L. et al. (2014). Distributions and sources of persistent organic pollutants (aliphatic hydrocarbons, PAHs, PCBs and pesticides) in surface sediments of an industrialized urban river (Huveaune), France. *Science of the Total Environment* 478: 141–151.

Karigar, C.S. and Rao, S.S. (2011). Role of microbial enzymes in the bioremediation of pollutants: a review. *Enzyme Research* 2011.

Karlaganis, G., Marioni, R., Sieber, I., and Weber, A. (2001). The elaboration of the 'Stockholm convention' on persistent organic pollutants (POPs): a negotiation process fraught with obstacles and opportunities. *Environmental Science and Pollution Research* 8 (3): 216–221.

Katagi, T. (2010). Bioconcentration, bioaccumulation, and metabolism of pesticides in aquatic organisms. In: *Reviews of Environmental Contamination and Toxicology*, 1–132. New York, NY: Springer.

Kennedy, R.S. and Finnerty, W.R. (1975). Microbial assimilation of hydrocarbons. I. The fine-structure of a hydrocarbon oxidizing Acinetobacter sp. *Archives of Microbiology* 102 (2): 75–83.

Koenig, S., Huertas, D., and Fernández, P. (2013). Legacy and emergent persistent organic pollutants (POPs) in NW Mediterranean deep-sea organisms. *Science of the Total Environment* 443: 358–366.

La Merrill, M., Emond, C., Kim, M.J. et al. (2013). Toxicological function of adipose tissue: focus on persistent organic pollutants. *Environmental Health Perspectives* 121 (2): 162–169.

Langenbach, T. (2013). Persistence and bioaccumulation of persistent organic pollutants (POPs). In: *Applied Bioremediation-Active and Passive Approaches*, 307. Rijeka: InTech.

Laxminarayan, R., Matsoso, P., Pant, S. et al. (2016). Access to effective antimicrobials: a worldwide challenge. *The Lancet* 387 (10014): 168–175.

Lily, M.K., Bahuguna, A., Dangwal, K., and Garg, V. (2010). Optimization of an inducible, chromosomally encoded benzo [a] pyrene (BaP) degradation pathway in Bacillus subtilis BMT4i (MTCC 9447). *Annals of Microbiology* 60 (1): 51–58.

Ludvik, J., Musk, V., and Dostaler, I. (1968). Pltrastractiiral changes in the yeast Candida lipolytica caused by penetration of hydrocarbons into the-cell. *Experientia* 24: 1066–1068.

Luna, J.D.S., De Carvalho, J.M., De Lima, M.R.F. et al. (2006). Acetogenins in Annona muricata L.(Annonaceae) leaves are potent molluscicides. *Natural Product Research* 20 (3): 253–257.

Luo, Q., Liu, Z.H., Yin, H. et al. (2018). Migration and potential risk of trace phthalates in bottled water: a global situation. *Water Research* 147: 362–372.

Maier, G., Nimmo-Smith, R.J., Glegg, G.A. et al. (2009). Estuarine eutrophication in the UK: current incidence and future trends. *Aquatic Conservation: Marine and Freshwater Ecosystems* 19 (1): 43–56.

Mansouri, A., Cregut, M., Abbes, C. et al. (2017). The environmental issues of DDT pollution and bioremediation: a multidisciplinary review. *Applied Biochemistry and Biotechnology* 181 (1): 309–339.

Maszenan, A.M., Liu, Y., and Ng, W.J. (2011). Bioremediation of wastewaters with recalcitrant organic compounds and metals by aerobic granules. *Biotechnology Advances* 29 (1): 111–123.

Matsumoto, E., Kawanaka, Y., Yun, S.J., and Oyaizu, H. (2009). Bioremediation of the organochlorine pesticides, dieldrin and endrin, and their occurrence in the environment. *Applied Microbiology and Biotechnology* 84 (2): 205–216.

Mattes, T.E., Alexander, A.K., and Coleman, N.V. (2010). Aerobic biodegradation of the chloroethenes: pathways, enzymes, ecology, and evolution. *FEMS Microbiology Reviews* 34 (4): 445–475.

Mazzarino, I., Piccinini, P., and Spinelli, L. (1999). Degradation of organic pollutants in water by photochemical reactors. *Catalysis Today* 48 (1–4): 315–321.

Mourier, B., Desmet, M., Van Metre, P.C. et al. (2014). Historical records, sources, and spatial trends of PCBs along the Rhône River (France). *Science of the Total Environment* 476: 568–576.

Muir, D.C. and Howard, P.H. (2006). Are there other persistent organic pollutants? A challenge for environmental chemists. *Environmental Science & Technology* 40 (23): 7157–7166.

Mulligan, C.N. and Yong, R.N. (2004). Natural attenuation of contaminated soils. *Environment International* 30 (4): 587–601.

Mullin, C.A., Frazier, M., Frazier, J.L. et al. (2010). High levels of miticides and agrochemicals in North American apiaries: implications for honey bee health. *PLoS One* 5 (3): e9754.

Murali, K.I.V. and Valli, M. (2017). *Environmental Management: Science and Engineering for Industry*. Oxford: Butterworth-Heinemann.

Naik, P., Abdellah, I.M., Abdel-Shakour, M. et al. (2018). An efficient aniline-based co-sensitizer for high performance N3-sensitized solar cells. *ChemistrySelect* 3 (43): 12297–12302.

Newman, J., Aucompaugh, A.G., Schell, L.M. et al. (2006). PCBs and cognitive functioning of Mohawk adolescents. *Neurotoxicology and Teratology* 28 (4): 439–445.

Nikolopoulou, M. and Kalogerakis, N. (2008). Enhanced bioremediation of crude oil utilizing lipophilic fertilizers combined with biosurfactants and molasses. *Marine Pollution Bulletin* 56 (11): 1855–1861.

Noutsopoulos, C., Koumaki, E., Sarantopoulos, V., and Mamais, D. (2019). Analytical and mathematical assessment of emerging pollutants fate in a river system. *Journal of Hazardous Materials* 364: 48–58.

Noziglia, A.J., Abbott, J.K., Polidoro, B., and Gerber, L.R. (2018). Flame retardant contamination and seafood sustainability. *Sustainability* 10 (4): 1070.

O'Brien, D., Lewis, S., Gallen, C. et al. (2014). Barron River pesticide monitoring and Cairns WWTP WQ assessment. Cairns: Centre for Tropical Water & Aquatic Ecosystem Research (TropWATER) Publication, James Cook University, Cairns, Australia, 49.

Ortiz-Vazquez, H., Shin, J., Soto-Valdez, H., and Auras, R. (2011). Release of butylated hydroxytoluene (BHT) from Poly (lactic acid) films. *Polymer Testing* 30 (5): 463–471.

Osumi, M. (2012). Visualization of yeast cells by electron microscopy. *Journal of Electron Microscopy* 61 (6): 343–365.

Otte, M.P., Gagnon, J., Comeau, Y. et al. (1994). Activation of an indigenous microbial consortium for bioaugmentation of pentachlorophenol/creosate contaminated soils. *Applied Microbiology and Biotechnology* 40 (6): 926–932.

Pathak, U., Roy, A., Mandal, D.D. et al. (2019). Bioattenuation of phenol and cyanide involving immobilised spent tea activated carbon with Alcaligenes faecalis JF339228: critical assessment of the degraded intermediates. *Asia-Pacific Journal of Chemical Engineering* 14 (1): e2278.

Pavela, R. and Benelli, G. (2016). Essential oils as ecofriendly biopesticides? Challenges and constraints. *Trends in Plant Science* 21 (12): 1000–1007.

Lorz, P.M., Towae, F.K., Enke, W. et al. (2007). *Phthalic Acid and Derivatives. Ullmann's Encyclopedia of Industrial Chemistry*. Weinheim: Wiley-VCH Verlag GmbH.

Pinero-Bonilla, J. (2019). Approach to knowledge of the microbial bioprospecting in Venezuela. *Journal of Applied Biotechnology & Bioengineering* 6 (1): 1–14.

Qu, B., Mu, Z., Liu, Y. et al. (2020). The synthesis of porous ultrathin graphitic carbon nitride for the ultrasensitive fluorescence detection of 2, 4, 6-trinitrophenol in environmental water. *Environmental Science: Nano*.

Raherison, C., Baldi, I., Pouquet, M. et al. (2019). Pesticides exposure by air in vineyard rural area and respiratory health in children: a pilot study. *Environmental Research* 169: 189–195.

Ramakrishnan, B., Megharaj, M., Venkateswarlu, K. et al. (2010). The impacts of environmental pollutants on microalgae and cyanobacteria. *Critical Reviews in Environmental Science and Technology* 40 (8): 699–821.

Remberger, M., Sternbeck, J., Palm, A. et al. (2004). The environmental occurrence of hexabromocyclododecane in Sweden. *Chemosphere* 54 (1): 9–21.

Riser-Roberts, E. (1998). *Remediation of Petroleum Contaminated Soils: Biological, Physical, and Chemical Processes*. CRC press.

Roze, E., Meijer, L., Bakker, A. et al. (2009). Prenatal exposure to organohalogens, including brominated flame retardants, influences motor, cognitive, and behavioral performance at school age. *Environmental Health Perspectives* 117 (12): 1953–1958.

Saari, L.L. and Mauvais, C.J. (2018). Sulfonylurea herbicide-resistant crops. In: *Herbicide-Resistant Crops*, 127–142. CRC Press.

Salam, J.A. and Das, N. (2012). Remediation of lindane from environment-an overview. *International Journal of Advanced Biotechnology and Research* 2: 9–15.

Salam, L.B., Obayori, O.S., and Hawa, O. (2016). Hydrocarbon degradation and biosurfactant production by an acenaphthene-degrading pseudomonas species. *Soil and Sediment Contamination: An International Journal* 25 (8): 837–856.

Salazar-Beltrán, D., Hinojosa-Reyes, L., Maya-Alejandro, F. et al. (2019). Automated on-line monitoring of the TiO_2-based photocatalytic degradation of dimethyl phthalate and diethyl phthalate. *Photochemical & Photobiological Sciences* 18 (4): 863–870.

Schell, L.M., Gallo, M.V., Denham, M. et al. (2008). Relationship of thyroid hormone levels to levels of polychlorinated biphenyls, lead, p, p'-DDE, and other toxicants in Akwesasne Mohawk youth. *Environmental Health Perspectives* 116 (6): 806–813.

Schrader, K.K. (2003). Natural algicides for the control of cyanobacterial-related off-flavor in catfish aquaculture. In: *Off-Flavors in Aquaculture* (eds. A.M. Rimando and K.K. Schrader), 195–208. Washington, DC: ACS Publications.

Schwarzenbach, R.P., Escher, B.I., Fenner, K. et al. (2006). The challenge of micropollutants in aquatic systems. *Science* 313 (5790): 1072–1077.

Schwedler, G., Rucic, E., Lange, R. et al. (2020). Phthalate metabolites in urine of children and adolescents in Germany. Human biomonitoring results of the German environmental survey GerES V, 2014–2017. *International Journal of Hygiene and Environmental Health* 225: 113444.

Shannon, M.A., Bohn, P.W., Elimelech, M. et al. (2010). Science and technology for water purification in the coming decades. In: *Nanoscience and Technology: A Collection of Reviews from Nature Journals*, 337–346.

Shorey, H.H. (2013). *Animal Communication by Pheromones*. Academic Press.

Silva, M.J., Wong, L.Y., Preau, J.L. et al. (2019). Urinary concentrations of diisoheptyl phthalate biomarkers in convenience samples of US adults in 2000 and 2018–2019. *Toxics* 7 (4): 53.

Smets, B.F. and Pritchard, P.H. (2003). Elucidating the microbial component of natural attenuation. *Current Opinion in Biotechnology* 14 (3): 283–288.

Stefan, M.I., Hoy, A.R., and Bolton, J.R. (1996). Kinetics and mechanism of the degradation and mineralization of acetone in dilute aqueous solution sensitized by the UV photolysis of hydrogen peroxide. *Environmental Science & Technology* 30 (7): 2382–2390.

Strong, P.J. and Claus, H. (2011). Laccase: a review of its past and its future in bioremediation. *Critical Reviews in Environmental Science and Technology* 41 (4): 373–434.

Sugiura, M. and Futagawa, S. (1984). Esophageal transection with paraesophagogastric devascularizations (the Sugiura procedure) in the treatment of esophageal varices. *World Journal of Surgery* 8 (5): 673–679.

Sverko, E.D., Tomy, G.T., Marvin, C.H. et al. (2008). Dechlorane plus levels in sediment of the lower Great Lakes. *Environmental Science & Technology* 42 (2): 361–366.

Teil, M.J., Tlili, K., Blanchard, M. et al. (2012). Occurrence of polybrominated diphenyl ethers, polychlorinated biphenyls, and phthalates in freshwater fish from the Orge River (Ile-de France). *Archives of Environmental Contamination and Toxicology* 63 (1): 101–113.

Tripathi, P., Singh, P.C., Mishra, A. et al. (2013). Trichoderma: a potential bioremediator for environmental cleanup. *Clean Technologies and Environmental Policy* 15 (4): 541–550.

Umar, M. and Aziz, H.A. (2013). Photocatalytic degradation of organic pollutants in water. *Organic Pollutants-Monitoring, Risk and Treatment* 8: 196–197.

Van der Heul, R.M. (2011). Environmental degradation of petroleum hydrocarbons (Master's thesis).

Vandevivere, P. and Kirchman, D.L. (1993). Attachment stimulates exopolysaccharide synthesis by a bacterium. *Applied and Environmental Microbiology* 59 (10): 3280–3286.

Vasquez-Murrieta, M.S., Hernández-Hernández, O.J., Cruz-Maya, J.A. et al. (2016). Approaches for removal of PAHs in soils: Bioaugmentation, biostimulation and bioattenuation. In: *Soil Contamination-Current Consequences and Further Solutions*, 329–342. Rijeka, Croatia: InTech.

Wang, Y., Ren, H., Pan, H. et al. (2015). Enhanced tolerance and remediation to mixed contaminates of PCBs and 2, 4-DCP by transgenic alfalfa plants expressing the 2, 3-dihydroxybiphenyl-1, 2-dioxygenase. *Journal of Hazardous Materials* 286: 269–275.

Wang, M., Safe, S., Hearon, S.E., and Phillips, T.D. (2019). Strong adsorption of polychlorinated biphenyls by processed montmorillonite clays: potential applications as toxin enterosorbents during disasters and floods. *Environmental Pollution* 255: 113210.

Wang, Y., He, L., Li, Y. et al. (2020). Ag NPs supported on the magnetic Al-MOF/PDA as nanocatalyst for the removal of organic pollutants in water. *Journal of Alloys and Compounds*: 154340.

Watanabe, K. (2001). Microorganisms relevant to bioremediation. *Current Opinion in Biotechnology* 12 (3): 237–241.

Watt, B.E., Proudfoot, A.T., Bradberry, S.M., and Vale, J.A. (2005). Anticoagulant rodenticides. *Toxicological Reviews* 24 (4): 259–269.

Weber, R., Bell, L., Watson, A. et al. (2019). Assessment of pops contaminated sites and the need for stringent soil standards for food safety for the protection of human health. *Environmental Pollution* 249: 703–715.

Wikoff, D.S., Bennett, D.C., Brorby, G.P., and Franke, K.S. (2020). Evaluation of potential human health risk associated with consumption of edible products from livestock fed ration supplemented with Red Lake Diatomaceous Earth. *Food Additives & Contaminants: Part A*: 1–11.

Wilkinson, S., Nicklin, S., and Faull, J.L. (2002). Biodegradation of fuel oils and lubricants: soil and water bioremediation options. *Progress in Industrial Microbiology* 36: 69–100.

Wolfaardt, G.M., Lawrence, J.R., Robarts, R.D., and Caldwell, D.E. (1995). Bioaccumulation of the herbicide diclofop in extracellular polymers and its utilization by a biofilm community during starvation. *Applied and Environmental Microbiology* 61 (1): 152–158.

Xing, Y., Lu, Y., Dawson, R.W. et al. (2005). A spatial temporal assessment of pollution from PCBs in China. *Chemosphere* 60 (6): 731–739.

Zahm, S.H. and Ward, M.H. (1998). Pesticides and childhood cancer. *Environmental Health Perspectives* 106 (suppl 3): 893–908.

Zamkowska, D., Karwacka, A., Jurewicz, J., and Radwan, M. (2018). Environmental exposure to non-persistent endocrine disrupting chemicals and semen quality: An overview of the current epidemiological evidence. *International Journal of Occupational Medicine and Environmental Health* 31 (4): 377–414.

Zhang, H., Lu, Y., Dawson, R.W. et al. (2005). Classification and ordination of DDT and HCH in soil samples from the Guanting Reservoir, China. *Chemosphere* 60 (6): 762–769.

Zhang, T., Tremblay, P.L., Chaurasia, A.K. et al. (2013). Anaerobic benzene oxidation via phenol in Geobacter metallireducens. *Applied and Environmental Microbiology* 79 (24): 7800–7806.

Zhang, F., Zhang, J., Shi, N., and Li, T. (2019). Growth-promoting effects of low-level butyl benzyl phthalate exposure on human neuroblastoma SH-SY5Y cells. *Journal of Applied Toxicology* 39 (8): 1181–1191.

Zhou, L., Song, W., Chen, Z., and Yin, G. (2013). Degradation of organic pollutants in wastewater by bicarbonate-activated hydrogen peroxide with a supported cobalt catalyst. *Environmental Science & Technology* 47 (8): 3833–3839.

Part IV

Removal of Water Pollutants by Nanotechnology

10

Detection and Removal of Heavy Metals from Wastewater Using Nanomaterials

Swati Chaudhary[1], Mohan Kumar[2*], Saami Ahmed[3*], and Mahima Kaushik[4]*

[1]*Department of Applied Sciences, MSIT, GGSIP University, New Delhi, India*
[2]*Department of Chemistry, Shri Varshney College, Aligarh, Uttar Pradesh, India*
[3]*Department of Chemistry, Zakir Husain Delhi College, University of Delhi, New Delhi, India*
[4]*Nano-bioconjugate Chemistry Lab, Cluster Innovation Centre, University of Delhi, New Delhi, India*
All three authors have contributed equally to this manuscript.

10.1 Introduction

The huge rise in the world's population and accelerated advances in industries has resulted in a prevalent menace to human life (Luvita et al. 2019). At present, water pollution has been one of the major environmental issues worldwide. The predominant source of water pollution is the dumping of wastewater from industries related to chemicals, mining, electroplating, metallurgy, agriculture waste, etc. (Breida et al. 2019; Wu et al. 2020). Heavy metal pollution of water has become an imperative concern in countries like India and China due to their huge populations (Ding 2019). The presence of heavy metals in water resources, even in trace quantities, could be disastrous to human life. However, a few of them at very low concentrations are required for the better functioning of various biological processes in the human body, e.g. iron, cobalt, copper, manganese, and zinc (De Acha et al. 2019), however, at higher concentrations, they can exert toxic effects on human health. The discharge of heavy metals into water resources could be through anthropogenic and natural resources (Parvin et al. 2019). All heavy metals are nonbiodegradable and show an adverse effect because they can be concentrated in the food chain (Azimi et al. 2017). Many dangerous health problems such as kidney failure, anemia, nervous system disorders, hypertension, damage of lungs, and loss of memory could be due to the presence of heavy metals in water resources (Kim et al. 2015; Engwa et al. 2019; Kabeer et al. 2019). The US Environmental Protection Agency (USEPA) and the Bureau of Indian Standards (BIS) has set the maximum permissible limit of heavy metals in drinking water (Kumar and Saravanan 2017). Substantial attention has been given to remove heavy metals from several water resources in the last two decades. Diverse methods have been utilized for the detoxification of heavy metals such as electrochemical, chemical coagulants, photocatalytic oxidation, reverse osmosis, ion exchange resin, and

Pollutants and Water Management: Resources, Strategies and Scarcity, First Edition. Edited by Pardeep Singh, Rishikesh Singh, Vipin Kumar Singh, and Rahul Bhadouria.

membrane filtration (Wu et al. 2017; Bolisetty et al. 2019; Elkacmi and Bennajah 2019). Every method has its built-in superiority and constraints in its application.

The World Health Organization (WHO) has established several standard protocols such as ion exchange (Hubicki and Kołodyńska 2012), precipitation (Fu et al. 2012), reverse osmosis (Thaçi and Gashi 2019), oxidation (Barakat et al. 2013), and membrane filtration (Qdais and Moussa 2004) to remove effluents from wastewater. Yet, these traditional treatment strategies have some disadvantages, including corrosion problems, sludge formation, high solution pH sensitivity, and some processes that are not considered economically feasible, which hinders the removal of heavy metals from wastewater. To overcome these issues, the adsorption technique was introduced, which is considered as the most significant technique to remove even trace amounts of heavy metal ions from wastewater due to its high efficiency, simple operation, and low cost (Lakherwal 2014; Ray and Shipley 2015). Although heavy metals can be successfully removed from wastewater with the help of conventional sorbents, they show their efficiency only in concentrated solutions. For this reason, advancements in the development of adsorbents are needed for the effective removal of heavy metal ions from dilute solutions. The use of nanomaterials for the removal of heavy metals is the most novel advancement, which addresses all the issues faced by conventional sorbents (Wang et al. 2012). Potential nanoadsorbents include carbon, metal oxides, polymers, etc., which can efficiently work for pollution control in environmental processes and industrial applications. Nanoadsorbents should possess some properties such as nontoxicity, high adsorption capacities, high sensitivity toward low concentrations of heavy metals, easy removal of adsorbed metals from the surface of nanoadsorbents, and these must be recycled infinitely (Wang et al. 2012). Metal-oxide nanomaterials studied so far include iron oxide, titanium oxide, cerium oxide, and zinc oxide, etc. They have some limitations of being unstable at a very small size, which inhibits their ability to adsorb heavy metals from wastewater (Thekkudan et al. 2016). However, carbon-based nanomaterials have been widely used as functional adsorbents in numerous applications. They contain high-energy binding sites, which are a significant and useful characteristic for adsorption (Tian et al. 2012). Other classes of nanoadsorbents include dendrimers and nanocomposites, which have shown high adsorption potential and are discussed later in this chapter.

10.2 Sources of Heavy Metals in Wastewater

Generally, the concentration of heavy metals in water resources is very low, however, they can still be poisonous to human life. A few heavy metals that could be toxic to human health are nickel, mercury, cadmium, arsenic, cobalt, and chromium (Basheer et al. 2020). A variety of human activities along with natural disasters are responsible for heavy metal water pollution. To fulfill the required daily need of the world, human activities have furnished more contamination to water resources. Natural sources such as volcanic eruptions, forest fires, weathering of rocks, sea salt sprays, and biogenic sources have also increased water pollution (Masindi and Muedi 2018). The existence of heavy metals in the environment can be found in the form of oxides, sulfides, hydroxides, sulfates, and phosphates. Various anthropogenic processes are responsible for the same, including industrial waste, agricultural waste, mining, and metallurgical processes. The prime source of heavy metal pollution of natural water resources is the mining industry (Wright and Ryan 2016; Ochieng et al. 2017). To extract metals from their ore, the use of an acid mine drainage system is very prevalent, as the acid solution can dissolve metals very well (Basir et al. 2019). Usually, the acid solution that contains a high level of metals is dispersed into the river after finishing with the drainage process. A few other sources of water pollution are fertilizers, lead-acid batteries, paint products, and microplastics (Sharma and Bhattacharya 2017).

10.3 Toxicity of Heavy Metals Concerning Human Health and Environment

The term heavy metal refers to the atomic weight and toxicity to the environment and human health. Heavy metals have a relatively higher density than water. Few heavy metals have the capacity to induce toxicity, even at minute concentrations, and they are lethal to human health in different forms and dosages. Generally, toxic heavy metals include Cd, Pb, Zn, Cr, Hg, and As. Hg is highly toxic in its elemental form (Hg^0), while the others are more dangerous in their cation forms (Pb^{2+}, Cd^{2+}, As^{3+}). This is because they become soluble in water in their ionic forms and are more easily transported in the body. They are typically involved in several biological processes when present in very low concentration, however, they become problematic when they exceed their permissible limit. They can also mimic other elements, and thus, interfere with several biological and metabolic processes (Jaishankar et al. 2014). The permissible limit of certain heavy metals in wastewater is presented in Table 10.1 (Alalwan et al. 2020). Hg is considered the most toxic heavy metal in the environment. Industries such as paper, pulp preservation, agriculture, and pharmaceuticals are responsible for Hg pollution (Morais et al. 2012). Several forms of Hg, including inorganic Hg (Hg^{1+}, Hg^{2+}), elemental Hg (Hg^0), and organic Hg compounds (MeHg), induce modifications in body tissues and cause adverse health problems. A disease known as acrodynia is caused due to Hg poisoning (Alina et al. 2012). Exposure to increased levels of Hg leads to the deterioration of kidneys, brain, lungs, memory loss, hair loss, and skin rashes. Hg depletes cellular

Table 10.1 Permissible limit of heavy metals according to the World Health Organization (WHO).

S. no	Heavy metal	Permissible limit in mg/l according to the WHO
1	As	0.01
2	Cd	0.05
3	Cu	1.0
4	Cr	0.05
5	Pb	0.005
6	Hg	0.001
7	Ni	0.02
8	Zn	5.0

antioxidants due to the formation of covalent bonds with cysteine residues once it enters inside the cell in the form of MeHg or Hg^{2+}. Reactive oxygen species (ROS) are accumulated, causing oxidative stress due to the interaction of Hg compounds (Valko et al. 2006; Tchounwou et al. 2012). Lead (Pb) is declared as a carcinogenic element by almost all environmental protection agencies (EPAs). Lead pollution causes birth defects such as allergies, autism, paralysis, dyslexia, mental retardation, kidney, brain damage, etc. (Baby et al. 2019). It also causes a decrease in sperm count, spontaneous abortions, and adversely affects vitamin D metabolism (Tchounwou et al. 2012). Cadmium (Cd) is usually present in small proportions in the environment, but its concentration has now increased due to industrial wastes. It occupies the seventh position in the list of most toxic heavy metals by the Agency for Toxic Substances and Disease Registry (ATSDR). Kidneys, skeleton, and respiratory systems are damaged due to Cd-induced toxicity (Jaishankar et al. 2014). Two forms of arsenic (As), i.e. arsenate and arsenite, are considered fatal to the environment and human beings. As toxicity mainly depends on the amount of exposure, frequency, and duration. It also depends on the age and gender of biological species, individual susceptibilities, and genetic factors (Tchounwou et al. 2004). They interact with the sulfhydryl group of the cells, which disturbs the protoplasm of the cells, thus causing malfunctioning in the respiration process and affecting mitosis and cell enzymes (Jaishankar et al. 2014). The toxicity of chromium (Cr) depends on its oxidation state, for instance, Cr (III) is an essential element for humans and animals, whereas Cr (VI) is extremely toxic and carcinogenic in nature (ATSDR 2012). It is an industrial pollutant and strong oxidizing agent, which, as a result, becomes corrosive in nature. The reduction of Cr (VI) can be caused by glutathione (GSH), hydrogen peroxide (H_2O_2), and ascorbic acid, which produces reactive intermediates such as Cr (IV), Cr (V), hydroxyl radicals, thiyl radicals, and finally, Cr (III). These reactive species can further attack lipid membranes, DNA, and proteins. This attack is responsible for the disruption of various cellular functions and integrity (O'Brien et al. 2003). Excessive absorption of zinc (Zn) can inhibit the absorption of iron (Fe) from being an essential trace element. Zn ions show high toxicity against plants, fish, vertebrates, invertebrates, etc. (Baby et al. 2019).

10.4 Role of Nanomaterials in the Removal of Heavy Metals

Nanomaterials have found a large variety of applications in the field of healthcare, electronic devices, energy, medicine, environmental protection, water purification, etc. in the last few decades. Nanomaterials possess unique properties such as small size, large surface area, customized functionalities, etc., which contribute extensively to their significant adsorption potential and reactivity. The extraordinary adsorption potential is responsible for the use of nanomaterials in the removal of heavy metals from wastewater. Various strategies have been developed to investigate the applications of nanomaterials for the removal of heavy metal ions. Different types of nanomaterials exhibiting distinctive characteristics have been employed for water treatment. The role of some nanomaterials, such as metal-based nanomaterials, carbon-based nanomaterials, dendrimers, and nanocomposites in heavy metal ion removal, are discussed in detail in the following sections.

10.5 Different Types of Nanoadsorbents

10.5.1 Carbon-Based Nanomaterials

Carbon nanomaterials found their application in the field of electronics, energy storage, drug delivery, sensors, diagnostics, and water purification due to their remarkable thermal and electrical properties, large surface area, facile modifications in their physical and chemical properties, and their potential to treat wastewater (Smith and Rodrigues 2015). They generally have high sorption capabilities. Carbon-based nanomaterials are generally available in the form of hollow spheres, tubes, and ellipsoids. Carbon-based cylindrical tubes are referred to as nanotubes, whereas spherical and ellipsoidal materials are called fullerenes. They also include graphene, graphene oxides, and activated carbons. Carbon nanotubes have the ability to form π-π electrostatic interactions, due to which they can effectively remove heavy metal pollutants from wastewater (Gangadhar et al. 2012). Fullerenes and quantum dots are considered zero-dimensional (0-D) nanomaterials because all three dimensions are less than 100 nm. On the other hand, carbon nanotubes and graphene are termed as one-dimensional (1-D) and two-dimensional (2-D) nanomaterials, respectively. This is due to two dimensions of carbon nanotubes and only one dimension of graphene is less than 100 nm. Similarly, graphite and other carbon nanomaterials have all the dimensions greater than 100 nm, due to which they are termed as three-dimensional (3-D) nanomaterials (Baby et al. 2019). At first, activated carbon was used as sorbent but it becomes difficult to remove heavy metals from activated carbon at parts per billion (ppb) levels. Later on, carbon nanotubes, graphene, and fullerene were developed and employed as nanoadsorbents (Wang et al. 2012).

10.5.1.1 Carbon Nanotubes

Carbon nanotubes (CNTs) were discovered by Sumio Iijima and widely used to remove heavy metals from wastewater due to their distinctive electronic, structural, chemical, physical, mechanical, semiconductor, and optoelectronic properties (Gupta et al. 2016). Initially, CNTs showed their effective adsorption properties for divalent metal ions. They constitute carbon molecules in a cylindrical shape. CNTs possess much higher adsorption potential as compared with that of activated carbon, which is due to their high surface area that ensures a strong interaction between CNTs and metal ions. Heavy metals such as lead (Pb), zinc (Zn), cadmium (Cd), copper (Cu), and nickel (Ni) can effectively be removed by CNTs (Mubarak et al. 2014). Depending on the number of layers present in each tube of CNTs, they are divided into two categories: single-walled CNTs (SWCNTs) and multi-walled CNTs (MWCNTs). SWCNTs comprise only a single graphite sheet of carbon atoms, whereas MWCNTs comprise multilayer graphene sheets rolled with dozens of concentric tubes. CNTs are considered better adsorbents for divalent metal ions present in water due to their ability to form π-π electrostatic interactions with the metal ion and also due to a large surface area. They exhibit high adsorption capacities for several heavy metal ions other than Hg (II). They can be effectively used for the removal of Pb (II), Cd (II), Cr (VI), and Cu (II) from wastewater. Hg (II) was effectively removed from aqueous solutions by MWCNTs. The adsorption rate was found to be highly affected by pH, contact time, and the initial concentration of the metal ion (Yaghmaeian et al. 2015). The probable adsorption active sites of CNTs majorly constitute internal sites, interstitial channels, outside surface, and external groove sites. Functional groups such as —OH, —NH_2, and —COOH are generally introduced on the surface of CNTs to improve the adsorption potential of CNTs toward heavy metals. These groups are added by chemical modification, endohedral filling, and heat treatment (Kumar et al. 2014). Few reports suggest that oxidants like H_2SO_4, HNO_3, $KMnO_4$, and NaOCl significantly enhance the adsorption capacity of CNTs by modifying their surface (El-Sheikh et al. 2011). The unique functionalized CNTs were prepared by sonicating the pre-oxidized CNTs with deep eutectic solvent, which was prepared by using glycerol and allyltriphenylphosphonium. This modified CNT was used to remove Hg (II) from wastewater (AlOmar et al. 2017; Xu et al. 2018). To make better nanocomposites, CNTs are combined with other compounds such as the preparation of amino-functionalized Fe_3O_4/carboxylic MWCNTs hybrid, which shows a high adsorption capacity for Cu (II) (Zhan et al. 2016). Even though CNTs could effectively remove heavy metals from wastewater, there are still some problems associated with them, which include high cost, separation of CNTs from wastewater after adsorption, and the risk of increasing pollution (Yang et al. 2019).

10.5.1.2 Fullerenes

Since 1985, fullerenes have been considered the third allotropic form of carbon. They comprise 12 pentagonal rings and an unspecified number of six-membered rings in a cage-like structure. For instance, C_{60} constitutes 32 faces with 12 pentagons and 20 hexagons of carbon atoms arranged at the 60 vertices of a truncated icosahedron (Shan et al. 2017). Fullerenes possess significant physicochemical properties such as high electron affinity, hydrophobic character, surface defects, and a high surface to volume ratio (Mohan Gokhale and Ravindra Somani 2015). Due to the defects/spaces present in the fullerenes, adsorbates penetrate these spaces and are adsorbed. Fullerenes are successfully used for the adsorption of heavy metals due to their unique properties (Kaneko et al. 1993; Baby et al. 2019). They show greater efficiency for the removal of Cu (II) as compared with nanocomposite-polystyrene (Alekseeva et al. 2016). The adsorption capacity of fullerenes is generally attributed to its surface defects. It is presumed that the Cu (II) ion penetrates the lattice of fullerenes for adsorption. In 1992, Rowe et al. performed spectroscopic and X-ray absorption experiments, which revealed that Cu (II) strongly adsorbs on the C60 surface through a charge transfer from Cu (4s) to C(2p) of C60. They also showed that Cu atoms chemically modify the environment of the interface of the C60 molecule (Rowe et al. 1992).

10.5.1.3 Graphene and Graphene Oxide

Graphene is the thinnest material known, consisting of a thick layered hexagonal lattice of carbon atoms. Graphene is extensively used in almost every field due to its 200 times more strength than steel. These have wide and significant properties such as high surface area, electronic properties, high mechanical strength, thermal mobility, and modified surface chemistry (Aïssa et al. 2015; Mohan et al. 2018). Graphene-based nanomaterials have emerged as a promising precursor in water treatment due to their unique and intriguing properties. Recently, Tabish et al. developed porous graphene, which showed 80% efficiency in the removal of As^{3+} from water (Tabish et al. 2018).

Graphene oxide (GO) is a 2-D sheet with a large number of oxygenated functional groups such as carbonyl, epoxide, hydroxyl, and carboxyl on their basal plane and edges

(Kim et al. 2012). GO becomes highly water dispersible in nature due to the breakage of the π-conjugation network, which occurs during the fabrication of GO with other compounds. The adsorption potential of GO can be increased by reducing the aggregation between individual GO sheets (Chowdhury and Balasubramanian 2014). Numerous metal oxides and polymers were integrated into the graphene sheets, which resulted in nanocomposites and exhibited multifunctional properties and outstanding performance (Shan et al. 2017). Reduced GO (RGO) contains more defects as compared with graphene and can be easily altered by functional groups including —COOH, —OH, etc. Graphene-based nanomaterials have been widely used to remove heavy metals from wastewater as shown in Figure 10.1. A nanocomposite produced by fabricating partially RGO with Fe_3O_4 shows exceptional potential in removing Pb^{2+} from water (Guo et al. 2018). Similarly, numerous heavy metal ions such as Pb^{2+}, Ni^{2+}, Cr^{3+}, Cd^{2+}, and Cu^{2+} are effectively removed by modified GO, which was produced by functionalizing the RGO with 4-sulfophenylazo compound (Zhang et al. 2018). A graphene-based self-propelled microbot system was developed consisting of nanosized multilayers of GO, Ni, and Pt, which can successfully remove 80% of Pb^{2+} from water (Vilela et al. 2016). A high absorption capacity of 56.49 mg/g was observed for the absorption of Mn^{2+} by a combination of graphene oxide and sodium alginate (GO-SA) hydrogen beads (Yang et al. 2018). Another nanocomposite containing zinc oxide and tea polyphenol with reduced GO (TPG-ZnO) was also designed for the removal of heavy metal ions. This GO-nanomaterial exhibited 98.9% adsorption efficiency in the removal of Pb^{2+} ions from water and also showed antibacterial properties against *Streptococcus mutans* gram-positive bacteria (Zheng et al. 2018). Although graphene-based nanomaterials found extensive applications in the removal of heavy metals, it is difficult to separate them from

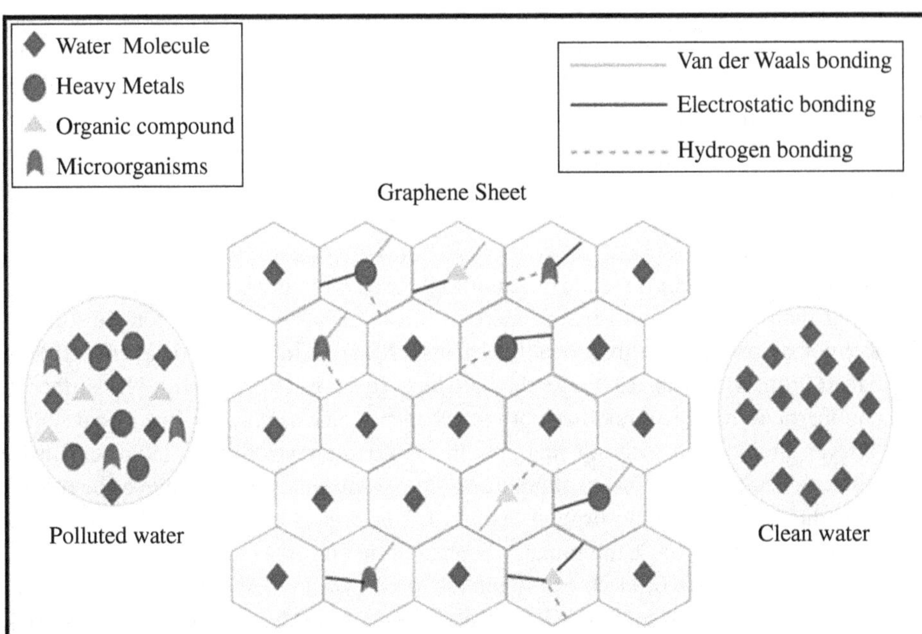

Figure 10.1 Removal of heavy metals from wastewater by using graphene-based nanomaterials.

aqueous solution due to the high water dispersible of GO (Yang et al. 2019). To overcome this problem, a modified graphene-based adsorbent was designed by embedding calcium alginate beads into GO, which was further reduced by polyethyleneimine. This modified adsorbent further enhanced the adsorption capacity for heavy metals (Arshad et al. 2019).

10.5.1.4 Activated Carbon

Activated carbon is also referred to as activated charcoal produced from coal, agricultural waste, and wood, and has a high surface area. Activated carbon also has high mechanical strength, due to which it finds wide application in capacitors, catalyst support, gas storage, electrodes, and more significantly, they are used as an adsorbent for the removal of heavy metal ions from water (Baby et al. 2019). The source of carbon, in particular, activated carbon, affects its adsorption potential. Activated carbon can be produced either by physical or chemical activation or by a combination of both. Its adsorption properties vary with the raw material and the process of activation. The quality of activated carbon depends on the type of biomass, final activation temperature impregnation rate of the chemical agent, and heating rate. Thus, it can be inferred that the type of raw material and the experimental conditions used for the preparation of activated carbon forms the basis of its adsorption properties and removal efficiency (Yeganeh et al. 2006). An efficiency of 92% was observed in the removal of Al^{3+} and Zn^{2+} ions from water by activated carbon, which was generated from apricot stone (El-Saharty et al. 2018). Another activated carbon, developed from sewage sludge was applied for the removal of Cu^{2+} and exhibited 50% adsorption capacity (Salihi et al. 2018). Sulfonated activated carbon was utilized for the removal of several heavy metal ions including Ni^{2+}, Cd^{2+}, Pb^{2+}, and Cu^{2+} (Li et al. 2018a,b). Agricultural waste of long-root *Eichhornia crassipes* was used to manufacture multipore activated carbon (MPAC) with a large surface area. This modified activated carbon was found efficient in the removal of heavy metal ions such as Zn^{2+}, Pb^{2+}, Cu^{2+}, Cd^{2+}, and Ni^{2+} (Cao et al. 2019). The synthesis and modification of activated carbon is easy, facile, and cheap. Also, they show higher adsorption potential for heavy metal ions.

10.5.2 Metal-Based Nanomaterials

10.5.2.1 Nanogold

Gold is considered a precious metal and is normally used for making jewelry and ornaments because of their shiny and anti-rust properties. But it has also been used for the study of nanotechnology-based water treatment (Shah et al. 2014). Gold nanoparticles (AuNPs) are less than 100 nm, which makes it a good tool to be used for water treatment. The surface of gold nanoparticles has been modified with many agents such as polyethylene glycol (PEG), lipid bilayer, and protein coating that has increased their stability in biological fluids (Guerrini et al. 2018). The removal of pollutant Hg^{2+} from water can be achieved by the use of AuNPs due to their high adsorption feature. It has already been very well studied that AuNPs having a size of 10–20 nm range, when supported on Al_2O_3, could not adsorb Hg^{2+} very well due to the lack of electrostatic interaction. But if the Hg^{2+} present in the water sample was reduced to Hg metal (Hg (II) to Hg (0)) by using a reducing agent such as $NaBH_4$, the adsorption of the Hg metal was potentially enhanced (Yang et al. 2019). The reduced Hg metal could adsorb very well in the form of Au_3Hg NPs which are present in amalgam form.

To regenerate the sorbent material, only 5% hydrochloric acid solution is required, which confirmed that the Hg metal adsorbed on the surface of the material. The filter membrane is commonly used for water treatment, which is still under study to enhance their efficacy and making them a cost-effective tool. The deposition of the membrane by various metal NPs has emerged as a very efficient technique and is being used at present (Li et al. 2019; Ye et al. 2019). The cost of water treatment using AuNPs after the reduction of the metal ion is very low along with the potential recovery of Au nanoparticles. Another study has been done by Ojea-Jimenez and their coworkers, where they have used citrate-coated AuNPs to adsorb Hg^{2+} from the water sample (Ojea-Jiménez et al. 2012). The Hg metal in a zero oxidation state was adsorbed in the form of Au_3Hg on the surface of AuNPs and could be easily removed from the water sample by using centrifugation. The membrane can be coated with AuNPs for its further application in water treatment. The efficiency of membrane coated AuNPs has been found to be much higher than the traditional membrane. The membrane can be coated with AuNPs by immersing it into a nanoparticle solution and then leaving it overnight. By using membrane coated AuNPs, the flux efficiency has also been increased significantly (Bharathi et al. 2016). Organohalides present in the water sample can also be effectively removed by using AuNPs. Bindhu and Umadevi prepared gold nanoparticles by a green method, where *Solanum lycopersicum* extract has played the role of reducing agent (Bindhu and Umadevi 2014a). The prepared nanoparticles were of 14 nm size range and have shown wonderful water purification. In another study, the same group has prepared AuNPs with *Ananas comosus* extract as a reducing agent. It can also be used as a potent antimicrobial agent (Bindhu and Umadevi 2014b). This method can be vital for the purification of the environmental water sample. Figure 10.2 depicts the removal of heavy metal ions by adsorption on gold NPs.

10.5.2.2 Nano Silver

Silver nanoparticles have also been explored for their applicability in removing heavy metal ions from wastewater samples. Several reports are available, which have explored the interaction of Hg (II) metal ions with silver nanoparticles (Morris et al. 2002; Ramana et al. 2013; Cheng et al. 2014). Sumesh et al. synthesized silver nanoparticles protected by

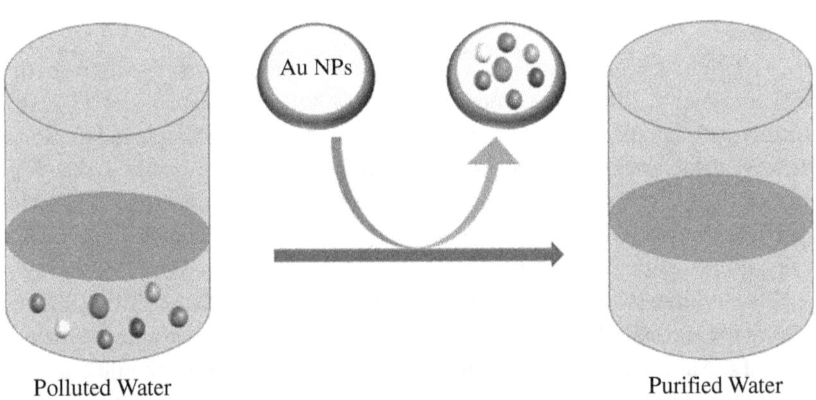

Figure 10.2 Heavy metal removal by adsorption on gold nanoparticles by adsorption method.

mercaptosuccinic acid (MSA) and supported on the surface of the alumina for the detoxification of the mercuric metal ion in the polluted water sample. Two different types of nanoparticles were prepared with a ratio of 1 : 6 and 1 : 3 Ag: MSA. The efficacy of 1 : 6 Ag: MSA was found to be 0.8 g of mercury per gram of Ag@MSA (Sumesh et al. 2011). The separation of Hg (II) and Ni (II) is much faster in comparison with metal ions such as Cr (III), Co (II), and Pb (II). Silver nanoparticles were dispersed on the surface of the carbon nanotubes for the efficient removal of Cu (II) and Cd (II) at an optimum pH of 6.0 and 7.0, respectively (Ramana et al. 2013). The silver nanoparticles have been studied for their spectacular properties such as ease of separation, catalytic activity, high adsorption activity, reusability, and surface modification (Ali et al. 2018).

10.5.2.3 Titanium Oxide

The use of titanium oxide (TiO_2) for the removal of heavy metals from aqueous solutions has gained much attention in the last few years due to its exclusive properties in comparison with other metal oxides. It has been observed that chemical behavior and catalytic activity is due to its different surface planes (Xiong et al. 2012). When titanium oxide is used in combination with zirconium, the adsorbent capacity is enhanced due to the change in the surface properties (Chen et al. 2012). Ahmed et al. performed a comparative analysis of heavy metal removal (Pb^{2+}, Cu^{2+}, Fe^{3+}, Cd^{2+}, and Zn^{2+}) from wastewater samples by synthesizing TiO_2 nanowires hydrothermally. The large surface area and good magnetic properties provided by the titanium nanowires make them a superior choice. The adsorption of lead was found to be the maximum, while zinc absorbed the least (Youssef and Malhat 2014). The comparative study of heavy metals (Cr, Cd, Pb) removal by using titanium oxide and wetland technique was performed and it was found that TiO_2 nanoparticles were able to reduce 90–96% impurities of heavy metals (Sudarsan et al. 2015). Peng et al. recently came up with mesoporous phosphated titanium oxide (PTO) for the specific removal of chromium ions from polluted water (Peng et al. 2020). They have synthesized 8-PTO (using 8% molar ratio phosphoric acid) having a surface area of 278 m^2/g with a very minor distribution size of less than 5 nm. The modified titanium oxide has shown outstanding adsorption capacity up to 92 mg/g in polluted water, which is much higher in comparison (10–83 mg/g) to TiO_2 alone.

10.5.2.4 Iron Oxide

Iron oxide (FeO) NPs are well-renowned adsorbents for the removal of toxic metals from the water sample (Parham et al. 2012; Dave and Chopda 2014; Martínez-Cabanas et al. 2016). This method is relatively inexpensive in comparison with other methods employed for water treatment. The easy detachment of the adsorbents from the system is only due to the magnetic properties of the FeO NPs that facilitate their reuse again (Guardia et al. 2011). The FeO NPs have always been a favorite choice for water treatment owing to their high surface area, small size, and magnetic property (Saif et al. 2016; Nizamuddin et al. 2019). The synthesis of FeO NPs has already been reviewed very well (Hasany et al. 2012, 2013; Campos et al. 2015). The shape and size of FeO NPs can be modified by changing the antecedent and the temperature of the system. The modification in the surface of FeO NPs could enhance the efficiency and stability of the nanomaterial for the decontamination of water

(Zhu et al. 2018). The surface modification of FeO NPs with the hydroxyl group offers a diverse synthetic tool to attach various functional groups (Wu et al. 2008). The modification of FeO NPs with fly ash could be used for the manufacture of cementitious (cement-based composites) composite that is used for the removal of heavy metals such as Pb, Cu, Cr, and Cd from water samples (Maiti et al. 2018). Long et al. worked on ultrafine mesoporous magnetite (Fe_3O_4) nanoparticles, which were used for the removal of a mixture of heavy metals from river water, which is difficult due to their adsorption (Fato et al. 2019). The iron nanoparticles modified with rice husk have been tested for the efficient removal of arsenic metal from a polluted water sample, where it could remove 95% of the arsenic (Pillai et al. 2020). Chatterjee and their coworkers have recently reported the manufacture of superparamagnetic nanoparticles by using fungus *Aspergillus Niger* BSC-1 for the detoxification of hexavalent chromium Cr (VI) metal ion (Chatterjee et al. 2020). Iron nanoparticles supported on hyperbranched polyglycerol polymer was used for the very fast adsorption of a mixture of heavy metals from aqueous solution. However, it was found that the efficiency of the adsorption power of nanoparticles was dependent on the concentration of nitrogen in the industrial wastewater (Almomani et al. 2020). Quite recently, Wang et al. fabricated molybdenum disulfide modified iron oxide nanoparticles (MoS_2/Fe_3O_4) that could easily remove Pb and Hg metal ions from water and soil samples (Wang et al. 2020).

10.5.2.5 Aluminum Oxide

Aluminum oxide (Al_2O_3) nanoparticles have emerged as a significant adsorbent for heavy metal ions such as Cd, Zn, and many more because of their high surface area volume ratio, low-cost material, and high decontamination power (Mahdavi et al. 2013; Jacukowicz-Sobala et al. 2015). These can exist in various crystalline forms, out of which γ-Al_2O_3 is widely used for water detoxification. The standard size of the Al_2O_3 nanoparticles was found to be 6–13 nm and their heavy metal removal efficacy was noted to be 97% and 87% for Pb and Cd, respectively. These are also used for the detoxification of various heavy metals such as Ni (II), Cr (VI), Hg (II), As (III), and Cu (II) with satisfactory capacity (Yang et al. 2019). The chemical modification of γ-Al_2O_3 NPs with various functional groups consisting of donor atoms has increased the adsorptive power comprehensibly. After the addition of a chemical to the alumina nanoparticles, heavy metals are excluded from water by not only the adsorption method but via chemical interaction as well. The efficiency of alumina nanoparticles has been remarkably enhanced in the presence of phosphate and humic acid (Stietiya and Wang 2014). In another study, nano γ-Al_2O_3 NPs were synthesized and their efficacy for the removal of heavy metals such as Cr (VI), Ni (II), Cd (II), and Pb (II) was investigated. The maximum adsorption capacity of nanomaterials for the heavy metals was found to be 13.3, 6.0, 1.1, and 0.33 (mg/g) for Cr (VI), Pb (II), Cd (II), and Ni (II), respectively (Poursani et al. 2015). It was also found that γ-Al_2O_3 NPs have provided much more surface area in comparison with α-Al_2O_3. Xia et al. have worked with thiol functionalized γ-aluminum oxide hydroxide (γ-AlOOH) and established that it could be used for the efficient removal of heavy metals such as Hg (II), Pb (II), and Cd (II) from wastewater samples (Xia et al. 2017). The synthesis of γ-Al_2O_3 by using the sol–gel method was done by Tabesh et al. and the heavy metal adsorption was tested for Pb (II) and Cd (II) metal ions (Tabesh et al. 2018).

10.5.3 Dendrimers

Dendritic is a highly branched three-dimensional polymer that can be classified as dendrimers, multi-arm star polymers, dendritic-linear block polymers, hypergrafted polymers, hyperbranched polymers, and dendronized polymers based on structural differences (Figure 10.3) (Fréchet and Tomalia 2001; Ma et al. 2016). For water purification, dendrimers are the most widely used class. Some authors often call further subgroups of dendritic polymers as "dendrimer." Dendrimers are three-dimensional synthetic polymers having wide-ranging reactive terminal groups (Shahbazi et al. 2011; Han et al. 2012). Composites are the unique properties bearing systems, which arise due to the combination of two or more materials having different properties. Whereas composite dendrimers are formed by synthetic dendritic polymers that are in coexistence with either organic or inorganic units. Dendrimers have a well-defined ordered structure of a monodisperse macromolecule. Due to the multifunctional framework and internal void, the dendrimer's end reaction occurs between the end and other molecules at the nanoscale. They are forecast to be an excellent adsorbent for the removal of water contamination. To figure out the role of adsorbents, it is mandatory to investigate its structure. The dendrimers compose three fundamental elements: (i) core part; (ii) repeating branching in the middle; (iii) and branching at the outer surface containing functional groups. In between these branches, pollutants are trapped inside the voids (Vunain et al. 2016). The following reasons make the dendrimers an excellent adsorbent:

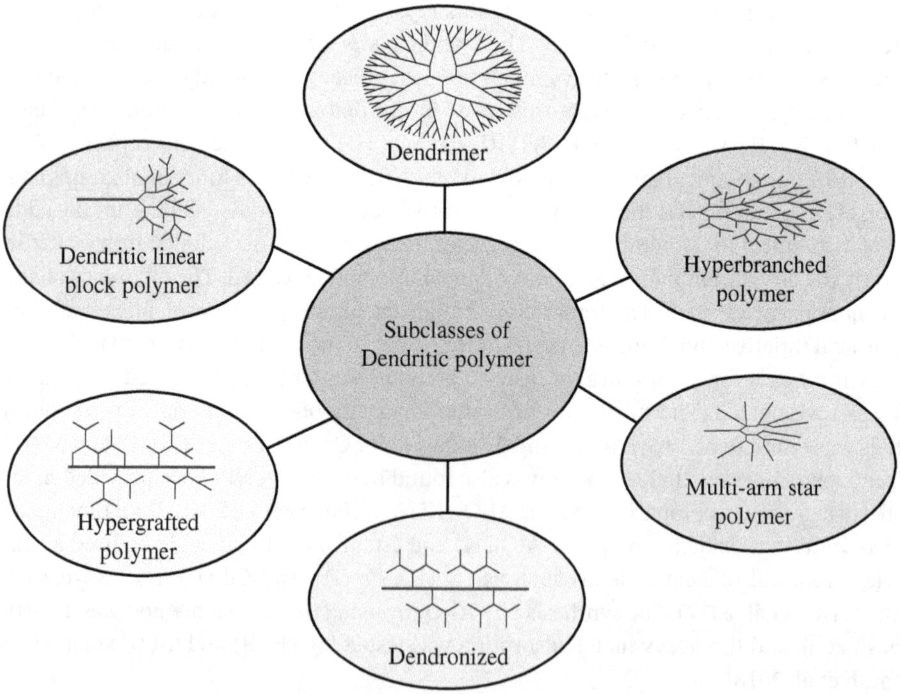

Figure 10.3 Different types of dendritic polymers. (*Source:* Sajid et al. 2018.)

i) Their 3-D highly branched architecture, modified surface with voids, and functionalized exterior makes them a suitable adsorbent.
ii) The large external and internal surface areas enhance their adsorbing capacity.
iii) The functional group at the exterior increases its selectivity.
iv) The physicochemical properties can be molded by modifying the structure of adsorbents.

In water purification, poly(amidoamine) (PAMAM) dendrimers are extensively employed, which consist of a core of ethylenediamine, repeating unit, and functionalized terminal. They are produced by Michael addition which results in methyl ester followed by amidation. The PAMAM dendrimer's diameter can be increased by 1 nm on increasing the repeating units (Yen et al. 2017). PAMAM dendrimers exhibit promising applications in catalytic reactions, molecular recognition, drug delivery, and water distillation. The PAMAM based dendrimers are cost-effective, nontoxic, synthesize from easily available raw material, and have a high binding affinity for pollutants (Barakat et al. 2013).

The adsorption of heavy metals onto dendrimers is based on physical and chemical adsorption. The eviction efficiency via adsorption of heavy metals on the altered dendrimer is a pH-dependent process as the functional group present at the molecular surface of dendrimers is differently effected by varying the pH. Besides, this adsorption efficiency is also determined by the physical and chemical nature of metal ions and other pollutants. In acidic medium, amine groups of dendrimers are protonated which assists in the removal of negatively charged impurities by electrostatically binding with them. For instance, at acidic pH 2, Cr^{6+} has been removed with the help of poly (amidoamine)-grafted cellulose nanofibril aerogel (PAMAM-g-CNF) (Figure 10.4). Contrary to that, at higher pH, the concentration of Cr^{6+} increases as at alkaline medium, and removal efficiency of PAMAM-g-CNF declines (Zhao et al. 2015). It has been reported that at lower pH, positive charge increases on the chitosan-PAMAM dendrimer surface, which impedes the removal of metal cations such as Pb (II) (Zarghami et al. 2016). Metal ions such as Cu^{2+}, Ni^{2+}, and Cr^{3+} are removed with the aid of dendrimer/titania composites from water (Figure 10.4) (Barakat et al. 2013). The tertiary amine in 4 hydroxyl-terminated PAMAM (G4-OH) dendrimer acts as a ligand and transfers its lone pair of electrons to Mn^{2+} and thereby leads to the eviction of metal from water. The 62 tertiary amine sites of dendrimer are coordinated with metal. The adsorption of positively charge Pb^{2+} on PAMAM dendrimer/titania composites is because of the development of a negative charge at the dendrimer's hydroxyl group (Barakat et al. 2014). Furthermore, Zn (II) ions are electrostatically attracted with dendrimer-based magnetic nanoparticles such as Gn-MNPs and thereby are removed from water (Chou and Lien 2011). Whereas metals like Au^{3+}, Pd^{4+}, Pd^{2+}, and Ag^{+} make a complex with MNP-G3, which is a third-generation dendrimer and is removed via the chemisorption process. On forming the complex with EDTA, the adsorption efficiency of dendrimer is enhanced (Yen et al. 2017).

Besides these, various mechanisms have been proposed such as complexation, electrostatic interaction, and physical encapsulation for the sorption of copper metal at the surface of poly(amidoamine) dendrimers (Xu and Zhao 2005). It has been reported that with the help of density functional theory (DFT), the adsorption of Hg (II) ions to silica gel supported modified PAMAM dendrimers can be explored (Niu et al. 2014). By involving nitrogen and oxygen groups, mercury ions may form a chelate with the dendrimers along with feeble $Hg^{2+}\pi$ interactions, which are also involved. As per the Dubinin–Radushkevich (DR) isotherm model, the Hg^{2+}eradication from water is spontaneous endothermic chemisorption.

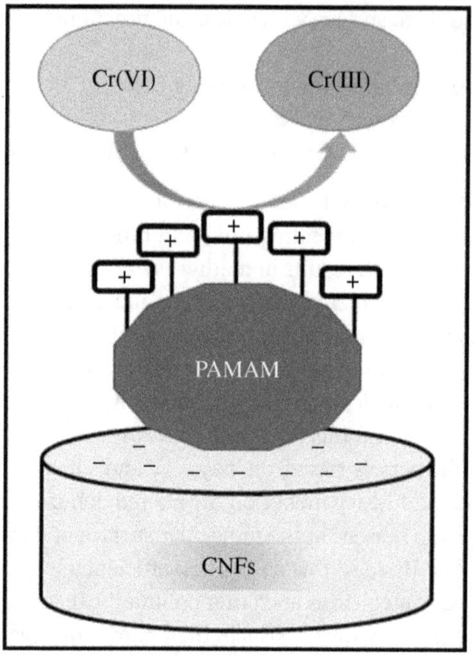

Figure 10.4 The removal of Cr^{6+} ions with the help of poly (amidoamine)-grafted cellulose nanofibril aerogel (PAMAM-g-CNF). (*Source:* Zhao et al. 2015.)

10.5.4 Nanocomposites

Nanocomposites are the composites in which various phases are present, out of which at least one of the phases contains nanometer range size particles (Li et al. 2018a,b). The preparation of nanocomposite shows a serious challenge in controlling the size and stoichiometry, but it emerges as a suitable substitution to micro-composites and monolithic. On considering the unique properties and design of nanocomposite, they are called a twenty-frst-century material. A composite is an aggregation of two or more different materials, segregated by an interface having different physical and chemical properties. In general, the composite comprises of two phases: a matrix phase (continuous phase) and a dispersed phase, whereas nanocomposite is a one in which one of the two phases constitutes nanoscale particles (Friedrich et al. 2005; Cinausero et al. 2008). The uniqueness in the property of nanocomposites arises from the material used, stoichiometry, size, and geometry of particles in various phases. The amalgamation of different materials having great mechanical and thermal strength, multifunctional potential, chemical functionality, and a large interphase area leads to the formulation of the more promising substance having a wide range of practical utility. Some of the properties exhibited by a nanocomposite are stiffness, strength, stability, corrosion resistance, low density, and thermal insulation. Nowadays, research has been carried out to synthesize effective polymeric-based nanocomposite membranes for water purification. For the treatment of water, membranes like polyethersulfone, polysulfone, polyvinylidene fluoride, and polypropylene seem to be effective (Saeki et al. 2016). Owing to its low cost, chemical resistivity, and thermal property these membranes are more in trend. For the treatment of water, membrane technology is supposed to be one of the most advanced methods employed so far. With the incorporation of nanomaterial in the membrane, this technology seems to be more promising. For the polymeric membrane synthesis, nanoparticles, nanotubes, or fiber of size ranging from 4 to 100 nm have been utilized (Mehwish et al. 2014). Generally, nanozeolites are preferred in the membrane of thin-film nanocomposites because of their better molecular sieve, enhanced membrane permeability, thickness of the active layer, hydrophilic nature, and increased negative charge. It was found that adding nanozeolites noticeably upgraded the membrane permeability and salt rejection capability in comparison with the reverse osmosis membrane (Qu et al. 2013). One of the prominent carbon-based nanomaterials used in membrane technology is graphene as it possesses high mechanical and thermal stability, high electrical conductivity, and resistance (Khan et al. 2017; Alghunaimi et al. 2019). Besides this, the large surface area of graphene-based nanomaterials enables them to combine with any active functional group and formulate a composite nanomembrane. A graphene-integrated membrane exhibits greater water permeability in comparison with the commercially available polymer-based membrane. In a computational finding, it has been indicated that the graphene-based membrane has 250 times more water flux than commercial reverse osmosis membrane in the same working conditions (Wu et al. 2019). Many researchers today employ graphene for the synthesis of an ultra-permeable membrane. Apart from graphene, highly permeable and selective carbon nanotubes (CNT) have been used in the development of membranes (Speltini et al. 2016; Song et al. 2017). The huge chemical, thermal, and mechanical stability of fullerenes make them an extensively employed carbon-based nanomaterial in membrane production. They show admirable electro-thermal conductivity and excellent optical qualities. The hydrophobic nature, large electron affinity, and surface/volume ratio makes fullerene an excellent adsorbing material of toxic waste from water. The super hydrophilic nature of TiO_2 nanoparticles enables them to be installed in the membrane filter for purification of wastewater (Lonnen et al. 2005). To meet the commercial scalability of TiO_2, its optimal composition and method of integrating it with the polymeric matrix are some of the serious challenges that still need to be answered. Microfiltration is best achieved by hollow fiber nanocomposite membranes. Nanocomposite catches the attention of many researchers because it is cheap, easy to fabricate, and has a high mechanical strength. In addition, they display the characteristics of both polymeric and inorganic material simultaneously. By introducing zinc oxide nanoparticles, the peculiar traits of the membrane such as hydrophilicity, permeability, and mechanical stability could be surpassed. Although the hollow fiber membranes lodging with nanocomposites sets a very good example for regulating organic in potable water, their high cost is its major drawback.

Based on the following aspects such as chemical composition, microstructures, fillers, matrix material, and applications, nanocomposites are classified into different groups such as polymeric matrix nanocomposites, metal-based matrix nanocomposites, polymer/layered silicate (PLS) nanocomposites, and polymer nanocomposites.

10.5.4.1 Polymeric Matrix Nanocomposites

In a way to improve the characteristics of composites by employing a large surface, area fillers led to the discovery of micro-composites, which eventually turned toward the development of nanocomposites. Polymeric matrix nanocomposites are composed of the polymeric matrix where high aspect ratio small size nanoparticles are reinforced (Gacitua et al. 2005; Saleh and Al-Absi 2017). Filler nanoparticles must be properly dispersed within the matrix by forming a proper interaction with the matrix. Polypyrrole/polyacrylonitrile (PANI/PPy) polymer nanofibers successfully adsorbed Co (II) ions from wastewater and showed 99.68% removal efficiency. Furthermore, the eviction of Cr (VI) was achieved by using PANI/PPy nanofibers, which were prepared by the polymerization of pyrrole and aniline monomers along with an oxidant, i.e. $FeCl_3$. Another hybrid nanocomposite containing poly(3,4-ethylenedioxythiophene)/polystyrene sulfonate (PEDOT/PSS) and biopolymer lignin (LG) were used for the removal of Pb^{2+} from wastewater (Zhao et al. 2018).

10.5.4.2 Metal Matrix Nanocomposites

When nano-dimension fillers are added to the large matrix of metal or an alloy, they lead to the formation of metal matrix-based nanocomposites. These nanocomposites possess the traits of both metals and ceramics. They are highly ductile and have great mechanical strength and thermal stability. A wider application can be achieved on combining the thermally and mechanically stable carbon nanotubes having a large surface area with the metal matrices. Polyvinyl alcohol (PVA) was fabricated with nanoscale Fe^0 in order to adsorb Sb^{3+} and Sb^{5+} from polluted water. PVA-Fe^0 nanocomposite exhibits rapid adsorption of antimony as compared with metal-based nanoparticles (Zhao et al. 2014). Fe^0 nanoparticles supported on PANI nanofibers showed rapid and efficient eviction of Cr^{6+} and As^{5+} metal ions from wastewater (Bhaumik et al. 2014). Recently, Cu-nanoparticles were effectively diffused on the chitosan-tripolyphosphate beads to fabricate nanocomposites, which was further utilized to intensify the adsorption efficiency for Cr (VI) as compared with chitosan-polyphosphate beads alone (Wu et al. 2009).

10.5.4.3 Polymer/Layered Silicate Nanocomposites

PLS nanocomposite materials are a blend of layered silicates into polymer matrices (Pavlidou and Papaspyrides 2008). They display significant improvement in their property as compared with conventional macro-composites. In a report by Greenland et al., the direct incorporation of polymer on clay was shown (Fornes et al. 2001). A nanocomposite consisting of silica gel microspheres that were encapsulated in imidazole functionalized polystyrene (SG-PS-azo-IM) was investigated for the eviction of several heavy metal ions like Hg (II), Ag (I), Au (III), Pb (II), Zn (II), Cu (II), Ni (II), Mn (II), etc. Out of all heavy metals, SG-PS-azo-IM nanocomposite showed great adsorption potential for Au (III) ions (Yin et al. 2010). Another nanocomposite poly(amidoxime)/SiO_2 (PAO/SiO_2) exhibited selective adsorption capability toward Cu (II) and Pb (II) ions. Recently, Cu (II) ions were made to adsorb on polyethylenimine (PEI)-tannins coated SiO_2 hybrids (SiO_2@PEI-TA). Apart from this, silica-sphere-poly(catechol hexamethylenediamine) (PCHA-SiO_2) nanocomposite was prepared for the selective eradication of Cd^{2+}, Pb^{2+}, and Cu^{2+} ions even at low concentrations (Zhao et al. 2018). By employing microporous layered silicate AMH-3 as sorbent, selective removal of heavy metals such as Pb^{2+}, Cu^{2+}, and Zn^{2+} was effectively achieved (Kim and Kwak 2017). Zeolite, a fabricated polymer layered silicate nanocomposite exhibited increment in the lattice d-spacing, which leads to a greater adsorption of lead and cadmium ions, i.e. 97.20% and 85.06%, respectively, from solution (Mbui et al. 2018).

10.6 Comparison of Adsorption Capacities of Various Heavy Metal Ions by Nanomaterials and Conventional Methods

Various conventional methods such as ion exchange, solvent extraction, chemical precipitation, electrochemical removal, membrane filtration, and coagulation were employed for the removal of heavy metals from aqueous media. However, these techniques are associated with some disadvantages, which include a requirement for high energy, incomplete removal, less efficiency, the generation of toxic sludge, costly disposal, and sensitive operating conditions. In order to overcome these problems, adsorption by nanomaterials has been explored. The adsorption capacity for heavy metal ions is greatly improved by using nanoadsorbents (Burakov et al. 2018). The adsorption capacity of some heavy metal ions by conventional methods is compared with nanoadsorbents in Table 10.2, whereas Table 10.3 depicts the maximum adsorption capacity of heavy metal ions by distinctive nanoadsorbents.

10.7 Conclusion and Future Trends

Numerous applications of nanomaterials have been exploited so far and the removal of heavy metals from wastewater on account of their extensive and unique properties is one of the most explored applications. In the present work, the role of different types of nanomaterials such as carbon- and metal-based nanomaterials, dendrimers, and nanocomposites are discussed in detail.

Table 10.2 Comparison of adsorption capacities of heavy metal ions by conventional methods and nanomaterials.

S. no	Heavy metal ion	Adsorption capacity by conventional method	Adsorption capacity by nanomaterials (mg/g)
1	Pb^{2+}	4.13 mg/g (pine wood char), 27.79 mg/g (*Cicer arientinum* biomass) Burakov et al. (2018), 0.53 mmol/g (activated carbon from eucalyptus bark)	49.71 mg/g (acidified MWCNTs), Wang et al. (2007), 137 mg/g (TiO_2-modified CNTs) Zhao et al. (2010)
2	Cu^{2+}	0.45 mmol/g (activated carbon from eucalyptus bark), 8.13 mg/g (zeolite), 37.17 mg/g (sawdust), 11.89 mg/g (bentonite clay) Burakov et al. (2018)	8.25 mg/g (As-produced CNTs), 47.39 mg/g (NaOCl-modified CNTs) Burakov et al. (2018)
3	Cd^{2+}	1.85 mg/g (activated carbon from olive stone), 26.24 mg/g (fly ash zeolite), 52.1 mg/g (red mud)	55.4 mg/g (GO/Fe_3O_4/sulfanilic acid), 166.7 mg/g ($GO/SiO_2/Fe_3O_4$)
4	Cr^{6+}	44 mg/g (activated carbon from rubber wood sawdust), 13.40 mg/g (activated carbon from sugarcane bagasse) Karthikeyan et al. (2005), Sharma and Forster (1994)	264.5 mg/g (CNTs/activated Al_2O_3), 638.56 mg/g (N-CNTs) Burakov et al. (2018), 2859.38 mg/g (graphene sand composite) Dubey et al. (2015)

Table 10.3 Adsorption capacities of heavy metal ions by various types of nanomaterials.

S· no	Heavy metal ion	Adsorption capacity (mg/g) by carbon-based nanoparticles	Adsorption capacity (mg/g) by metal-based nanoparticles	Adsorption capacity (mg/g) by dendrimers	Adsorption capacity (mg/g) by nanocomposites
1	Hg^{2+}	186.97 mg/g (CNTs) Yang et al. (2019)	237.60 mg/g (mercaptoethylamino monomer-modified maghemite) Madrakian et al. (2015); 4065.0 mg/g (gold nanoparticles) Lisha and Pradeep (2009)	The adsorption capacity of PAMAM-ATP adsorbents followed the order of G2.0 (PAMAM-ATP, 200.8 mg/g) > G1.0 (polyacrylamine modified ATP, PAM-ATP, 192.5 mg/g) > G3.0 (amine modified ATP, M-ATP, 90 mg/g) > G4.0 (raw ATP, 5 mg/g) Qin et al. (2019)	376.2 mg/g (silver/quartz nanocomposite) El-Tawil et al. (2019)
2	Pb^{2+}	373.14 mg/g (rGO-Fe_3O_4), 689 mg/g (rGO-sulfophenylazo), 238.1 mg/g (activated carbon) Baby (2019); 461.3 mg/g (chitosan/GO) Yang (2019)	68.9 mg/g (maghemite) Rajput et al. (2017)	568.18 mg/g Graphene oxide/polyamidoamine dendrimers Zhang et al. (2014)	887.98 mg/g polymer-based graphene oxide nanocomposite Musico et al. (2013)
3	Cd^{2+}	96.2 mg/g (activated carbon) Baby (2019); 106.3 mg/g (GO) Shan et al. (2017)	91.55 mg/g (mercaptoethylamino monomer-modified maghemite) Madrakian et al. (2015)	1225 mg/g poly (propyleneimine) (PPI) dendrimer Ebrahimi et al. (2020)	76.67 mg/g magnetite/carbon nanocomposites Andelescu et al. (2018)
4	Co^{2+}	68.2 mg/g (GO) Shan et al. (2017)		355 mg/g (xanthate functionalized dendrimer) Ilaiyaraja et al. (2014)	18.76 mg/g. Fe_3O_4/bentonite nanocomposite Hashemian et al. (2015)

5	Cu^{2+}	30.49 mg/g (magnetic amino-functionalized Fe_3O_4/carboxylic MWCNTs) Yang (2019); 14.6 mmol/g (fullerenes) Baby (2019)	84.46 mg/g (nano-hematite) Chen et al. (2009); 34.0 mg/g (maghemite) Rajput et al. (2017)	3333 mg/g (PAMAM, G4) Hayati et al. (2017)	48.78 mg/g (nanocomposite of magnetic hydroxyapatite nanorods) Thanh et al. (2018)
6	Zn^{2+}	246 mg/g (GO) Yang (2019)		24.3 mg/g (dendrimer-conjugated magnetic nanoparticles) Chou and Lien (2011)	24.3 mg/g (polyaniline nanocomposite coated on rice husk) Ghorbani et al. (2012)
7	Cr^{6+}	310.4 mg/g (chitosan/GO) Yang (2019)	232.51 mg/g (amino-functionalized Fe_3O_4) Baghani et al. (2016)	377.36 mg/g (dendrimer poly(amidoamine)-grafted cellulose nanofibril (PAMAM-g-CNF) aerogel) Zhao et al. (2015)	149.25 mg/g (polypyrrole-coated halloysite nanotube clay nanocomposite) Ballav et al. (2014)

These nanomaterials inherit significant properties, and thus, behave as an excellent adsorbent for heavy metals. The treatment of heavy metal pollution in wastewater is of utmost importance both from the perspective of the environment as well as human health. Due to the distinctive properties of nanomaterials, these have emerged as a promising alternative to the conventional adsorbents for removing heavy metals from wastewater. Carbon and metal-based nanomaterials exhibit great potential as nanoadsorbents toward heavy metals but they also have some limitations, which are needed to be overcome. They are unstable and lead to aggregation, resulting in a reduced potential for the removal of heavy metals. Also, it sometimes becomes hard to smoothly and efficiently separate nanomaterials from the aqueous solution due to their small size. Nevertheless, these problems were efficiently solved by the development of nanocomposites. The market for nanocomposites has been forecasted in numerous fields because of their distinctive mechanical, thermal, electrical, magnetic, optical properties, etc. Recently, magnetic nanoparticles have been utilized for the removal of heavy metals from wastewater due to their unique strength, less toxicity, and simple separation. In addition, they can be easily and effectively recovered after adsorption. Several magnetic nanoparticles require laboratory research despite their significant properties. The interaction between nanomaterials and heavy metals still requires a thorough understanding to synthesize more specific and selective nanomaterials. The utilization of nanomaterials for the removal of heavy metals from wastewater in terms of their potential, reusability, and cost of treatment should be extensively explored to find out robust and feasible solutions.

Acknowledgments

The authors wish to acknowledge the research and development grant of the University of Delhi, Delhi.

Conflict of Interest

None to declare.

References

Aïssa, B., Memon, N.K., Ali, A., and Khraisheh, M.K. (2015). Recent progress in the growth and applications of graphene as a smart material: a review. *Frontiers in Materials* 2: 58.

Alalwan, H.A., Kadhom, M.A., and Alminshid, A.H. (2020). Removal of heavy metals from wastewater using agricultural byproducts. *Journal of Water Supply: Research and Technology – AQUA* 69 (2): 99–112.

Alekseeva, O.V., Bagrovskaya, N.A., and Noskov, A.V. (2016). Sorption of heavy metal ions by fullerene and polystyrene/fullerene film compositions. *Protection of Metals and Physical Chemistry of Surfaces* 52 (3): 443–447.

Alghunaimi, F.I., Alsaeed, D.J., Harith, A.M., and Saleh, T.A. (2019). Synthesis of 9-octadecenoic acid grafted graphene modified with polystyrene for efficient light oil removal from water. *Journal of Cleaner Production* 233: 946–953.

Ali, A., Mannan, A., Hussain, I. et al. (2018). Effective removal of metal ions from aqueous solution by silver and zinc nanoparticles functionalized cellulose: isotherm, kinetics and statistical supposition of process. *Environmental Nanotechnology, Monitoring & Management* 9: 1–11.

Alina, M., Azrina, A., Mohd Yunus, A.S. et al. (2012). Heavy metals (mercury, arsenic, cadmium, plumbum) in selected marine fish and shellfish along the Straits of Malacca. *International Food Research Journal* 19 (1): 135–140.

Almomani, F., Bhosale, R., Khraisheh, M., and Almomani, T. (2020). Heavy metal ions removal from industrial wastewater using magnetic nanoparticles (MNP). *Applied Surface Science* 506: 144924.

AlOmar, M.K., Alsaadi, M.A., Hayyan, M. et al. (2017). Allyl triphenyl phosphonium bromide-based DES-functionalized carbon nanotubes for the removal of mercury from water. *Chemosphere* 167: 44–52.

Andelescu, A., Nistor, M.A., Muntean, S.G., and Rădulescu-Grad, M.E. (2018). Adsorption studies on copper, cadmium, and zinc ion removal from aqueous solution using magnetite/carbon nanocomposites. *Separation Science and Technology* 53 (15): 2352–2364.

Arshad, F., Selvaraj, M., Zain, J. et al. (2019). Polyethylenimine modified graphene oxide hydrogel composite as an efficient adsorbent for heavy metal ions. *Separation and Purification Technology* 209: 870–880.

Atsdr, S. (2012). Toxicological Profile for Chromium. Agency for Toxic Substances and Disease Registry. Washington, DC: Public Health Service, US Department of Health and Human Services. www.atsdr.cdc.gov/toxprofiles/tp.asp

Azimi, A., Azari, A., Rezakazemi, M., and Ansarpour, M. (2017). Removal of heavy metals from industrial wastewaters: a review. *ChemBioEng Reviews* 4 (1): 37–59.

Baby, R., Saifullah, B., and Hussein, M.Z. (2019). Carbon nanomaterials for the treatment of heavy metal-contaminated water and environmental remediation. *Nanoscale Research Letters* 14 (1): 341.

Baghani, A.N., Mahvi, A.H., Gholami, M. et al. (2016). One-Pot synthesis, characterization and adsorption studies of amine-functionalized magnetite nanoparticles for removal of Cr (VI) and Ni (II) ions from aqueous solution: Kinetic, isotherm and thermodynamic studies. *Journal of Environmental Health Science and Engineering* 14 (1): 1–12.

Ballav, N., Choi, H.J., Mishra, S.B., and Maity, A. (2014). Polypyrrole-coated halloysite nanotube clay nanocomposite: synthesis, characterization and Cr (VI) adsorption behaviour. *Applied Clay Science* 102: 60–70.

Barakat, M.A., Ramadan, M.H., Alghamdi, M.A. et al. (2013). Remediation of Cu (II), Ni (II), and Cr (III) ions from simulated wastewater by dendrimer/titania composites. *Journal of Environmental Management* 117: 50–57.

Barakat, M.A., Ramadan, M.H., Kuhn, J.N., and Woodcock, H.L. (2014). Equilibrium and kinetics of Pb2+ adsorption from aqueous solution by dendrimer/titania composites. *Desalination and Water Treatment* 52 (31–33): 5869–5875.

Basheer, A., Attiq-ur-Rehman, S., Khan, N., and Hayatullah, A.B. (2020). 7. Physico-chemical and heavy metals analysis of drinking water and their effect on human health: a review. *Pure and Applied Biology (PAB)* 9 (1): 587–594.

Basir, N.A.M., Yaacob, W.Z.W., Mohammed, N.H. et al. (2019). Prediction and Remediation of Water Quality in Monitoring Potential of Acid Mine Drainage. *American Journal of Engineering and Applied Sciences* 12: 173–184.

Bharathi, P., Balasubramanian, N., Anitha, S., and Vijayabharathi, V. (2016). Development of gold nanoparticles supported membrane as an efficient material for protein detection. *Journal of Environmental Biology* 37 (6): 1415.

Bhaumik, M., Choi, H.J., McCrindle, R.I., and Maity, A. (2014). Composite nanofibers prepared from metallic iron nanoparticles and polyaniline: high performance for water treatment applications. *Journal of Colloid and Interface Science* 425: 75–82.

Bindhu, M.R. and Umadevi, M. (2014a). Antibacterial activities of green synthesized gold nanoparticles. *Materials Letters* 120: 122–125.

Bindhu, M.R. and Umadevi, M. (2014b). Silver and gold nanoparticles for sensor and antibacterial applications. *Spectrochimica Acta Part A: Molecular and Biomolecular Spectroscopy* 128: 37–45.

Bolisetty, S., Peydayesh, M., and Mezzenga, R. (2019). Sustainable technologies for water purification from heavy metals: review and analysis. *Chemical Society Reviews* 48 (2): 463–487.

Breida, M., Younssi, S.A., Ouammou, M. et al. (2019). Pollution of water sources from agricultural and industrial effluents: special attention to NO 3 $^-$, Cr (VI), and Cu (II). In: *Water Chemistry* (ed. M. Eyvaz). London: IntechOpen https://doi.org/10.5772/intechopen.86921.

Burakov, A.E., Galunin, E.V., Burakova, I.V. et al. (2018). Adsorption of heavy metals on conventional and nanostructured materials for wastewater treatment purposes: a review. *Ecotoxicology and Environmental Safety* 148: 702–712.

Campos, E.A., Pinto, D.V.B.S., Oliveira, J.I.S.D. et al. (2015). Synthesis, characterization and applications of iron oxide nanoparticles-A short review. *Journal of Aerospace Technology and Management* 7 (3): 267–276.

Cao, F., Lian, C., Yu, J. et al. (2019). Study on the adsorption performance and competitive mechanism for heavy metal contaminants removal using novel multi-pore activated carbons derived from recyclable long-root Eichhorniacrassipes. *Bioresource Technology* 276: 211–218.

Chatterjee, S., Mahanty, S., Das, P. et al. (2020). Biofabrication of iron oxide nanoparticles using manglicolous fungus Aspergillus niger BSC-1 and removal of Cr (VI) from aqueous solution. *Chemical Engineering Journal* 385: 123790.

Chen, D., Cao, L., Hanley, T.L., and Caruso, R.A. (2012). Facile synthesis of monodisperse mesoporous zirconium titanium oxide microspheres with varying compositions and high surface areas for heavy metal ion sequestration. *Advanced Functional Materials* 22 (9): 1966–1971.

Cheng, Z., Li, G., Zhang, N., and Liu, H.O. (2014). A novel functionalized silver nanoparticles solid chemosensor for detection of Hg (II) in aqueous media. *Dalton Transactions* 43 (12): 4762–4769.

Chou, C.M. and Lien, H.L. (2011). Dendrimer-conjugated magnetic nanoparticles for removal of zinc (II) from aqueous solutions. *Journal of Nanoparticle Research* 13 (5): 2099–2107.

Chowdhury, S. and Balasubramanian, R. (2014). Recent advances in the use of graphene-family nanoadsorbents for removal of toxic pollutants from wastewater. *Advances in Colloid and Interface Science* 204: 35–56.

Cinausero, N., Howell, B., Schmaucks, G. et al. (2008). *Fire Retardancy of Polymers: New Strategies and Mechanisms*. Royal Society of Chemistry.

Dave, P.N. and Chopda, L.V. (2014). Application of iron oxide nanomaterials for the removal of heavy metals. *Journal of Nanotechnology* 2014: 1–14.

De Acha, N., Elosúa, C., Corres, J.M., and Arregui, F.J. (2019). Fluorescent sensors for the detection of heavy metal ions in aqueous media. *Sensors* 19 (3): 599.

Ding, Y. (2019). Heavy metal pollution and transboundary issues in ASEAN countries. *Water Policy* 21 (5): 1096–1106.

Dubey, R., Bajpai, J., and Bajpai, A.K. (2015). Green synthesis of graphene sand composite (GSC) as novel adsorbent for efficient removal of Cr (VI) ions from aqueous solution. *Journal of Water Process Engineering* 5: 83–94.

Ebrahimi, R., Hayati, B., Rezaee, R. et al. (2020). Adsorption of cadmium and nickel from aqueous environments using a dendrimer. *Journal of Advances in Environmental Health Research* 8 (1): 19–24.

Elkacmi, R. and Bennajah, M. (2019). Advanced oxidation technologies for the treatment and detoxification of olive mill wastewater: a general review. *Journal of Water Reuse and Desalination* 9 (4): 463–505.

El-Saharty, A., Mahmoud, S.N., Manjood, A.H. et al. (2018). Effect of apricot stone activated carbon adsorbent on the removal of toxic heavy metals ions from aqueous solutions. *International Journal of Ecotoxicology and Ecobiology* 3 (2): 13.

El-Sheikh, A.H., Al-Degs, Y.S., Al-As' ad, R.M., and Sweileh, J.A. (2011). Effect of oxidation and geometrical dimensions of carbon nanotubes on Hg (II) sorption and preconcentration from real waters. *Desalination* 270 (1–3): 214–220.

El-Tawil, R.S., El-Wakeel, S.T., Abdel-Ghany, A.E. et al. (2019). Silver/quartz nanocomposite as an adsorbent for removal of mercury (II) ions from aqueous solutions. *Heliyon* 5 (9): e02415.

Engwa, G.A., Ferdinand, P.U., Nwalo, F.N., and Unachukwu, M.N. (2019). Mechanism and health effects of heavy metal toxicity in humans. In: *Poisoning in the Modern World-New Tricks for an Old Dog?* IntechOpen https://doi.org/10.5772/intechopen.82511.

Fato, F.P., Li, D.W., Zhao, L.J. et al. (2019). Simultaneous removal of multiple heavy metal ions from river water using ultrafine mesoporous magnetite nanoparticles. *ACS Omega* 4 (4): 7543–7549.

Fornes, T.D., Yoon, P.J., Keskkula, H., and Paul, D.R. (2001). Nylon 6 nanocomposites: the effect of matrix molecular weight. *Polymer* 42 (25): 09929–09940.

Fréchet, J.M. and Tomalia, D.A. (eds.) (2001). *Dendrimers and Other Dendritic Polymers*, 647. New York: Wiley.

Friedrich, K., Fakirov, S., and Zhang, Z. (eds.) (2005). *Polymer Composites: From Nano-to Macro-Scale*. Springer Science & Business Media.

Fu, F., Xie, L., Tang, B. et al. (2012). Application of a novel strategy-advanced Fenton-chemical precipitation to the treatment of strong stability chelated heavy metal containing wastewater. *Chemical Engineering Journal* 189: 283–287.

Gacitua, W., Ballerini, A., and Zhang, J. (2005). Polymer nanocomposites: synthetic and natural fillers a review. *Maderas. Ciencia y Tecnologia* 7 (3): 159–178.

Gangadhar, G., Maheshwari, U., and Gupta, S. (2012). Application of nanomaterials for the removal of pollutants from effluent streams. *Nanoscience & Nanotechnology-Asia* 2 (2): 140–150.

Ghorbani, M., Eisazadeh, H., and Ghoreyshi, A.A. (2012). Removal of zinc ions from aqueous solution using polyaniline nanocomposite coated on rice husk. *Iranica Journal of Energy & Environment* 3 (1): 83–88.

Guardia, P., Labarta, A., and Batlle, X. (2011). Tuning the size, the shape, and the magnetic properties of iron oxide nanoparticles. *The Journal of Physical Chemistry C* 115 (2): 390–396.

Guerrini, L., Alvarez-Puebla, R.A., and Pazos-Perez, N. (2018). Surface modifications of nanoparticles for stability in biological fluids. *Materials* 11 (7): 1154.

Guo, T., Bulin, C., Li, B. et al. (2018). Efficient removal of aqueous Pb (II) using partially reduced graphene oxide-Fe3O4. *Adsorption Science & Technology* 36 (3–4): 1031–1048.

Gupta, V.K., Moradi, O., Tyagi, I. et al. (2016). Study on the removal of heavy metal ions from industry waste by carbon nanotubes: effect of the surface modification: a review. *Critical Reviews in Environmental Science and Technology* 46 (2): 93–118.

Han, K.N., Yu, B.Y., and Kwak, S.Y. (2012). Hyperbranched poly (amidoamine)/polysulfone composite membranes for Cd (II) removal from water. *Journal of Membrane Science* 396: 83–91.

Hasany, S.F., Ahmed, I., Rajan, J., and Rehman, A. (2012). Systematic review of the preparation techniques of iron oxide magnetic nanoparticles. *Nanoscience and Nanotechnology* 2 (6): 148–158.

Hasany, S.F., Abdurahman, N.H., Sunarti, A.R., and Jose, R. (2013). Magnetic iron oxide nanoparticles: chemical synthesis and applications review. *Current Nanoscience* 9 (5): 561–575.

Hashemian, S., Saffari, H., and Ragabion, S. (2015). Adsorption of cobalt (II) from aqueous solutions by Fe 3 O 4/bentonite nanocomposite. *Water, Air, and Soil Pollution* 226 (1): 2212.

Hayati, B., Maleki, A., Najafi, F. et al. (2017). Super high removal capacities of heavy metals (Pb2+ and Cu2+) using CNT dendrimer. *Journal of Hazardous Materials* 336: 146–157.

Hubicki, Z. and Kołodyńska, D. (2012). Selective removal of heavy metal ions from waters and waste waters using ion exchange methods. *Ion Exchange Technologies* 7: 193–240.

Ilaiyaraja, P., Singha Deb, A.K., Ponraju, D., and Venkatraman, B. (2014). Removal of cobalt from aqueous solution using xanthate functionalized dendrimer. *Desalination and Water Treatment* 52 (1-3): 438–445.

Jacukowicz-Sobala, I., Ociński, D., and Kociołek-Balawejder, E. (2015). Iron and aluminium oxides containing industrial wastes as adsorbents of heavy metals: application possibilities and limitations. *Waste Management & Research* 33 (7): 612–629.

Jaishankar, M., Tseten, T., Anbalagan, N. et al. (2014). Toxicity, mechanism and health effects of some heavy metals. *Interdisciplinary Toxicology* 7 (2): 60–72.

Kabeer, A., Mailafiya, M.M., Danmaigoro, A. et al. (2019). Therapeutic potential of curcumin against lead-induced toxicity. *Biomedical Research and Therapy* 6: 3053–3066.

Kaneko, K., Ishii, C., Arai, T., and Suematsu, H. (1993). Defect-associated microporous nature of fullerene C60 crystals. *The Journal of Physical Chemistry* 97 (26): 6764–6766.

Karthikeyan, T., Rajgopal, S., and Miranda, L.R. (2005). Chromium (VI) adsorption from aqueous solution by HeveaBrasilinesis sawdust activated carbon. *Journal of Hazardous Materials* 124 (1-3): 192–199.

Khan, A., Wang, J., Li, J. et al. (2017). The role of graphene oxide and graphene oxide-based nanomaterials in the removal of pharmaceuticals from aqueous media: a review. *Environmental Science and Pollution Research* 24 (9): 7938–7958.

Kim, J. and Kwak, S.Y. (2017). Efficient and selective removal of heavy metals using microporous layered silicate AMH-3 as sorbent. *Chemical Engineering Journal* 313: 975–982.

Kim, J., Cote, L.J., and Huang, J. (2012). Two dimensional soft material: new faces of graphene oxide. *Accounts of Chemical Research* 45 (8): 1356–1364.

Kim, H.C., Jang, T.W., Chae, H.J. et al. (2015). Evaluation and management of lead exposure. *Annals of Occupational and Environmental Medicine* 27 (1): 30.

Kumar, P.S. and Saravanan, A. (2017). Sustainable wastewater treatments in textile sector. In: *Sustainable Fibres and Textiles* (ed. S. Senthilkannan Muthu), 323–346. Cambridge: Woodhead Publishing.

Kumar, R., Khan, M.A., and Haq, N. (2014). Application of carbon nanotubes in heavy metals remediation. *Critical Reviews in Environmental Science and Technology* 44 (9): 1000–1035.

Lakherwal, D. (2014). Adsorption of heavy metals: a review. *International Journal of Environmental Research and Development* 4 (1): 41–48.

Li, J., Wang, X., Zhao, G. et al. (2018a). Metal–organic framework-based materials: superior adsorbents for the capture of toxic and radioactive metal ions. *Chemical Society Reviews* 47 (7): 2322–2356.

Li, J., Xing, X., Li, J. et al. (2018b). Preparation of thiol-functionalized activated carbon from sewage sludge with coal blending for heavy metal removal from contaminated water. *Environmental Pollution* 234: 677–683.

Li, B., Chen, X., Li, K. et al. (2019). Coupling membrane and Fe–Pd bimetallic nanoparticles for trichloroethene removing from water. *Journal of Industrial and Engineering Chemistry* 78: 198–209.

Lisha, K.P. and Pradeep, T. (2009). Towards a practical solution for removing inorganic mercury from drinking water using gold nanoparticles. *Gold Bulletin* 42 (2): 144–152.

Lonnen, J., Kilvington, S., Kehoe, S.C. et al. (2005). Solar and photocatalytic disinfection of protozoan, fungal and bacterial microbes in drinking water. *Water Research* 39 (5): 877–883.

Luvita, V., Darmayanti, N.E., Zaid, G., and Setyarini, D. (2019). Testing instrument for water quality and drinking water using oxidation and electromagnetic methods (Case Study: Local Water Company at Bangka Barat). *IOP Conference Series: Materials Science and Engineering* 543 (1): 012048.

Ma, Y., Mou, Q., Wang, D. et al. (2016). Dendritic polymers for theranostics. *Theranostics* 6 (7): 930.

Madrakian, T., Afkhami, A., Zadpour, B., and Ahmadi, M. (2015). New synthetic mercaptoethylamino homopolymer-modified maghemite nanoparticles for effective removal of some heavy metal ions from aqueous solution. *Journal of Industrial and Engineering Chemistry* 21: 1160–1166.

Mahdavi, S., Jalali, M., and Afkhami, A. (2013). Heavy metals removal from aqueous solutions using TiO2, MgO, and Al2O3 nanoparticles. *Chemical Engineering Communications* 200 (3): 448–470.

Maiti, M., Sarkar, M., Malik, M.A. et al. (2018). Iron oxide NPs facilitated a smart building composite for heavy-metal removal and dye degradation. *ACS Omega* 3 (1): 1081–1089.

Martínez-Cabanas, M., López-García, M., Barriada, J.L. et al. (2016). Green synthesis of iron oxide nanoparticles. Development of magnetic hybrid materials for efficient As (V) removal. *Chemical Engineering Journal* 301: 83–91.

Masindi, V. and Muedi, K.L. (2018). Environmental contamination by heavy metals. *Heavy Metals* 19: 2019.

Mbui, D.W., Andala, D.M., Abong'o, D.A., and Mmbaga, J.N. (2018). Design and fabrication of polymer layered silicate nanocomposites for water purification. *MRS Advances* 3 (36): 2099–2107.

Mehwish, N., Kausar, A., and Siddiq, M. (2014). Advances in polymer-based nanostructured membranes for water treatment. *Polymer-Plastics Technology and Engineering* 53 (12): 1290–1316.

Mohan Gokhale, M. and Ravindra Somani, R. (2015). Fullerenes: chemistry and its applications. *Mini-Reviews in Organic Chemistry* 12 (4): 355–366.

Mohan, V.B., Lau, K.T., Hui, D., and Bhattacharyya, D. (2018). Graphene-based materials and their composites: a review on production, applications and product limitations. *Composites Part B: Engineering* 142: 200–220.

Morais, S., Costa, F.G., and Pereira, M.D.L. (2012). Heavy metals and human health. *Environmental Health–Emerging Issues and Practice* 10: 227–246.

Morris, T., Copeland, H., McLinden, E. et al. (2002). The effects of mercury adsorption on the optical response of size-selected gold and silver nanoparticles. *Langmuir* 18 (20): 7261–7264.

Mubarak, N.M., Sahu, J.N., Abdullah, E.C., and Jayakumar, N.S. (2014). Removal of heavy metals from wastewater using carbon nanotubes. *Separation & Purification Reviews* 43 (4): 311–338.

Musico, Y.L.F., Santos, C.M., Dalida, M.L.P., and Rodrigues, D.F. (2013). Improved removal of lead (II) from water using a polymer-based graphene oxide nanocomposite. *Journal of Materials Chemistry A* 1 (11): 3789–3796.

Niu, Y., Qu, R., Chen, H. et al. (2014). Synthesis of silica gel supported salicylaldehyde modified PAMAM dendrimers for the effective removal of Hg (II) from aqueous solution. *Journal of Hazardous Materials* 278: 267–278.

Nizamuddin, S., Siddiqui, M.T.H., Mubarak, N.M. et al. (2019). Iron oxide nanomaterials for the removal of heavy metals and dyes from wastewater. In: *Nanoscale Materials in Water Purification* (eds. S. Thomas, D. Pasquini, D.A. Gopakumar and S.-Y. Leu), 447–472. Amsterdam: Elsevier.

O'Brien, T.J., Ceryak, S., and Patierno, S.R. (2003). Complexities of chromium carcinogenesis: role of cellular response, repair and recovery mechanisms. *Mutation Research/Fundamental and Molecular Mechanisms of Mutagenesis* 533 (1–2): 3–36.

Ochieng, G.M., Seanego, E.S., and Nkwonta, O.I. (2017). Impacts of mining on water resources in South Africa: A review. *Scientific Research and Essays* 5 (22): 3351–3357.

Ojea-Jiménez, I., López, X., Arbiol, J., and Puntes, V. (2012). Citrate-coated gold nanoparticles as smart scavengers for mercury (II) removal from polluted waters. *ACS Nano* 6 (3): 2253–2260.

Parham, H., Zargar, B., and Shiralipour, R. (2012). Fast and efficient removal of mercury from water samples using magnetic iron oxide nanoparticles modified with 2-mercaptobenzothiazole. *Journal of Hazardous Materials* 205: 94–100.

Parvin, F., Rikta, S.Y., and Tareq, S.M. (2019). Application of nanomaterials for the removal of heavy metal from wastewater. In: *Nanotechnology in Water and Wastewater Treatment*, 137–157. Elsevier https://doi.org/10.1016/B978-0-12-813902-8.00008-3.

Pavlidou, S. and Papaspyrides, C.D. (2008). A review on polymer–layered silicate nanocomposites. *Progress in Polymer Science* 33 (12): 1119–1198.

Peng, W., Du, S., Shaoning, Z. et al. (2020). Constructing mesoporous phosphated titanium oxide for efficient Cr (III) removal. *Journal of Hazardous Materials* 384: 121278.

Pillai, P., Kakadiya, N., Timaniya, Z. et al. (2020). Removal of arsenic using iron oxide amended with rice husk nanoparticles from aqueous solution. *Materials Today: Proceedings* https://doi.org/10.1016/j.matpr.2019.12.307.

Poursani, A.S., Nilchi, A., Hassani, A.H. et al. (2015). A novel method for synthesis of nano-γ-Al_2O_3: study of adsorption behavior of chromium, nickel, cadmium and lead ions. *International Journal of Environmental Science and Technology* 12 (6): 2003–2014.

Qdais, H.A. and Moussa, H. (2004). Removal of heavy metals from wastewater by membrane processes: a comparative study. *Desalination* 164 (2): 105–110.

Qin, W., Qian, G., Tao, H. et al. (2019). Adsorption of Hg (II) ions by PAMAM dendrimers modified attapulgite composites. *Reactive and Functional Polymers* 136: 75–85.

Qu, X., Alvarez, P.J., and Li, Q. (2013). Applications of nanotechnology in water and wastewater treatment. *Water Research* 47 (12): 3931–3946.

Rajput, S., Singh, L.P., Pittman, C.U. Jr., and Mohan, D. (2017). Lead (Pb2+) and copper (Cu2+) remediation from water using superparamagnetic maghemite (γ-Fe2O3) nanoparticles synthesized by Flame Spray Pyrolysis (FSP). *Journal of Colloid and Interface Science* 492: 176–190.

Ramana, D.V., Yu, J.S., and Seshaiah, K. (2013). Silver nanoparticles deposited multiwalled carbon nanotubes for removal of Cu (II) and Cd (II) from water: surface, kinetic, equilibrium, and thermal adsorption properties. *Chemical Engineering Journal* 223: 806–815.

Ramesh, G.V. and Radhakrishnan, T.P. (2011). A universal sensor for mercury (Hg, HgI, HgII) based on silver nanoparticle-embedded polymer thin film. *ACS Applied Materials & Interfaces* 3 (4): 988–994.

Ray, P.Z. and Shipley, H.J. (2015). Inorganic nano-adsorbents for the removal of heavy metals and arsenic: a review. *RSC Advances* 5 (38): 29885–29907.

Rowe, J.E., Rudolf, P., Tjeng, L.H. et al. (1992). Synchrotron radiation and low energy electron diffraction studies of ultrathin C 60 films deposited on Cu (100), Cu (111) and Cu (110). *International Journal of Modern Physics B* 6 (23–24): 3909–3913.

Saeki, D., Nagashima, Y., Sawada, I., and Matsuyama, H. (2016). Effect of hydrophobicity of polymer materials used for water purification membranes on biofilm formation dynamics. *Colloids and Surfaces A: Physicochemical and Engineering Aspects* 506: 622–628.

Saif, S., Tahir, A., and Chen, Y. (2016). Green synthesis of iron nanoparticles and their environmental applications and implications. *Nanomaterials* 6 (11): 209.

Sajid, M., Nazal, M.K., Baig, N., and Osman, A.M. (2018). Removal of heavy metals and organic pollutants from water using dendritic polymers based adsorbents: a critical review. *Separation and Purification Technology* 191: 400–423.

Saleh, T.A. and Al-Absi, A.A. (2017). Kinetics, isotherms and thermodynamic evaluation of amine functionalized magnetic carbon for methyl red removal from aqueous solutions. *Journal of Molecular Liquids* 248: 577–585.

Salihi, I. U., Kutty, S. R. M., and Ismail, H. H. M. (2018). Copper Metal Removal using Sludge Activated Carbon Derived from Wastewater Treatment Sludge. In *MATEC Web of Conferences*, 203, 03009. Les Ulis: EDP Sciences. https://doi.org/10.1051/matecconf/201820303009

Shah, M., Badwaik, V., Kherde, Y. et al. (2014). Gold nanoparticles: various methods of synthesis and antibacterial applications. *Frontiers in Bioscience* 19 (8): 1320–1344.

Shahbazi, A., Younesi, H., and Badiei, A. (2011). Functionalized SBA-15 mesoporous silica by melamine-based dendrimer amines for adsorptive characteristics of Pb (II), Cu (II) and Cd (II) heavy metal ions in batch and fixed bed column. *Chemical Engineering Journal* 168 (2): 505–518.

Shan, S. J., Zhao, Y., Tang, H., and Cui, F. Y. (2017). A mini-review of carbonaceous nanomaterials for removal of contaminants from wastewater. *IOP Conference Series: Earth and Environmental Science*, 68(1), 012003. Bristol: IOP Publishing.

Sharma, S. and Bhattacharya, A. (2017). Drinking water contamination and treatment techniques. *Applied Water Science* 7 (3): 1043–1067.

Sharma, D.C. and Forster, C.F. (1994). A preliminary examination into the adsorption of hexavalent chromium using low-cost adsorbents. *Bioresource Technology* 47 (3): 257–264.

Smith, S.C. and Rodrigues, D.F. (2015). Carbon-based nanomaterials for removal of chemical and biological contaminants from water: a review of mechanisms and applications. *Carbon* 91: 122–143.

Song, H., Qiu, Y., Wang, Y. et al. (2017). Polymer/carbon nanotube composite materials for flexible thermoelectric power generator. *Composites Science and Technology* 153: 71–83.

Speltini, A., Sturini, M., Maraschi, F., and Profumo, A. (2016). Recent trends in the application of the newest carbonaceous materials for magnetic solid-phase extraction of environmental pollutants. *Trends in Environmental Analytical Chemistry* 10: 11–23.

Stietiya, M.H. and Wang, J.J. (2014). Zinc and cadmium adsorption to aluminum oxide nanoparticles affected by naturally occurring ligands. *Journal of Environmental Quality* 43 (2): 498–506.

Sudarsan, J.S., Prasanna, K., Baskar, G., and George, R.B. (2015). Role of titanium oxide nanoparticle on heavy metal reduction in electroplating waste water treatment. *International Journal of ChemTech Research* 7 (2): 547–553.

Sumesh, E., Bootharaju, M.S., and Pradeep, T. (2011). A practical silver nanoparticle-based adsorbent for the removal of Hg2+ from water. *Journal of Hazardous Materials* 189 (1–2): 450–457.

Tabesh, S., Davar, F., and Loghman-Estarki, M.R. (2018). Preparation of γ-Al2O3 nanoparticles using modified sol-gel method and its use for the adsorption of lead and cadmium ions. *Journal of Alloys and Compounds* 730: 441–449.

Tabish, T.A., Memon, F.A., Gomez, D.E. et al. (2018). A facile synthesis of porous graphene for efficient water and wastewater treatment. *Scientific Reports* 8 (1): 1–14.

Tchounwou, P.B., Centeno, J.A., and Patlolla, A.K. (2004). Arsenic toxicity, mutagenesis, and carcinogenesis–a health risk assessment and management approach. *Molecular and Cellular Biochemistry* 255 (1–2): 47–55.

Tchounwou, P.B., Yedjou, C.G., Patlolla, A.K., and Sutton, D.J. (2012). Heavy metal toxicity and the environment. In: *Molecular, Clinical and Environmental Toxicology*, 133–164. Basel: Springer https://doi.org/10.1007/978-3-7643-8340-4_6.

Thaçi, B.S. and Gashi, S.T. (2019). Reverse osmosis removal of heavy metals from wastewater effluents using biowaste materials pretreatment. *Polish Journal of Environmental Studies* 28 (1) https://doi.org/10.15244/pjoes/81268.

Thanh, D.N., Novák, P., Vejpravova, J. et al. (2018). Removal of copper and nickel from water using nanocomposite of magnetic hydroxyapatite nanorods. *Journal of Magnetism and Magnetic Materials* 456: 451–460.

Thekkudan, V.N., Vaidyanathan, V.K., Ponnusamy, S.K. et al. (2016). Review on nanoadsorbents: a solution for heavy metal removal from wastewater. *IET Nanobiotechnology* 11 (3): 213–224.

Tian, Y., Gao, B., Morales, V.L. et al. (2012). Methods of using carbon nanotubes as filter media to remove aqueous heavy metals. *Chemical Engineering Journal* 210: 557–563.

Valko, M., Rhodes, C., Moncol, J. et al. (2006). Free radicals, metals and antioxidants in oxidative stress-induced cancer. *Chemico-Biological Interactions* 160 (1): 1–40.

Vilela, D., Parmar, J., Zeng, Y. et al. (2016). Graphene-based microbots for toxic heavy metal removal and recovery from water. *Nano Letters* 16 (4): 2860–2866.

Vunain, E., Mishra, A.K., and Mamba, B.B. (2016). Dendrimers, mesoporous silicas and chitosan-based nanosorbents for the removal of heavy-metal ions: a review. *International Journal of Biological Macromolecules* 86: 570–586.

Wang, H.J., Zhou, A.L., Peng, F. et al. (2007). Adsorption characteristic of acidified carbon nanotubes for heavy metal Pb (II) in aqueous solution. *Materials Science and Engineering: A* 466 (1–2): 201–206.

Wang, X., Guo, Y., Yang, L. et al. (2012). Nanomaterials as sorbents to remove heavy metal ions in wastewater treatment. *Journal of Environmental & Analytical Toxicology* 2 (7): 154–158.

Wang, Z., Zhang, J., Wen, T. et al. (2020). Highly effective remediation of Pb (II) and Hg (II) contaminated wastewater and soil by flower-like magnetic MoS2 nanohybrid. *Science of the Total Environment* 699: 134341.

Wright, I.A. and Ryan, M.M. (2016). Impact of mining and industrial pollution on stream macroinvertebrates: importance of taxonomic resolution, water geochemistry and EPT indices for impact detection. *Hydrobiologia* 772 (1): 103–115.

Wu, W., He, Q., and Jiang, C. (2008). Magnetic iron oxide nanoparticles: synthesis and surface functionalization strategies. *Nanoscale Research Letters* 3 (11): 397.

Wu, S.J., Liou, T.H., and Mi, F.L. (2009). Synthesis of zero-valent copper-chitosan nanocomposites and their application for treatment of hexavalent chromium. *Bioresource Technology* 100 (19): 4348–4353.

Wu, P., Jiang, L.Y., He, Z., and Song, Y. (2017). Treatment of metallurgical industry wastewater for organic contaminant removal in China: status, challenges, and perspectives. *Environmental Science: Water Research & Technology* 3 (6): 1015–1031.

Wu, Y., Pang, H., Liu, Y. et al. (2019). Environmental remediation of heavy metal ions by novel-nanomaterials: a review. *Environmental Pollution* 246: 608–620.

Wu, G., Wang, J., and Geng, J. (2020). Chemical HRPs in wastewater. In: *High-Risk Pollutants in Wastewater*, 5–39. Elsevier https://doi.org/10.1016/B978-0-12-816448-8.00002-2.

Xia, Z., Baird, L., Zimmerman, N., and Yeager, M. (2017). Heavy metal ion removal by thiol functionalized aluminum oxide hydroxide nanowhiskers. *Applied Surface Science* 416: 565–573.

Xiong, L.B., Li, J.L., Yang, B., and Yu, Y. (2012). Ti^{3+} in the surface of titanium dioxide: generation, properties and photocatalytic application. *Journal of Nanomaterials* 2012: 1–13.

Xu, Y. and Zhao, D. (2005). Removal of copper from contaminated soil by use of poly (amidoamine) dendrimers. *Environmental Science & Technology* 39 (7): 2369–2375.

Xu, J., Cao, Z., Zhang, Y. et al. (2018). A review of functionalized carbon nanotubes and graphene for heavy metal adsorption from water: preparation, application, and mechanism. *Chemosphere* 195: 351–364.

Yaghmaeian, K., Mashizi, R.K., Nasseri, S. et al. (2015). Removal of inorganic mercury from aquatic environments by multi-walled carbon nanotubes. *Journal of Environmental Health Science and Engineering* 13 (1): 55.

Yang, X., Zhou, T., Ren, B. et al. (2018). Removal of Mn (II) by sodium alginate/graphene oxide composite double-network hydrogel beads from aqueous solutions. *Scientific Reports* 8 (1): 1–16.

Yang, J., Hou, B., Wang, J. et al. (2019). Nanomaterials for the removal of heavy metals from wastewater. *Nanomaterials* 9 (3): 424.

Ye, C.C., An, Q.F., Wu, J.K. et al. (2019). Nanofiltration membranes consisting of quaternized polyelectrolyte complex nanoparticles for heavy metal removal. *Chemical Engineering Journal* 359: 994–1005.

Yeganeh, M.M., Kaghazchi, T., and Soleimani, M. (2006). Effect of raw materials on properties of activated carbons. Chemical Engineering and Technology: Industrial Chemistry-Plant Equipment-Process Engineering. *Biotechnology* 29 (10): 1247–1251.

Yen, C.H., Lien, H.L., Chung, J.S., and Yeh, H.D. (2017). Adsorption of precious metals in water by dendrimer modified magnetic nanoparticles. *Journal of Hazardous Materials* 322: 215–222.

Yin, P., Xu, Q., Qu, R. et al. (2010). Adsorption of transition metal ions from aqueous solutions onto a novel silica gel matrix inorganic–organic composite material. *Journal of Hazardous Materials* 173 (1–3): 710–716.

Youssef, A.M. and Malhat, F.M. (2014). Selective removal of heavy metals from drinking water using titanium dioxide nanowire. *Macromolecular Symposia* 337 (1): 96–101.

Zarghami, Z., Akbari, A., Latifi, A.M., and Amani, M.A. (2016). Design of a new integrated chitosan-PAMAM dendrimer biosorbent for heavy metals removing and study of its adsorption kinetics and thermodynamics. *Bioresource Technology* 205: 230–238.

Zhan, Y., Hu, H., He, Y. et al. (2016). Novel amino-functionalized Fe 3 O 4/carboxylic multi-walled carbon nanotubes: one-pot synthesis, characterization and removal for Cu (II). *Russian Journal of Applied Chemistry* 89 (11): 1894–1902.

Zhang, F., Wang, B., He, S., and Man, R. (2014). Preparation of graphene-oxide/ polyamidoamine dendrimers and their adsorption properties toward some heavy metal ions. *Journal of Chemical and Engineering Data* 59 (5): 1719–1726.

Zhang, C.Z., Chen, B., Bai, Y., and Xie, J. (2018). A new functionalized reduced graphene oxide adsorbent for removing heavy metal ions in water via coordination and ion exchange. *Separation Science and Technology* 53 (18): 2896–2905.

Zhao, X., Jia, Q., Song, N. et al. (2010). Adsorption of Pb (II) from an aqueous solution by titanium dioxide/carbon nanotube nanocomposites: kinetics, thermodynamics, and isotherms. *Journal of Chemical and Engineering Data* 55 (10): 4428–4433.

Zhao, X., Dou, X., Mohan, D. et al. (2014). Antimonate and antimonite adsorption by a polyvinyl alcohol-stabilized granular adsorbent containing nanoscale zero-valent iron. *Chemical Engineering Journal* 247: 250–257.

Zhao, J., Zhang, X., He, X. et al. (2015). A super biosorbent from dendrimer poly (amidoamine)-grafted cellulose nanofibril aerogels for effective removal of Cr (VI). *Journal of Materials Chemistry A* 3 (28): 14703–14711.

Zhao, G., Huang, X., Tang, Z. et al. (2018). Polymer-based nanocomposites for heavy metal ions removal from aqueous solution: a review. *Polymer Chemistry* 9 (26): 3562–3582.

Zheng, S., Hao, L., Zhang, L. et al. (2018). Tea polyphenols functionalized and reduced graphene oxide-ZnO composites for selective Pb2+ removal and enhanced antibacterial activity. *Journal of Biomedical Nanotechnology* 14 (7): 1263–1276.

Zhu, N., Ji, H., Yu, P. et al. (2018). Surface modification of magnetic iron oxide nanoparticles. *Nanomaterials* 8 (10): 810.

11

Spinel Ferrite Magnetic Nanoparticles

An Alternative for Wastewater Treatment

Sanjeet Kumar Paswan[1], *Pawan Kumar*[2], *Ram Kishore Singh*[1], *Sushil Kumar Shukla*[3], *and Lawrence Kumar*[1]

[1] *Department of Nanoscience and Technology, Central University of Jharkhand, Ranchi, Jharkhand, India*
[2] *Department of Physics, Mahatma Gandhi Central University, Motihari, Bihar, India*
[3] *Department of Transport Science and Technology, Central University of Jharkhand, Ranchi, Jharkhand, India*

11.1 Introduction

In the present scenario, water pollution caused by domestic, industrial, commercial, and agricultural activities has emerged as one of the major environmental challenges in both developed and developing countries. The consumption of available freshwater is almost 70%, 22%, and 8% in agricultural, industrial, and domestic sectors, respectively. The current trend of a growing population might lead to an acute shortage of water and it is anticipated that by the end of 2027, around four billion people will encounter the problem of clean drinking water due to groundwater depletion and water pollution (Nemerow and Dasgupta 1991; Helmer and Hespanhol 1997; Lehr and DeMarre 1980; Ali and Aboul-Enein 2004; Murray et al. 2015). Freshwater is essential for the survival of life. River and water resource contamination around the world is directly increasing due to rapid expansion in population and urbanization, as well as the rapid pace of industrialization leading to direct disposal of untreated noxious industrial waste, sanitary waste, and excess from agricultural fields. The scarcity of freshwater and its treatment cost impose paying for clean water and water tariffs in developing countries (Nemerrow 1978; Forgacs et al. 2004). Wastewater contains several pollutants such as biological pollutants, undesirable inorganic and organic chemicals that include heavy metal ions, dyes, and medication waste, which makes water unsafe for drinking purposes. Treatment of this contaminated water is a large challenge at present (Rai et al. 2005; Reddy and Yun 2016; Santhosh et al. 2016). These pollutants are not only harmful to living beings but also directly affect the ecosystem. Hence, removing these kinds of pollutants completely from wastewater or to reduce it to below a certain level as per World Health Organization (WHO) guidelines is an urgent need to ensure human health and environmental safety. Considering these hazardous effects,

Pollutants and Water Management: Resources, Strategies and Scarcity, First Edition. Edited by Pardeep Singh, Rishikesh Singh, Vipin Kumar Singh, and Rahul Bhadouria.

recycling of freshwater from contaminated water has brought the attention of various researchers across the world. Over a few decades, various conventional methodologies (primary or physical treatment, secondary or biological, and tertiary or chemical treatment) have been employed for the treatment of wastewater (Ali et al. 2011; Gupta et al. 2011; Saleh et al. 2011; Saleh and Gupta 2011; Gupta and Nayak 2012; Saleh and Gupta 2012). Sedimentation and gravity separation, evaporation, flotation, solvent extraction, micro and ultra-filtration, ion exchange, reverse osmosis (RO), coagulation, distillation, electrodialysis, precipitation, oxidation, electrolysis, etc. are some examples of methods (Shannon et al. 2008; Kurniawan et al. 2012; Bora and Dutta 2014; Kumar et al. 2014). The rigorous requirements of chemicals, infrastructure, and engineering skills make the above treatment methods troublesome, time-consuming, and costly. Biological treatment under aerobic or anaerobic environments removes suspended solids by microorganisms (algae, fungi, or bacteria). Further, chemical treatment of water involves various chemicals such as hydrochloric acid, alum, ammonia, ozone, coagulants, permanganate, ferric salts, chlorine, and corrosion control chemicals (Shannon et al. 2008).

The current situation demands an operative, robust, and low-cost promising technique that improves water quality. The recent encroachment of nanotechnology in the area of water purification has emerged as one of the finest and advanced ways to treat wastewater (Bora and Dutta 2014; Kumar et al. 2014). Nanotechnology offers a variety of promising solutions to filter out inorganic and organic water contaminants such as heavy metals like mercury, lead, arsenic, and cadmium, and biological toxins that cause water-borne diseases such as cholera and typhoid. Nanotechnology has advantages over existing water purification technology by providing scaffold related rapid and low-cost detection (Doroodmand et al. 2012; Bora and Dutta 2014; Kumar et al. 2014). Nanotechnology helps to create novel structures, devices, and systems by altering the properties of materials at the nanoscale. Nanotechnology encompasses materials at least one dimension below 100 nm and that exhibit a high degree of functionalization and size-dependent characteristics. Large surface area and high ratio of surface to volume of nanomaterials with large surface energy, high catalytic ability, and high reacting ability make them more suitable adsorption materials than conventional ones (Lindsay 2010; Pradeep 2012). A greater number of active sites for the interaction of nanomaterials possess properties of colloidal behavior to mix with the aqueous suspension of different chemical species. Due to the small size of nanomaterials, there is a strong ability of energy conservation and they may chemically regenerate after being exhausted (Li et al. 2006). Nanoparticles have the ability to treat pollutants at various depth levels located in water bodies, which is generally left out by other conventional technologies. Nanosensor, a nanotechnology-based application may enable us to boost drinking water by detection of impurities and pathogens (Bora and Dutta 2014; Kumar et al. 2014). A synthesis method of nanomaterials has been used by researchers for the removal of aquatic pollutants, which is discussed extensively in the synthesis section.

The promising techniques for water treatment employed by a nanotechnologist using different types of nanomaterials are *photocatalysis*, *nanofiltration*, and *nanoadsorbents* (Bora and Dutta 2014; Kumar et al. 2014). Various water pollutants have been treated by using photocatalysis techniques where chemical reaction rate change occurs by illuminating nanostructured catalysts using electromagnetic radiation such as infrared, visible, and ultraviolet. These nanostructured catalysts are commonly known as the photocatalyst. In photocatalysis, reaction partners comprise the chemical transformation by the absorption of light. Employing semiconductor material as a catalyst gives rise to light energy absorption phenomena generating a pair of electrons and a hole known as an exciton. These phenomena occur when the energy of light is comparable with its band-gap energy. The photogenerated electron–hole pair produces highly reactive reducing and/or oxidizing radicals in water, like hydroxyl ions (OH^-) superoxides (O_2^-) or other radicals. The generated radical also participates in secondary reactions which enable the degradation of organic and inorganic molecules of pollutant presence in contaminated water (McNaught and Wilkinson 1997; Bora and Dutta 2014). Photocatalysis shows surface phenomenon with the complex process having the following steps: (i) reactant diffusion at the catalyst surface; (ii) reactant adsorption at the catalyst surface; (iii) occurrence of reaction at the catalyst surface; (iv) product desorption at the catalyst surface taking place; and (v) product diffusion from the catalyst surface. The possible degradation steps through photocatalysis for organic pollutants present in water bodies are expressed through the following chemical equation (Pareek and Adesina 2003; Bora and Dutta 2014).

$$Catalyst \xrightarrow{Light} h_{VB}^{+} + e_{CB}^{-} \qquad (e-h\ pair\ generation) \tag{11.1}$$

$$OP \rightarrow OA^{-}(aq) + OC^{+}(aq)$$
$$(aqoueous\ dissociation\ of\ the\ pollutant) \tag{11.2}$$

$$H_2O + h_{VB}^{+} \rightarrow OH^{\bullet} + H^{+}(aq)$$
$$(photo-spliting\ of\ water) \tag{11.3}$$

$$O_2 + e_{CB}^{-} \rightarrow O_2^{-}$$
$$(electrophilic\ adsorption\ of\ dissolved\ O_2) \tag{11.4}$$

$$H^{+} + O_2^{-} \rightarrow HO_2^{\bullet}$$
$$(protonation\ of\ superoxide\ anion) \tag{11.5}$$

$$OA^{-} + h_{VB}^{+} \rightarrow OA^{\bullet} \tag{11.6}$$

$$OH^{-} + h_{VB}^{+} \rightarrow OH^{\bullet} \tag{11.7}$$

$$OA^{\bullet} + OH^{\bullet} / OH_2^{\bullet} \rightarrow CO_2 + intermediate + mineral\ acids + neutral\ sites \tag{11.8}$$

$$OC^{+} + OH^{-} \rightarrow hydroxylated\ products \tag{11.9}$$

$$h_{VB}^{+} + e_{CB}^{-} \rightarrow catalyst + heat$$
$$(e-h\ pair\ recombination) \tag{11.10}$$

where *OP* represents organic pollutant, *OA* stands for okadaic acid, *OC* is organic contaminants, and *VB* and *CB* are valence band and conduction band, respectively.

Semiconductor nanoparticles are widely used as photocatalysts for wastewater treatment. In order to activate these nanoparticles as an efficient photocatalyst, it must exhibit a wide energy band gap to produce an electron–hole pair (exciton) with low recombination ability. In addition, there must be sufficient energy that could carry out secondary reactions. An ideal photocatalyst should demonstrate the following properties: (i) photoactivity of the catalyst should be high; (ii) it should show inertness in a biological and chemical environment; (iii) it should be photostable; (iv) and it must be nontoxicity and cost-effective (Bhatkhande et al. 2001). Prominent nanostructured semiconductor photocatalysts such as zinc sulfide (ZnS), ferric oxide (Fe_2O_3), zinc oxide (ZnO), cadmium sulfide (CdS), and titanium dioxide (TiO_2) have shown great potential and ability in the degradation of dyes present in the water released by the textile industry (Baruah et al. 2008, 2009; Chan et al. 2011). TiO_2 photocatalysts have shown humic acid degradation ability from drinking water. They also show Hg^{2+}, CH_3Hg^{2+}, and chloride toxicity removal ability from wastewater (Eggins et al. 1997; Serpone et al. 1987). Potassium cyanide and chromium toxicity present in water have been removed through ZnO nanoparticles (Domenech and Peral 1988; Khalil et al. 1998). As a wide application of a nano photocatalyst, it helps in the conversion of organic contaminants (alcohols, chlorinated aromatic contaminants, phenolic derivatives, and carboxylic acids) into carbon dioxide and water as a by-product (Mills et al. 1993; Bhatkhande et al. 2001) and decomposition of harmful inorganic contaminants (cyanide, thiocyanate, halide ions ammonia, nitrites, and nitrates) (Hoffmann et al. 1995; Mills et al. 1996). The antibacterial properties of photocatalysts make them a suitable material for inhibiting microbial growth, preventing the growth of harmful bacteria and viruses (*Streptococcus cricetus, Lactobacillus acidophilus, Escherichia coli, Saccharomyces cerevisiae, Streptococcus mutans, Streptococcus natuss,* etc.) by destructing their cell wall (Mills and Hunte 1997).

Nanofiltration is another technique where filter media or a membrane separating the solid part from the liquid is made of newly developed nanomaterials like carbon nanomaterials, metal oxides, and zeolites. The high permeation rate and exemplary charge-dependent repulsion properties bring advancement in the pressure-driven membrane separation nanofiltration technique. Nanofiltration is also gaining widespread importance due to the requirements of low pressure (7–30 atm) in comparison with the RO method with a pressure range of 20–100 atm. In addition, it is also a low energy consumption technique (Cadotte et al. 1988; Drewes et al. 2008). In advanced water filtration technology, membrane processes are employed. But the function of the membrane generally becomes weak due to membrane fouling (Mohmood et al. 2013; Lutchmiah et al. 2014). The membrane performance could be enhanced by embedding spinel ferrite nanoparticles in the membrane. Incorporation of spinel ferrite nanoparticles enhances the membrane permeability, fouling resistance, mechanical and thermal stability execution for degradation of the contaminant, and self-cleaning (Pendergast et al. 2010). Spinel ferrite nanoparticle incorporation enhances surface hydrophilicity by decreasing the fouling capacity of the membrane (Kefeni et al. 2017). For example, Fe_3O_4-polyethersulfonate (PES) nanocomposite membrane exhibited a 99.4% elimination of humic acid during 48 hours of filtration (Ng et al. 2015).

Adsorption technology using nanoadsorbents offers high performance, affordability, and environmentally friendly application for wastewater treatment and water purification. Due to easy operation and cost-effectiveness, adsorption is a widely used physical treatment technique with great potential to eradicate soluble and insoluble inorganic, organic, and biological pollutants from wastewater. Adsorption depends on various factors like temperature, contact duration, adsorbent and adsorbate nature, chemical environment, particle size, etc. (Bora and Dutta 2014). A lack of distinguished adsorption capacity of bulk adsorbent materials limits its commercial application. Single adsorbents cannot be used for removing all kinds of pollutants. Several investigations and various aspects of water and waste treatment by adsorption techniques using nanoparticles have been carried out by the research community over a decade (Bora and Dutta 2014; Kumar et al. 2014; Reddy and Yun 2016; Santhosh et al. 2016). Nanoparticles with higher specific surface areas than bulk particles, easily functionalization properties with various chemical groups, and nanosized pores lead to increases in their adsorbent affinity toward the target contaminants (Santhosh et al. 2016; Reddy and Yun 2016). Nanoadsorbents can also be reprocessed by getting rid of the absorbed pollutants, thereby restoring them. For the commercial application of an ideal adsorbent, it should possess the following characteristics: (i) adsorption capacity should be rapid; (ii) must be nontoxic in the environment; (iii) must be reusable; (iv) must be cost-efficient; (v) performance should be very high; (vi) and after treatment of wastewater, its separation must be very easy (Gómez-Pastora et al. 2014). Getting separation of the adsorbed materials from the aqueous solution in a batch solution presents a challenge. After adsorption, retrieval of the nonbiodegradable sorbent is crucial if one wants to avoid secondary pollution problems. The widespread literature survey exposes the extensive use of nonmagnetic nanomaterials such as TiO_2, ZnO, ZnS, CdS, carbon nanomaterials, zeolites, etc. for treatment of wastewater (Santhosh et al. 2016). However, separation of nonmagnetic nanomaterials and recovery after treatment is tough, time-consuming, and expensive. As an effect, the problem of insufficient recovery might cause an additional environmental problem. This problem could be overcome using magnetic nanoadsorbents. Magnetic nanoadsorbents are new generation adsorbents materials that can easily bind with chemical contaminants and are currently used for the magnetic separation of chemical pollutants. Arsenic or other oil contamination having a binding ability to magnetic nanoparticle surfaces can be eliminated easily by using a magnet, which results in an affordable method for wastewater treatment. Hence, tunable morphology, large surface area, easy separation after sorption, and large efficiency make magnetic nanoadsorbents a prominent method for water purification and wastewater treatment (Reddy and Yun 2016; Kefeni et al. 2017, Kefeni and Mamba 2020).

11.2 Spinel Ferrite Nanoparticles

For water and wastewater treatment, the role of various iron oxide based magnetic nanoparticles like hematite (Fe_2O_3), maghemite (Fe_2O_3), and magnetite (Fe_3O_4) have been extensively studied (Ngomsik et al. 2005; Li et al. 2006; Xu et al. 2012; Tang and Lo 2013; Zhu et al. 2013; Gómez-Pastora et al. 2014). Fe_3O_4 nanoparticles have a phase changing ability and they transform into other oxides depending on environmental conditions. They show a significant impact on magnetic properties and may lead to aggregation. Moreover, by reducing the pH, stability also reduces for Fe_3O_4 nanoparticles (Gómez-Lopera et al. 2006; Parsons et al. 2009; Ramimoghadam et al. 2015). Recently, spinel ferrites (SFs) nanoparticles emerged as prominent materials for the treatment of wastewater owing to specific characteristics like large adsorption capacity, excellent tunable magnetic properties, excellent chemical reactivity, superparamagnetic character, surface active sites, large specific surface area, and excellent chemical resistance to oxidation as compared with iron oxide based materials (Reddy and Yun 2016; Kefeni et al. 2017; Kefeni and Mamba 2020). Spinel ferrite materials show a wide range of stability in acidic medium with pH 2.0–6.0, which provides a range of applications. Synthesis of spinel ferrite nanoparticles with tuned shape and size and its functionalization by attaching various ligands can be made effortlessly (Zhang et al. 2010; Shaikh et al. 2016). Recovering property of spinel ferrite nanoparticles after treatment of wastewater employing an external magnetic field makes it reusable for many cycles and assures their long-lasting performance in water treatment. The natural abundance of spinel ferrite based magnetic adsorbents and their inexpensive manufacturing can provide commercial profit from the applications. Spinel ferrites MFe_2O_4 (where M = Fe, Mg, Ni, Mn, Zn, Co, and Cu) has a similar crystal structure and composition to the naturally occurring spinel $MgAl_2O_4$, making them reasonable to manufacture. They provide a wide range of solid solutions by changing chemical composition without altering crystal structures (Zhang et al. 2010; Shaikh et al. 2016).

Spinel ferrite nanoparticles with spinel structure and cubic symmetry are represented by the general formula AB_2O_4, where B and A are trivalent and divalent metallic cations situated at the tetrahedral site and octahedral site, respectively. The four oxygen anions surround the tetrahedral site (A-site) whereas six oxygen anions surround the octahedral site (B-site). Its unit cell consists of eight formula units, i.e. 56 ions. Out of 56 ions, 32 oxygen anions form a cubic close-packed array giving rise to 96 interstitial sites. The 96 interstitial sites consist of 64 tetrahedral sites and 32 octahedral sites. Out of 64 tetrahedral and 32 octahedral sites, only 8 tetrahedral sites and 16 octahedral sites are filled by metal cations to satisfy the charge neutrality. A total of 24 cations occupy the available 96 interstitial sites in the unit cell. Hence, there are significant numbers of vacant interstitial sites inside the unit cell, which is conducive for cation migration between the interstitial sites. Cationic distribution depends on site preference, stabilization energy of crystal field, cationic ionic radius, size of the interstitial sites, synthesis conditions, and method (Valenzuela 1994; Sickafus and Wills 1999; Goldman 2006; Henderson et al. 2007). The stimulating physical and chemical behavior of spinel ferrite nanoparticles strongly depends on the types, concentration, and metal cation distribution over interstitial sites (Goldman 2006).

Spinel ferrite crystal structure is categorized into normal, inverse, and mixed spinel structure on the basis of the distribution of cations over A-site and B-site with preferable occupancy of divalent and trivalent cations, respectively. In the case of inverse spinel structure, site A is the occupancy position of half of the trivalent cations, whereas site B is occupied by half trivalent and the entire divalent cations. Both A-sites and B-sites are occupied by divalent cations in the case of the mixed spinel structure (Sickafus and Wills 1999). $ZnFe_2O_4$ falls under the category of normal spinel ferrite, $CoFe_2O_4$, and $NiFe_2O_4$ belong to inverse spinel, and $MnFe_2O_4$ exhibits a mixed ferrite structure (Reddy and Yun 2016). Possible cationic distribution in spinel ferrite can be signified by the general formula $(M^{2+}{}_{1-x}Fe^{3+}{}_{x})$ $[M^{2+}{}_{x}Fe^{3+}{}_{2-x}]$ where cations in the () and [] brackets represent A-sites and B-sites cations. In the above formula, x denotes the degree of inversion with existing value x = 0, 1 and between 0 and 1 corresponding to normal spinel, inverse spinel, and mixed spinel ferrite. Cation distribution variation over A-sites and B- sites results in materials with different magnetic properties with a similar chemical composition (Goldman 2006). For example, $ZnFe_2O_4$ in its bulk form is a normal spinel structure with antiferromagnetic property whereas at the nanoscale, it shows ferrimagnetic behavior due to different cation distribution over its interstitial sites (Valenzuela 1994; Henderson et al. 2007). A wide range of cationic substitution ability results in a variety of solid solutions without altering its crystal structure (Culity and Graham 2009). The intense chemical reactivity, high adsorption capacity, and reasonable saturation magnetization of ferrite materials provide a rapid elimination ability of water and wastewater pollutants. Spinel ferrite nanoparticles as a catalyst facilitate organic and inorganic reaction rates and reduce the required reaction temperature for wastewater treatment. In addition, it could easily be separated out by applying an external magnetic field to the solution mixture once water treatment is over (Zhang et al. 2010; Reddy and Yun 2016; Shaikh et al. 2016; Kefeni et al. 2017; Kefeni and Mamba 2020). In addition to ferrite nanoparticles, their nanocomposites are also utilized for wastewater treatment, where organic pollutants are subjected to photodegradation. An extensive literature review using the Database of Scopus with the key search term "spinel ferrite and waste water" revealed the increasing trend of using spinel ferrite nanomaterials, its derivatives compounds, and nanocomposites for the treatment of water and wastewater bodies.

Various research has reported on spinel ferrite nanoparticle-based catalysts for organic and inorganic contamination removal from wastewater due to their strong photodegradation ability (Reddy and Yun 2016; Kefeni et al. 2017; Kefeni and Mamba 2020). Transition metal substitution in spinel ferrite nanoparticles provides antibacterial properties for wastewater treatment (Sanpo et al. 2014). Higher antibacterial efficiency of cobalt ferrite with the substitution of zinc, manganese, and copper over a bacterial resistance drug has been reported (Maksoud et al. 2018, 2019). A better cationic and anionic adsorption ability of spinel ferrite nanoparticles is essential to act as photocatalytic materials for wastewater treatment. Ferrite nanoparticle surfaces adsorb anions due to positively charged surfaces below the pH having zero-point charge (pH_{zpc}) and adsorb cations due to negative charge surfaces above the pH having zero-point charge (pH_{zpc}) (Kefeni and Mamba 2020). For example, the modified surfaces of Fe_3O_4 nanoparticles using ascorbic acid remove about 45% of arsenate (AsO_4^{3-}) and arsenite (AsO_3^{3-}) at pH less than 7, whereas at pH above than 7, there is a 15% reduction in solution removal ability (Feng et al. 2012). There is an increase of arsenic removal at higher pH of positively charged surfaces of $CoFe_2O_4$, Fe_3O_4, and $MnFe_2O_4$ nanoparticles (Zhang et al. 2010). There is a wide range of applications of the photocatalytic behavior of spinel ferrite nanoparticles like $CoFe_2O_4$, $CuFe_2O_4$, $MnFe_2O_4$, $NiFe_2O_4$, and $ZnFe_2O_4$ and their derivative composites for water contaminants degradation (Reddy and Yun 2016). For example, $CoFe_2O_4$@ZnS shows better photodegradation of methylene blue (MB) contamination removal than bare $CoFe_2O_4$ nanoparticles and ZnS (Farhadi et al. 2017). There is photodegradation enhancement of MB in the presence of $CoFe_2O_4$/GO than the pure $CoFe_2O_4$ and GO (graphene oxide), which is due to improvement in adsorption capacity and easy transfer of negatively charged electrons from cobalt ferrite nanoparticles to the sheet of graphene, creating hindrance electron recombination (Zhang et al. 2013). $CuFe_2O_4$ nanoparticles have the highest photodegradation ability for Malachite green (MG) irradiating with visible light in comparison with $NiFe_2O_4$ and $ZnFe_2O_4$, whereas $NiFe_2O_4$ and $ZnFe_2O_4$ nanoparticles show the complete removal of 4-chlorophenol by photodegradation process in the presence of hydrogen peroxide under similar synthesis condition (Kurian et al. 2014; Shetty et al. 2017). Enhanced photodegradation activity has been observed for core-shell type $NiFe_2O_4$@ TiO_2 nanomaterials toward Rhodamine B (RhB) dyes, whereas $ZnFe_2O_4$ nanoparticles show photodegradation activity toward RR198 (reactive red) and RR 120 dyes (reactive red) (Mohmood et al. 2013; Wang et al. 2017). $ZnFe_2O_4$/rGO also shows excellent photocatalytic activity toward the degradation of RhB dyes, which may arise due to the band-gap reduction of $ZnFe_2O_4$ due to GO incorporation (Wang et al. 2019). Thus, numerous investigations advocate the highly efficient photodegradation ability of a variety of spinel ferrite nanoparticles and their composites due to the energy band-gap reduction and improvement in charge transportation and separation.

11.3 Synthesis of Spinel Ferrite Magnetic Nanoparticles

In order to utilize the spinel ferrite nanoparticles for wastewater treatment, one has to be aware of different synthesis methods to produce it. The different synthesis methods give rise to different morphology and variation in porosity, which greatly affects the adsorption properties for a variety of pollutants. There are two broad approaches to the synthesis of ferrite nanoparticles, namely top-down where chemical combination occurs to form particles, and the bottom-up approach, where larger size particles are crushed into minute particles (Pradeep 2012). The different routes of the synthesis process alter the properties of prepared nano ferrites significantly. Various methods can be used for ferrite nanoparticle preparation. Some of the methods are highlighted here.

11.3.1 Co-precipitation Method

The co-precipitation method of nano ferrites and their composite synthesis is the easiest and most effective pathway to synthesize various kinds of spinel ferrite nanomaterials and their composites. It has advantages such as less time-consuming and high mass production to synthesize uniform size particles. The aqueous solution has been prepared by the uniform mixing of divalent and trivalent transition metal salts in 1:2 mol ratios in an alkaline medium. Throughout the process, basic nature is maintained by optimizing pH through NaOH/NH_4OH. The resulting precipitates are filtered and washed many times with acetone and distilled water. Thereafter it is dried in a hot air oven. The obtained sample has been calcinated at the required temperature to get a crystalline sample with the desired phase. Several spinel ferrite nanoparticles such as $NiFe_2O_4$, $CoFe_2O_4$, $MnFe_2O_4$ (Kumar and Kar 2011; Pereira et al. 2012; Ahmad et al. 2017), etc. have been synthesized using this method. A general reaction mechanism for spinel nickel ferrite nanoparticles formation is given as follows:

$$Ni(NO_3)_2.6H_2O + 2NaOH \rightarrow Ni(OH)_2 + 2NaNO_3 + 6H_2O \quad (11.11)$$

$$Fe(NO_3)_3.9H_2O + 3NaOH \rightarrow Fe(OH)_3 + 2NaNO_3 + 9H_2O \quad (11.12)$$

$$2Fe(OH)_3 + Ni(OH)_2 \rightarrow NiFe_2O_4 + 4H_2O \quad (11.13)$$

11.3.2 Citrate Precursor Method

A clear solution is prepared through nitrate precursors or other soluble salts of metals in the citrate precursor method. There is a mixing of citric acid with metal nitrate solutions. Here, citric acid provides heat for the chemical reaction and prevents precipitation of hydroxylated compounds by forming a complex with metal ions. Better complex-forming ability, a low temperature required for ignition (i.e. 200–250 °C), and a combustion reaction in a controlled way with nitrates make citric acid preferable (Kumar and Kar 2011). A molar ratio 1 : 3 of metal nitrates to citric acid is maintained. The raise molar amount of citric acid minimizes precipitate production by arresting the free movement of metal ions. An 80–90 °C continuous stirring forms a viscous solution by evaporation, and finally, a highly viscous colored gel formation occurs. The prepared gel is kept in a hot air oven to remove excess water. The obtained samples have to be heat treated at the required temperature to obtain crystalline phase materials. A lack of precipitation and the probability of phase segregation provide advantages to the citrate precursor method over other methods. For example, $CoFe_2O_4$ spinel ferrites can be synthesized through the citrate precursor method and following the chemical reaction (Kumar and Kar 2011);

$$\begin{aligned} &\mathrm{Co(NO_3)_2\ldots 6H_2O + 2Fe(NO_3)_3\ldots 9H_2O + 3C_6H_8O_7\ldots H_2O + \frac{7}{2}O_2} \\ &\quad \rightarrow \mathrm{CoFe_2O_4 + 39H_2O + 4N_2 + 18CO_2} \end{aligned} \tag{11.14}$$

11.3.3 Hydrothermal Method

In the case of hydrothermal synthesis, an aqueous-alcoholic solution of metal salts is prepared. The basic pH nature of the solution is maintained by adding bases. The prepared solution is moved to a pressurized autoclave and is kept in temperature within the range of 150–180 °C for a fixed time interval. The metal concentration, solvent composition, temperature, and reaction time play a crucial role in the size and shape distribution of prepared NPs. Addition of surfactants like cetyltrimethylammonium bromide (CTAB) and polyethylene glycol (PEG) avoid agglomeration, control the growth of NPs, and change the shape. Synthesis of $ZnFe_2O_4$, $NiFe_2O_4$, $CoFe_2O_4$, and $MnFe_2O_4$ spinel ferrite nanoparticles using the hydrothermal method has been elucidated in the literature (Komarneni et al. 1998).

11.3.4 Sol–Gel Method

In sol–gel synthesis, metal alkoxide solutions have been used, which follow hydrolysis and the condensation polymerization process to form gels. Perfectly crystalline materials have been obtained by the removal of volatile by-products through heating. Low-temperature requirements and cost-effectiveness provide advantages to sol-gel synthesis compared with other methods. Various research has reported uniformity, composition control, and particle size distribution ability by the sol–gel method during various ferrite nanoparticles like $CuFe_2O_4$, $ZnFe_2O_4$, $MnFe_2O_4$, $CoFe_2O_4$, $NiFe_2O_4$, etc. (Mathew and Juang 2007; Carta et al. 2009; Sharma et al. 2015; Tadjarodi et al. 2015).

11.3.5 Solvothermal Method

The solvothermal method (if the water is solvent then it may be called hydrothermal) is a promising approach of aqueous or nonaqueous solvents with improved particle size and morphology. It is the most eco-friendly method. In this process, experimental conditions like precursor materials, solvent, time, temperature, and surfactant play a major role in the determination of morphology size and shape of spinel ferrite nanoparticles. Synthesis of various spinel ferrite nanoparticles like metal-doped $MgFe_2O_4$, Ni—Zn ferrite, $MnFe_2O_4$, Fe_3O_4, and $CoFe_2O_4$ with enhanced physical and chemical properties have been investigated (Li et al. 2015; Ni et al. 2015; Yan et al. 2015; Yin et al. 2016).

11.3.6 Microemulsion Method

In the microemulsion method, two relatively immiscible and thermodynamically feasible liquids are dispersed with surfactant. It diversifies ferrite nanoparticles by varying the nature of co-surfactant and surfactant, ratio of oil to water, and reaction conditions that regulate particle size. Low-temperature requirement and reuse of surfactant make it eco-friendly and favorable. Poor crystallinity, high poly dispersion, and a higher amount of solvent obligation due to the slow rate of nucleation at low temperature are the major disadvantages of this method. It is of two types, namely normal oil-in-water and reverse water-in-oil where the dispersed phase comprises monodisperse droplets of 2–100 nm size range. Synthesis of $CoFe_2O_4$, $NiFe_2O_4$, and Fe_3O_4 stable nanosized spinel ferrite particles has been done through this method (Kale et al. 2004; Mathew and Juang 2007; Ai et al. 2010).

11.3.7 Sonochemical Method

The Sonochemical method is extensively used for the preparation of SF nanoparticles owing to the good control of conditions of the reaction, which achieve homogeneous mixing and control distribution of particle size. Here, the intensity and temperature of ultrasonic waves play a major role in the particle size of ferrites, resulting in bubble formation and bringing in situ calcination due to heat developed by high energy collisions between the particles. Atomic-level mixing can be achieved through the sonochemical method, which reduces the required annealing temperature for crystalline phase formation. Some spinel ferrites synthesized by this technique take in $CuFe_2O_4$ and Fe_3O_4 (Lai et al. 2004; Lv et al. 2008).

11.3.8 Microwave-Assisted Method

This is a new methodology of spinel ferrite nanoparticle synthesis with advantages such as speed, good yield, high-energy efficiency, cost-effective, lower reaction time, and appropriate dielectric loss, making it commercially viable. But the yield obtained during this method is lower in comparison with other methods. In this method, microwave energy is used to combust the precursors in order to form SFs, whereas heating is used by other conventional combustion reactions. The microwave energy is converted to thermal energy, which raises the temperature from 100 to 200 °C for a short interval of time. A Teflon vessel helps as an exhaust drain for gas removal during the reaction. Several SF nanoparticles like Fe_3O_4, $Co_2Fe_2O_4$, $Mn_{1-x}Ni_xFe_2O_4$, and $ZnFe_2O_4$ have been synthesized through the microwave-assisted method (Manikandan et al. 2014; Tadjarodi et al. 2015; Jesudoss et al. 2016).

11.3.9 Thermal Decomposition

The thermal decomposition method encompasses thermal decomposition of precursors of organometallics such as carbonyls and metallic acetylacetonates in the presence of organic surfactants (hexadecyl amine and oleic acid) and solvents for ferrite nanoparticle synthesis. Monodispersed highly crystalline NPs have been obtained. Various factors act as controlling parameters for change in size and morphology of SF nanoparticles like aging, temperature, kind and concentration of surfactant, reaction time, and solvent nature. Generally, this method can be employed in the production of high-quality crystalline ferrite nanoparticles on a large-scale (per each synthesis at the level of 40 g) with controlled size and shape output (Kefeni et al. 2017).

11.3.10 Mechanical Milling Method

This technique falls under the category of a top-down approach scheme for ferrite nanoparticle synthesis by using a planetary ball mill. The outcome of this technique is in the form of a random shell structure and an ordered ferromagnetic core. This technique's advantages are short duration, simple, low-cost, and production on a large-scale. But milling for a long duration results in frequent contamination which changes the stoichiometry of the as-obtained ferrite particles. It is the major disadvantage of this technique. Synthesis of some SFs $CoFe_2O_4$, $CuFe_2O_4$, $NiFe_2O_4$, and $Ni_{1-x}Mn_xFe_2O_4$ has been reported through this method (Kefeni et al. 2017).

Attachment of several functional groups on the surfaces of spinel ferrite nanoparticles leads to the modification of properties of nanoparticles and enhances flexibility in these magnetic systems. To raise the contamination removal efficacy of spinel ferrite nanoparticles, a favorable interaction of SFs and desired contaminants can be made by the selective attachment of functional groups on SFNP surfaces. Some commonly utilized functional groups in the modification of spinel ferrite nanoparticles are shown in Figure 11.1 (Kefeni et al. 2017).

In addition to surface functionalization, SFNPs could also be utilized as nanofiller in making nanocomposite materials for wastewater treatment. Most recently, carbon nanotubes or graphene has been used with spinel ferrite nanoparticles to upsurge the adsorptive surface and achieve performances unattainable by ferrite nanoparticles alone. The utilization of spinel ferrite nanoparticles and their composites for the remediation of water bodies mainly takes place via two approaches: (i) direct adsorption of contaminants and (ii) photodegradation. Both of these processes are discussed in the next section.

11.4 Adsorption and Photocatalytic Degradation Mechanisms

11.4.1 Adsorption

As a result of huge contaminant occurrences in wastewater, the adsorption method has emerged as an efficient and cost-effective way of removing contaminants from water without generating secondary waste in the process (Bora and Dutta 2014; Kumar et al. 2014). Adsorption has been divided into two categories: chemisorption and physisorption. Strong

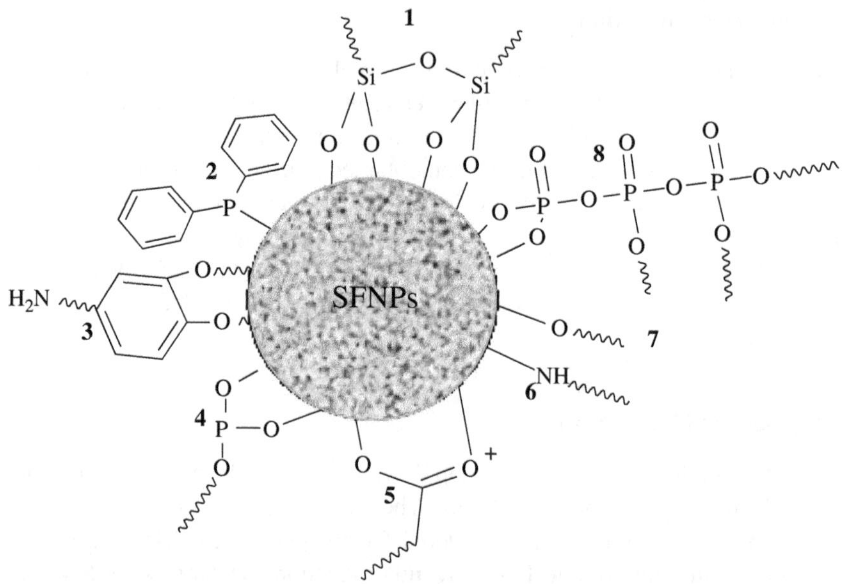

Figure 11.1 Representation of the functional groups utilized in the functionalization of spinel ferrite nanoparticles: 1, trimethoxy silane; 2, diphenylphosphine; 3, dopamine 1, 2-diol; 4, phosphoric acid; 5, carboxylic acids; 6, 7, and 8 represent amines, alcohols, and triphosphines, respectively. (*Source:* Kefeni et al. 2017. Reprinted with permission from Elsevier.)

chemical association between the surfaces of adsorbate and adsorbent takes place by electron pair sharing in chemisorption, while a weak attractive interaction exists between solid surfaces and pollutants (Sanghi and Verma 2013). The adsorption behavior of selected metals and organic contaminants on the surface of spinel ferrite magnetic nanoparticles is listed in Table 11.1 (Venturini 2019).

Several isotherm and kinetic models like isotherm equations proposed by Langmuir and Freundlich and kinetic equations of pseudo-first and pseudo-second order are employed to estimate the adsorption capacity of different spinel ferrite nanoparticles and their nanocomposites (Kumar et al. 2014; Reddy and Yun 2016; Sigdel et al. 2016). The equations help to design the adsorption system as well as to evaluate the capacity of adsorption and the affinity of adsorbent toward adsorbate.

11.4.1.1 Adsorption Mechanisms

Various adsorbent and adsorbate interactions govern the adsorption mechanism which mainly occurs due to surface charge and surface hydroxyl group behavior (Sun et al. 2015; Reddy and Yun 2016). The spinel ferrite (MFe_2O_4) surface makes its arrangement with plentiful hydroxyl groups (M-OH, Fe-OH), affording main binding positions for numerous cationic and anionic types (Reddy and Yun 2016). Various physiochemical interactions like complex formation with inner or outer spheres and weak forces interaction (dipole–dipole interactions, van der Waals forces, hydrogen bonding, and Pi-Pi interactions) and ion-exchange take part in the elimination of different contaminants from wastewater (Tombácz 2009; Hou et al. 2010; Zhang et al. 2010; Wang et al. 2012;

Table 11.1 Some examples of adsorption of organic and inorganic compounds on the surface of spinel ferrite magnetic nanoparticles.

Adsorbent	Size (nm)	Specific area (m^2/g)	Pollutant	Capacity (mg/g)	Reference
$CuFe_2O_4$/DC	17.91	16.96	Pb (II)	921.1	Khan et al. (2019)
$MnFe_2O_4$/GO	20	67.5	As (V)	240.4	Huong et al. (2018)
$CoFe_2O_4$	20	—	Cs (I)	75	Hassan and Aly (2020)
Fe_3O_4	—	109.6	U (IV)	52.63	Li et al. (2019)
$NiFe_2O_4$/rGO	32.2	167.3	Th (IV)	100	Lingamdinne et al. (2017)
$CoFe_2O_4$	9–21	31.2	RR 195 dye	91.7	Nassar and Khatab (2016)
$MnFe_2O_4$/GO	20	67.5	Methylene blue dye	177.3	Lan et al. (2018)
$Ni_{0.5}Zn_{0.5}Fe_2O_4$	9	—	Alizarin dye	250	Afkhami et al. (2015)
$CoFe_{1.9}Sm_{0.1}O_4$	11	—	Congo red dye	178.6	Wu et al. (2016)
$CaFe_2O_4$	15–30	41.8	Congo red dye	40.9	An et al. (2015)

Bao et al. 2013; Yang et al. 2014; Zhou et al. 2014). Hydroxyl groups (S-OH) form an outer layer because of aqueous dispersion of spinel oxide surfaces in water molecules (Tombácz 2009; McCafferty 2015). Solution pH alters the metal oxide surface charge. For example, a positive charge (S-OH^{2+}) develops on the spinel ferrite nanoparticle surface at low pH owing to the upsurge in H^+ ions, whereas, a negative charge develops at high pH owing to hydroxyl groups deprotonation. The protonation or deprotonation of SF active surface sites (S-OH) governed by pH_{zpc} in aqueous solution are given below (Jia et al. 2012; Kosmulski 2012; Ren et al. 2012; Jia et al. 2013; Reddy and Yun 2016):

$$\equiv S-OH_{(surface)}+H^{+}_{aq}\rightarrow\equiv S-OH_2^{+}(surface)(\text{pH}<\text{pH}_{\text{zpc}}) \tag{11.15}$$

$$\equiv S-OH_{(surface)}+OH^{-}_{aq}\rightarrow\equiv S-O^{-}(surface)+H_2O(\text{pH}>\text{pH}_{\text{zpc}}) \tag{11.16}$$

The term pH_{zpc} represents a pH at which net charge of particle surface is zero. The surface of $ZnFe_2O_4$ helps in the adsorption of anionic acid red 88 (AR88) at a lower pH, whereas adsorption property decreases with increasing pH, which may arise due to electrostatic repulsion between the negatively charged $ZnFe_2O_4$ and the anionic dye (Konicki et al. 2013). The adsorption of cationic species onto spinel ferrite nanoparticles takes place via a reaction with OH groups present on the surface (Ren et al. 2012; Kumar et al. 2014).

$$=S-OH+M^{2+}\rightarrow=S-O-M^{+}+H^{+} \tag{11.17}$$

$$\equiv S-O^{-}+M^{2+}\rightarrow\equiv S-O-\ldots M^{2+} \tag{11.18}$$

$$\equiv(S-OH)_2+M^{2+}\rightarrow\equiv(S-O^{-})_2-M^{2+}+2H^{+} \tag{11.19}$$

In the case of oxyanions adsorption, spinel ferrite nanoparticles adsorption capacity declines with an increase in pH and upsurges with the decrease in pH. Increased protonation raises the adsorption behavior of MFe_2O_4 by creating surface positive charges. In oxyanions adsorption, a ligand exchange process occurs through inner and outer-sphere surface complex formations (Kumar et al. 2014; Sun et al. 2015).

$$\equiv S - OH + H^+ + M^- \leftrightarrow \equiv S - M + H_2O(\text{inner} - \text{sphere}) \tag{11.20}$$

$$= S - OH + H^+ + M^{2-} \leftrightarrow = S - OH_2^+ - M^{2-}(\text{outer} - \text{sphere}) \tag{11.21}$$

The adsorption of Pb (II) on the $Co_{0.6}Fe_{2.4}O_4$ surface occurs through the chemical ion-exchange mechanism where ion exchange and outer-sphere complex formation take place at pH < 7, while for pH > 7, the formation of an inner-sphere surface complex takes place (Duan et al. 2015).

Various physiochemical interactions exhibit the contamination removal behavior of wastewater. Copper contamination is removed by EDTA functionalized silica-coated Fe_3O_4 surface through ion exchange and electrostatic force of attractions (Neyaz and Siddiqui 2015). Chromium contamination is removed from wastewater using $NiFe_2O_4$ nanoparticles by the physisorption process (Jia et al. 2012). The adsorption of cationic dyes on several spinel ferrite nanoparticles occurs because of electrostatic interaction and ion-exchange where surfactant groups present on the spinel ferrite nanoparticle, which is an important factor (Vîrlan et al. 2013). Elimination of Congo red (CR) and MB takes place through hydrogen bond association between the hydroxyl groups and dye existing with $MnFe_2O_4$ nanoparticles (Yang et al. 2014). Arsenate contamination removal favors through inner-sphere complex formation on γ-Fe_2O_3 NPs (Tuutijärvi et al. 2010). The adsorption mechanism result is governed by various factors like surface charge, porous structure, oxidation state, and functional groups. Hence, variation in adsorption mechanism occurs due to the type of pollutant existing in wastewater and ferrite nanoparticles and their nanocomposites are employed for the treatment of wastewater.

11.4.1.2 Factors Influencing Adsorption Capacity

Several studies highlighted the role of various factors like the type of synthesis, charge, morphology, system temperature, adsorbent particle size and dosage, contaminant concentration, wastewater pH, etc. on the adsorption capacity of spinel ferrite nanoparticles and their nanocomposites for the treatment of water (Yagub et al. 2014; Lata and Samadder 2016; Lofrano et al. 2016; Reddy and Yun 2016). Augmented surface area to volume ratio arises due to the nano-size effect, which raises the adsorption capacity. For example, $MnFe_2O_4$ nanoparticles, having an average diameter of 10 nm and a surface area of 208 m^2/mg, eliminate Cr (VI) with an exhibiting adsorption ability of 31.6 mg/g, whereas a diameter of ~20 nm and surface area of 180 m^2/g exhibited enhanced adsorption capacity (Hu et al. 2005, 2007; Auffan et al. 2008). The morphology of the nanoparticle influences the adsorption behavior of ferrite nanoparticles. For example, cobalt ferrite nanoparticles with spherical morphology show uranium (VI) removal ability with an adsorption capacity of 170 mg/g, whereas nanorod morphology of zinc ferrite shows an adsorption capacity of 34.2 mg/g for Cr (VI) elimination (Wei et al. 2014, Jia et al. 2015). Functionalization of ferrite nanoparticles with aminol, anhydride, and thiol groups exhibited higher adsorption capacity (Kraus et al. 2009; Rooygar et al. 2014;

Pirouz et al. 2015). There is an upsurge in adsorption of Cu (II) possessing an adsorption capacity of 22.6 mg/g, and 15.4 mg/g adsorption capacity for Cr (VI) adsorption using chitosan-coated $MnFe_2O_4$ nanoparticles (CCMNPs) (Xiao et al. 2013). Change in structural composition by rare-earth substitution into MFe_2O_4 results in structural disorders that enhance the adsorption capacity due to increased pore size and pore volume, reduced particle size, active surface binding sites, and high surface area (Cheng et al. 1999; Sharma et al. 2015). For example, an upsurge adsorption capacity of 57.2 mg/g has been reported for Cr (VI) contamination removal for Ce^{3+} doped $ZnFe_2O_4$ clusters (Kuai et al. 2013). The surface charge of nanoparticles/nanocomposites becomes positively and negatively charged at a low value and high value of pH, respectively, which favors adsorption of anions at a low value of pH while cations adsorption at a high value of pH. At a pH range of 4–6, γ-Fe_2O_3 nanoparticles show optimum adsorption of Mo (VI) with the value of adsorption capacity at 33.4 mg/g (Afkhami and Norooz-Asl 2009). $CuFe_2O_4$ has increased the elimination efficacy of Mo (VI) at a lower pH while increasing the pH above 10 leads to the disappearance of the removal ability. It may develop due to increased attractive interaction between the $(MoO_4)^{2-}$ anion and the positively charged surface of spinel ferrite nanoparticles as pH drops (Tu et al. 2014). Proper investigations of Cu (II) removal highlight the effect of initial concentration, adsorbent dose, and pH. There is an increase in copper removal from wastewater solution by increasing the pH from 2 to 5.3 with Fe_3O_4 nanoparticles. The 66% reduction in copper removal activity has been observed by enhancing the early solution concentration of Fe_3O_4 from 10 to 100 mg/l. There is an increase in copper removal from 41.65 to 97.05% by raising Fe_3O_4 adsorbent quantity from 0.1 to 1 g (Davarnejad and Panahi 2016). The impact of pH on contamination removal by using various spinel ferrites has been widely reported. The increase in Pb^{2+} adsorption has been observed at zero-point charge (zpc) 4.35 by using $NiFe_2O_4$ and $MgFe_2O_4$ nanoparticles. The adsorption ability of lead cations (87%) and mercury cations (88%) has been accomplished at pH of zero-point charge (zpc) 4 with $CoFe_2O_4$-rGO (Zhang et al. 2014; Fang et al. 2016). Functionalized surface cobalt ferrite nanoparticles employing the amine functional group has been observed for the removal of Direct green 6 (DG6), Direct red 80 (DR80), and Acid blue 92 (AB92) dyes with an adsorption capacity of 384.61, 333.3, and 625 mg/g, respectively, from aqueous solutions (Reddy and Yun 2016). Elimination of reactive blue 5 (RB5) dye from wastewater is enhanced by above 90% at pH = 1 and 25 °C by using $NiFe_2O_4$ nanoparticles with particle size 17 nm (Kefeni et al. 2017; Kefeni and Mamba 2020). Temperature also influences adsorption abilities by enhancing adsorption for endothermic reactions and decreasing for exothermic reactions. The adsorption of Pb (II) on $Co_{0.6}Fe_{2.4}O_4$ nanoparticle surface synthesized by the thermal decomposition method exhibited elevated removal efficacy (>85%) within 20 minutes of reaction time and the reaction was revealed to be endothermic and spontaneous in nature (Reddy and Yun 2016; Kefeni et al. 2017; Kefeni and Mamba 2020). Calcination temperature has a direct effect on spinel ferrite nanoparticle surface area due to controlling the growth of nanoparticles. For example, $NiFe_2O_4$ hollow fiber calcinated at three different temperatures, 500, 600, and 700 °C, revealed a decreasing trend in BET surface area, but the 600 °C calcinated sample exhibited highest pore volume with a higher adsorption ability of 89.85 mg/g for CR elimination from wastewater (Kefeni and Mamba 2020). Similarly, 300 °C calcined manganese ferrite nanoparticles showed greater

Cr (VI) elimination behavior in comparison with 900 °C (Kefeni and Mamba 2020). Hence, an appropriate understanding is essential to establish a comprehensive knowledge for the complete elimination of every targeted wastewater contaminant.

11.4.1.3 Adsorptions of Dye, Pharmaceuticals, and Pesticides

A major part of water pollution is contributed by textile industries, which discharge dyes as waste and create critical health-related issues and pollute the environment. Due to complex and stable molecular structures, dye degradation is difficult, and their small concentrations generate toxic chemicals through various chemical reactions (hydrolysis, oxidation, etc.). Dyes have a carcinogenic and mutation causing effect on humans and aquatic organisms. With the aim of the removal of dyes, recently, bare and modified MFe_2O_4 have been used owing to their enhanced adsorption capability and smooth retrieval and recycle. Chitosan-glutamic-Fe_3O_4@SiO_2 and chitosan-Fe_3O_4@SiO_2 based nanocomposites have been widely used for the elimination of three categories of cationic dyes, viz. MB, crystal violet (CV), and cationic light yellow (7GL) from contaminated water solutions. Chitosan-glutamic-Fe_3O_4@SiO_2 with corresponding adsorption capacities values 375.4, 180.1, and 217.3 mg/g at 25 °C and pH 7, and chitosan-Fe_3O_4@SiO_2 with values 78.8, 28.8, and 17.6 mg/g have been reported (Yan et al. 2013). Various spinel ferrite derivatives applied for the elimination of diverse dyes from wastewater. The activated carbon-$NiFe_2O_4$ composite has shown high adsorption capacity for the 90% removal of methyl orange. The porous $Ni_{0.6}Fe_{2.4}O_4$ nanoparticles exhibited removal of 92.04% CR dye and $ZnFe_2O_4$ nanoparticles removed acid red 88 (AR88) dye (Konicki et al. 2013; Kefeni et al. 2017). Various pharmaceutical contaminants, specifically chlortetracycline (CTC), tetracycline (TCN), and oxytetracycline (OTC) from wastewater can be removed by using spinel ferrite nanoparticles (Kefeni et al. 2017; Kefeni and Mamba 2020). Pesticides widely used for increasing agriculture yield disseminate pesticides such as organochlorine and atrazine in the environment and are noticed in contaminated water. Pesticides produce adverse effects on human health even at low concentrations. Presently, spinel ferrite nanoparticles, mainly Fe_3O_4 based nanocomposites, have gained attention for pesticide elimination from wastewater (Kefeni et al. 2017; Kefeni and Mamba 2020). The photodegradation ability of spinal ferrite can also be helpful in organic pollutants removal.

11.4.2 Photocatalytic Degradation

Industrial wastewater discharge is a major concern around the world. Harmful organic pollutants, for example, dyes and phenols derivatives released as industrial waste, create a challenge for treatment under visible light irradiation. In this manner, a possible photocatalytic approach has been made by using spinel ferrite nanoparticles and their corresponding nanocomposites due to their durability and stability under photoirradiation. Without producing secondary waste, the organic pollutants may photocatalytically degrade into carbon dioxide gas, water, and other gaseous products and chemicals (Kefeni et al. 2017). The unique property of spinel ferrite nanoparticles in terms of the photocatalyst is that it can be used both in bare form and composite (semiconductor-coated surfaces) form with oxidants. For example, H_2O_2 is widely used in organic pollutant degradation using the photocatalytic approach. Degradation of organic contaminants via the photocatalytic approach in wastewater is enhanced by incorporating semiconductors because of reduction in electron–hole recombination, corrosion, and band gap.

11.4.2.1 Mechanism of Contaminant Degradation Using the Photocatalytic Approach

A wide range of band-gap energy arises due to various spinel ferrite materials like $CuFe_2O_4$ (1.89 eV), $CoFe_2O_4$ (2.31 eV), $ZnFe_2O_4$ (1.91 eV) Fe_3O_4 (1.92 eV), $NiFe_2O_4$ (2.2 eV), and γ-Fe_2O_3 (2.03 eV) (Kefeni et al. 2017) make it appropriate for visible light absorption. The energy band gap of spinel ferrite nanoparticles is low in comparison with semiconductors such as zinc oxide (3.2), titanium dioxide (3.2 eV), WO_3 (2.5 eV), zinc sulfide (3.7 eV), and cadmium sulfide (2.62 eV), which utilize visible light energy effectively and simply change light energy into chemical energy to sustain oxidation and reduction for organic pollutant degradation (Kefeni et al. 2017). An excitation of electron (e−) in the valence band takes place which moves to the conduction band, creating a photogenerated hole (h+) captured by dissolved oxygen gas presented in the contaminated water after mixing the ferrite photocatalyst. Free radical reactive oxygen is produced. Thereafter, a further reaction takes place with hydrogen ions so that an active free hydroxyl radical and hydroxyl ion could form (Eq. 11.23). An active hydroxyl free radical is produced after the reaction of the photogenerated hole (h+) (produced in valence band) with water (Eq. 11.24).

The contaminants present in wastewater are in close vicinity of spinel ferrite nanoparticles. Due to the large adsorption capacity of spinel ferrite nanoparticles and the creation of hydroxyl radicals on the ferrite photocatalyst surface, the contaminants are attacked easily. As an effect, it is easily degraded (reaction 11.25). The active surface sites of the ferrite photocatalyst continue the desorption of degraded products. In addition to organic compound degradation through the radical attack, the degradation of organic contaminants could also be possible by directly reducing the electrons present in the conduction band (reaction 11.26) and the oxidation of holes occupying the valence band (reaction 11.27) (Tseng et al. 2010; Henderson 2011; Lee et al. 2014; Mamba and Mishra 2016). The chemical reaction for the photocatalytic approach is expressed as follows (Kefeni et al. 2017):

$$e^- + O_2 \rightarrow O_2^- \quad (11.22)$$

$$O_2^- + 2H^+ + 2e^- \rightarrow OH + OH^- \quad (11.23)$$

$$h^+ + H_2O \rightarrow OH + H^+ / OH \quad (11.24)$$

$$OH + \text{Organic Contamination (OC)} \rightarrow CO_2(g) + H_2O + \text{others} \quad (11.25)$$

$$e^- + \text{Organic Contaminant (OC)} \rightarrow OC^- \rightarrow\rightarrow \text{products degraded} \quad (11.26)$$

$$h^+ + \text{Organic Contaminant (OC)} \rightarrow OC^+ \rightarrow\rightarrow \text{products degraded} \quad (11.27)$$

The schematic picture for the formation of hydroxyl radicals and reactive oxygen is depicted in Figure 11.2.

Textile industries widely use dyes for coloring purposes of cloths. The complex structures of dyes provide stability, which makes them less prone to chemical and biological

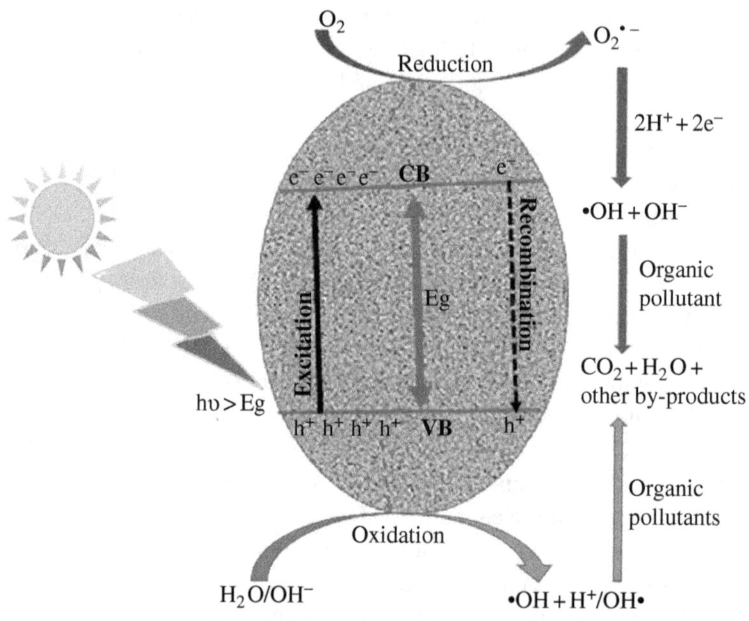

Figure 11.2 Schematic diagram showing the formation of oxygen and hydroxyl radicals (which are reactive) under illumination of visible light and possible degraded product. (*Source:* Kefeni and Mamba 2020. Reprinted with permission from Elsevier.)

degradation (Yagub et al. 2014). Recent developments have been made by using several spinel ferrite materials such as $MnFe_2O_4$@PANI@Ag and $Co_{0.53}Mn_{0.3}Fe_{2.16}O_4$@ TiO_2 with the upsurge in degradation ability for azo dyes in comparison with pure spinel ferrite nanoparticles (Amir et al. 2016; Neris et al. 2018). Against indigo carmine synthetic dye degradation, Ni doped $MnFe_2O_4$ with particular composition $Mn_{0.5}Ni_{0.5}Fe_2O_4$ has revealed superior photocatalytic activity (Jesudoss et al. 2016).

In the photodegradation method of water treatment, Fenton and photo-Fenton methodology are commonly used, which comprises the reagent H_2O_2 and a ferrous ion (Fe^{2+}) source. When the ferric ion (Fe^{3+}) is generated, the decomposition of the peroxide is induced by metal ions giving rise to OH radicals, resulting in a chemical reaction with an organic compound and eventually decomposition. The high recombination rate of the produced free radicals can be overcome by accelerating the breakdown of the peroxide by the exploitation of UV photons. Nevertheless, in the liquid current, the ferric ions remain present, which requires another processing so that complete purification of the water after processing can be achieved. In this regard, the application of a magnetic-based heterogeneous catalyst comprising iron such as ferrites makes the treatment process very easy. A rate of MB discoloration under the influence of neutral pH is high using Fe_3O_4@SiO_2 nanocomposite behaving like a Fenton catalyst in order to decompose H_2O_2 in comparison with bare Fe_3O_4 (Yang et al. 2015). A possible mechanism of hydroxyl ion formation due to divalent cations has been reported (He et al. 2016). A list of the catalytic systems using a heterogeneous Fenton process with spinel ferrite nanoparticles from various available research is listed in Table 11.2.

Table 11.2 Typical examples of spinel ferrite nanoparticles using heterogeneous Fenton processes.

Name of adsorbent	Size of nanoparticle (nm)	Specific area exhibited by nanoparticle surface (m^2/g)	Compound decomposed during treatment	Apparent reaction rate during decomposition	Reference
$ZnFe_2O_4$	30.06	151	Methylene blue	0.267	Sharma et al. (2015)
$MgFe_2O_4$	20	14	Methylene blue	0.117	Ivanetsa et al. (2019)
$NiFe_2O_4$ /CNT	—	54	Amaranth	0.017	Rigo et al. (2017)
Doped $MgFe_2O_4$	52	141.5	Rhodamine B	0.0197	Diao et al. (2018)
$CoFe_2O_4$	25.3	48.6	2,4-dichlorophenol	0.0273	Nair and Kurian (2017)

Phenol and its derivatives are considered as the largest groups of environmental pollutants. This is mainly due to their wide industrial application as antibacterial and antifungal agents. The US Environmental Protection Agency (USEPA) has put them in the category of main organic pollutants in wastewater due to their toxicity, carcinogenicity, and mutagenic nature (Boruah et al. 2017). They can cause severe harm to human and animal liver, lungs, and red blood cells, even at very low concentrations (Anku et al. 2017). In addition, the degradation of these pollutants is difficult due to their stability. Consequently, they remain in the environment for longer periods (Wang et al. 1999). The application of Fe_3O_4 mixed with H_2O_2 in wastewater containing phenol has removed 85% of phenol in three hours at a temperature of 16 °C without the formation of any secondary pollutant (Zhang et al. 2008). The Fe_3O_4 (ferrite)–ZnO (semiconductor) hybrid nanoparticles are reported to degrade the phenol through enhanced photocatalytic activities. It has been observed that 89% of the photocatalyst used was recovered after three cycles, with phenol removal of 82.8, 72.4, and 65.1% in cycles one, two, and three, respectively (Feng et al. 2014). These values are much greater than that of freshly prepared ZnO, in which only 52% degradation of phenol was observed. This enhanced photocatalytic performance could be attributed to the synergic effects between the SFNPs and semiconductors that reduced the fast recombination of photogenerated electrons and holes, thereby increasing the efficiency of charge separation and allowing more electrons and holes to be available for the reduction and oxidation of contaminants (Kefeni and Mamba 2020). The spinel ferrite nanoparticles and their composites are promising materials for the effective degradation of antibiotics through the photocatalytic approach. For example, graphitic carbon sand composite (GSC) and bentonite (BT) supported superparamagnetic $MnFe_2O_4$ nanoparticles have been used effectively for the photodegradation of ampicillin (AMP) and oxytetracycline (OTC) antibiotics under solar light. The reported results have shown 96 and 83% of AMP and 99 and 90% OTC degradation in 60 and 120 minutes under solar irradiation by using $MnFe_2O_4$/GSC and $MnFe_2O_4$/BT (Reddy and Yun 2016; Kefeni et al. 2017; Kefeni and Mamba 2020).

11.5 Recovery and Reuse

Recovery and reuse of spinel ferrite nanoparticles and their composites after being used for water treatment is one of the key steps. Separation of spinel ferrite nanoparticles after treatment of water with the external magnetic field is superior to the commonly used filtration and centrifugation process because it is simple, selective, and rapid. The presence of these materials as a core nanomaterial aids the possible recovery of the adsorbent along with the pollutant removed from water using an external magnetic field. The schematic illustration of spinel ferrite nanoparticles and their composite adsorption, desorption, recovery, and reuse is presented in Figure 11.3. The aqueous solutions of sodium hydroxide and strong acids are widely used for the regeneration of adsorbent and desorption of toxic metals from loaded ferrites nanoparticles and its derivative compounds. The amount of recovery of initial used ferrite nanoparticles and composites is dependent on the stability of the adsorbent and efficiency of the chemicals used for desorption (Kefeni and Mamba 2020). The main purpose of the regenerating process is to restore the adsorption capacity of an exhausted adsorbent. The desorption efficiency of contaminants is dependent on many factors, such as the type of adsorbent and desorbate, pH, temperature, eluent, and contact time between the solid and liquid phases. The desorption process of contaminants from ferrite nanoparticles is relatively easy, mainly due to its stability under basic and acidic conditions (particularly $pH > 3$) (Zhang et al. 2010). Hence, ferrite nanoparticles can be easily regenerated by using low concentrated acid or base solutions or alcohols. The anions, metallic cations, and anion and cation dyes are effectively desorbed using 0.001 to 0.2 HCl or HNO3, 0.001 to 2 M NaOH (aq.), and 4-6% (v/v) acetic acid in methanol, respectively (Gómez-Pastora et al. 2014).

After regeneration, SFNPs can be used over and over for several cycles, and this makes the process cost-effective. The effective regeneration of the adsorbent used and the

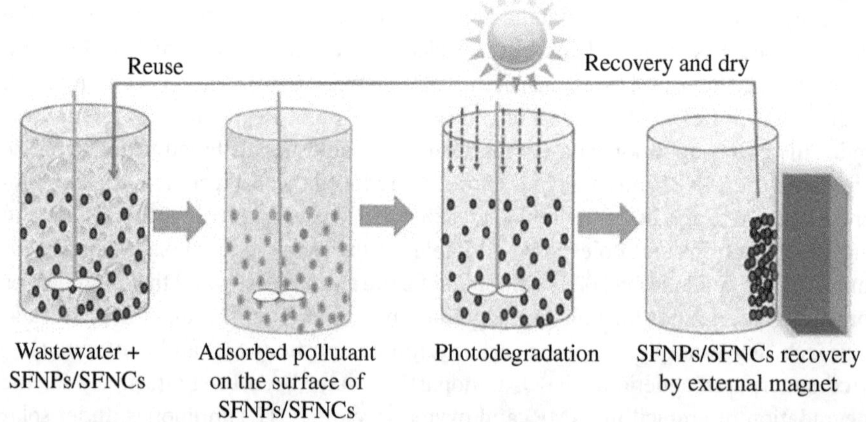

Figure 11.3 Schematic diagram for wastewater treatment using spinel ferrite nanoparticles and their composites and possible recovery and reuse. (*Source:* Kefeni and Mamba 2020. Reprinted with permission from Elsevier.)

desorption of contaminants may vary depending on the aforementioned factors, which require proper optimization. The research conducted on adsorption and photocatalysis has demonstrated that they can be reused for several cycles before the catalytic activities start to diminish. For example, $CuFe_2O_4$ nanoparticles have been used for the removal of MG in aqueous solution and exhibited removal efficiency of about 60% after four cycles of reusing $CuFe_2O_4$ nanoparticles. The adsorption stability is reported over 15 cycles when modified Fe_3O_4 nanoparticles are applied in the removal of Cu (II) (Kefeni et al. 2017).

11.6 Future Perspectives

Proper quality control is required for the complete removal of spinel ferrite nanoparticles and toxic contaminants after wastewater treatment. Proper quality control monitors the impact of treated water on human health to avoid unintended consequences. Hence, expertise is a must in order to properly control all the process for which rapid technological developments is required. Despite the strong capacity of spinel ferrite nanoparticles for removing varieties of toxic chemicals from wastewater, studies on the impact of spinel ferrite nanoparticles on human health and their environmental behavior and ecological risk have not been fully addressed. Hence, an in-depth toxicity study is required before the commercialization and application of these ferrite nanoparticles on a large scale (Kefeni et al. 2017; Kefeni and Mamba 2020). The stability, sensitivity, selectivity, adsorption capacity, and ease of recovery of ferrite nanoparticles to be used for wastewater treatment should be thoroughly evaluated in actual wastewater on a small scale before they are commercialized or used on a large scale (Kefeni et al. 2017). All the published articles deal with the photocatalysis application of spinel ferrite nanoparticles for water treatment on a small scale and laboratory level where superb performances have been reported by these ferrite nanoparticles. However, their practical applicability on an industrial scale is not yet reported. Information on the assessment of the complete degradation of pollutants is a research gap observed in the literature. In the case of dyes, it is difficult to interpret that the disappearance of color has completely removed the organic compounds. Incomplete degradation may introduce secondary pollutants that might be more toxic than the parent pollutants. The presence of higher pollutant concentrations in wastewater should always be considered when evaluating the degradation capacity of photocatalytic materials. This is because the pollutant can cover the active surface of the photocatalyst and inhibit its proper function. Therefore, predetermination of the concentration of the pollutant and optimizing the catalyst dosage need to be considered, in fact, such tests are rarely available in the literature. Theoretical software is required in predicting and guiding the selection of the appropriate type of dopant, surface functionalization materials, and the amount of photocatalytic loading. Consequently, it would help in reducing the number of tedious experimental trials. However, no effort has been made to develop powerful software. Furthermore, coating spinel ferrite nanoparticles is important in order to increase their lifetime and stability but there is no standard technique yet available in the literature in order to uniformly functionalize or coat the surface of spinel ferrite nanoparticles. The thickness of the coating affects the magnetization values of the spinel ferrite nanoparticles, thereby decreasing possible separation and reuse of ferrite nanoparticles. Therefore, proper optimization of coating techniques requires detailed future study. Furthermore, the higher the surface coating may mask the photosensitive part and reduce the photocatalytic effect. In addition to the type of photocatalyst material, the size of nanoparticles, crystallinity, accessibility of the active surface to the pollutant, and diffusion resistance of organic pollutants are very important characteristics for enhancing photocatalytic properties. With this respect, small particle sizes with high crystallinity are important due to their higher specific surface area and several active sites that favor higher photocatalytic activity. In order to synthesize particles with high crystallinity, high-temperature synthesis is required but a high temperature increases the particle size (Kefeni et al. 2017; Kefeni and Mamba 2020). Thus, one has to think about the optimization of crystallinity without compromising the particle size.

11.7 Conclusion

The topics discussed in this chapter demonstrate the flexibility of spinel ferrite nanoparticles in the purification of contaminated water. The attractive physical and chemical properties of spinel ferrite nanoparticles and their composites with tailored size, composition, magnetic characteristics, high chemical stability, and easy modification of their surface with a suitable semiconductor and organic species make them promising nanomaterials for wastewater treatment. Factors such as surface area, surface charge, annealing temperature, and functional groups influence the adsorption properties of SFs and their composites. The high adsorption capacity is due to their higher specific surface area and greater active sites for interaction with contaminants available in the solution. Besides their inherent low toxicity, these materials are also economically interesting due to their ease of separation by the external magnetic field from the purified stream. These materials have a higher photodegradation capacity where pollutants are attacked by active radicals generated by the spinel photocatalyst during the reaction and they easily degrade. These nanoparticles show great promise for future applications in water treatment systems.

References

Afkhami, A. and Norooz-Asl, R. (2009). Removal, preconcentration and determination of Mo(VI) from water and wastewater samples using maghemite nanoparticles. *Colloids Surf. A Physicochem. Eng. Asp.* 346: 52–57. https://doi.org/10.1016/j.colsurfa.2009.05.024.

Afkhami, A., Sayari, S., Moosavi, R., and Madrakian, T. (2015). Magnetic nickel zinc ferrite nanocomposite as an efficient adsorbent for the removal of organic dyes from aqueous solutions. *J. Ind. Eng. Chem.* 21: 920–924. https://doi.org/10.1016/j.jiec.2014.04.033.

Ahmad, N., Alam, M., Ansari, A.A. et al. (2017). Nickel ferrite nanomaterials: synthesis, characterization and properties. *Nanosci. Nanotechnol. Lett.* 9: 1688–1695. https://doi.org/10.1166/nnl.2017.2545.

Ai, Z., Deng, K., Wan, Q. et al. (2010). Facile microwave-assisted synthesis and magnetic and gas sensing properties of Fe_3O_4 nanoroses. *J. Phys. Chem. C* 114 (14): 6237–6242. https://doi.org/10.1021/jp910514f.

Ali, I. and Aboul-Enein, H.Y. (2004). *Chiral Pollutants: Distribution, Toxicity and Analysis by Chromatography and Capillary Electrophoresis*. Chichester: Wiley http://gen.lib.rus.ec/book/index.php?md5=53c7665cae18ac504df8c2cadaeb2b12.

Ali, I., Khan, T.A., and Asim, M. (2011). Removal of arsenic from water by electro-coagulation and electrodialysis techniques. *Sep. Purif. Rev.* 40: 25–42. https://doi.org/10.1080/15422119.2011.542738.

Amir, M., Kurtan, U., Baykal, A., and Sözeri, H. (2016). $MnFe_2O_4$@PANI@Ag heterogeneous nanocatalyst for degradation of industrial aqueous organic pollutants. *J. Mater. Sci. Technol.* 32: 134–141. https://doi.org/10.1016/j.jmst.2015.12.011.

An, S., Liu, X., Yang, L., and Zhang, L. (2015). Enhancement removal of crystal violet dye using magnetic calcium ferrite nanoparticle: study in single- and binary-solute systems. *Chem. Eng. Res. Des.* 94: 726–735. https://doi.org/10.1016/j.cherd.2014.10.013.

Anku, W.W., Mamo, M.A., and Govender, P.P. (2017). Phenolic compounds in water: sources, reactivity, toxicity and treatment methods. *INTECH*: 419–443. https://doi.org/10.5772/66927.

Auffan, M., Rose, J., Proux, O. et al. (2008). Enhanced adsorption of arsenic onto maghemites nanoparticles: As(III) as a probe of the surface structure and heterogeneity. *Langmuir* 24: 3215–3222. https://doi.org/10.1021/la702998x.

Bao, X., Qiang, Z., Ling, W., and Chang, J.H. (2013). Sonohydrothermal synthesis of MFe_2O_4 magnetic nanoparticles for adsorptive removal of tetracyclines from water. *Sep. Purif. Technol.* 117: 104–110. https://doi.org/10.1016/j.seppur.2013.03.046.

Baruah, S., Rafique, R.F., and Dutta, J. (2008). Visible light photocatalysis by tailoring crystal defects in zinc oxide nanostructures. *Nano* 3: 399–407. https://doi.org/10.1142/S179329200800126X.

Baruah, S., Sinha, S.S., Ghosh, B. et al. (2009). Photoreactivity of ZnO nanoparticles in visible light: effect of surface states on electron transfer reaction. *J. Appl. Phys.* 105: 074308. https://doi.org/10.1063/1.3100221.

Bhatkhande, D.S., Pangarkar, V.G., and Beenackers, A.A.C.M. (2001). Photocatalytic degradation for environmental applications – a review. *J. Chem. Technol. Biotechnol.* 77: 102. https://doi.org/10.1002/jctb.532.

Bora, T. and Dutta, J. (2014). Applications of nanotechnology in wastewater treatment—a review. *J. Nanosci. Nanotechnol.* 14: 613–626. https://doi.org/10.1166/jnn.2014.8898.

Boruah, P.K., Sharma, B., Karbhal, I. et al. (2017). Ammonia-modified graphene sheets decorated with magnetic Fe_3O_4 nanoparticles for the photocatalytic and photo-Fenton degradation of phenolic compounds under sunlight irradiation. *J. Hazard. Mater.* 325: 90–100. https://doi.org/10.1016/j.jhazmat.2016.11.023.

Cadotte, J., Forester, R., Kim, M. et al. (1988). Nanofiltration membranes broaden the use of membrane separation technology. *Desalination* 70: 77–88. https://doi.org/10.1016/0011-9164(88)85045-8.

Carta, D., Casula, M.F., Falqui, A. et al. (2009). A structural and magnetic investigation of the inversion degree in ferrite nanocrystals MFe_2O_4 (M=Mn, Co, Ni). *J. Phys. Chem. C* 113: 8606–8615. https://doi.org/10.1021/jp901077c.

Chan, S.H.S., Wu, T.Y., Juan, J.C., and Teh, C.Y. (2011). Recent developments of metal oxide semiconductors as photocatalysts in advanced oxidation processes (AOPs) for treatment of dye waste-water. *J. Chem.Technol. Biotechnol.* 86: 1130–1158. https://doi.org/10.1002/jctb.2636.

Cheng, F.X., Jia, J.T., Xu, Z.G. et al. (1999). Microstructure, magnetic, and magneto-optical properties of chemical synthesized Co–RE (RE= Ho, Er, Tm, Yb, Lu) ferrite nanocrystalline films. *J. Appl. Phys.* 86: 2727–2732. https://doi.org/10.1063/1.371117.

Culity, B.D. and Graham, C.D. (2009). *Introduction to Magnetic Materials*. Hoboken, NJ: Wiley http://gen.lib.rus.ec/book/index.php?md5=db0800c7d344419a0c04bdd557ac41df.

Davarnejad, R. and Panahi, P. (2016). Cu(II) removal from aqueous wastewaters by adsorption on the modified Henna with Fe_3O_4 nanoparticles using response surface methodology. *Sep. Purif. Technol.* 158: 286–292. https://doi.org/10.1016/j.seppur.2015.12.018.

Diao, Y., Yan, Z., Guo, M., and Wang, X. (2018). Magnetic multi-metal co-doped magnesium ferrite nanoparticles: An efficient visible light-assisted heterogeneous Fenton-like catalyst synthesized from saprolite laterite ore. *J. Hazard. Mater.* 344: 829–838. https://doi.org/10.1016/j.jhazmat.2017.11.029.

Domenech, J. and Peral, J. (1988). Removal of toxic cyanide from water by heterogeneous photocatalytic oxidation over ZnO. *Sol. Energy* 41: 55–59. https://doi.org/10.1016/0038-092X(88)90115-6.

Doroodmand, M.M., Tahvildar, Z., and Sheikhi, M.H. (2012). Multi-walled carbon nanotubes/polycrylonitrile composite as novel semi-permeable filter for water treatment process. *Sci. Adv. Mater.* 4: 1085–1095. https://doi.org/10.1166/sam.2012.1395.

Drewes, J.E., Bellona, C.L., and Xu, P. (2008). *Comparing Nanofiltrationand Reverse Osmosis for Treating Recycled Water*. London: IWA Publishing (International Water Assoc). https://doi.org/10.1002/j.1551-8833.2008.tb09724.x.

Duan, S., Tang, R., Xue, Z. et al. (2015). Effective removal of Pb(II) using magnetic $Co_{0.6}Fe_{2.4}O_4$ micro-particles as the adsorbent: synthesis and study on the kinetic and thermodynamic behaviors for its adsorption. *Colloids Surf. A Physicochem. Eng. Asp.* 469: 211–223. https://doi.org/10.1016/j.colsurfa.2015.01.029.

Eggins, B.R., Palmer, F.L., and Byrne, J.A. (1997). Photocatalytic treatment of humic substances in drinking water. *Water Res.* 31: 1223–1226. https://doi.org/10.1016/S0043-1354(96)00341-7.

Fang, B., Yan, Y., Yang, Y. et al. (2016). Adsorption of Pb^{2+} from aqueous solution using spinel ferrite prepared from steel pickling sludge. *Water Sci. Technol.* 73: 1112–1121. https://doi.org/10.2166/wst.2015.580.

Farhadi, S., Siadatnasab, F., and Khataee, A. (2017). Ultrasound-assisted degradation of organic dyes over magnetic $CoFe_2O_4$@ZnS core-shell nanocomposites. *Ultrason. Sonochem.* 37: 298–309. https://doi.org/10.1016/j.ultsonch.2017.01.019.

Feng, L., Cao, M., Ma, X. et al. (2012). Superparamagnetic high-surface-area Fe_3O_4 nanoparticles as adsorbents for arsenic removal. *J. Hazard. Mater.* 217–218: 439–446. https://doi.org/10.1016/j.jhazmat.2012.03.073.

Feng, X., Guo, H., Patel, K. et al. (2014). High performance, recoverable Fe_3O_4-ZnO nanoparticles for enhanced photocatalytic degradation of phenol. *Chem. Eng. J.* 244: 327–334. https://doi.org/10.1016/j.cej.2014.01.075.

Forgacs, E., Cserháti, T., and Oros, G. (2004). Removal of synthetic dyes from wastewaters: a review. *Environ. Int.* 30: 953–971. https://doi.org/10.1016/j.envint.2004.02.001.

Goldman, A. (2006). *Modern Ferrite Technology*. New York: Springer http://gen.lib.rus.ec/book/index.php?md5=a405089044aaa302587fce26cc28ca31.

Gómez-Lopera, S.A., Arias, J.L., Gallardo, V., and Delgado, Á.V. (2006). Colloidal stability of magnetite/poly(lactic acid) Core/Shell nanoparticles. *Langmuir* 22: 2816–2821. https://doi.org/10.1021/la0530079.

Gómez-Pastora, J., Bringas, E., and Ortiz, I. (2014). Recent progress and future challenges on the use of high performance magnetic nano-adsorbents in environmental applications. *Chem. Eng. J.* 256: 187–204. https://doi.org/10.1016/j.cej.2014.06.119.

Gupta, V.K. and Nayak, A. (2012). Cadmium removal and recovery from aqueous solutions by novel adsorbents prepared from orange peel and Fe_2O_3 nanoparticles. *Chem. Eng. J.* 180: 81–90. https://doi.org/10.1016/j.cej.2011.11.006.

Gupta, V.K., Agarwal, S., and Saleh, T.A. (2011). Synthesis and characterization of alumina coated carbon nanotubes and their application for lead removal. *J. Hazard. Mater.* 185: 17–23. https://doi.org/10.1016/j.jhazmat.2010.08.053.

Hassan, M.R. and Aly, M.I. (2020). Adsorptive removal of cesium ions from aqueous solutions using synthesized Prussian blue/magnetic cobalt ferrite nanoparticles. *Part Sci. Technol.* 38: 236–246. https://doi.org/10.1080/02726351.2018.1539799.

He, J., Yang, X., Men, B., and Wang, D. (2016). Interfacial mechanisms of heterogeneous Fenton reactions catalyzed by iron-based materials: a review. *J. Environ. Sci.* 39: 97–109. https://doi.org/10.1016/j.jes.2015.12.003.

Helmer, R., Hespanhol, I., and World Health Organization (1997). *Water Pollution Control: A Guide to the Use of Water Quality Management Principles*. London: E & FN.

Henderson, M.A. (2011). A surface science perspective on TiO_2 photocatalysis. *Surf. Sci. Rep.* 66: 185–297. https://doi.org/10.1016/j.surfrep.2011.01.001.

Henderson, C.M.B., Charnock, J.M., and Plant, D.A. (2007). Cation occupancies in Mg, Co, Ni, Zn, Al ferrite spinels: a multi-element EXAFS study. *J. Phys. Condens. Matter* 19: 076214. https://doi.org/10.1088/0953-8984/19/7/076214.

Hoffmann, M.R., Martin, S.T., Choi, W., and Bahnemann, D.W. (1995). Environmental applications of semiconductor photocatalysis. *Chem. Rev.* 95: 69–96. https://doi.org/10.1021/cr00033a004.

Hou, X., Feng, J., Ren, Y. et al. (2010). *Colloids Surf. A Physicochem. Eng. Asp.* 363: 1–7.

Hu, J., Lo, I.M., and Chen, G. (2005). Fast removal and recovery of Cr(VI) using surface-modified jacobsite (MnFe2O4) nanoparticles. *Langmuir* (21): 11173–11179. https://doi.org/10.1021/la051076h.

Hu, J., Lo, I.M.C., and Chen, G. (2007). Performance and mechanism of chromate (VI) adsorption by δ-FeOOH-coated maghemite (γ-Fe_2O_3) nanoparticles. *Sep. Purif. Technol.* 56: 249–256. https://doi.org/10.1016/j.seppur.2007.07.023.

Huong, L., Thi, T.P., Lan, N. et al. (2018). Functional manganese ferrite/ graphene oxide nanocomposites: effects of graphene oxide on the adsorption mechanisms of organic MB dye and inorganic As(v) ions from aqueous solution. *RSC Adv.* 8: 12376–12389. https://doi.org/10.1039/C8RA00270C.

Ivanetsa, A., Roshchinaa, M., Srivastavab, V. et al. (2019). Effect of metal ions adsorption on the efficiency of methylene blue degradation onto $MgFe_2O_4$ as Fenton-like catalysts. *Colloids Surf. A Physicochem. Eng. Asp.* 571: 17–26. https://doi.org/10.1016/j.colsurfa.2019.03.071.

Jesudoss, S.K., Vijaya, J.J., Kennedy, L.J. et al. (2016). Studies on the efficient dual performance of $Mn_{1-x}Ni_xFe_2O_4$ spinel nanoparticles in photo-degradation and antibacterial activity. *J. Photochem. Photobiol. B* 165: 121–132. https://doi.org/10.1016/j.jphotobiol.2016.10.004.

Jia, Z., Peng, K., and Xu, L. (2012). Preparation, characterization and enhanced adsorption performance for Cr(VI) of mesoporous $NiFe_2O_4$ by twice pore-forming method. *Mater. Chem. Phys.* 136: 512–519. https://doi.org/10.1016/j.matchemphys.2012.07.019.

Jia, Z., Wang, Q., Liu, J. et al. (2013). Effective removal of phosphate from aqueous solution using mesoporous rodlike NiFe2O4 as magnetically separable adsorbent. *Colloids Surf., A: Physicochem. Eng. Asp.* 436: 495–503.

Kale, A.S., Gubbala, S., and Misra, R.D.K. (2004). Magnetic behavior of nanocrystalline nickel ferrite synthesized by the reverse micelle technique. *J. Magn. Magn. Mater.* 277: 350–358. https://doi.org/10.1016/j.jmmm.2003.11.015.

Kefeni, K.K. and Mamba, B.B. (2020). Photocatalytic application of spinel ferrite nanoparticles and nanocomposites in wastewater treatment: review. *Sustain. Mater. Technol.* 23: e00140. https://doi.org/10.1016/j.susmat.2019.e00140.

Kefeni, K.K., Mamba, B.B., and Msagati, T.A.M. (2017). Application of spinel ferrite nanoparticles in water and wastewater treatment: a review. *Sep. Purif. Technol.* 188: 399–422. https://doi.org/10.1016/j.seppur.2017.07.015.

Kefeni, K.K., Msagati, T.A.M., and Mamba, B.B. (2017). Ferrite nanoparticles: synthesis, characterisation and applications in electronic device. *Mater. Sci. Eng. B* 215: 37–55. https://doi.org/10.1016/j.mseb.2016.11.002.

Khalil, L.B., Mourad, W.E., and Rophael, M.W. (1998). Photocatalytic reduction of environmental pollutant Cr(VI) over some semiconductors under UV/visible light illumination. *Appl. Catal. B Environ.* 17: 267–273. https://doi.org/10.1016/S0926-3373(98)00020-4.

Khan, M.A., Otero, M., Kazi, M., and Alqadami, A.A. (2019). Unary and binary adsorption studies of lead and malachite green onto a nanomagnetic copper ferrite/drumstick pod biomass composite. *J. Hazard. Mater.* 365: 759–770. https://doi.org/10.1016/j.jhazmat.2018.11.072.

Komarneni, S., D'Arrigo, M.C., Leonelli, C., and Pellacani, G.C. (1998). Microwave-hydrothermal synthesis of nanophase ferrites. *J. Am. Ceram. Soc.* 81 (11): 3041–3043. https://doi.org/10.1111/j.1151-2916.1998.tb02738.x.

Konicki, W., Sibera, D., Mijowska, E. et al. (2013). Equilibrium and kinetic studies on acid dye Acid Red 88 adsorption by magnetic $ZnFe_2O_4$ spinel ferrite nanoparticles. *J. Colloid Interface Sci.* 398: 152–160. https://doi.org/10.1016/j.jcis.2013.02.021.

Kosmulski, M. (2012). IEP as a parameter characterizing the pH-dependent surface charging of materials other than metal oxides. *Adv. Colloid Interf. Sci.* 171–172: 77–86. https://doi.org/10.1016/j.cis.2012.01.005.

Kraus, A., Jainae, K., Unob, F., and Sukpirom, N. (2009). Synthesis of MPTS-modified cobalt ferrite nanoparticles and their adsorption properties in relation to Au(III). *J. Colloid Interface Sci.* 338: 359–365. https://doi.org/10.1016/j.jcis.2009.06.045.

Kuai, S., Zhang, Z., and Nan, Z. (2013). Synthesis of Ce^{3+} doped $ZnFe_2O_4$ self- assembled clusters and adsorption of chromium(VI). *J. Hazard. Mater.* 250–251: 229–237. https://doi.org/10.1016/j.jhazmat.2013.01.074.

Kumar, L. and Kar, M. (2011a). Effect of annealing temperature and preparation condition on magnetic anisotropy in Nanocrystalline cobalt ferrite. *IEEE Trans. Magn.* 47: 3645–3648. https://doi.org/10.1109/TMAG.2011.2151841.

Kumar, L. and Kar, M. (2011b). Influence of Al^{3+} ion concentration on the crystal structure and magnetic anisotropy of Nanocrystalline spinel cobalt ferrite. *J. Magn. Magn. Mater.* 323: 2042–2048. https://doi.org/10.1016/j.jmmm.2011.03.010.

Kumar, S., Ahlawat, W., Bhanjana, G. et al. (2014a). Nanotechnology-based water treatment strategies. *J. Nanosci. Nanotechnol.* 14: 1838–1858. https://doi.org/10.1166/jnn.2014.9050.

Kumar, S., Nair, R.R., Pillai, P.B. et al. (2014b). Graphene oxide–$MnFe_2O_4$ magnetic nanohybrids for efficient removal of lead and arsenic from water. *ACS Appl. Mater. Interfaces* 6: 17426–17436. https://doi.org/10.1021/am504826q.

Kurian, M., Nair, D.S., and Rahnamol, A.M. (2014). Influence of the synthesis conditions on the catalytic. *React. Kinet. Mech. Catal.* 111: 591–604. https://link.springer.com/article/10.1007/s11144-013-0667-x.

Kurniawan, T.A., Sillanpaa, M.E.T., and Sillanpaa, M. (2012). Nanoadsorbents for remediation of aquatic environment: local and practical solutions for global water pollution problem. *Crit. Rev. Environ. Sci. Technol.* 42: 1233. https://doi.org/10.1080/10643389.2011.556553.

Lai, J., Shafi, K.V.P.M., Ulman, A., and Loos, K. (2004). Mixed iron manganese oxide nanoparticles. *J. Phys. Chem. B* 108: 14876–14883. https://doi.org/10.1021/jp049913w.

Lan, H., Pham, T.T., Lan, N., and Hong, L. (2018). Functional manganese ferrite/graphene oxide nanocomposites: effects of graphene oxide on the adsorption mechanisms of organic MB dye and inorganic As(v) ions from aqueous solution. *RSC Adv.* 8: 12376–12389. https://doi.org/10.1039/C8RA00270C.

Lata, S. and Samadder, S.R. (2016). Removal of arsenic from water using nano adsorbents and challenges: a review. *J. Environ. Manag.* 166: 387–406. https://doi.org/10.1016/j.jenvman.2015.10.039.

Lee, T., Jeon, E., and Kim, B. (2014). Mussel-inspired nitrogen-doped graphene nanosheet supported manganese oxide nanowires as highly efficient electro-catalysts for oxygen reduction reaction. *J. Mater. Chem.* 2: 6167–6173. https://doi.org/10.1039/C3TA14147K.

Lehr, T.E.G.J.H. and DeMarre, J.P. (1980). *Domestic Water Treatment*. New York: McGraw Hill.

Li, L., Fan, M., Brown, R. et al. (2006). Synthesis, properties, and environmental applications of nanoscale iron based materials: a review. *Crit. Rev. Environ. Sci. Technol.* 36: 405–431. https://doi.org/10.1080/10643380600620387.

Li, Z., Gao, K., Han, G. et al. (2015). Solvothermal synthesis of $MnFe_2O_4$ colloidalnanocrystal assemblies and their magnetic and electrocatalytic properties. *New J. Chem.* 39: 361–368. https://doi.org/10.1039/C4NJ01466A.

Li, M., Liu, H., Chen, T. et al. (2019). Synthesis of magnetic biochar composites for enhanced uranium (VI) adsorption. *Sci. Total Environ.* 651: 1020–1028. https://doi.org/10.1016/j.scitotenv.2018.09.259.

Lindsay, S.M. (2010). *Introdution to Nanoscience*. New York: Oxford University Press http://gen.lib.rus.ec/book/index.php?md5=4812c7bbb97cecb0ab423c72fc75b5af.

Lingamdinne, L.P., Choi, Y.L., Kim, I.S. et al. (2017). Preparation and characterization of porous reduced graphene oxide based inverse spinel nickel ferrite nanocomposite for adsorption removal of radionuclides. *J. Hazard. Mater.* 326: 145–156. https://doi.org/10.1016/j.jhazmat.2016.12.035.

Liu, X., An, S., Wang, Y. et al. (2015). Rapid selective separation and recovery of a specific target dye from mixture consisted of different dyes by magnetic Ca-ferrites nanoparticles. *Chem. Eng. J.* 262: 517–526. https://doi.org/10.1016/j.cej.2014.10.002.

Lofrano, G., Carotenuto, M., Libralato, G. et al. (2016). Polymer functionalized nanocomposites for metals removal from water and wastewater: An overview. *Water Res.* 92: 22–37. https://doi.org/10.1016/j.watres.2016.01.033.

Lutchmiah, K., Verliefde, A.R.D., Roest, K. et al. (2014). Forward osmosis for application in wastewater treatment: a review. *Water Res.* 58: 179–197. https://doi.org/10.1016/j.watres.2014.03.045.

Lv, W.Z., Liu, B., Luo, Z.K. et al. (2008). XRD studies on the nanosized copper ferrite powders synthesized by sonochemical method. *J. Alloys Compd.* 465: 261–264. https://doi.org/10.1016/j.jallcom.2007.10.049.

Mahmoodi, N.M. (2013). Zinc ferrite nanoparticle as amagnetic catalyst: synthesis and dyedegradation. *Mater. Res. Bull.* 48: 4255–4260. https://doi.org/10.1016/j.materresbull.2013.06.070.

Maksoud, M.I.A., El-Sayyad, G.S., Ashour, A.H. et al. (2018). Synthesis and characterization of metals-substituted cobalt ferrite [$M_xCo_{(1-x)}Fe_2O_4$; (M = Zn, Cu and Mn; x = 0 and 0.5)] nanoparticles as antimicrobial agents and sensors for Anagrelide determination in biological samples. *Mater. Sci. Eng. C* 92: 644–656. https://doi.org/10.1016/j.msec.2018.07.007.

Maksoud, M.I.A.A., El-Sayyad, G.S., Ashour, A.H. et al. (2019). Antibacterial, antibiofilm, and photocatalytic activities of metals-substituted spinel cobalt ferrite nanoparticles. *Microb. Pathog.* 127: 144–158. https://doi.org/10.1016/j.micpath.2018.11.045.

Mamba, G. and Mishra, A.K. (2016). Graphitic carbon nitride (g-C_3N_4) nanocomposites: a new and exciting generation of visible light driven photocatalysts for environmental pollution remediation. *Appl. Catal. B Environ.* 198: 347–377. https://doi.org/10.1016/j.apcatb.2016.05.052.

Manikandan, A., Sridhar, R., Antony, S.A., and Ramakrishna, S. (2014). A simple aloe vera plant-extracted microwave and conventional combustion synthesis: morphological, optical, magnetic and catalytic properties of $CoFe_2O_4$ nanostructures. *J. Mol. Struct.* 1076: 188–200. https://doi.org/10.1016/j.molstruc.2014.07.054.

Mathew, D.S. and Juang, R.S. (2007). An overview of the structure and magnetism of spinel ferrite nanoparticles and their synthesis in microemulsions. *Chem. Eng. J.* 129: 51–65. https://doi.org/10.1016/j.cej.2006.11.001.

McCafferty, E. (2015). Acid–base properties of surface oxide films. In: *Surface Chemistry of Aqueous Corrosion Processes*, 1–54. Cham: Springer International Publishing https://link.springer.com/chapter/10.1007/978-3-319-15648-4_1.

McNaught, A.D. and Wilkinson, A. (1997). *IUPAC Gold Book*. Oxford: Blackwell Scientific Publications.

Mills, A. and Hunte, S.L. (1997). An overview of semiconductor photocatalysis. *J. Photochem. Photobiol. A Chem.* 108: 1–35. https://doi.org/10.1016/S1010-6030(97)00118-4.

Mills, A., Davies, R.H., and Worsley, D. (1993). Water purification by semiconductor photocatalysis. *Chem. Soc. Rev.* 22: 417–425. https://doi.org/10.1039/CS9932200417.

Mills, A., Belghazi, A., and Rodman, D. (1996). Bromate removal from drinking water by semiconductor photocatalysis. *Water Res.* 30: 1973–1978. https://doi.org/10.1016/0043-1354(96)00012-7.

Mohmood, I., Lopes, C.B., and Lopes, I. (2013). Nanoscale materials and their use in water contaminants removal- a review. *Environ. Sci. Pollut. Res.* 20: 1239–1260. https://link.springer.com/article/10.1007/s11356-012-1415-x.

Murray, B.P., Walsh, S.N., and O'Sullivan, D.T.J. (2015). The water energy nexus, an ISO50001 water case study and the need for a water value system. *Water Resour. Ind.* 10: 15–28. https://doi.org/10.1016/j.wri.2015.02.001.

Nair, D.S. and Kurian, M. (2017). Heterogeneous catalytic oxidation of persistent chlorinated organics over cobalt substituted zinc ferrite nanoparticles at mild conditions: reaction kinetics and catalyst reusability studies. *J. Environ. Chem. Eng.* 5: 964–974. https://doi.org/10.1016/j.jece.2017.01.021.

Nassar, M.Y. and Khatab, M. (2016). Cobalt ferrite nanoparticles via a template-free hydrothermal route as an efficient nano-adsorbent for potential textile dye removal. *RSC Adv.* 6: 79688–79705. https://doi.org/10.1039/C6RA12852A.

Nemerow, N.L. and Dasgupta, A. (1991). *Industrial and Hazardous Waste Treatment*. New York: Van Nostrand Reinhold https://www.osti.gov/biblio/6829221.

Nemerrow, N.L. (1978). *Industrial Water Pollution: Origins. Characteristics, and Treatment*. Massachusetts: Addison-Wesley https://cmc.marmot.org/Record/.b16104614.

Neris, A.M., Schreiner, W.H., Salvador, C. et al. (2018). Photocatalytic evaluation of the magnetic core@shell system (Co,Mn) Fe_2O_4 @TiO_2 obtained by the modified Pechini method. *Mater. Sci. Eng. B* 229: 218–226. https://doi.org/10.1016/j.mseb.2017.12.029.

Neyaz, N. and Siddiqui, W.A. (2015). Removal of Cu(II) by modified magnetite nanocomposite as a nanosorbent. *Int. J. Sci. Res.* 4: 1868–1873. https://pdfs.semanticscholar.org/5f66/f7271b4bb36a3182dd68b54312bd4f335009.pdf.

Ng, Q.H., Lim, J.K., Ahmad, A.L. et al. (2015). Magnetic nanoparticles augmented composite membranes in removal of organic foulant through magnetic actuation. *J. Membr. Sci.* 493: 134–146. https://doi.org/10.1016/j.memsci.2015.06.045.

Ngomsik, A.-F., Bee, A., Draye, M. et al. (2005). Magnetic nano- and microparticles for metal removal and environmental applications: a review. *C. R. Chim.* 8: 963–970. https://doi.org/10.1016/j.crci.2005.01.001.

Ni, D., Lin, Z., Xiaoling, P. et al. (2015). Preparation and characterization of nickel-zinc ferrites by a solvothermal method. *Rare Metal Mater. Eng.* 44: 2126–2131. https://doi.org/10.1016/S1875-5372(16)30010-8.

Pareek, V. and Adesina, A.A. (2003). *Nanotechnlogy for sustainable water resources, in: Handbook of Photochemistry and Photobiology: Inorganic Photochemistry* (ed. H.S. Nalwa). Los Angeles: American Scientific Publishers http://www.aspbs.com/photochemistry.html.

Parsons, J.G., Lopez, M.L., Peralta-Videa, J.R., and Gardea-Torresdey, J.L. (2009). Determination of arsenic(III) and arsenic(V) binding to microwave assisted hydrothermal synthetically prepared Fe_3O_4, Mn_3O_4, and $MnFe_2O_4$nanoadsorbents. *Microchem. J.* 91: 100–106. https://doi.org/10.1016/j.microc.2008.08.012.

Pendergast, M.T.M., Nygaard, J.M., Ghosh, A.K., and Hoek, E.M.V. (2010). Using nanocomposite materials technology to understand and control reverse osmosis membrane compaction. *Desalination* 261: 255–263. https://doi.org/10.1016/j.desal.2010.06.008.

Pereira, C., Pereira, A., Fernandes, C., and Rocha, M. (2012). Superparamagnetic MFe_2O_4 (M = Fe, Co, Mn) nanoparticles: tuning the particle size and magnetic properties through a novel one-step coprecipitation route. *Chem. Mater.* 24: 1496–1504. https://doi.org/10.1021/cm300301c.

Pirouz, M.J., Beyki, M.H., and Shemirani, F. (2015). Anhydride functionalised calcium ferrite nanoparticles: a new selective magnetic material for enrichment of lead ions from water and food samples. *Food Chem.* 170: 131–137. https://doi.org/10.1016/j.foodchem.2014.08.046.

Pradeep, T. (2012). *A Text Book of Nanoscience and Nanotechnology*. New Delhi: Tata McGraw Hill Book.

Rai, H.S., Bhattacharyya, M.S., Singh, J. et al. (2005). Removal of dyes from the effluent of textile and dyestuff manufacturing industry: a review of emerging techniques with reference to biological treatment. *Crit. Rev. Environ. Sci. Technol.* 35: 219–238. https://doi.org/10.1080/10643380590917932.

Ramimoghadam, D., Bagheri, S., and Hamid, S.B.A. (2015). Stable monodispersenanomagnetic colloidal suspensions: an overview. *Colloids Surf. B: Biointerfaces* 133: 388–411. https://doi.org/10.1016/j.colsurfb.2015.02.003.

Reddy, D.H.K. and Yun, Y.S. (2016). Spinel ferrite magnetic adsorbents: alternative future materials for water purification. *Coord. Chem. Rev.* 315: 90–111. https://doi.org/10.1016/j.ccr.2016.01.012.

Ren, Y., Li, N., Feng, J. et al. (2012). Adsorption of Pb(II) and Cu(II) from aqueous solution on magnetic porous ferrospinel MnFe2O4. *J. Colloid Interface Sci.* 367: 415–421. https://doi.org/10.1016/j.jcis.2011.10.022.

Richard, H. and Ivanildo, H. (1997). Water Pollution Control: A Guide to the Use of Water A: ontrolCollution P to tuide Ghe Water Quality Management Principles. London: World Health Organization, E & FN. https://apps.who.int/iris/handle/10665/41967.

Rigo, C., Severo, E.D.C., Mazutti, M.A. et al. (2017). Preparation of nickel ferrite/ carbon nanotubes composite by microwave irradiation technique for use as catalyst in photo-fenton reaction. *Mater. Res.* 20: 311–316. https://doi.org/10.1590/1980-5373-mr-2016-0672.

Rooygar, A.A., Mallah, M.H., Abolghasemi, H., and Safdari, J. (2014). Magmolecular process for the separation of antimony(III) from aqueous solution. *J. Chem. Eng. Data* 59: 3545–3554.

Saleh, T.A. and Gupta, V.K. (2011). Functionalization of tungsten oxide into MWCNT andits application for sunlight-induced degradation of Rhodamine B. *J. Colloid Interface Sci.* 362: 337–344. https://doi.org/10.1016/j.jcis.2011.06.081.

Saleh, T.A. and Gupta, V.K. (2012). Column with CNT/magnesium oxide composite for lead (II) removal from water. *Environ. Sci. Pollut. Res.* 19: 1224–1228. https://link.springer.com/article/10.1007/s11356-011-0670-6.

Saleh, T.A., Agarwal, S., and Gupta, V.K. (2011). Synthesis of MWCNT/MnO_2 and their application for simultaneous oxidation of arsenite and sorption of arsenate. *Appl. Catal., B* 106: 46–53. https://doi.org/10.1016/j.apcatb.2011.05.003.

Sanghi, R. and Verma, P. (2013). Decolorisation of aqueous dye solutions by low-cost adsorbents: a review. *Color. Technol.* 129: 85–108. https://doi.org/10.1111/cote.12019.

Sanpo, N., Tharajak, J., and Li, Y. (2014). Biocompatability of transition metal-substituted cobalt ferrite nanoparticle. *J. Nanopart. Res.* 16: 2510. https://link.springer.com/article/10.1007/s11051-014-2510-3.

Santhosh, C., Velmurugan, V., Jacob, G. et al. (2016). Role of nanomaterials in water treatment applications: a review. *Chem. Eng. J.* 306: 1116–1137. https://doi.org/10.1016/j.cej.2016.08.053.

Serpone, N., Ah-You, Y.K., Tran, T.P. et al. (1987). AMI simulated sunlight photoreduction and elimination of Hg(II) and CH_3Hg(II) chloride salts from aqueous suspension of titanium dioxide. *Sol. Energy* 39: 491–498. https://doi.org/10.1016/0038-092X(87)90056-9.

Shaikh, M., Sahu, M., Atyam, K.K., and Ranganath, K.V.S. (2016). Surface modification of ferrite nanoparticles with dicarboxylic acids for the synthesis of 5-hydroxymethylfurfural: a novel and green protocol. *RSC Adv.* 6: 76795–76801. https://doi.org/10.1039/C6RA13354A.

Shannon, M.A., Bohn, P.W., Elimelech, M. et al. (2008). Science and technology for water purification in the coming decades. *Nature* 452: 301–310. https://www.nature.com/articles/nature06599.

Sharma, R., Bansal, S., and Singhal, S. (2015). Tailoring the photo-Fenton activity of spinel ferrites (MFe_2O_4) by incorporating different cations (M = Cu, Zn, Ni and Co) in the structure. *RSC Adv.* 5: 6006–6018. https://doi.org/10.1039/C4RA13692F.

Shetty, K., Renuka, L., Nagaswarupa, H.P. et al. (2017). A comparative study on $CuFe_2O_4$, $ZnFe_2O_4$and $NiFe_2O_4$: morphology, impedance and photocatalytic studies. *Mater. Today Proc.* 4: 11806–11815. https://doi.org/10.1016/j.matpr.2017.09.098.

Sickafus, K.E. and Wills, J.M. (1999). Spinel coumpounds: structure and property relations. *J. Am. Ceram. Soc.* 82: 3279–3292. https://doi.org/10.1111/j.1151-2916.1999.tb02241.x.

Sigdel, A., Park, J., Kwak, H., and Park, P. (2016). Arsenic removal from aqueous solutions by adsorption onto hydrous iron oxide-impregnated alginate beads. *J. Ind. Eng. Chem.* 35: 277–286. https://doi.org/10.1016/j.jiec.2016.01.005.

Sun, W., Pan, W., Wang, F., and Xu, N. (2015). Removal of Se(IV) and Se(VI) by MFe_2O_4 nanoparticles from aqueous solution. *Chem. Eng. J.* 273: 353–362. https://doi.org/10.1016/j.cej.2015.03.061.

Tadjarodi, A., Imani, M., and Salehi, M. (2015). $ZnFe_2O_4$ nanoparticles and a clay encapsulated $ZnFe_2O_4$ nanocomposite: synthesis strategy, structural characteristics and the adsorption of dye pollutants in water. *RSC Adv.* 5: 56145–56156. https://doi.org/10.1039/C5RA02163D.

Tang, S.C.N. and Lo, I.M.C. (2013). Magnetic nanoparticles: Essential factors for sustainable environmental applications. *Water Res.* 47: 2613–2632. https://doi.org/10.1016/j.watres.2013.02.039.

Tombácz, E. (2009). pH-dependent surface charging of metal oxides. *Period. Polytech., Chem. Eng.* 53: 77–86. https://doi.org/10.3311/pp.ch.2009-2.08.

Tseng, T.K., Lin, Y.S., Chen, Y.J., and Chu, H. (2010). A review of photocatalysts prepared by sol-gel method for VOCs removal. *Int. J. Mol. Sci.* 11: 2336–2361. https://doi.org/10.3390/ijms11062336.

Tu, Y.J., You, C.F., Chang, C.K. et al. (2014). XANES evidence of molybdenum adsorption onto novel fabricated nano-magnetic $CuFe_2O_4$. *Chem. Eng. J.* 244: 343–349. https://doi.org/10.1016/j.cej.2014.01.084.

Tuutijärvi, T., Lu, J., Sillanpaa, M., and Chen, G. (2010). Adsorption mechanism of arsenate on crystal γ-Fe_2O_3 nanoparticles. *J. Environ. Eng.* 136: 897–905. https://ascelibrary.org/doi/full/10.1061/%28ASCE%29EE.1943-7870.0000233.

Valenzuela, R. (1994). *Magnetic Ceramics*. Cambridge: Cambridge University Press http://gen.lib.rus.ec/book/index.php?md5=bab1a29c09d1a7592240382535da49d1.

Venturini, J. (2019). *Application of Ferrite Nanoparticles in Wastewater Treatment, Nanomaterials for Eco-Friendly Applications*, 141–151. Engineering Materials, Springer https://link.springer.com/chapter/10.1007/978-3-030-26810-7_9.

Vîrlan, C., Ciocârlan, R.G., Roman, T. et al. (2013). Studies on adsorption capacity of cationic dyes on several magnetic nanoparticles. *Acta Chem.* 30: 19–30. https://doi.org/10.2478/achi-2013-0003.

Wang, K., Hsieh, Y., Chou, M., and Chang, C. (1999). Photocatalytic degradation of 2-chloro and 2-nitrophenol by titanium dioxide suspensions in aqueous solution. *Appl. Catal. B Environ.* 21: 3–10. https://doi.org/10.1016/S0926-3373(98)00116-7.

Wang, L., Li, J., Wang, Y. et al. (2012). Adsorption capability for Congo red on nanocrystalline MFe_2O_4 (M = Mn, Fe, Co, Ni) spinel ferrites. *Chem. Eng. J.*: 181–182 72–79. https://doi.org/10.1016/j.cej.2011.10.088.

Wang, Y., Li, C., Tan, J. et al. (2017). Synthesis of magnetically separable coreshell structured $Ni_xFe_{1-x}Fe_2O_4$@TiO_2 nanoparticles photocatalysts for the degradation of organic dyes. *J. Porous. Mater.* 24: 639–646. https://link.springer.com/article/10.1007/s10934-016-0300-2.

Wang, Y., Yan, H., and Zhang, Q. (2019). Graphene-magnetic spinel ferrite nanocomposite: facile synthesis and excellent photocatalytic performance. *Aust. J. Chem.* 74: 267–275. https://doi.org/10.1071/CH18432.

Wei, J., Zhang, X., Liu, Q. et al. (2014). Magnetic separation of uranium by $CoFe_2O_4$ hollow spheres. *Chem. Eng. J.* 241: 228–234. https://doi.org/10.1016/j.cej.2013.12.035.

Wu, X., Ding, Z., Song, N., and Li, L. (2016). Effect of the rare-earth substitution on the structural, magnetic and adsorption properties in cobalt ferrite nanoparticles. *Ceram. Int.* 42: 4246–4255. https://doi.org/10.1016/j.ceramint.2015.11.100.

Xiao, Y., Liang, H., Chen, W., and Wang, Z. (2013). Synthesis and adsorption behavior of chitosan -coated $MnFe_2O_4$ nanoparticles for trace heavy metal ions removal. *Appl. Surf. Sci.* 285 (Part B): 498–504. https://doi.org/10.1016/j.apsusc.2013.08.083.

Xu, P., Zeng, G.M., Huang, D.L. et al. (2012). Use of iron oxide nanomaterials in wastewater treatment: a review. *Sci. Total Environ.* 424: 1–10. https://doi.org/10.1016/j.scitotenv.2012.02.023.

Yagub, M.T., Sen, T.K., Afroze, S., and Ang, H.M. (2014). Dye and its removal from aqueous solution by adsorption: a review. *Adv. Colloid Interf. Sci.* 209: 172–184. https://doi.org/10.1016/j.cis.2014.04.002.

Yan, H., Li, H., Yang, H. et al. (2013). Removal of various cationic dyes from aqueous solutions using a kind of fully biodegradable magnetic composite microsphere. *Chem. Eng. J.* 223: 402–411. https://doi.org/10.1016/j.cej.2013.02.113.

Yan, Z., Gao, J., Li, Y., and Zhang, M. (2015). Hydrothermal synthesis and structure evolution of metal-doped magnesium ferrite from saprolite laterite. *RSC Adv.* 5: 92778–92787. https://doi.org/10.1039/C5RA17145H.

Yang, L., Zhang, Y., Liu, X. et al. (2014). The investigation of synergistic and competitive interaction between dye Congo red and methyl blue on magnetic $MnFe_2O_4$. *Chem. Eng. J.* 246: 88–96.

Yang, S.T., Zhang, W., Xie, J. et al. (2015). Fe_3O_4@SiO_2 nanoparticles as a high-performance Fenton-like catalyst in a neutral environment. *RSC Adv.* 5: 5458–5463. https://doi.org/10.1039/C4RA10207J.

Yin, Y., Liu, W., Huo, N., and Yang, S.T. (2016). Synthesis of vesicle-like $MgFe_2O_4$/Graphene 3D network anode material with enhanced lithium storage performance. *ACS Sustain. Chem. Eng.* 5 (1): 563–570. https://doi.org/10.1021/acssuschemeng.6b01949.

Zhang, J., Zhuang, J., Gao, L. et al. (2008). Decomposing phenol by the hidden talent of ferromagnetic nanoparticles. *Chemosphere* 73: 1524–1528. https://doi.org/10.1016/j.chemosphere.2008.05.050.

Zhang, S., Niu, H., Cai, Y. et al. (2010). Arsenite and arsenate adsorption onco- precipitated bimetal oxide magnetic nanomaterials $MnFe_2O_4$ and $CoFe_2O_4$. *Chem. Eng. J.* 158: 599–607. https://doi.org/10.1016/j.cej.2010.02.013.

Zhang, D., Pu, X., Gao, Y. et al. (2013). One-step combustion synthesis of $CoFe_2O_4$-graphene hybrid materialsfor photodegradation of methylene blue. *Mater. Lett.* 113: 179–181. https://doi.org/10.1016/j.matlet.2013.09.088.

Zhang, Y., Yan, L., Xu, W. et al. (2014). Adsorption of Pb(II) and Hg(II) from aqueous solution using magnetic $CoFe_2O_4$-reduced graphene oxide. *J. Mol. Liq.* 191: 177–182. https://doi.org/10.1016/j.molliq.2013.12.015.

Zhou, L., Ji, L., Ma, P.C. et al. (2014). Development of carbon nanotubes/CoFe2O4 magnetic hybrid material for removal of Tetrabromobisphenol A and Pb(II). *J. Hazard. Mater.* 265: 104–114.

Zhu, J., Wei, S., Chen, M. et al. (2013). Magnetic nanocomposites for environmental remediation. *Adv. Powder Technol.* 24: 459–467. https://doi.org/10.1016/j.apt.2012.10.012.

12

Biocompatible Cellulose-Based Sorbents for Potential Application in Heavy Metal Ion Removal from Wastewater

Shashikant Shivaji Vhatkar, Kavita Kumari, and Ramesh Oraon

Department of Nanoscience and Technology, Central University of Jharkhand, Ranchi, Jharkhand, India

12.1 Introduction

Anthropogenic activities, including industrial activity, unsustainable agricultural practices, and inefficient waste management, have led to the intrusion of toxic heavy metal ions (HMi) (Pb^{2+}, Hg^{2+}, Cu^{2+}, Cd^{2+}, etc.) into aquatic eco-systems and bioreserves (Holmes et al. 2019). Their mobility and bioaccumulation have made their presence in food chains result in various health hazards (Ranieri et al. 2020). Thus, the removal and seclusion of HMi by using appropriate biomaterials both ecologically and economically have become urgent tasks. At present, various comprehensive treatments are in practice to remediate the affected environment (Maniquiz-Redillas and Kim 2016). To date, among all reported remediation methods (filtration, ion-exchange, precipitation, sorption/desorption), the sorption method has been proven to effectively control mobility and seclusion of these HMi (Wu et al. 2019b). In addition to this, their efficacy, low cost, design-flexibility, convenient operations with no secondary pollution, and noble regenerations make them a viable choice over conventional absorbent-based technologies (Maruyama et al. 2014). Efficacies of sorption are proportional to metal ion concentration, pH, and temperature, and thus, prominence is laid on the sorption properties of biomaterials. First, mechanism and kinetics of the sorption of HMi are examined through various adsorption isotherms like Dubinin-Radushkevich, Linear, Langmuir, Freundlich, Temkin, etc., followed by a selection of appropriate sorbent material in the sorption system (Akpomie et al. 2015; Mustapha et al. 2019). To date, a number of sorbents have been reported for HMi removal, mainly with activated carbons, agricultural biomass, clay minerals, industrial by-products, polymers, etc. (Akpomie et al. 2015; Ge and Li 2018; Khulbe and Matsuura 2018; Zhao et al. 2018; Koohzad et al. 2019).

However, with advances in technology, these traditional materials suffer lower sorption rates and thus present lower efficiencies (Saha et al. 2016). Several studies have been made to address the issue of lower sorption efficiency by developing new types of sorbents with improved characteristics. With advances in nanotechnology, the past two decades have

Pollutants and Water Management: Resources, Strategies and Scarcity, First Edition. Edited by Pardeep Singh, Rishikesh Singh, Vipin Kumar Singh, and Rahul Bhadouria.

witnessed an increasing trend in deploying 1D (i.e. graphene quantum dots) and 2D (i.e. graphene oxide) structured nanomaterials as adsorbent for HMi removal owing to their better physical, chemical, and mechanical properties (B. Chen et al. 2015). Apart from this, secondary forces arising from intermolecular interactions provided by high specific surfaces make these nanomaterials a better sorptionable candidate (Koohzad et al. 2019; Özsin et al. 2019).

In recent times, the world has perceived a paradigm shift toward sustainability with a focus on the use of biomaterials including sorption applications (Wang and Chen 2009). Of all biomaterials, cellulose is the most abundant material on the planet that exists today (Muhd Julkapli and Bagheri 2017). It is a biodegradable polymer, has high tensile strength, strong structural resistance, rich nano-fibrillar, and a porous structure making cellulose a good sorption candidate imparted by the presence of hydroxyl groups in cellulose chains (Thomas et al. 2018). Additionally, the presence of hydroxyl groups allows the functionalization of cellulose to further boost their overall physicochemical property. With such intriguing properties of cellulose, it has become the biomaterial of the present era for multifunctional applications (Ding et al. 2018). Recently, cellulose-based adsorbents for HMi removal have been attracting more and more attention due to their low cost, ease of availability, and biocompatibility (Jamshaid et al. 2017). Numerous cellulose-based materials have been reported for the past few years as a sorptionable candidate along with metal-organo composites, thiols, nanoparticles, grafting, etc. for HMi removal (Kumar et al. 2012; Cruz-Tato et al. 2017; Yao et al. 2019).

The present chapter highlights the significant role of cellulose and its derivatives (as biosorbents) toward sorption characteristics in environmental remediation and waste treatment. In view of rapid industrial commercialization coupled with the issue of global warming, cellulose-based sorbents can pave the way for a clean environment and greener eco-systems.

12.2 Heavy Metal Ions as Pollutants

Transition metals (Ru, Cu, Cd, Hg, etc.) and some metalloids (As, Sb, Pb, etc.) in freshwater reservoirs are exclusively classified as heavy metals due to their toxic, nonbiodegradable, and persistent nature, which are biologically hazardous or environmentally unfriendly (Azimi et al. 2017). Rapid industrial growth is the major source of HMi introduction into different segments of the environment (including air, water, soil, and the biosphere). Their electronic configuration makes them highly reactive along with a tendency to form undesirable and harmful complexations. HMi are mobile and bioaccumulate into the soil and are taken up by plants, thus entering animal and human food chains, thereby negatively affecting the health and vital activity of all organisms. Consequently, they disrupt metabolic functions by affecting vital organs in humans, and through the transposition of nutritional minerals, they impede biological functions in plants. Thus, HMi are often regarded as synonyms to pollution and toxicity (Hodson 2004). Table 12.1 demonstrates the source and their toxicologic effects on human health caused by heavy metals along with permissible limits as defined by the United Nations Environment Programme (UNEP) (based on the joint group of experts on the scientific aspects of marine pollution: reports of the 17th session, Geneva, Switzerland: World Health Organization 1987).

Table 12.1 UNEP reports on HMi and their toxicological effects.

Heavy metal	Major source	Effects	Permissible levels (mg/l)
Cd	Welding, electroplating, pesticide, fertilizers, Cd- batteries, nuclear fission plants	Renal dysfunction, lung cancer, bone defects (like osteoporosis), blood pressure, kidney damage, bronchitis, gastrointestinal disorder, bone marrow, cancer	0.06
Cr	Mines, mineral sources	Fatigue, irritability, damaged nervous system	0.05
Cu	Mining, pesticide production	Anemia, liver and kidney damage, stomach and intestinal irritation	0.1
Hg	Pesticides, batteries, paper industries	Protoplasm poisoning, tremors, gingivitis, minor psychological changes	0.01
Mn	Welding, fuel addition, ferromanganese production	Contact/inhalation direct damage to the central nervous system	0.26
Pb	Paint, pesticides, auto emissions, mining, coal burn	Congenital paralysis, liver, kidney, gastrointestinal damage	0.1
Zn	Refineries, brass manufacture, metal plating, plumbing	Corrosive effect on the skin, damaged nervous membrane	15

12.3 Cellulose as Biosorbents: Fundamentals to the Mechanistic Approach

Cellulose, a biopolymer, is the most abundant material on the planet to date. Its renewability, biodegradability, and nontoxicity make cellulose an attractive choice as a sorbent (Burba and Willmer 1983). Cellulose consists of several thousand units of glucose units linked end to end together in the β-(1–4) configuration (see Figure 12.1). The basic chemical formula can be represented as:

$$C_{6.P}H_{10(P+2)}O_{5(P+1)} \approx \left(C_6H_{10}O_5\right)_P$$

Where, P = degree of polymerization and n = number of glucose monomers in the cellulose chain. The elemental composition of cellulose is carbon (44–45%), hydrogen (6–6.5%), and oxygen (48.5–50%).

Apart from being a most abundant natural biopolymer (40–60% of wood mass), cellulose is a low-cost versatile material, thermodynamically stable, presenting a crystalline structure, and offers numerous hydrogen bonding opportunities. Using these advantages, hydroxyl groups (-OH) on cellulose influence the adsorption of HMi through electrostatic interaction (see Figure 12.2).

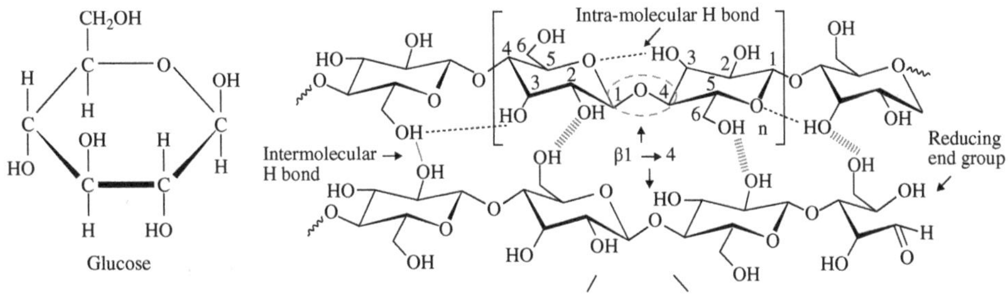

Figure 12.1 The chemical structure of cellulose.

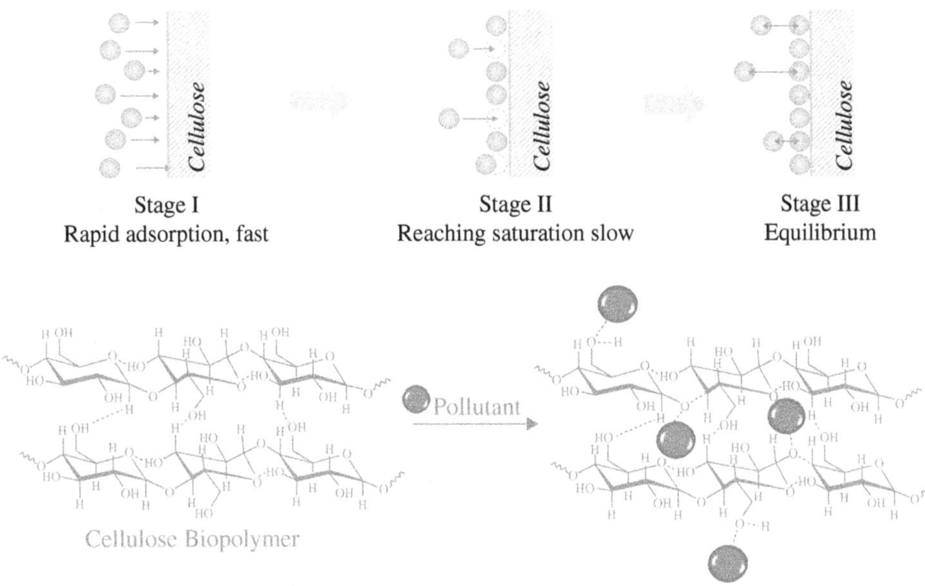

Figure 12.2 The sorption mechanism in cellulose for pollutants. (*Source:* Modified from Barathi et al. 2013.)

Cellulose, in its natural form, has limited sites (hydroxyl groups) for sorption activity, thus, it exhibits low sorption rates (Abbas et al. 2017). Many methods have been reported with respect to cellulose modification, either chemically or through grafting to further enhance the sorption characteristics of cellulose. The presence of the hydroxyl (–OH) group makes these modifications possible. Thus, efforts are still being channeled toward development through the optimization of various analytical parameters, functionalization, etc.

12.4 Modification of Cellulose

Recently, surface modification of cellulose through the hydroxyl group was exclusively carried out via simple reactions (viz. oxidation, acetylation, grafting of functional material or polymers onto the surface) (Abushammala and Mao 2019). The modification aimed at modulating the surface property of cellulose to improve its affinity for polar/nonpolymer molecules to place functional groups to target certain applications (Ding et al. 2018). Modified cellulose renders a higher surface area to volume ratio, better pore size distribution, mechanical rigidity, tunable interfacial surface chemistry, and viable regeneration under ambient conditions (Baiardo et al. 2002). Functionalization is a simple way to enhance hydrophilicity which leads to larger mass transfer, elimination ability, and selectivity. Cellulose modified with chelated groups (esterification, –COOH, amines, etc.) was extensively used for the removal of HMi. Different techniques involved in the surface modification of cellulose is represented in Figure 12.3.

In addition to chemical functionalization/grafting, there are five important parameters that should also be considered when investigating sorbent behavior with respect to the optimized sorption rate. They are as follows:

- **Effect of initial metal concentration**: The metal uptake mechanism is solely dependent on HMi concentration. When initial metal concentration increases, the percentage adsorption of heavy metals decreases. This could be attributed to saturated specific sites due to which exchange sites are filled, and thus, percent metal removal decreases, which means adsorption occurs only on the surface (Mouni et al. 2010; Al-Senani and Al-Fawzan 2018).
- **Effect of sorbent dosage**: Since increased dose is consistent with greater area and larger number of desorption sites, adsorption percentage increases with increasing adsorbent dose (Al-Senani and Al-Fawzan 2018).

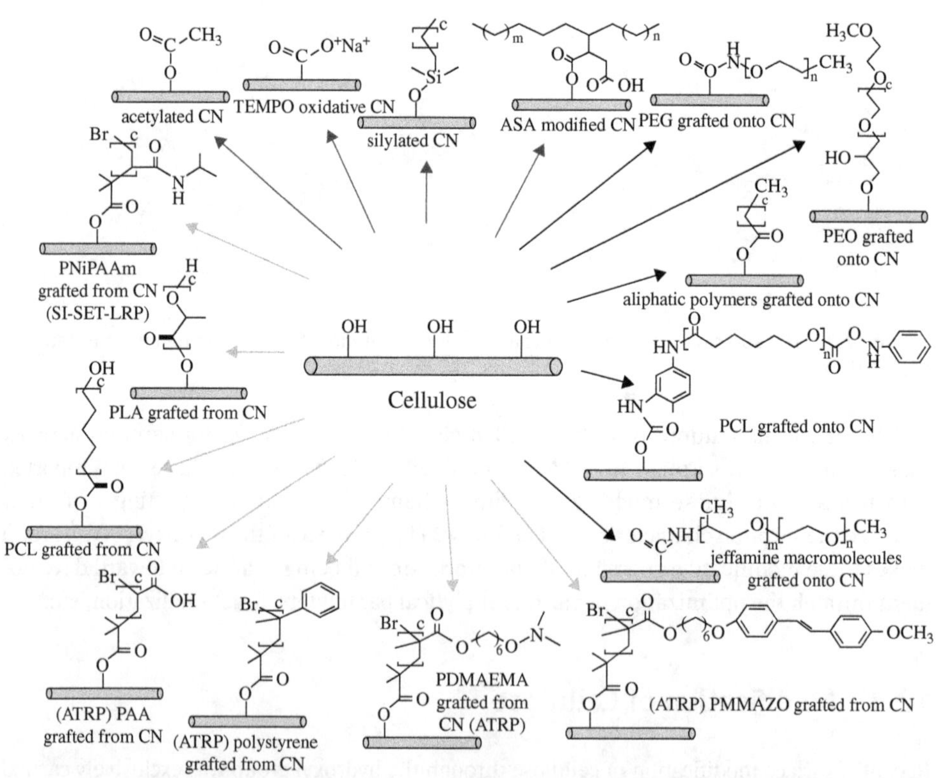

Figure 12.3 Different techniques for the surface modification of cellulose. (*Source:* Adapted from Dufresne 2013.)

- **Effect of pH on metal-uptake**: The influence of pH on sorption capacity could be attributed to the chemical form of heavy metals in the solution at specific pH. At low pH, the sorbent surface becomes more positively charged, thus, reducing the attraction with metal ions (due to effective competition between H^+ and H_3O^+). Contrarily, as the pH increases (pH ~ 9), a more negatively charged surface facilitates greater metal uptake (Mouni et al. 2010). In addition to this, the effect of pH on sorption can vary substantially in the presence of different functional groups.
- **Effect of temperature**: The adsorption capacity of HMi depends on the nature of the adsorbent. Depending on the temperature of the system (from room temperature to higher temperature), the rate of diffusion of adsorbate across the external boundary layer and internal pores also increases or decreases, thus, leading to a variable sorption efficiency of the sorbent (Mouni et al. 2010; Al-Senani and Al-Fawzan 2018).
- **Effect of contact time**: Contact time of HMi with the adsorbent depends on the nature of the adsorbent used, its exposed surface area, and their corresponding diffusion (Mouni et al. 2010).

12.4.1 Cellulose-Derived Material

Considerable work has been carried out on the sorption of HMi by cellulose-based materials due to its relatively lower cost and abundance. Bagasse, agricultural by-products, waste material from wood processing, wood sawdust, rice hull, etc. have been extensively used as biosorbents for HMi removal (Jamshaid et al. 2017). Waste materials with high cellulosic composition were further chemically modified using minerals, organic acid/base, organic–inorganic compound, oxidizing agents, etc. to further enhance the HMi sorption potential. A significant share of effort has been devoted to environmental remediation viz. wastewater treatment, purification, adsorption of HMi, air and soil purification, etc.

12.5 Removal of Various HMi

Lead, chromium, cobalt, copper, and nickel are all found to be nondegradable and nonessential environmental contaminants. Their toxicity has been found to cause anemia, renal dysfunctions, and damaged nervous systems in humans. To tackle such issues, Abbas et al. (2017) reported highly efficient, regenerable supersorbent succinylated hydroxyethylcellulose (HEC-Suc-Na) to evaluate multiple HMi viz. Pb(II), Cr(VI), Co(II), Cu(II), and Ni(II) from aqueous solutions. The presence of a single functional group, i.e. Na-succinate on HEC backbone and the difference in ionic charge and redox potential resulted in enhanced sorption capacities of 174, 163.64, 102.3, 97.6, 82.6, and 68.0 mg/g for Pb (II), Cr (VI), Co (II), Cu (II), and Ni (II), respectively. Apart from this, HMi (like Pb, Cd, Hg, Cu, etc.) has also been reported to be adsorbed by several cellulose-derived materials. Cadmium, which is another nondegradable HMi, has been reported to be removed by cellulose-based sorbent selectively as well as collectively. Chen et al. (2019) modified cellulose using carboxyl, amides, and secondary amines for cadmium only adsorption, resulting in an optimum sorption of 401.1 mg/g. Mercury ions are a serious environmental hazard. They have been listed as a precedence-controlled pollutant by the United Nations Environmental Protection agency (UNEP). Thus, effective removal by using an environmentally friendly approach is of great significance. In this regard, Bansal et al. (2018) extracted and esterified cellulose nanofibers from bagasse and functionalized them by incorporation of thiol (-SH) groups. The thiol-mercury interaction was such that adsorption was effective even at the lowest concentration (1 ppm). In a similar approach, Sun et al. (2018) grafted steric acid onto sugarcane bagasse fiber (SBF) through esterification, with sorption as high as 178 mg/g. This was higher than the pristine SBF. Also, the lack of functional groups to complex with HMi in virgin cellulose limits adsorption rate. Diethylenetriaminepentaacetic acid (DTPA), a potential ligand with five carboxylate groups, was grafted onto cellulose by B. Li et al. (2019b), achieving maximum adsorption of 476.2 mg/g. It is observed that cellulose when powdered, results in microcrystalline cellulose (MCC) with particle diameter ranging 20–100 μm, making them ineligible for sorption related activities. This limitation was crossed by Li et al. (2019b) using modified cellulose powders (MCPs), prepared by precipitating the mixture solution of cellulose with glutaric anhydride in ionic liquid, followed by grafting with aminoguanidine hydrochloride, and recorded a maximum adsorption of 625 mg/g. However, another instance of similar interaction where Li et al. (2018c) implemented in situ graft polymerization of glycidyl methacrylate (GMA) onto cellulose resulted in a coupling between modified cellulose and the chelating receptor of the Hg^{2+} ions. The maximum adsorption capacity was found to be 657.9 mg/g. Bacterial cellulose has gained popularity in the past five years. One of the group reported by Tamahkar et al. (2018), attached Cibacron Blue F3GA (as a dye ligand) to bacterial cellulose for selective Hg^{2+} removal, resulting in an adsorption capacity as high as 322 mg/g. This high rate was accounted for by the large number of functional groups incorporated into the structure and the high surface area to volume ratio exhibited by bacterial cellulose. Similarly, other HMi that have been reported to be sorbed by various cellulose-based materials in the past three years have been tabulated in Table 12.2.

Table 12.2 Different cellulose-based material used as sorbent for HMi.

HMi	Sorbent	ARR (mg/g)	References
Pb(II), Cr(VI), Co(II), Cu(II), Ni(II)	Hydroxyethylcellulose succinate-Na (HEC-Suc-Na)	1000, 909.09, 666.6, 588, and 500	Abbas et al. (2017)
Cd(II), Pb(II), Hg(II)	Nano-TiO_2-glu-CMC	2440 μmol/g (pH 7), 2880 μmol/g (pH 6), and 1160 μmol/g (pH 3)	Mahmoud et al. (2017)
Pb(II), Co(II)	Cellulosic adsorbent (PDA–MCC)	177.75 and 122.70	Sun et al. (2017)
Cr(VI)	Ion imprinted polymer (IIP) from sodium carboxymethyl cellulose	177.62 at 25 °C	Velempini et al. (2017)

Table 12.2 (Continued)

HMi	Sorbent	ARR (mg/g)	References
Cu(II), Cd(II), Pb(II)	Cellulose/chitosan (1 : 1) composite	0.53 mmol/g Cu^{2+}, 0.28 mmol/g Cd^{2+}, and 0.16 mmol/g Pb^{2+} ions at Ph 5.0.	Xiao and Hu (2017)
Pb(II), Cd(II)	Cellulose nanofibers (CNFs) tethered with sulfur	1.16 mmol/g and 0.82 mmol/g	Abu-Danso et al. (2018)
Co(II)	NP/cotton nanocomposite	125.0 mmol/g on ZnCt	Ali et al. (2018)
Hg(II)	Amine functionalized green material (Cys-CNFs)	116.822	Bansal et al. (2018)
Cr(VI)	Hybrid cellulose aerogels (named as ZIF-8@CA)	41.8	Bo et al. (2018)
Cu(II), Pb(II)	Natural CNFs cross-linked with acrylic acid (AA)	40.01 and 130.36	Tian et al. (2018)
Fe(III), Cu(II), Cd(II)	Amidoxime polyacrylonitrile/ regenerated cellulose (AOPAN/ RC) composite nanofibrous membranes (AOPAN/RC	7.47, 4.26, and 1.13 mmol/g	Feng et al. (2018)
Cd(II), Pb(II)	Jute/PAA hydrogel	401.7 and 542.9	Zhou et al. (2018a)
Cu(II), Cr(VI)	Zeolitic imidazolate framework-67 (ZIF-67)-based "pearl-necklace-like" 2-methylimidazole/cellulose acetate (MIM/CA) composite membranes	18.9 and 14.5	Hou et al. (2018)
Cd(II)	Magnetic bioadsorbent (MCIA)	40.8, 49.4, 54.6, and 56.6 at 293, 303, 313, and 323 K, respectively	Zhou et al. (2018b)
Cu(II), Zn(II), Cd(II), Pb(II)	Guanyl-modified cellulose (Gu-MC)	83, 78, 68, 52, and 48	Kenawy et al. (2018)
Cu(II)	Lignocellulose-based composite hydrogel	541	Zhang et al. (2018a)
Cu(II), Pb(II)	Aerogel adsorbent based on nanofibrillated cellulose (NFC) and polyethyleneimine (PEI)	175.44 and 357.44	Li et al. (2018a)
Hg (II)	Aminosalicylic acid-modified cellulose composite	657.9	M. Li et al. (2018b)
Pb(II)	CF/poly(NIPAM-co-AAC)	80.8, higher than that of CF (45.6) at 293 K	Zhang et al. (2018b)
Pb(II)	Purely extracted cellulose from oil palm empty fruit bunches (OPEFB) and EDTA modified - OPEFB	232.9 and 236.7	Nazir et al. (2018)

(Continued)

Table 12.2 (Continued)

HMi	Sorbent	ARR (mg/g)	References
Cu(II), Fe(II), Zn(II), Mn(II)	Modified lettuce roots, nonmodified lettuce roots, and modified sugarcane bagasse, nonmodified sugarcane bagasse	24.61, 2.64, 23.04, and 5.92 for NLR; 2.29, 16.89, 1.97, and 2.88 for MLR; 0.81, 0.06, 0.83, and 0.46 for NSB; and 1.35, 2.89, 20.76, and 1.56 for MSB.	Milani et al. (2018)
Cu(II) Pb(II), Fe(III)	CMC-g-AMPS hydrogel	86.21, 102.04, and 33.56	Panchan et al. (2018)
Cu(II), Fe(III), Cr(III), Co(III), Ni(II)	Poly(acrylonitrile)/cellulose block copolymer (PAN-b-cell)	272, 242, 219, 201, and 195, respectively, at pH 6	Rahman et al. (2018)
Ni(II), Co(II), Zn(II)	Hydroxyethyl cellulose-graft-poly(N-vinyl imidazole-co-acrylic acid)	165, 122, and 68	Ray et al. (2018)
Hg(II)	Esterification of sugarcane bagasse cellulose with excess stearic acid and the reaction of grafting polyacrylamide brush by ultraviolet radiation initiation	178	Sun et al. (2018)
Hg(II), Cd(II), Pb(II)	Bacterial cellulose nanofibers (BC) were prepared using a dye, namely Cibacron Blue F3GA (CB)	322.4, 48.5, and 41.9	Tamahkar et al. (2018)
Cu(II), Zn(II), Pb(II)	Crosslinked cellulose/sodium alginate (SA) was modified with PEI as an adsorbent (PEI-RCSA)	177.1, 110.2, and 234.2	Zhan et al. (2018)
Cd(II)	Electrospun nonwoven chitosan (CS) and phosphorylated nanocellulose (PNC) sorbent	232.55 at 25 °C	Brandes et al. (2019)
Pb(II)	Activated carbon fibers and cellulose/polyester (ACFs) (ACFCP)	385.77	Carraro et al. (2019)
Cd(II)	Novel carboxyl, amide, carbonyl sulfide, and secondary amino group grafted cellulose derivative adsorbent (modified cellulose)	401.1	Chen et al. (2019)
Cr(IV)	NC-based nanoadsorbents with excellent Cr(VI) adsorption performance through polydopamine (PDA) (CNC@PDA)	205	Dong et al. (2019)
Cu(II), Ni (II)	Cellulose-based adsorbent-glycidyl methacrylate/cellulose microspheres/iminodiacetic acid disodium (IDA)	96.90 and 111.11	Du et al. (2019)
Pb(II)	Carboxymethyl cellulose-immobilized Fe_3O_4 nanoparticles (CMC-Fe_3O_4)	152.0	Fan et al. (2019)

Table 12.2 (Continued)

HMi	Sorbent	ARR (mg/g)	References
Pb(II), Cu(II), Cd(II)	Agroindustrial waste biomass	86.2, 35.8, and 24.4	Fernández-López et al. (2019)
Cs(I)	CMC-kcufc-PEG carboxymethyl cellulose sodium (CMC) potassium copper ferrocyanide (KCuFC) using polyethylene glycol 6000 (PEG6000)	149.8	Hu et al. (2019)
Cr(VI), methylene blue dye (MB)	CS-GO/CMC-1/1 composite aerogel	127.4, 3190	Huang et al. (2019)
Pb(II)	Zero-valent iron (nZVI) with simultaneously modified with CMC	1376.07	Jiao et al. (2019)
Pb(II), Cu(II)	Uio-66-NH2@CA metal–organic frameworks (uio-66 and uio-66-NH2) on the flexible cellulose aerogels as metal –organic frameworks@cellulose aerogels composite materials	89.40	B. Li et al. (2019b)
Hg(II), Cu(II)	Modified porous cellulose beads (MCBs)	625 and 98.52	Li et al. (2019b)
Hg(II)	Diethylenetriaminepentaacetic acid (DTPA)-modified cellulose adsorbent	476.2	B. Li et al. (2019b)
Pb(II), MB	Carboxylated cellulose fabrics	81.30 and 76.92	Li et al. (2019c)
Hg(II)	Fluorogenic bioadsorbent	143.88	M. Li et al. (2019d)
Cu(II), Zn(II)	TEMPO-oxidized cellulose nanofibers (TOCNFs)	102.9 and 73.9	Li et al. (2019e)
Pb(II)	CS/NFC	252.6	Li et al. (2019f)
Pb(II) Cd(II), Cu(II)	Acrylic acid grafted onto cellulose nanocrystal (AA-g-CNC)	1026, 898.8, and 872.4	Liu et al. (2019)
Rhodamine B, Cr(IV), DMF	Porous zeolitic imidazolate frameworks-8@cellulose nanofiber@cellulose foam (ZIF-8@CNF@cellulose foam)	24.6, 35.6, and 45.2	Ma et al. (2019)
Cu(II)	2,2,6,6-tetramethylpiperidine-1-oxyl (TEMPO) oxidized cellulose nanofibril (TO-CNF) crosslinked trimethylolpropane-tris-(2-methyl-1-aziridine) propionate (TMPTAP) and PEI TO-CNF/TMPTAP/PEI	485.44	Mo et al. (2019)
Arsenate	3-mercapto-propanoic acid (MPA) modified cellulose filter paper (MPA-cell paper)	92.59	Pramanik et al. (2019)
Cu(II)	Carboxymethylated CNFs (CMCNFs)	115.3 at pH 5.0	Qin et al. (2019)

(Continued)

Table 12.2 (Continued)

HMi	Sorbent	ARR (mg/g)	References
Pb(II), Cu(II)	Hydrogel composites based on chitosan-g-poly(acrylic acid) matrices	818.4 and 325.5	Rodrigues et al. (2019)
Cd(II)	Nano mgs decorated cellulose nanofibers (mgs@CNF)	333.33	Sankararamakrishnan et al. (2019)
Cu(II), methyl orange (MO)	CNFs, which were cross-linked with PEI aerogels.	103.5 and 265.9	Tang et al. (2019)
Cr(VI)	Synthetic iron sulfide nanoparticles (FeS NPs) stabilized with CMC	1046.1 mg Cr(VI) per gram of FeS NPs	Wang et al. (2019)
Zn(II), Cd(II)	Cellulose acetate (CA) /natural rectorite (REC) nanofiber mat	104.31 and 99.33	Wu et al. (2019a)
Cr(IV)	Cellulose-based solid amine adsorbent (MCC/TEPAA)	327.72	Xue et al. (2019)

12.6 Adsorption and Kinetic Studies

Adsorption is a result of various interactions that take place between sorbent sites and HMi. Among many interactions that occur on the surface of a sorbent, cellulose sorbents favor swelling, ion-exchange, and metal-ion interactions (Öztürk et al. 2009). Adsorption isotherms provide sorption insight that allows for the determination of various sorbent surface morphology such as surface area and pore behavior. They provide a basis for thermodynamics as well as sorbed molecule distribution in a sorbate (liquid/gas phase)-sorbent (solid phase) environment, once equilibrium is achieved. Typically, a sorbent is subjected to batch experiments, which provide data like HMi adsorbed, rate constant, etc. These are then fitted to available isotherms (summarized in Table 12.3) (Gunay et al. 2013). A model agreement with respect to experimental data is determined by the value of correlation coefficients (R^2). The higher the value, the better conformity can be assured. Maximum adsorption (q_e), which is predefined for each model, can be calculated using various equations given in Table 12.3. A separation factor, R_L valued between 0 and 1 indicates favorable adsorption ($0 < R_L < 1$). Any outbound value beyond this limit could be considered unfavorable.

12.6.1 Kinetics

The following kinetics have been exclusively considered for isotherm modeling.

12.6.1.1 Pseudo-first Order Model

$$\text{i) } log\left(q_e - q_t\right) = log_e - \frac{k_2}{2.303}t : \text{pseudo} - \text{first order} \tag{12.1}$$

Table 12.3 Two- and three-parameter equations for isotherm modeling.

	Formula	Parameters
Two-parameter models		
Freundlich	$q_c = K_F C_e^{1/n}$	K_F, n
Langmuir	$q_e = \dfrac{q_m a_L C_e}{1 + a_L C_e}$	q_m, a_L
Dubinin–Radushkevich	$q_e = q_m \exp\left(\dfrac{RT \ln(1 + 1/C_e))^2}{-2E^2}\right)$	q_s, E
Temkin	$q_e = \left(\dfrac{RT}{b}\right) \ln(K_{Te} C_e)$	b, K_{Te}
Three-parameter models		
Redlich-Peterson	$q_e = \dfrac{K_{RP} C_e}{1 + a_{RP} C_e^{\beta}}$	K_{RP}, a_{RP}, β
Sips	$q_e = \dfrac{q_s a_S C_e^{1/sp}}{1 + a_s C_e^{1/sp}}$	q_s, a_S, sp
Toth	$q_e = \dfrac{q_m C_e}{\left(K_{To} + C_e^{tn}\right)^{1/tn}}$	q_m, K_{To}, tn
Khan	$q_e = \dfrac{q_k b_K C_e}{\left(1 + b_K C_e\right)^{a_K}}$	q_m, b_K, a_K

Source: Gunay et al. (2013).

12.6.1.2 Pseudo-second Order Model

$$\text{ii)}\quad \frac{t}{q_t} = \frac{1}{kq_e^2} + \frac{t}{q_e} : \text{pseudo} - \text{second order} \tag{12.2}$$

q_e: metal adsorbed at equilibrium;
q_t: metal adsorbed at equilibrium at time t;
k: rate constant (g/ min).

12.6.1.3 Weber and Morris Intraparticle Diffusion Model

$$\text{iii)}\quad Q = kt^{\frac{1}{2}} + C : \text{Wb and Morris Intraparticle diffusion model} \tag{12.3}$$

C: Intercept

12.6.1.4 Ion Exchange Mechanism

$$\text{iv) } \log(1-F) = -\frac{S}{2.303}t : \text{ion exchange mechanism} \tag{12.4}$$

$$S : \min^{1}$$

$$F = \frac{q_t}{q_e}$$

$$q_t = \frac{C_0V_0 - C_tV_t}{m}$$

C_0 is the initial concentration of HMi in the solution (mg/l);
V_0 is the initial volume of the solution (L);
C_t is the concentration of ion at time t;
V_t is the volume of solution at time t;
m is the sorbent mass (g).

12.7 The Role of Thermodynamics

Thermodynamics in adsorption plays an important role in determining phase equilibria between sorbent and HMi. It should be noted that these calculations hold good only over the equilibrium model. Thermodynamic parameters like free energy change (ΔG°, kJ/mol K), entropy change (ΔS°, kJ/ mol K), and enthalpy (ΔH°) can be estimated by using the following relations:

$$\Delta G^0 = -RTlnK_L \tag{12.5}$$

$$\ln K_L = \frac{\Delta S^0}{R} - \frac{\Delta H^0}{RT} \tag{12.6}$$

Where, T is the absolute temperature (K); R = 8.314 J/ mol K, the universal gas constant; K_L: Langmuir constant. A sign convention is observed sorbents as usual for certain parameters like negative values of ΔG°, ΔS°, and ΔH°. The negative value is a good indication that sorption activity would occur spontaneously and exothermically. Positive values of ΔH° on the sorbents represent an endothermic activity.

12.8 Prospects Toward Sustainability

Derived cellulose (plants, bacterial) hydrogels either natively or with composites have a tremendous deployment in the field of protecting technologies related to environmental problems owing to their surface morphologies (Chang and Zhang 2011). Powdered cellulose (Ivanova et al. 2019) and wood cellulose (Zadorecki and Michell 1989) could be potential reinforcement agents in synthetic organic polymers employed in several thermoplastics and elastomers. Moreover, the past decade has witnessed many reports related

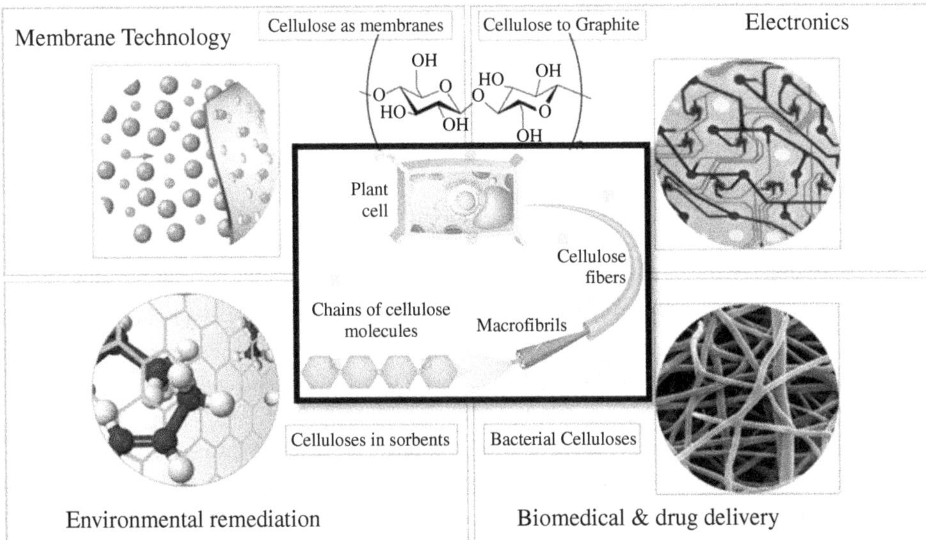

Figure 12.4 Versatile applications of cellulose-based materials.

to microbial cellulose's unique nanostructures making it a suitable candidate for numerous medical and tissue engineering applications. With advances in nanotechnology, cellulose and nanocellulose along with numerous composites (Lin and Dufresne 2014) are promising candidates as bioscaffolds. Their biocompatibility, biodegradability, and lower toxicity make them a sustainable choice over other different materials. Increasing sustainability demands must bring cellulose into focus in various technological areas. The present scenario projects a promising market dynamic worldwide, for example, in India and China at present, where textile industries are already implementing cellulose acetate over silk fibers. With an expected market scale of over $10 billion/year, cellulose could be, with no doubt, the next generation material in many areas, including automobile components, electronic materials, packaging, construction materials, and high-performance fillers (see in Figure 12.4).

12.9 Summary

In this chapter, sorption efficacy of cellulose-based sorbents has highlighted fundamental and mechanical methods toward HMi removal from wastewater. Among various wastewater treatment processes, cellulose-based biosorbents are considered to be efficient, cost-effective, and environmentally friendly methods in the purification process. This chapter provides a basic understanding of cellulose chemistry, functionalization, development, and optimization through analytical parameters. Due to the presence of hydroxyl groups in the cellulose surface, several modification routes are offered via oxidation, grafting chemical functionalization, etc., which allow larger mass transfer, elimination ability, and stability.

This chapter also reviews the different cellulose-based sorbents for the removal of harmful toxic pollutants from wastewater and their modified forms. A brief overview on sorption, kinetic studies, and thermodynamics has also been discussed. The chapter concludes with the prospects for cellulose-based sorbents in various areas, along with forecasts for the market share of cellulose.

References

Abbas, A., Hussain, M.A., Sher, M. et al. (2017). *Design, characterization and evaluation of hydroxyethylcellulose based novel regenerable supersorbent for heavy metal ions uptake and competitive adsorption. International Journal of Biological Macromolecules*; Elsevier B.V. doi: https://doi.org/10.1016/j.ijbiomac.2017.04.024.

Abu-Danso, E., Peräniemi, S., Leiviskä, T., and Bhatnagar, A. (2018). Synthesis of S-ligand tethered cellulose nanofibers for efficient removal of Pb(II) and Cd(II) ions from synthetic and industrial wastewater. *Environmental Pollution* 242: 1988–1997. https://doi.org/10.1016/j.envpol.2018.07.044.

Abushammala, H. and Mao, J. (2019). A review of the surface modification of cellulose and nanocellulose using aliphatic and aromatic mono- and di-isocyanates. *Molecules* 24 (15) https://doi.org/10.3390/molecules24152782.

Akpomie, K.G., Dawodu, F.A., and Adebowale, K.O. (2015). Mechanism on the sorption of heavy metals from binary-solution by a low cost montmorillonite and its desorption potential. *Alexandria Engineering Journal* 54 (3): 757–767. https://doi.org/10.1016/j.aej.2015.03.025.

Ali, A., Gul, A., Mannan, A., and Zia, M. (2018). Efficient metal adsorption and microbial reduction from Rawal Lake wastewater using metal nanoparticle coated cotton. *Science of the Total Environment* 639: 26–39. https://doi.org/10.1016/j.scitotenv.2018.05.133.

Al-Senani, G.M. and Al-Fawzan, F.F. (2018). Adsorption study of heavy metal ions from aqueous solution by nanoparticle of wild herbs. *The Egyptian Journal of Aquatic Research* 44 (3): 187–194. https://doi.org/10.1016/j.ejar.2018.07.006.

Azimi, A., Azari, A., Rezakazemi, M., and Ansarpour, M. (2017). Removal of heavy metals from industrial wastewaters: a review. *ChemBioEng Reviews* 4 (1): 37–59. https://doi.org/10.1002/cben.201600010.

Baiardo, M., Frisoni, G., Scandola, M., and Licciardello, A. (2002). Surface chemical modification of natural cellulose fibers. *Journal of Applied Polymer Science* 83 (1): 38–45. https://doi.org/10.1002/app.2229.

Bansal, M., Ram, B., Chauhan, G.S., and Kaushik, A. (2018). L-Cysteine functionalized bagasse cellulose nanofibers for mercury(II) ions adsorption. *International Journal of Biological Macromolecules* 112: 728–736. https://doi.org/10.1016/j.ijbiomac.2018.01.206.

Barathi, M., Krishna, S., and Rajesh, K.N. (2013). Efficacy of novel Al-Zr impregnated cellulose adsorbent prepared using microwave irradiation for the facile defluoridation of water. *Journal of Environmental Chemical Engineering* 1 (4): 1325–1335.

Bo, S., Ren, W., Lei, C. et al. (2018). Flexible and porous cellulose aerogels/zeolitic imidazolate framework (ZIF-8) hybrids for adsorption removal of Cr(IV) from water. *Journal of Solid State Chemistry* 262: 135–141. https://doi.org/10.1016/j.jssc.2018.02.022.

Brandes, R., Belosinschi, D., Brouillette, F., and Chabot, B. (2019). A new electrospun chitosan/phosphorylated nanocellulose biosorbent for the removal of cadmium ions from aqueous solutions. *Journal of Environmental Chemical Engineering* 7 (6) https://doi.org/10.1016/j.jece.2019.103477.

Burba, P. and Willmer, P.G. (1983). Cellulose: a biopolymeric sorbent for heavy-metal traces in waters. *Talanta* 30 (5): 381–383. https://doi.org/10.1016/0039-9140(83)80087-3.

Carraro, P.S., Spessato, L., Crespo, L.H.S. et al. (2019). Activated carbon fibers prepared from cellulose and polyester–derived residues and their application on removal of Pb2+ ions from aqueous solution. *Journal of Molecular Liquids* 289 https://doi.org/10.1016/j.molliq.2019.111150.

Chang, C. and Zhang, L. (2011). Cellulose-based hydrogels: present status and application prospects. *Carbohydrate Polymers* 84 (1): 40–53. https://doi.org/10.1016/j.carbpol.2010.12.023.

Chen, B., Ma, Q., Tan, C. et al. (2015). Carbon-based sorbents with three-dimensional architectures for water remediation. *Small* 11 (27): 3319–3336. https://doi.org/10.1002/smll.201403729.

Chen, Q., Zheng, J., Wen, L. et al. (2019). A multi-functional-group modified cellulose for enhanced heavy metal cadmium adsorption: performance and quantum chemical mechanism. *Chemosphere* 224: 509–518. https://doi.org/10.1016/j.chemosphere.2019.02.138.

Cruz-Tato, P., Ortiz-Quiles, E.O., Vega-Figueroa, K. et al. (2017). Metalized Nanocellulose composites as a feasible material for membrane supports: design and applications for water treatment. *Environmental Science and Technology* 51 (8): 4585–4595. https://doi.org/10.1021/acs.est.6b05955.

Ding, Q., Xu, X., Yue, Y. et al. (2018). Nanocellulose-mediated electroconductive self-healing hydrogels with high strength, plasticity, viscoelasticity, stretchability, and biocompatibility toward multifunctional applications. *ACS Applied Materials and Interfaces* 10 (33): 27987–28002. https://doi.org/10.1021/acsami.8b09656.

Dong, L., Deng, R., Xiao, H. et al. (2019). Hierarchical polydopamine coated cellulose nanocrystal microstructures as efficient nanoadsorbents for removal of Cr(VI) ions. *Cellulose* 26 (11): 6401–6414. https://doi.org/10.1007/s10570-019-02529-3.

Du, J. F., Dong, Z., & Zhao, L. (2019). Radiation synthesis and heavy metal ions removal of cellulose microsphere based adsorbent. *5th International Conference on Water Resource and Environment, WRE 2019, 344*, Macau University. Macau. Bristol: IOP Publishing Ltd. doi: https://doi.org/10.1088/1755-1315/344/1/012072

Dufresne, A. (2013). Nanocellulose: a new ageless bionanomaterial. *Materials Today* 16 (6): 220–227. https://doi.org/10.1016/j.mattod.2013.06.004.

Fan, H., Ma, X., Zhou, S. et al. (2019). Highly efficient removal of heavy metal ions by carboxymethyl cellulose-immobilized Fe 3 O 4 nanoparticles prepared via high-gravity technology. *Carbohydrate Polymers* 213: 39–49. https://doi.org/10.1016/j.carbpol.2019.02.067.

Feng, Q., Wu, D., Zhao, Y. et al. (2018). Electrospun AOPAN/RC blend nanofiber membrane for efficient removal of heavy metal ions from water. *Journal of Hazardous Materials* 344: 819–828. https://doi.org/10.1016/j.jhazmat.2017.11.035.

Fernández-López, J.A., Angosto, J.M., Roca, M.J., and Doval Miñarro, M. (2019). Taguchi design-based enhancement of heavy metals bioremoval by agroindustrial waste biomass from artichoke. *Science of the Total Environment* 653: 55–63. https://doi.org/10.1016/j.scitotenv.2018.10.343.

Ge, Y. and Li, Z. (2018). Application of lignin and its derivatives in adsorption of heavy metal ions in water: a review. *ACS Sustainable Chemistry & Engineering* 6 (5): 7181–7192. https://doi.org/10.1021/acssuschemeng.8b01345.

Gunay, A., Ersoy, B., Dikmen, S., and Evcin, A. (2013). Investigation of equilibrium, kinetic, thermodynamic and mechanism of basic blue 16 adsorption by montmorillonitic clay. *Adsorption* 19: 757–768. https://doi.org/10.1007/s10450-013-9509-4.

Hodson, M.E. (2004). Heavy metals - geochemical bogey men? *Environmental Pollution* 129 (3): 341–343. https://doi.org/10.1016/j.envpol.2003.11.003.

Holmes, R.R., Hart, M.L., and Kevern, J.T. (2019). Reuse of drinking water treatment waste for remediation of heavy metal contaminated groundwater. *Groundwater Monitoring and Remediation* 39 (4): 69–79. https://doi.org/10.1111/gwmr.12348.

Hou, X., Zhou, H., Zhang, J. et al. (2018). High adsorption pearl-necklace-like composite membrane based on metal–organic framework for heavy metal ion removal. *Particle and Particle Systems Characterization* 35 (6) https://doi.org/10.1002/ppsc.201700438.

Hu, J., Chen, S., Zhang, N. et al. (2019). Porous composite CMC-KCuFC-PEG spheres for efficient cesium removal from wastewater. *New Journal of Chemistry* 43 (24): 9658–9665. https://doi.org/10.1039/c9nj01697j.

Huang, T., Shao, Y.W., Zhang, Q. et al. (2019). Chitosan-cross-linked graphene oxide/carboxymethyl cellulose aerogel globules with high structure stability in liquid and extremely high adsorption ability. *ACS Sustainable Chemistry & Engineering* 7 (9): 8775–8788. https://doi.org/10.1021/acssuschemeng.9b00691.

Ivanova, V.N., Makhotina, L.G., and Berestovaya, E.D. (2019). Prospects of using commercial cellulose types to make products with a high added value. *Fibre Chemistry* 51 (1): 14–17. https://doi.org/10.1007/s10692-019-10038-z.

Jamshaid, A., Hamid, A., Muhammad, N. et al. (2017). Cellulose based materials for the removal of heavy metals from wastewater-An overview. *Chemical and Biochemical Engineering* 4: 240–256. https://doi.org/10.1002/cben.201700002.

Jiao, W., Song, Y., Zhang, D. et al. (2019). Nanoscale zero-valent iron modified with carboxymethyl cellulose in an impinging stream-rotating packed bed for the removal of lead(II). *Advanced Powder Technology* 30 (10): 2251–2261. https://doi.org/10.1016/j.apt.2019.07.005.

Kenawy, I.M., Ismail, M.A., Hafez, M.A.H., and Hashem, M.A. (2018). Synthesis and characterization of novel ion-imprinted guanyl-modified cellulose for selective extraction of copper ions from geological and municipality sample. *International Journal of Biological Macromolecules* 115: 625–634. https://doi.org/10.1016/j.ijbiomac.2018.04.100.

Khulbe, K.C. and Matsuura, T. (2018). Removal of heavy metals and pollutants by membrane adsorption techniques. *Applied Water Science* 8 (1) https://doi.org/10.1007/s13201-018-0661-6.

Koohzad, E., Jafari, D., and Esmaeili, H. (2019). Adsorption of lead and arsenic ions from aqueous solution by activated carbon prepared from Tamarix leaves. *ChemistrySelect* 4 (42): 12356–12367. https://doi.org/10.1002/slct.201903167.

Kumar, A.S.K., Kalidhasan, S., Rajesh, V., and Rajesh, N. (2012). Application of cellulose-clay composite biosorbent toward the effective adsorption and removal of chromium from industrial wastewater. *Industrial and Engineering Chemistry Research* 51 (1): 58–69. https://doi.org/10.1021/ie201349h.

Li, J., Zuo, K., Wu, W. et al. (2018a). Shape memory aerogels from nanocellulose and polyethyleneimine as a novel adsorbent for removal of Cu(II) and Pb(II). *Carbohydrate Polymers* 196: 376–384. https://doi.org/10.1016/j.carbpol.2018.05.015.

Li, M., Li, B., Pan, Y. et al. (2018b). An Aminosalicylic acid-modified cellulose composite used for mercury (II) removal from single and quarternary aqueous solutions. *ChemistrySelect* 3 (35): 10096–10102. https://doi.org/10.1002/slct.201802254.

Li, M., Liu, Z., Wang, S. et al. (2018c). Fluorescence detection and removal of copper from water using a biobased and biodegradable 2D soft material. *Chemical Communications* 54 (2): 184–187. https://doi.org/10.1039/c7cc08035b.

Li, B., Li, M., Zhang, J. et al. (2019a). Adsorption of Hg (II) ions from aqueous solution by diethylenetriaminepentaacetic acid-modified cellulose. *International Journal of Biological Macromolecules* 122: 149–156. https://doi.org/10.1016/j.ijbiomac.2018.10.162.

Li, B., Pan, Y., Zhang, Q. et al. (2019b). Porous cellulose beads reconstituted from ionic liquid for adsorption of heavy metal ions from aqueous solutions. *Cellulose* 26 (17): 9163–9178. https://doi.org/10.1007/s10570-019-02687-4.

Li, C., Ma, H., Venkateswaran, S., and Hsiao, B.S. (2019c). Highly efficient and sustainable carboxylated cellulose filters for removal of cationic dyes/heavy metals ions. *Chemical Engineering Journal* https://doi.org/10.1016/j.cej.2019.123458.

Li, M., Li, B., Zhou, L. et al. (2019d). Fluorescence-sensitive adsorbent based on cellulose using for mercury detection and removal from aqueous solution with selective "on-off" response. *International Journal of Biological Macromolecules* 132: 1185–1192. https://doi.org/10.1016/j.ijbiomac.2019.04.048.

Li, M., Messele, S.A., Boluk, Y., and Gamal El-Din, M. (2019e). Isolated cellulose nanofibers for Cu (II) and Zn (II) removal: performance and mechanisms. *Carbohydrate Polymers* 221: 231–241. https://doi.org/10.1016/j.carbpol.2019.05.078.

Li, Y., Guo, C., Shi, R. et al. (2019f). Chitosan/ nanofibrillated cellulose aerogel with highly oriented microchannel structure for rapid removal of Pb (II) ions from aqueous solution. *Carbohydrate Polymers* 223 https://doi.org/10.1016/j.carbpol.2019.115048.

Lin, N. and Dufresne, A. (2014). Nanocellulose in biomedicine: current status and future prospect. *European Polymer Journal* 59: 302–325. https://doi.org/10.1016/j.eurpolymj.2014.07.025.

Liu, X., Xu, M., An, B. et al. (2019). A facile hydrothermal method–fabricated robust and ultralight weight cellulose nanocrystal-based hydro/aerogels for metal ion removal. *Environmental Science and Pollution Research* 26 (25): 25583–25595. https://doi.org/10.1007/s11356-019-05810-x.

Ma, S., Zhang, M., Nie, J. et al. (2019). Lightweight and porous cellulose-based foams with high loadings of zeolitic imidazolate frameworks-8 for adsorption applications. *Carbohydrate Polymers* 208: 328–335. https://doi.org/10.1016/j.carbpol.2018.12.081.

Mahmoud, M.E., Abdou, A.E.H., Sobhy, M.E., and Fekry, N.A. (2017). Solid–solid crosslinking of carboxymethyl cellulose nanolayer on titanium oxide nanoparticles as a novel biocomposite for efficient removal of toxic heavy metals from water. *International Journal of Biological Macromolecules* 105: 1269–1278. https://doi.org/10.1016/j.ijbiomac.2017.07.156.

Maniquiz-Redillas, M.C. and Kim, L.H. (2016). Evalulation of the capability of low-impact development practices for the removal of heavy metal from urban stormwater runoff.

Environmental Technology 37 (18): 2265–2272. https://doi.org/10.1080/09593330.2016.1147610.

Maruyama, T., Terashima, Y., Takeda, S. et al. (2014). Selective adsorption and recovery of precious metal ions using protein-rich biomass as efficient adsorbents. *Process Biochemistry* 49 (5): 850–857. https://doi.org/10.1016/j.procbio.2014.02.016.

Milani, P.A., Consonni, J.L., Labuto, G., and Carrilho, E.N.V.M. (2018). Agricultural solid waste for sorption of metal ions, part II: competitive assessment in multielemental solution and lake water. *Environmental Science and Pollution Research* 25 (36): 35906–35914. https://doi.org/10.1007/s11356-018-1726-7.

Mo, L., Pang, H., Tan, Y. et al. (2019). 3D multi-wall perforated nanocellulose-based polyethylenimine aerogels for ultrahigh efficient and reversible removal of Cu(II) ions from water. *Chemical Engineering Journal* 378 https://doi.org/10.1016/j.cej.2019.122157.

Mouni, L., Merabet, D., Bouzaza, A., and Belkhiri, L. (2010). Removal of Pb2+ and Zn2+ from the aqueous solutions by activated carbon prepared from dates stone. *Desalination and Water Treatment* 16 (1–3): 66–73. https://doi.org/10.5004/dwt.2010.1106.

Muhd Julkapli, N. and Bagheri, S. (2017). Nanocellulose as a green and sustainable emerging material in energy applications: a review. *Polymers for Advanced Technologies* 28 (12): 1583–1594. https://doi.org/10.1002/pat.4074.

Mustapha, S., Shuaib, D.T., Ndamitso, M.M. et al. (2019). Adsorption isotherm, kinetic and thermodynamic studies for the removal of Pb(II), Cd(II), Zn(II) and Cu(II) ions from aqueous solutions using Albizia lebbeck pods. *Applied Water Science* 9 (6): 142. https://doi.org/10.1007/s13201-019-1021-x.

Nazir, M.S., Ajab, H., Raza, M.R., and Abdullah, M.A. (2018). Surface modification of cellulose fibers from oil palm empty fruit bunches for heavy metal ion sorption and diesel desulphurization. *Desalination and Water Treatment* 107: 241–256. https://doi.org/10.5004/dwt.2018.22170.

Özsin, G., Kılıç, M., Apaydın-Varol, E., and Pütün, A.E. (2019). Chemically activated carbon production from agricultural waste of chickpea and its application for heavy metal adsorption: equilibrium, kinetic, and thermodynamic studies. *Applied Water Science* 9 (3) https://doi.org/10.1007/s13201-019-0942-8.

Öztürk, H., Vu-manh, H., and Bechtold, T. (2009). Interaction of cellulose with alkali metal ions and complexed heavy metals. *Lenzinger Berichte* 87.

Panchan, N., Niamnuy, C., Dittanet, P., and Devahastin, S. (2018). Optimization of synthesis condition for carboxymethyl cellulose-based hydrogel from rice straw by microwave-assisted method and its application in heavy metal ions removal. *Journal of Chemical Technology and Biotechnology* 93 (2): 413–425. https://doi.org/10.1002/jctb.5370.

Pramanik, K., Sarkar, P., and Bhattacharyay, D. (2019). 3-Mercapto-propanoic acid modified cellulose filter paper for quick removal of arsenate from drinking water. *International Journal of Biological Macromolecules* 122: 185–194. https://doi.org/10.1016/j.ijbiomac.2018.10.065.

Qin, F., Fang, Z., Zhou, J. et al. (2019). Efficient removal of Cu2+ in water by Carboxymethylated Cellulose Nanofibrils: performance and mechanism. *Biomacromolecules* 20 (12): 4466–4475. https://doi.org/10.1021/acs.biomac.9b01198.

Rahman, M.L., Sarkar, S.M., Farid, E.M. et al. (2018). Synthesis of tapioca cellulose-based poly(amidoxime) ligand for removal of heavy metal ions. *Journal of Macromolecular Science, Part B: Physics* 57 (2): 83–99. https://doi.org/10.1080/00222348.2018.1432179.

Ranieri, E., Moustakas, K., Barbafieri, M. et al. (2020). Phytoextraction technologies for mercury- and chromium-contaminated soil: a review. *Journal of Chemical Technology and Biotechnology* 95 (2): 317–327. https://doi.org/10.1002/jctb.6008.

Ray, J., Jana, S., Bhanja, S.K., and Tripathy, T. (2018). Efficient removal of Co(II), Ni(II), and Zn(II) metal ions from binary and ternary solutions using a pH responsive bifunctional graft copolymer. *Colloid and Polymer Science* 296 (8): 1275–1291. https://doi.org/10.1007/s00396-018-4345-4.

Rodrigues, F.H.A., de Magalhães, C.C.E., Medina, A.L., and Fajardo, A.R. (2019). Hydrogel composites containing nanocellulose as adsorbents for aqueous removal of heavy metals: design, optimization, and application. *Cellulose* 26 (17): 9119–9133. https://doi.org/10.1007/s10570-019-02736-y.

Saha, D., Barakat, S., Van Bramer, S.E. et al. (2016). Noncompetitive and competitive adsorption of heavy metals in sulfur-functionalized ordered mesoporous carbon. *ACS Applied Materials and Interfaces* 8 (49): 34132–34142. https://doi.org/10.1021/acsami.6b12190.

Sankararamakrishnan, N., Singh, R., and Srivastava, I. (2019). Performance of novel MgS doped cellulose nanofibres for Cd(II) removal from industrial effluent – mechanism and optimization. *Scientific Reports* 9 (1) https://doi.org/10.1038/s41598-019-49076-2.

Sun, C., Ni, J., Zhao, C. et al. (2017). Preparation of a cellulosic adsorbent by functionalization with pyridone diacid for removal of Pb(II) and Co(II) from aqueous solutions. *Cellulose* 24 (12): 5615–5624. https://doi.org/10.1007/s10570-017-1519-z.

Sun, N., Wen, X., and Yan, C. (2018). Adsorption of mercury ions from wastewater aqueous solution by amide functionalized cellulose from sugarcane bagasse. *International Journal of Biological Macromolecules* 108: 1199–1206. https://doi.org/10.1016/j.ijbiomac.2017.11.027.

Tamahkar, E., Türkmen, D., Akgönüllü, S. et al. (2018). Bacterial cellulose nanofibers for efficient removal of Hg 2+ from aqueous solutions. In: *Nanotechnology for Sustainable Water Resources* (eds. E. Tamahkar, D. Türkmen, S. Akgönüllü, et al.), 501–522. Wiley Blackwell https://doi.org/10.1002/9781119323655.ch16.

Tang, J., Song, Y., Zhao, F. et al. (2019). Compressible cellulose nanofibril (CNF) based aerogels produced via a bio-inspired strategy for heavy metal ion and dye removal. *Carbohydrate Polymers* 208: 404–412. https://doi.org/10.1016/j.carbpol.2018.12.079.

Thomas, B., Raj, M.C., Athira, B.K. et al. (2018). Nanocellulose, a versatile green platform: from biosources to materials and their applications. *Chemical Reviews* 118 (24): 11575–11625. https://doi.org/10.1021/acs.chemrev.7b00627.

Tian, C., She, J., Wu, Y. et al. (2018). Reusable and cross-linked cellulose nanofibrils aerogel for the removal of heavy metal ions. *Polymer Composites* 39 (12): 4442–4451. https://doi.org/10.1002/pc.24536.

Velempini, T., Pillay, K., Mbianda, X.Y., and Arotiba, O.A. (2017). Epichlorohydrin crosslinked carboxymethyl cellulose-ethylenediamine imprinted polymer for the selective uptake of Cr(VI). *International Journal of Biological Macromolecules* 101: 837–844. https://doi.org/10.1016/j.ijbiomac.2017.03.048.

Wang, J. and Chen, C. (2009). Biosorbents for heavy metals removal and their future. *Biotechnology Advances* 27 (2): 195–226. https://doi.org/10.1016/j.biotechadv.2008.11.002.

Wang, T., Liu, Y., Wang, J. et al. (2019). In-situ remediation of hexavalent chromium contaminated groundwater and saturated soil using stabilized iron sulfide nanoparticles. *Journal of Environmental Management* 231: 679–686. https://doi.org/10.1016/j.jenvman.2018.10.085.

Wu, Y., Qiu, X., Cao, S. et al. (2019a). Adsorption of natural composite sandwich-like nanofibrous mats for heavy metals in aquatic environment. *Journal of Colloid and Interface Science* 539: 533–544. https://doi.org/10.1016/j.jcis.2018.12.099.

Wu, Y., Pang, H., Liu, Y. et al. (2019b). Environmental remediation of heavy metal ions by novel-nanomaterials: a review. *Environmental Pollution* 246: 608–620. https://doi.org/10.1016/j.envpol.2018.12.076.

Xiao, M. and Hu, J. (2017). Cellulose/chitosan composites prepared in ethylene diamine/potassium thiocyanate for adsorption of heavy metal ions. *Cellulose* 24 (6): 2545–2557. https://doi.org/10.1007/s10570-017-1287-9.

Xue, F., He, H., Zhu, H. et al. (2019). Structural Design of a Cellulose-Based Solid Amine Adsorbent for the complete removal and colorimetric detection of Cr(VI). *Langmuir* 35 (39): 12636–12646. https://doi.org/10.1021/acs.langmuir.9b01788.

Yao, M., Wang, Z., Liu, Y. et al. (2019). Preparation of dialdehyde cellulose graftead graphene oxide composite and its adsorption behavior for heavy metals from aqueous solution. *Carbohydrate Polymers* 212: 345–351. https://doi.org/10.1016/j.carbpol.2019.02.052.

Zadorecki, P. and Michell, A.J. (1989). Future prospects for wood cellulose as reinforcement in organic polymer composites. *Polymer Composites* 10 (2): 69–77. https://doi.org/10.1002/pc.750100202.

Zhan, W., Xu, C., Qian, G. et al. (2018). Adsorption of Cu(ii), Zn(ii), and Pb(ii) from aqueous single and binary metal solutions by regenerated cellulose and sodium alginate chemically modified with polyethyleneimine. *RSC Advances* 8 (33): 18723–18733. https://doi.org/10.1039/c8ra02055h.

Zhang, L., Lu, H., Yu, J. et al. (2018a). Synthesis of lignocellulose-based composite hydrogel as a novel biosorbent for Cu2+ removal. *Cellulose* 25 (12): 7315–7328. https://doi.org/10.1007/s10570-018-2077-8.

Zhang, M., Li, Y., Yang, Q. et al. (2018b). Temperature and pH responsive cellulose filament/poly (NIPAM-co-AAc) hybrids as novel adsorbent towards Pb(II) removal. *Carbohydrate Polymers* 195: 495–504. https://doi.org/10.1016/j.carbpol.2018.04.082.

Zhao, G., Huang, X., Tang, Z. et al. (2018). Polymer-based nanocomposites for heavy metal ions removal from aqueous solution: a review. *Polymer Chemistry* 9 (26): 3562–3582. https://doi.org/10.1039/c8py00484f.

Zhou, G., Luo, J., Liu, C. et al. (2018a). Efficient heavy metal removal from industrial melting effluent using fixed-bed process based on porous hydrogel adsorbents. *Water Research* 131: 246–254. https://doi.org/10.1016/j.watres.2017.12.067.

Zhou, J., Zhou, X., Ren, J. et al. (2018b). Preparation of bio-based magnetic adsorbent and application for efficient removal of Cd(II) from water. *Water Science and Technology* 77 (5): 1313–1323. https://doi.org/10.2166/wst.2018.007.

Part V

Advances in Remediation of Water Pollution

13

Advances in Membrane Technology Used in the Wastewater Treatment Process

Naresh K. Sethy[1], Zeenat Arif[1], K.S. Sista[2], P.K. Mishra[1], Pradeep Kumar[1], and Avinash K. Kushwaha[3]

[1] *Department of Chemical Engineering and Technology, IIT (BHU), Varanasi, Uttar Pradesh, India*
[2] *Research and Development, Tata Steel, Jamshedpur, Uttar Pradesh, India*
[3] *Department of Botany, BHU, Varanasi, Uttar Pradesh, India*

13.1 Introduction

Water is an important resource and life on Earth cannot be assumed without it. About 97% of water is salty, which is not fit for consumption, and only less than 3% is available as freshwater (Hong et al. 2017; Sun et al. 2018). Growth in the population combined with industrialization and urbanization led to an increase in the demand for freshwater, creating alarming situations globally, and in the future, will lead to drastic water shortages, even in regions currently known as water-rich (Adeleye et al. 2016). In addition, existing freshwater supplies are being slowly degraded to wastewater due to the accumulation of numerous synthetic organic and inorganic contaminants associated with various human or industrial activities threatening its direct usage, which, if not taken care of, will lead to various health issues (Judd and Jefferson 2003). Thus, it is a necessity to filter and purify the available water before its usage in multiple human utilities (Santhosh et al. 2016). In the last few decades, various technologies for handling wastewater have been explored. Traditional methods such as biological treatment, adsorption, sedimentation, coagulation, and distillation are not very effective in removing organic contaminants from wastewater, These techniques usually require complex equipment with high energy consumption and high operating costs, some of which also require large quantities of chemicals, leading to waste by-products and sludge (Helali et al. 2020). A pictographic representation of various challenges and limitations of conventional wastewater treatment techniques is shown in Figure 13.1.

Besides conventional techniques, membrane filtration has been considered as one of the most promising methods for water decontamination due to its advantages of high separation, selectivity, low energy consumption, no additional chemicals requirements, simple scale-up, and continuous operation (Thakur and Voicu 2016; Zhang et al. 2016). This method imparts the usage of semipermeable membranes to separate harmful contaminants from wastewater.

Pollutants and Water Management: Resources, Strategies and Scarcity, First Edition. Edited by Pardeep Singh, Rishikesh Singh, Vipin Kumar Singh, and Rahul Bhadouria.

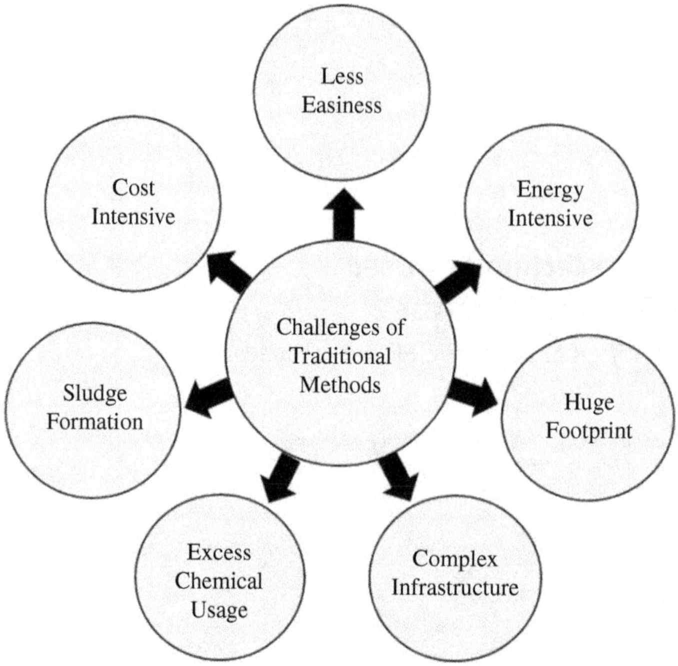

Figure 13.1 Challenges of traditional methods.

Membranes designed for water treatment need to attain adequate water permeability, successful rejection or selectivity of pollutants, proper mechanical integrity, and reasonable fouling and scaling resistance (Jiang and Cheng-Yin 2003; Cao et al. 2006; Jiang and Zhang 2007). Based on the targeted contaminants for removal and mode of application, various membrane-based processes are developed, including particle filtration (PF), microfiltration (MF), ultrafiltration (UF), nanofiltration (NF), reverse osmosis (RO), membrane distillation (MD), electrodialysis (ED), pervaporation (PV), and membrane bioreactor (MBR) (Erkanlı et al. 2017). Although they have major advantages, membrane systems also suffer from a few disadvantages. Throughout filtration, the trapping and aggregation of contaminants on the membrane surface and/or inside the membrane pores result in membrane fouling, which would eventually deteriorate the membrane performance. Factors like material of membrane, microorganisms, and feed water chemistry influence membrane fouling (Ahmed et al. 2016; Zinadini and Gholami 2016). Generally, fouling is of two modes, reversible, when the fouling caused by the anchoring of solute particles on the membrane surface, and irreversible, when the particles are firmly bound to the surface. Once a strong matrix of the fouled layer is created, the solute can transform reversible fouling into irreversible during the continuous filtration cycle causing severe damage (Zhao et al. 2017). It is also important to understand the water characteristics before employing suitable membrane filtration technology. A pictorial representation of various measurements used to understand wastewater characteristics is shown in Figure 13.2. Thus, the physical, chemical, and biological characteristics of water generally drive the treatment of wastewater by any of the conventional or membrane filtration techniques (Muttamara 1996; Brannon et al. 2017; Wigginton et al. 2018). Details on membrane technologies for wastewater treatment are shown in the following sections.

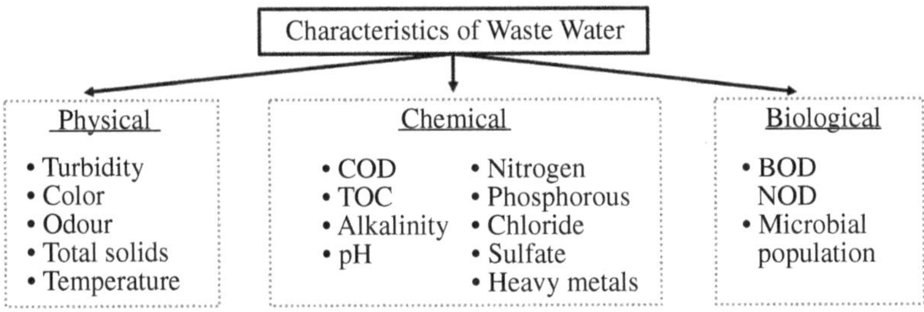

Figure 13.2 Characteristics of wastewater.

13.2 Membrane Technologies for Wastewater Treatment

Membrane separation technology is a process in which materials are selectively separated from targeted media by the use of permeable continuous molecular structures called membranes. Special features for membrane technologies that make them attractive among various separation techniques available for wastewater treatment are high efficiency, easy operation, space saving, easy scale-up, and environmentally friendly.

13.2.1 Membranes Classification

Membranes are typically classified based on nature, structure, transport mechanism, and geometry. An overview of classifications is given in Figure 13.3, Classification remains a guide for the choice of membrane for a specific application.

13.2.2 Preparation Methods

13.2.2.1 Solution Casting Method

In this method, polymeric membranes are created by pouring casting solutions onto a thin-layer chromatograph applicator and then drawing a blade across a glass plate. The solvent is allowed to evaporate at defined temperatures and time, and post drying, the

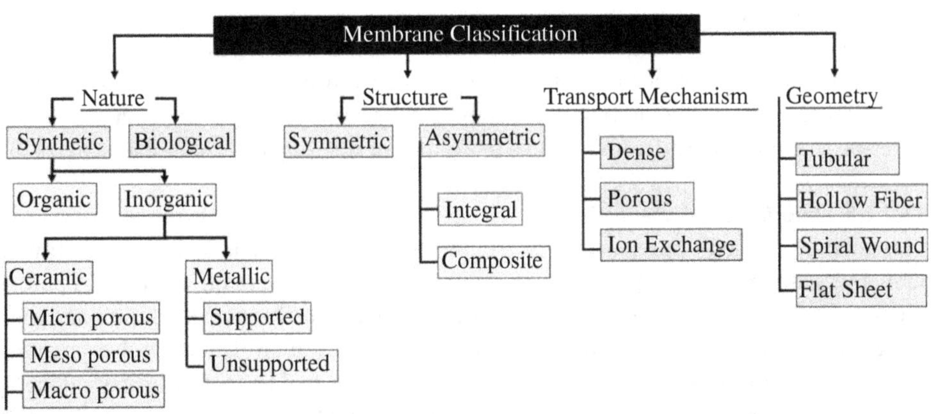

Figure 13.3 Classification of membranes used for wastewater treatment.

membrane formed is removed from the glass plate. The membranes resulting from this method are asymmetric structures consisting of a dense, thin skin layer and rough, porous support layers. Besides, because any gelating medium is not used and the casting solvent is completely evaporated in the process, the resulting membranes have a dense symmetrical structure. A variety of industrial applications such as MF, UF, NF, RO, gas separation, and PV have been applied to polymer membranes prepared by the solution casting method (Sethy et al. 2019).

13.2.2.2 Phase Inversion

This is the primary method used in polymeric membrane preparation. In the inversion of the thermal process, a polymer is dissolved in a hot solvent and cast to create a film. A demixing method is done by cooling, and a porous structure is produced. This production process is used both as a film and as a hollow fiber to manufacture membranes. A polymer is dissolved in a solvent in the wet phase inversion process and then assembled to form a film, which is then dipped into a coagulation bath where the membrane precipitates. To boost the membrane's mechanical properties, a nonwoven cloth is often used as casting support. The method of phase inversion can be used for a great variety of polymers. Using this method, it is possible to acquire membranes with various shapes (symmetrical and asymmetrical), configuration (flat surface, circle, "spaghetti," hollow fiber), porosity, and separation properties by properly working on the preparatory parameters. During membrane forming in the coagulation bath, contaminants are also added into the casting solution to promote pore creation by their dissolution (Arif et al. 2019).

13.2.2.3 Sol–Gel and Slip Casting

These techniques are used to prepare asymmetrical UF ceramic membranes (or NF ones at experimental production level) by the deposition of a mesoporous film on the surface of macroporous ceramic support. This technique is based on inorganic polymerization reaction where a precursor's hydrolysis and condensation contribute to a macromolecular network that is first dried and then fired or sintered through the membrane formation (Mallada et al. 2006). The precursor is typically an alkoxide or a salt of silicon. Film deposition can also be made on the macroporous support, starting from the slurry of oxide particles.

13.2.2.4 Sintering

This method is used primarily for the preparation of ceramic or metallic membranes used in UF and MF. Particles of the substance are pushed and heated at or below the melting temperature. The arrangement of the membrane is symmetrical and the thickness of the pores varies from some micrometers to a few tenths of micrometers (Abdullah et al. 2016).

13.2.2.5 Interfacial Polymerization

This method is used to prepare polymeric membranes for NF and RO. Two substituted monomers in immiscible liquids (e.g. water and an organic compound) are used. A porous membrane (UF) is first impregnated with the first monomer's aqueous solution and then placed into contact with the second monomer's organic solution. The fast polymerization reaction at the interface between the two solutions (corresponding to the transparent membrane surface) produces a thin polymer film (a few tenths of a nanometer). Different monomer groups are used to create a range of thin polymer films with very good salt retention and permeability (Gohil and Choudhury 2019).

13.2.2.6 Surface Modifications

Surface modification techniques are also used to offer any desirable properties to the polymeric membrane surface, such as a specified degree of hydrophilicity, which increases the membrane's fouling resistance. There are many methods of surface alteration (chemical oxidation, plasma injection, polymer grafting, and so on) (Zare and Kargari 2018).

13.2.3 Membrane Characterization

One of the key aspects of membrane applications is to understand membrane chemistry, structure, and properties. Various characterization techniques, including spectroscopy, microscopy, porometry, and performance testing, which are important for acquiring knowledge on membrane performance, are discussed in this section. The characterization of the membrane is the basis for establishing the relationship between chemistry – structure – properties, which is a crucial aspect in the optimization of the membrane to achieve excellent efficiency in separation and resistance to fouling.

13.2.3.1 Membrane Porometry

Different methods of porometry were used to assess the distribution of the MF and UF membranes by pore size. The bubble point method can be used to describe pores, which calculates the pressure at which air bubbles begin to pass through a wetted MF membrane. This method is strictly applicable when the pore is much larger than the solvent molecule. Another widely used method for characterizing the MF membrane pore is mercury intrusion, which is ideal for pores ranging from 3.5 to 1000 nm. In this method, high pressure needs to be used that may compress the membrane and distort the pore mouth opening. In mercury porosimetry, the membrane is tested dry. Along with previous methods, certain methods include gas adsorption, permporometry, and thermoporometry to identify smaller pores (Nakao 1994).

13.2.3.2 Performance Test

This test determines the complete rejection characteristics of the membrane. Generally, this method is a size-based rejection of solutes by UF and MF membrane pores. Performance data (e.g. rejection of probe molecules like dextran or polyethylene glycol) can be used to determine a membrane's pore-size distribution (Zhao et al. 2012).

13.2.3.3 Spectroscopic Method

Spectroscopic methods provide important information about a membrane's structure and chemistry. Techniques like Fourier transform infrared spectroscopy (FTIR), X-ray photoelectron spectroscopy (XPS), dispersive energy spectroscopy (EDX), and electrical impedance spectroscopy (EIS) are widely used. FTIR is able to recognize various chemical bonds in a membrane based on its infrared irradiation adsorption (Sethy et al. 2019). Both XPS and EDX are commonly used for the detection of elements present in the membrane. XPS is a highly surface-sensitive technique capable of measuring elemental composition and chemical binding details for the surface area's top 1–5 nm depth (Freger et al. 2005). EDX is less sensitive, compared with XPS. Nonetheless, EDX provides a unique benefit, as it can be paired with a scanning electron microscopy (SEM) or electron microscopy transmission (TEM), which allows it to evaluate micro-scale features at specific locations and create elemental mapping. EDX is commonly used for the study of chemical compositions of membrane foulants in membrane autopsies (Arif et al. 2019). The layered structures of the thin-film composite (TFC)-RO and NF membranes have recently been described by EIS (Fontananova et al. 2017).

13.2.3.4 Microscopic Method

Microscopic methods are useful for visualizing the morphology and function of the membranes. They can be used to analyze a membrane's surface characteristics as well as its cross-section. Microscopic methods are also used for the characterization of membrane pore size, In the presence of a fouling cake layer, great information about the composition, structure, and properties of the layer can be given by microscopic investigation. Due to its low cost and simple operation, conventional visible light microscopy is widely used for membrane visualization, but due to the relatively broad wavelength of visible light, its resolution is typically limited. On the other hand, electron microscopic methods such as SEM and TEM provide much greater resolutions. In polymeric materials, SEM has been commonly used to characterize the surface and cross-section of both clean and fouled thin (submicron) membranes. Whereas for ultrathin membranes (< 100 nm), TEM can provide a great deal of detail on membrane and foulant structural material. Other widely used microscopic methods include atomic force microscopy (AFM) and confocal laser scanning microscopy (CLSM). AFM can provide useful information about a membrane's surface characteristics, and it has become the standard approach for characterizing membrane roughness (Arif et al. 2019; Sethy et al. 2020). In comparison, CLSM is an effective tool for characterizing biofilms.

13.2.4 Pressure-Driven Membrane Separation Technologies

Membrane processes in which separation is carried out by means of driving force resulting from a gradient in pressure, concentration, temperature, and electrical field are classified as pressure-driven membrane processes, concentration-driven membrane processes, thermal-driven membrane processes, and electrical potential-driven membrane processes, respectively. Among the aforementioned separation processes, the pressure-driven membrane process is technically and commercially the most relevant process. A pictorial representation of size exclusion contaminant removal by various pressures driven membrane separation processes used is shown in Figure 13.4.

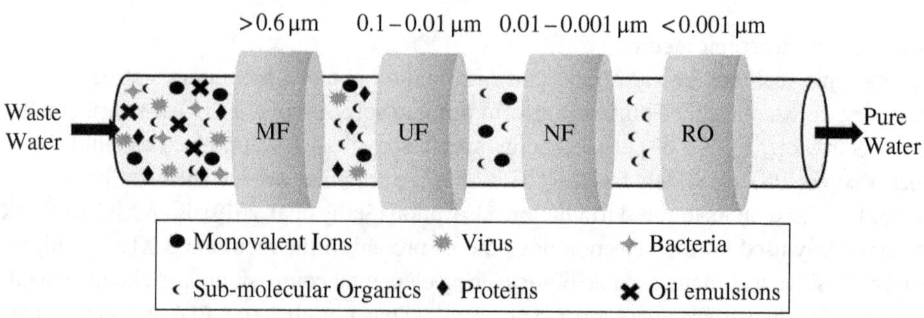

Figure 13.4 Pressure-driven membrane processes for wastewater treatment.

13.2.4.1 Microfiltration

MF is a pressure-driven process in which colloidal and suspended particles in the range of 0.6–10 μm are removed. It is operated at 0.2–3.5 bar pressure with varying membrane thickness from 10 to 150 μm (Qu et al. 2016; Torki et al. 2017). Membrane materials like cellulose acetate (CA), polysulfone (PSF), polyamide (PA), and polyethersulfone (PES) are used. MF is typically used as a pretreatment step for RO or NF to reduce the risk of fouling. The main advantages of MF membranes are uniform pore size, excellent separation performance, operate at low pressure, and they do not require chemical additives in the separation process (Saja et al. 2018).

13.2.4.2 Ultrafiltration

The ultrafiltration membrane process can separate compounds between 0.01 and 0.1 μm, which is between MF and RO (Qu et al. 2014). It is operated at 1–10 bar pressures. Membrane materials like PSF, PES, and polyvinylidene fluoride (PVDF) are used. This separation process is operated at room temperature and no solvent or other precipitating agents are required for the concentrating process. These membranes are extremely popular water filters with low energy usage for the elimination of pathogenic microorganisms, macromolecules, and suspended matters (Krüger et al. 2016). UF also has certain drawbacks, and these limitations are addressed by new developments like the hybrid ultrafiltration membrane (Zhan et al. 2016) and the nanocomposite membrane (Mocanu et al. 2017; Gebru and Das 2018). Nanocomposite UF membranes have been used for the separation of oil–water emulsion (Moeinzadeh et al. 2019) and wastewater treatment (Igbinigun et al. 2016) with enhanced antibacterial and antifouling properties (Nasrollahi et al. 2019; Cihanoglu and Altinkaya 2020).

13.2.4.3 Nanofiltration

NF membranes are also called loose RO membranes due to their potential to separate inorganic salts like divalent ions in the range 0.01–0.001 μm. NF has better water quality compared with RO and needs lower basic energy consumption (Abdel 2018). NF is capable of eliminating ions that greatly contribute to the osmotic effect, thereby requiring lower operating pressures than those in RO. For highly contaminated water, successful pretreatment with NF is needed (Wang et al. 2016). Membrane materials like polysulfone, sulfonated-polysulfone, and polyimide are used for the fabrication of the NF membranes, To increase the flux and selectivity along with the antifouling properties, composite NF membranes have been used (Xu et al. 2016; Abdel 2018).

13.2.4.4 Reverse Osmosis

RO is the process of pressure-driven elimination of liquid solids and smaller particles generally operating at high-pressure gradients of 15–100 bar. The membrane is only permeable to water molecules and has the ability to concentrate dissolved as well as suspended solids. The RO process is widely used in the desalination of brackish water. Membrane materials like cellulose acetate, polyamide and polyether urea, etc. are used in the fabrication of the RO membrane. The pores of RO membranes are finer than UF and are basically capable of extracting debris, bacteria, and organics (Yan et al. 2016; Wu et al. 2020), The surface of the membrane used in the RO process is modified to prevent fouling or biofouling in the membrane (Mohammadtabar et al. 2019).

Research utilizing pressure-driven membrane processes for water purification application is shown in Table 13.1.

Table 13.1 Pressure-driven membranes used for wastewater treatment.

S. no	Process	Membrane material	Target molecule/ions in wastewater	Application	Reference
1	MF	Polysulfone (PSF)	Pb^{2+}, Zn^{2+}, and Hg^{2+}	Removal of heavy metals	Gao et al. (2019)
		Pozzolan	Aluminum chloride suspension, tannery wastewater	Clarification of aluminum chloride suspension and treatment of wastewater from the tannery industry	Beqqour et al. (2019)
		Polyvinyl alcohol (PVA) and polyacrylonitrile (PAN)	Chromium (VI) and cadmium (II)	Removal of heavy metals from wastewater	Liu et al. (2020)
		Polyamide-imide (PAI)	Oily materials	Oil/water emulsion separation	Helali et al. (2020)
2	UF	Polyacrylate sodium (PAAS)	Ni (II) and Cr (III)	Removal of heavy metals	Zhang et al. (2019)
		–	Soya oil	Treatment of vegetable oil refinery wastewater	Khouni et al. (2020)
		Copolymer of maleic acid and acrylic acid (PMA)	Sn^{2+}	Removal of heavy metal ions from wastewater	Le et al. (2019)
3	NF	Polyethersulfone (PES)	Cr(VI))	Heavy metal removal	Wei et al. (2019)
		Polyvinyl chloride (PVC)	$MgCl_2$	Water softening and multivalent ion removal	Fanga et al. (2019)
		Polyethylenimine (PEI)	Congo red (CR), magnesium chloride ($MgCl_2$), magnesium sulfate ($MgSO_4$), sodium sulfate (Na_2SO_4), sodium chloride (NaCl)	Dye/salt mixture separation	Li et al. (2019)
		Polyvinyl alcohol (PVA)	Carbofuran	Pesticide wastewater treatment	Qin et al. (2020)
		PSF	Sodium sulfate (Na_2SO_4)	Water purification application	Zhai et al. (2020)
		PES	Direct red 23, Congo red, and Direct red 243, $MgSO_4$, Na_2SO_4, $MgCl_2$, NaCl	Separation of dye and inorganic salts	Tavangar et al. (2020)
4	RO	PSF	NaCl	Desalination	Li et al. (2019); Wang et al. (2019)
		Polyvinylidene fluoride (PVDF)	Cr (VI)	Heavy metal ion removal	Liu et al. (2019)
		PSF	NaCl, Na_2SO_4, $MgSO_4$	Salt removal from water	Hu et al. (2020)

13.3 Advancements in Membrane Technology for Wastewater Treatment

Recent studies by membrane researchers incline toward integrating nanomaterials with polymeric materials resulting in the creation of nanocomposite membranes, otherwise termed as nano-improved membranes, with improved mechanical and physicochemical properties (Khorshidi et al. 2019). Specific nanomaterials such as carbon nanotubes (CNTs), nano clays, nano silver, and nano-sized TiO_2, ZnO, Al_2O_3, Fe_3O_4, and SiO_2 are being used. More information on nanomaterials, their effect on nanocomposite membrane technologies, and their success in wastewater treatment are elucidated in the forthcoming sections.

13.3.1 Carbon Nanotubes

CNTs have gained a lot of attention between the different nanomaterials examined. CNTs deliver antibacterial properties along with strong hydrophilicity, reasonable chemical stability, large surface area, and increase porosity when used in nanocomposite membranes. Because of their high aspect ratio and high axis strength, they are suitable for strengthening membranes. Potential and a wide variety of applications of CNTs have been a guiding factor in the use of high-quality single-walled (SW), double-walled (DW), and multiwalled (MW) CNTs for integration into CNT-polymer composites (Vatanpour et al. 2011), Several methods have been implemented to create CNT-polymer membranes, such as in situ polymerization, interfacial polymerization, and immersion of nanoparticles into the casting fluid (Celik et al. 2011), CNT is incorporated into the polymer to exhibit high efficiency in terms of tolerance to fouling, permeability, and salt rejection for nanofiltration membranes (Vatanpour et al. 2019), During the separation of the oil-in-water emulsion, the superoleophobic surface of the hybrid membrane having CNT allowed rapid water transport, thus achieving 99% separation efficiency (Yan et al. 2019).

13.3.2 Silver Nanoparticles

The silver nanoparticle (AgNP) is one of the first nanomaterials to receive interest in composite membrane manufacturing. A new avenue of mixed matrix membranes with bactericidal properties is driven by silver nanoparticles. AgNP provides a broad surface-to-volume ratio which promotes their use as a safe local source for ions and prolonging bacterial adhesion protection (Cao et al. 2010). Many experts claim that silver is nonallergic, nontoxic, biologically safe, and less detrimental to human cells. Due to its low toxicity against humans and their antibacterial capacity, silver nanoparticles are widely used in membrane formation (Cui et al. 2019; Li et al. 2019). Reported works have shown that the inclusion of AgNP on the surface or in the membrane matrix is successful in applications against gram-positive and gram-negative bacteria (Li et al. 2019). AgNP along with other metallic nanoparticles were added in polymers to fabricate the composite membranes with enhanced antibacterial and antibiofouling properties. In addition to the antibactericidal and antifouling properties of silver, an improvement in the hydrophilicity and selectivity of membranes mixed with Ag nanoparticles was also noticed (Haghighat et al. 2020). On the other hand, an interesting use of AgNP incorporated PVA and starch composite membranes in a biomedical application for wound dressing was reported by Batool et al. (2019).

13.3.3 Titanium Dioxide Nanoparticles

Titanium dioxide (TiO_2), a special semiconductor material, is widely used in various industries due to its outstanding photocatalytic and hydrophilicity properties. Membranes incorporating nano TiO_2 are superior in the absorption of UV rays, catalytic performance, durability, cheapness, and industrial usability. The major advantage of improved antifouling resistance of membranes is delivered by using TiO_2 nanoparticles. The surface of TiO_2 nanoparticles is modified to prevent aggregation and increase the dispersion of TiO_2 nanoparticles into an organic solvent (Li et al. 2007). Incorporation of TiO_2 particles on PVDF polymer matrix to study the qualitative and quantitative assessment of the antifouling property of membranes is reported by Arif et al. (2019). It was concluded that the addition of TiO_2 particles increases the repulsive interactive energy barrier between the membrane surface and foulant, thereby reducing the fouling tendency of the membrane. Also, the permeation results showed a high % rejection value for bovine serum albumin (BSA) and humic acid (HA) (foulants) by the PVDF/TiO_2 membrane compared with pure PVDF. Besides antifouling application, nano TiO_2 added membranes also play an important role in removing toxic metals from wastewater using a low-cost UF process. A recent study to remove Cr (VI) from the model solution using PVDF/TiO_2 membrane by a UF process was carried out by Arif et al. 2020a, Higher removal of Cr (VI) for PVDF/TiO_2 compared with pristine PVDF is attributed to ionic–ionic interaction at acidic pH between Cr ion and nano TiO_2 particles. Another study on the usage of nano TiO_2 incorporated membrane for Cr (VI) removal from real tannery wastewater with a greater percentage of rejection was also reported Arif et al. (2020b). Researchers also prepared TiO_2-mixed matrix membranes to boost membrane efficiency by growing hydrophilicity and minimizing fouling (Yang and Wang 2006; Yang et al. 2007). In another study, PVDF-TNT (titania nanotubes) UF membrane was used to treat palm oil mill wastewater. An increase in membrane hydrophilicity, flux, fouling resistance, and reusability was observed in the presence of titania nanotubes due to the presence of abundant hydroxyl groups contained on the TNT surface (Subramaniam et al. 2018).

13.3.4 Zinc Oxide Nanoparticles

Nano-sized zinc oxide (ZnO NP) is another important nanoproduct used to alter membranes. ZnO nanoparticles have recently gained growing interest as a multifunctional inorganic nanoparticle due to its prominent physical and chemical properties, such as strong catalytic activity and efficient antibacterial and bactericidal capabilities (Balta et al. 2012). ZnO nanoparticles are also economical than many other nanoparticles, such as TiO_2 and Al_2O_3 nanoparticles. ZnO NP is used either alone or in combination with other nano oxide particles to improve various properties of the membrane and its application. Researchers reported that the presence of ZnO NP in the nanocomposite UF membrane enhanced the antifouling performance with the support of other metallic oxide nanoparticles (Ayyaru et al. 2020). In another work, without affecting the contact angle, ZnO NP presence enhanced the oil–water separation with added benefit of good antibacterial property (Manaf et al. 2019). Usage of a nanocomposite membrane with ZnO and CuO enhancing the adsorption, catalytic and antibacterial properties in an application for water purification was reported (He et al. 2019). Recent studies done by Rajakumaran et al. (2019) and Nasrollahi et al. (2019) proved the improvement in permeability and antifouling properties by the addition of ZnO NP in the nanocomposite membrane used for UF and RO applications, respectively. Another interesting application for the use of ZnO based thin-film nanocomposite membrane for water desalination application was reported (Amini et al. 2020).

13.3.5 Aluminum Oxide Nanoparticles

The durability, quality, hydrophilicity, and sufficient mechanical intensity of nanoparticles made of aluminum oxide (Al_2O_3) makes this inorganic material an excellent choice for nanocomposite membrane preparation. Due to these properties, membranes when blended with Al_2O_3 NPs show unique benefits in terms of thermal resistance, chemical tolerance, separation efficiency, and adaptability to extreme wastewater conditions. Use of Al_2O_3 NPs along with polymers like cellulose acetate (CAc), cellulose acetate phthalate (CAP), polyaniline (PA), and PVDF to modify membrane properties such as hydrophilicity, permeability, and fouling resistance was reported (Yan et al. 2005; Mukherjee and De 2013). Improved properties like flux, hydrophilicity, fouling resistance for PVDF-Al_2O_3 nanocomposite polymeric membrane due to surface activity, adsorptive ability, and reactive hydroxyl groups of Al_2O_3 NP was reported (Yan et al. 2009; Liu et al. 2011). Another work on the usage of Al_2O_3 nanowires as a filler in polyvinyl butyral nonwoven membranes to improve the mechanical and thermal properties with high swelling resistance was reported (Liu et al. 2020). Usage of alumina in the nanocomposite membrane to separate metal ions like arsenic (Razmgar et al. 2019), copper (Ayaz et al. 2019), and cadmium (Hussein et al. 2020) from the aqueous environments is also reported by many researchers. Usage of Al_2O_3 incorporated cellulose triacetate (CTA) membrane for desalination application was also reported (Prihatiningtyas et al. 2020).

13.3.6 Iron Oxide Nanoparticles

In recent years, magnetic nanoparticles have received significant research interest, especially in drug and gene tracking, cell isolation, and hyperthermia applications due to their intrinsic superparamagnetic properties (Ye et al. 2009). Among the magnetic nanoparticles, iron oxide (Fe_3O_4) has outstanding thermal resistance, chemical resilience, strong magnetic strength, superior adsorption, and reasonable biocompatibility (Dudek et al. 2012). Iron oxide nanoparticles (IONP) as adsorbents were used to produce a nanocomposite membrane with improved sensitivity in the PES matrix to extract copper ions

(Daraei et al. 2012). Another work on the usage of IONP in the nanofiltration membrane for separation of dye from wastewaters was reported. It was observed that higher dye retention is achieved due to the strong repulsive forces between the dye functional groups and the nanomaterials (Karimnezhad et al. 2019). It was also observed that along with dye removal, IONP also improved membrane properties like hydrophilicity, permeability, and antifouling resistance. Similarly, works on modified polysulfone membranes using IONP for various applications like protein purification (Said et al. 2019), seawater purification (Lakhotia et al. 2019), and removal of phenolic compounds from wastewater was reported (Modia and Bellarea 2020).

13.3.7 Silicon Dioxide Nanoparticles

Silicon dioxide (SiO_2) is another useful nanoparticle commonly used in the manufacture of nanocomposite membranes owing to its mild reactivity and distinguished capacity to enhance hydrophilicity and boost thermal and mechanical stability and fouling resistance of membranes (Wu et al. 2003; Jin et al. 2012). Silica has been extensively used for polymeric membranes such as PMMA (Zulfikar et al. 2006), PSF (Zhang et al. 2013), PA (Jin et al. 2012), PVA (Nagarale et al. 2005), and PVDF (Ogoshi and Chujo 2005). A basic approach for producing a SiO_2 hybrid membrane is to combine an organic polymer with a metal alkoxide such as tetraethoxysilane (TEOS) accompanied by a sol–gel technique. Another technique of phase inversion for easy and appropriate preparation of polymer-silica mixed matrix membranes was also chosen (Cui et al. 2010; Shen et al. 2011). Membrane properties like mechanical strength, thermal resilience, hydrophilicity, permeability, and fouling resistance dramatically improved with an addition of TEOS into composites (Wu et al. 2003). Superhydrophilic nanocomposite membranes using SiO_2 NPs as filler was used for oil–water emulsion separation application (Cui et al. 2019). Another work on the usage of SiO_2 NPs incorporated nanofiltration membrane applications like heavy metal (Cd II) ion recovery and dye (methylene red) removal from wastewater was reported (Kumari et al. 2020). Similarly, usage of the nanocomposite UF membrane (PSF/PVP/SiO_2) for removal of amoxicillin (antibiotics) from aqueous solutions with improved flux due to SiO_2 nanoparticle presence was reported (Shakak et al. 2020). Another work on the usage of SiO_2 NPs to enhance RO membrane properties was reported (Wang et al. 2019).

13.3.8 Bioinspired Membranes

Bioinspired membranes impart usage of biological materials or natural biopolymers in their manufacturing to enhance the structural and functional properties of membranes (Morones-Ramırez 2013). Experimental conditions for the synthesis of these membranes are mild with aqueous, temperature regulated, moderate pressure, and neutral or near-neutral pH conditions. Bioinspired membranes have great advantages, such as high permeability, catalytic reactivity, and resistance to fouling, showcasing potential to use in water treatment applications (Rawlings et al. 2012; Cui et al. 2019; Kumar et al. 2020). Because of various advantages, bioinspired membranes have the potential to become the next generation of membranes.

13.4 Conclusions

Freshwater is needed for life and the purification of available water before its use for daily applications has been mandated since the inception of industrialization and subsequent contamination of available water resources with harmful pollutants. Conventional wastewater treatment techniques like coagulation, flocculation, adsorption, distillation, etc. are not very efficient in treating wastewater. Given the numerous advantages of membranes, such as relatively high selectivity, good permeability, low energy consumption, operational simplicity, energy efficiency, easiness to scale up, adaptability, and satisfactory stability under different conditions, membrane treatment technologies have been dominating the wastewater treatment industry over the years. Pressure-driven membrane technologies like, MF, UF, NF, and RO are used for wastewater treatment. However, to address certain particle limitations and improve membrane usability in various kinds of wastewaters, the interest of the research fraternity inclined toward improvements in fouling resistance, thermal resistance, chemical resistance, permeability, durability, and decreasing energy consumption of membranes and membrane-based technologies. One such important advancement toward superior membrane synthesis is the use of nanoparticles like CNT, Ag, ZnO, Al_2O_3, TiO_2, SiO_2, Fe_3O_4, etc., and the advent of bioinspired membranes. Further, research into the generation of new polymers with special properties for use in composite membranes for water treatment is also in focus.

References

Abdel, M. (2018). Nanofiltration systems and applications in wastewater treatment. Review article. *Ain Shams Engineering Journal* 9: 3077–3092.

Abdullah, N., Rahman, M.A., Othman, M.H.D. et al. (2016). Preparation and characterization of self-cleaning alumina hollow fiber membrane using the phase inversion and sintering technique. *Ceramics International* 42: 12312–12322.

Adeleye, A.S., Conway, J.R., Garner, K. et al. (2016). Engineered nanomaterials for water treatment and remediation: costs, beneits, and applicability. *Chemical Engineering Journal* 286: 640–662.

Ahmed, F., Lalia, B.S., Kochkodan, V. et al. (2016). Electrically conductive polymeric membranes for fouling prevention and detection. *A review. Desalination* 391: 1–15.

Amini, M., Seifi, M., Akbari, A., and Hosseinifard, M. (2020). Polyamide-zinc oxide-based thin film nanocomposite membranes: towards improved performance for forward osmosis. *Polyhedron* 179: 114362. https://doi.org/10.1016/j.poly.2020.114362.

Arif, Z., Sethy, N.K., Kumari, L. et al. (2019). Antifouling behaviour of PVDF/TiO_2 composite membrane: a quantitative and qualitative assessment. *Iranian Polymer Journal* 28: 301–312. https://doi.org/10.1007/s13726-019-00700-y.

Arif, Z., Sethy, N.K., Mishra, P.K., and Verma, B. (2020a). Development of eco-friendly, self-cleaning, antibacterial membrane for the elimination of chromium (VI) from tannery wastewater. *International Journal of Environmental Science and Technology* https://doi.org/10.1007/s13762-020-02753-6.

Arif, Z., Sethy, N.K., Mishra, P.K., and Verma, B. (2020b). Green approach for the synthesis of ultrafiltration photocatalytic membrane for tannery wastewater: modeling and optimization. *International Journal of Environmental Science and Technology* https://doi.org/10.1007/s13762-020-02719-8.

Ayaz, M., Muhammad, A., Younas, M. et al. (2019). Enhanced water flux by fabrication of polysulfone/alumina nanocomposite membrane for copper(II) removal. *Macromolecular Research* 27: 565–571.

Ayyaru, S., Dinh, T.T.L., and Ahn, Y.H. (2020). Enhanced antifouling performance of PVDF ultrafiltration membrane by blending zinc oxide with support of graphene oxide nanoparticle. *Chemosphere* 241: 125068. https://doi.org/10.1016/j.chemosphere.2019.125068.

Balta, S., Sotto, A., Luis, P. et al. (2012). A new outlook on membrane enhancement with nanoparticles: the alternative of ZnO. *Journal of Membrane Science* 389: 155–161.

Batool, S., Hussain, Z., Niazi, M.B.K. et al. (2019). Biogenic synthesis of silver nanoparticles and evaluation of physical and antimicrobial properties of Ag/PVA/starch nanocomposites hydrogel membranes for wound dressing application. *Journal of Drug Delivery Science and Technology* 52: 403–414.

Beqqour, D., Achiou, B., Bouazizi, A. et al. (2019). Enhancement of microfiltration performances of pozzolan membrane by incorporation of micronized phosphate and its application for industrial wastewater treatment. *Journal of Environmental Chemical Engineering* 7: 102981. https://doi.org/10.1016/j.jece.2019.102981.

Brannon, E., Moseman-Valtierra, S., Lancellotti, B. et al. (2017). Comparison of N_2O emissions and gene abundances between wastewater nitrogen removal systems. *Journal of Environmental Quality* 46: 931–938.

Cao, G.P., Zhao, P., and Wen-Jie, L.I. (2006). Study progress and development trend in water treatment technology of the membrane separation technology. *Water Conservancy Science & Technology & Economy* https://doi.org/10.2991/mmme-16.2016.202.

Cao, X., Tang, M., Liu, F. et al. (2010). Immobilization of silver nanoparticles onto sulfonated polyethersulfone membranes as antibacterial materials. *Colloids and Surfaces B: Biointerfaces* 81: 555–562.

Celik, E., Park, H., Choi, H., and Choi, H. (2011). Carbon nanotube blended polyethersulfone membranes for fouling control in water treatment. *Water Resources* 45: 274–282.

Cihanoglu, A. and Altinkaya, S.A. (2020). A facile route to the preparation of antibacterial polysulfone-sulfonated polyethersulfone ultrafiltration membranes using a cationic surfactant cetyltrimethylammonium bromide. *Journal of Membrane Science* 594: 117438. https://doi.org/10.1016/j.memsci.2019.117438.

Cui, A., Liu, Z., Xiao, C., and Zhang, Y. (2010). Effect of micro-sized SiO_2 -particle on the performance of PVDF blend membranes via TIPS. *Journal of Membrane Science* 360: 259–264.

Cui, Y., Yang, L., Meng, M. et al. (2019). Facile preparation of antifouling g-C_3N_4/Ag_3PO_4 nanocomposite photocatalytic polyvinylidene fluoride membranes for effective removal of rhodamine B. *Korean Journal of Chemical Engineering* 36: 236–247.

Daraei, P., Madaeni, S.S., Ghaemi, N. et al. (2012). Novel polyethersulfone nanocomposite membrane prepared by PANI/Fe_3O_4 nanoparticles with enhanced performance for Cu(II) removal from water. *Journal of Membrane Science* 415-416: 250–259.

Dudek, G., Turczyn, R., Strzelewicz, A. et al. (2012). Preparation and characterization of iron oxides-polymer composite membranes. *Separation Science and Technology* 47: 1390–1394.

Erkanlı, M., Yilmaz, L., Çulfaz-Emecen, P.Z., and Yetis, U. (2017). Brackish water recovery from reactive dyeing wastewater via ultrafiltration. *Journal of Cleaner Production* 165: 1204–1214.

Fanga, L.F., Zhou, M.Y., Cheng, L. et al. (2019). Positively charged nanofiltration membrane based on cross-linked polyvinyl chloride copolymer. *Journal of Membrane Science* 572: 28–37.

Fontananova, E., Profio, G.D., Giorno, L., and Drioli, E. (2017). Membrane and interfaces characterization by impedance spectroscopy. *Comprehensive Membrane Science and Engineering* 2: 393–410. https://doi.org/10.1016/B978-0-12-409547-2.12238-3.

Freger, V., Bottino, A., Capannelli, G. et al. (2005). Characterization of novel acid-stable NF membranes before and after exposure to acid using ATR-FTIR, TEM and AFM. *Journal of Membrane Science* 256: 134–142.

Gao, B., Zhang, D., and Li, Y. (2019). Preparation of PSSS-grafted polysulfone microfiltration membrane and its rejection and removal properties towards heavy metal ions. *Polymers Advanced Technologies* 30: 1096–1105.

Gebru, K.A. and Das, C. (2018). Removal of chromium (VI) ions from aqueous solutions using amine-impregnated TiO_2 nanoparticles modified cellulose acetate membranes. *Chemosphere* 191: 673–684.

Gohil, J.M. and Choudhury, R.R. (2019). Introduction to nanostructured and nano-enhanced polymeric membranes. Preparation, function, and application for water purification. *Nanoscale Materials in Water Purification*: 25–57. https://doi.org/10.1016/B978-0-12-813926-4.00038-0.

Haghighat, N., Vatanpour, V., Sheydaei, M., and Nikjavan, Z. (2020). Preparation of a novel polyvinyl chloride (PVC) ultrafiltration membrane modified with Ag/TiO_2 nanoparticle with enhanced hydrophilicity and antibacterial activities. *Separation and Purification Technology* 237: 116374. https://doi.org/10.1016/j.seppur.2019.116374.

He, X., Yang, D.P., Zhang, X. et al. (2019). Waste eggshell membrane-templated CuO-ZnO nanocomposites with enhanced adsorption, catalysis and antibacterial properties for water purification. *Chemical Engineering Journal* 369: 621–633. https://doi.org/10.1016/j.cej.2019.03.047.

Helali, N., Rastgar, M., Ismail, M.F., and Sadrzadeh, M. (2020). Development of underwater superoleophobic polyamide-imide (PAI) microfiltration membranes for oil/water emulsion separation. *Separation and Purification Technology* 238: 116451. https://doi.org/10.1016/j.seppur.2019.116451.

Hong, H., Yan, X., Song, X. et al. (2017). Bromine incorporation into five DBP classes upon chlorination of water with extremely low SUVA values. *Science of the Total Environment* 590-591: 720–728. https://doi.org/10.1016/j.scitotenv.2017.03.032.

Hu, X., Sun, J., Peng, R. et al. (2020). Novel thin-film composite reverse osmosis membrane with superior water flux using parallel magnetic field induced magnetic multiwalled carbon nanotubes. *Journal of Cleaner Production* 242: 118423. https://doi.org/10.1016/j.jclepro.2019.118423.

Hussein, M.A., Shahzad, H.K., Patel, F. et al. (2020). Porous Al_2O_3-CNT nanocomposite membrane produced by spark plasma sintering with tailored microstructure and properties for water treatment. *Nanomaterials* 10: 845. https://doi.org/10.3390/nano10050845.

Igbinigun, E., Fennell, Y., Malaisamy, R. et al. (2016). Graphene oxide functionalized polyethersulfone membrane to reduce organic fouling. *Journal of Membrane Science* 514: 518–526.

Jiang, S.J. and Cheng-Yin, D.U. (2003). Application of membrane separation technology in water & wastewater treatment and its relative priority. *Journal of Chongqing Architecture University* https://doi.org/10.2991/mmme-16.2016.202.

Jiang, C.C. and Zhang, J.F. (2007). Progress and prospect in electro Fenton process for wastewater treatment. *Journal of Zhejiang University - Science A: Applied Physics & Engineering* 8: 1118–1125.

Jin, L., Shi, W., Yu, S. et al. (2012). Preparation and characterization of a novel PA-SiO_2 nanofiltration membrane for raw water treatment. *Desalination* 298: 34–41.

Judd, S. and Jefferson, B. (2003). *Membranes for Industrial Wastewater Recovery and Re-Use*. Oxford, United Kingdom: Elsevier https://doi.org/10.1016/B978-1-85617-389-6.X5000-0.

Kamari, S. and Shahbazi, A. (2020). Biocompatible Fe_3O_4@SiO_2-NH_2 nanocomposite as a green nanofiller embedded in PES-nanofiltration membrane matrix for salts, heavy metal ion and dye removal: Long-term operation and reusability tests. *Chemosphere* 243: 125282.

Karimnezhad, H., Navarchian, A.H., Gheinani, T.T., and Zinadini, S. (2019). Incorporation of iron oxyhydroxide nanoparticles in polyacrylonitrile nanofiltration membrane for improving water permeability and antifouling property, Reactive and Functional. *Polymers* 135: 77–93.

Khorshidi, B., Hosseini, S.A., Ma, G. et al. (2019). Novel nanocomposite polyethersulfone-antimony tin oxide membrane with enhanced thermal, electrical and antifouling properties. *Polymer* 163: 48–56.

Khouni, I., Louhichi, G., Ghrabi, A., and Moulin, P. (2020). Efficiency of a coagulation/flocculation–membrane filtration hybrid process for the treatment of vegetable oil refinery wastewater for safe reuse and recovery. *Process Safety and Environmental Protection* 135: 323–341.

Krüger, R., Vial, D., Ariin, D. et al. (2016). Novel ultrafiltration membranes from low-fouling copolymers for RO pretreatment applications. *Desalination and Water Treatment* 57: 23185–23195.

Lakhotia, S.R., Mukhopadhyay, M., and Kumari, P. (2019). Iron oxide (FeO) nanoparticles embedded thin-film nanocomposite nanofiltration (NF) membrane for water treatment. *Separation and Purification Technology* 211: 98–107.

Le, H., Qiu, Y., and Tang, S. (2019). Removal of Sn^{2+} using copolymer of maleic acid-acrylic acid (PMA) by complexation-ultrafiltration and regeneration of PMA using shear induced dissociation and ultrafiltration. *Desalination and Water Treatment* 160 (2019): 41–49.

Li, J.B., Zhu, J.W., and Zheng, M.S. (2007). Morphologies and properties of poly(phthalazinone ethersulfoneketone) matrix ultrafiltration membranes with entrapped TiO_2 nanoparticles. *Journal of Applied Polymer Science* 103: 3623–3629.

Li, Q., Liao, Z., Fang, X. et al. (2019a). Tannic acid-polyethyleneimine crosslinked loose nanofiltration membrane for dye/salt mixture separation. *Journal of Membrane Science* 584: 324–332.

Li, S., Gao, B., Wang, Y. et al. (2019b). Antibacterial thin film nanocomposite reverse osmosis membrane by doping silver phosphate loaded graphene oxide quantum dots in polyamide layer. *Desalination* 464: 94–104.

Li, W., Li, B., Meng, M. et al. (2019c). Bimetallic Au/Ag decorated TiO_2 nanocomposite membrane for enhanced photocatalytic degradation of tetracycline and bactericidal efficiency. *Applied Surface Science* 487: 1008–1017.

Li, J., Yuan, S., Zhu, J., and Bruggen, B.V.D. (n.d.). High-flux, antibacterial composite membranes via polydopamine-assisted PEI-TiO_2/Ag modification for dye removal. *Chemical Engineering Journal* 373: 275–284.

Liu, F., Moghareh Abed, M.R., and Li, K. (2011). Preparation and characterization of poly(vinylidene fluoride) (PVDF) based ultrafiltration membranes using nano g- Al_2O_3. *Journal of Membrane Science* 366: 97–103.

Liu, L., Xub, Y., Wang, K. et al. (2019). Fabrication of a novel conductive ultrafiltration membrane and its application for electrochemical removal of hexavalent chromium. *Journal of Membrane Science* 584: 191–201.

Liu, M., Turcheniuk, K., Fu, W. et al. (2020a). Scalable, safe, high-rate supercapacitor separators based on the Al_2O_3 nanowire polyvinyl butyral nonwoven membranes. *Nano Energy* 71: 104627. https://doi.org/10.1016/j.nanoen.2020.104627.

Liu, X., Jiang, B., Yin, X. et al. (2020b). Highly permeable nanofibrous composite microfiltration membranes for removal of nanoparticles and heavy metal ions. *Separation and Purification Technology* 233: 115976. https://doi.org/10.1016/j.seppur.2019.115976.

Mallada, R. and Mene'ndez, M. (2008). *Inorganic Membranes: Synthesis, Characterization and Applications*, vol. 13, 1–460. Oxford, United Kingdom: Elsevier.

Manaf, O., Anjana, K., Prasanth, R. et al. (2019). ZnO decorated anti-bacterial electrospun ABS nanocomposite membrane for oil-water separation. *Materials Letters* 25: 126626. https://doi.org/10.1016/j.matlet.2019.126626.

Mocanu, A., Rusen, E., Diacon, A. et al. (2017). Electrochemical deposition of zinc oxide on the surface of composite membrane polysulfone-graphene-polystyrene in the presence of water soluble polymers. *Journal of Nanomaterials*: 1401503. https://doi.org/10.1155/2017/1401503.

Modia, A. and Bellarea, J. (2020). Efficient removal of 2,4-dichlorophenol from contaminated water and alleviation of membrane fouling by high flux polysulfone-iron oxide/graphene oxide composite hollow fiber membranes. *Journal of Water Process Engineering* 33: 101113. https://doi.org/10.1016/j.jwpe.2019.101113.

Moeinzadeh, R., Ghadam, A.G.J., Lau, W.J., and Emadzadeh, D. (2019). Synthesis of nanocomposite membrane incorporated with amino-functionalized nanocrystalline cellulose for refinery wastewater treatment. *Carbohydrate Polymers* 225: 115–212.

Mohammadtabar, F., Pillai, R.G., Khorshidi, B. et al. (2019). Efficient treatment of oil sands produced water: process integration using ion exchange regeneration wastewater as a chemical coagulant. *Separation and Purification Technology* 221: 166–174.

Morones-Ramırez, J.R. (2013). Bioinspired synthesis of optically and thermally responsive nanoporous membranes. *NPG Asia Materials* 5: 1–9.

Mukherjee, R. and De, S. (2013). Adsorptive removal of phenolic compounds using cellulose acetate phthalate alumina nanoparticle mixed matrix membrane. *Journal of Hazardous Materials* 265: 8–19.

Muttamara, S. (1996). Wastewater characteristics. *Resources, Conservation and Recycling* 16: 1–4.

Nakao, S. (1994). Determination of pore size and pore size distribution. 3. Filtration membranes. *Journal of Membrane Science* 96: 131–165.

Nasrollahi, N., Aber, S., Vatanpour, V., and Mahmoodi, N.M. (2019a). Development of hydrophilic microporous PES ultrafiltration membrane containing CuO nanoparticles with improved antifouling and separation performance. *Materials Chemistry and Physics* 222: 338–350.

Nasrollahi, N., Vatanpour, V., and Aber, S. (2019b). Improving the permeability and antifouling property of PES ultrafiltration membranes using the drying method and incorporating the

CuO-ZnO nanocomposite. *Journal of Water Process Engineering* 31: 100891. https://doi.org/10.1016/j.jwpe.2019.100891.

Ogoshi, T. and Chujo, Y. (2005). Synthesis of poly(vinylidene fluoride)(PVDF)/silica hybrids having interpenetrating polymer network structure by using crystallization between PVDF chains. *Journal of Polymer Science A: Polymer Chemistry* 43: 3543–3550.

Prihatiningtyas, I., Gebreslase, G.A., and Bruggen, B.V.D. (2020). Incorporation of Al_2O_3 into cellulose triacetate membranes to enhance the performance of pervaporation for desalination of hypersaline solutions. *Desalination* 474: 114198. https://doi.org/10.1016/j.desal.2019.114198.

Qin, H., Guo, W., Huang, X. et al. (2020). Preparation of yttria-stabilized ZrO2 nanofiltration membrane by reverse micelles-mediated sol-gel process and its application in pesticide wastewater treatment. *Journal of the European Ceramic Society* 40: 145–154.

Qu, F., Liang, H., Zhou, J. et al. (2014). Ultrafiltration membrane fouling caused by extracellular organic matter (EOM) from Microcystisaeruginosa: Effects of membrane pore size and surface hydrophobicity. *Journal of Membrane Science* 449: 58–66.

Qu, X., Alvarez, P., and Werber, J.R. (2016). The critical need for increased selectivity, not increased water permeability, for desalination membranes. *Environmental Science & Technology Letters* 3: 112–120.

Rajakumaran, R., Boddu, V., Kumara, M. et al. (2019). Effect of ZnO morphology on GO-ZnO modified polyamide reverse osmosis membranes for desalination. *Desalination* 467: 245–256. https://doi.org/10.1016/j.desal.2019.06.018.

Rawlings, A.E., Bramble, J.P., and Staniland, S.S. (2012). Innovation through imitation: biomimetic, bioinspired and biokleptic research. *Soft Matter* 8: 2675–2679.

Razmgar, K., Saljoughi, E., and Mousavi, S.M. (2019). Preparation and characterization of a novel hydrophilic PVDF/PVA/Al2O3 nanocomposite membrane for removal of As(V) from aqueous solutions. *Polymer Composites* 2453 https://doi.org/10.1002/pc.25115.

Said, N., Hasbullah, H., Abidin, M.N.Z. et al. (2019). Facile modification of polysulfone hollow-fiber membranes via the incorporation of well-dispersed iron oxide nanoparticles for protein purification. *Journal of Applied Polymer Science* https://doi.org/10.1002/APP.47502.

Saja, S., Bouazizi, A., Achiou, B. et al. (2018). Elaboration and characterization of low-cost ceramic membrane made from natural Moroccan perlite for treatment of industrial wastewater. *Journal of Environmental Chemical Engineering* 6: 451–458.

Santhosh, C., Velmurugan, V., and Jacob, G. (2016). Role of nanomaterials in water treatment applications. A review. *Chemical Engineering Journal* 306: 1116–1137.

Sethy, N.K., Arif, Z., Mishra, M.K., and Kumar, P. (2019). Synthesis of SiO_2 nanoparticle from bamboo leaf and its incorporation in PDMS membrane to enhance its separation properties. *Journal of Polymeric Engineering* https://doi.org/10.1515/polyeng-2019-0120.

Sethy, N.K., Arif, Z., Mishra, M.K., and Kumar, P. (2020). Nanocomposite film with green synthesized TiO2 nanoparticles and hydrophobic polydimethylsiloxane polymer: synthesis, characterization, and antibacterial test. *Journal of Polymeric Engineering* 40: 211–220.

Shakak, M., Rezaee, R., Maleki, A. et al. (2020). Synthesis and characterization of nanocomposite ultrafiltration membrane (PSF/PVP/SiO2) and performance evaluation for the removal of amoxicillin from aqueous solutions. *Environmental Technology & Innovation* 17: 100529. https://doi.org/10.1016/j.eti.2019.100529.

Shen, J., Ruan, W.L., and Gao, C. (2011). Preparation and characterization of PES-SiO_2 organic-inorganic composite ultrafiltration membrane for raw water pretreatment. *Chemical Engineering Journal* 168: 1272–1278.

Subramaniam, M.N., Goh, P.S., Lau, W.J. et al. (2018). AT-POME colour removal through photocatalytic submerged filtration using antifouling PVDF-TNT nanocomposite membrane. *Separation and Purification Technology* 191: 266–275.

Sun, H., Song, X., Ye, T. et al. (2018). Formation of disinfection by-products during chlorination of organic matter from phoenix tree leaves and Chlorella vulgaris. *Environmental Pollution* 243: 1887–1893.

Tavangar, T., Karimi, M., Rezakazemi, M. et al. (2020). Textile waste, dyes/inorganic salts separation of cerium oxide-loaded loose nanofiltration polyethersulfone membranes. *Chemical Engineering Journal* 385: 123787. https://doi.org/10.1016/j.cej.2019.123787.

Thakur, V.K. and Voicu, S.I. (2016). Recent advances in cellulose and chitosan based membranes for water purification. A concise review. *Carbohydrate Polymers* 146: 148–165.

Torki, M., Nazari, N., and Mohammadi, T. (2017). Evaluation of biological fouling of RO/MF membrane and methods to prevent it. *European Journal of Advances in Engineering and Technology* 4: 707–710.

Vatanpour, V., Madaeni, S.S., Moradian, R. et al. (2011). Fabrication and characterization of novel antifouling nanofiltration membrane prepared from oxidized multiwalled carbon nanotube/polyethersulfone nanocomposite. *Journal of Membrane Science* 375: 284–294.

Vatanpour, V., Esmaeili, M., Safarpour, M. et al. (2019). Synergistic effect of carboxylated MWCNTs on the performance of acrylic acid UV-grafted polyamide nanofiltration membranes. *Reactive and Functional Polymers* 134: 74–84.

Wang, N., Liu, T., Shen, H. et al. (2016). Ceramic tubular MOF hybrid membrane fabricated through in situ layer-by-layer self-assembly for nanofiltration. *AIChE Journal* 62: 538–546.

Wang, Y., Gao, B., Li, S. et al. (2019). Cerium oxide doped nanocomposite membranes for reverse osmosis desalination. *Chemosphere* 218: 974–983.

Wei, X.Z., Gan, Z.Q., Shen, Y.J. et al. (2019). Negatively-charged nanofiltration membrane and its hexavalent chromium removal performance. *Journal of Colloid and Interface Science* 553: 475–483.

Wigginton, S., Brannon, E., Kearns, P.J. et al. (2018). Nitrifying and denitrifying bacterial communities in advanced N-removal onsite wastewater treatment systems. *Journal of Environmental Quality* 47: 1163–1171.

Wu, C., Xu, T., and Yang, W. (2003). A new inorganic-organic negatively charged membrane: membrane preparation and characterizations. *Journal of Membrane Science* 224: 117–125.

Wu, H., Zhang, X., Zhao, X.T. et al. (2020). High flux reverse osmosis membranes fabricated with hyperbranched polymers via novel twice-crosslinked interfacial polymerization method. *Journal of Membrane Science* 595: 117480. https://doi.org/10.1016/j.memsci.2019.117480.

Xu, Y.C., Wang, Z.X., Cheng, X.Q. et al. (2016). Positively charged nanofiltration membranes via economically mussel-substance-simulated co-deposition for textile wastewater treatment. *Chemical Engineering Journal* 303: 555–564.

Yan, L., Li, Y.S., and Xiang, C.B. (2005). Preparation of poly(vinylidene fluoride)(pvdf) ultrafiltration membrane modified by nano-sized alumina (Al_2O_3) and its antifouling research. *Polymer* 46: 7701–7706.

Yan, L., Hong, S., Li, M.L., and Li, Y.S. (2009). Application of the Al_2O_3-PVDF nanocomposite tubular ultrafiltration (UF) membrane for oily wastewater treatment and its antifouling research. *Separation and Purification Technology* 66: 347–352.

Yan, L., Zhang, G., Zhang, L. et al. (2019). Robust construction of underwater superoleophobic CNTs/nanoparticles multifunctional hybrid membranes via interception effect for oily wastewater purification. *Journal of Membrane Science* 569: 32–40.

Yang, Y. and Wang, P. (2006). Preparation and characterizations of new PS/TiO2 hybrid membranes by sol-gel process. *Polymer* 47: 2683–2688.

Yang, Y., Zhang, H., Wang, P. et al. (2007). The influence of nano-sized TiO_2 fillers on the morphologies and properties of PSF UF membrane. *Journal of Membrane Science* 288: 231–238.

Ye, X.Y., Liu, Z.M., Wang, Z.G. et al. (2009). Preparation and characterization of magnetic nanofibrous composite membranes with catalytic activity. *Materials Letters* 63: 1810–1813.

Zare, S. and Kargari, A. (2018). Membrane properties in membrane distillation. In: *Emerging Technologies for Sustainable Desalination Handbook* (ed. V.G. Gude), 107–156. Oxford: Elsevier.

Zhai, Z., Zhao, N., Liu, J. et al. (2020). Advanced nanofiltration membrane fabricated on the porous organic cage tailored support for water purification application. *Separation and Purification Technology* 230: 115845. https://doi.org/10.1016/j.seppur.2019.115845.

Zhan, L., Zhang, P., Wang, M. et al. (2016). Research on the experiment of reservoir water treatment applying ultrailtration membrane technology of diferent processes. *Journal of Environmental Biology* 37: 1007.

Zhang, Y., Liu, Y., Zhao, L., and Song, L. (2013). Investigation of phosphorylated TiO_2- SiO_2 particles/polysulfone composite membrane for wastewater treatment. *Desalination* 324: 118–126.

Zhang, F., Gao, S., and Zhu, Y. (2016). Alkaline-induced super hydrophilic/underwater superoleophobicpolyacrylonitrile membranes with ultralow oil-adhesion for high-efficient oil/water separation. *Journal of Membrane Science* 513: 67–73.

Zhang, Q., Gao, J., and Qiu, Y.R. (2019). Removal of Ni (II) and Cr (III) by complexation-ultrafiltration using rotating disk membrane and the selective separation by shear induced dissociation. *Chemical Engineering & Processing: Process Intensification* 135: 236–244.

Zhao, Z., Zheng, J., Wang, M. et al. (2012). High performance ultrafiltration membrane based on modified chitosan coating and electrospun nanofibrous PVDF scaffolds. *Journal of Membrane Science* 394–395: 209–217.

Zhao, F., Chu, H., Zhang, Y. et al. (2017). Increasing the vibration frequency to mitigate reversible and irreversible membrane fouling using an axial vibration membrane in microalgae harvesting. *Journal of Membrane Science* 529: 215–223.

Zinadini, S. and Gholami, F. (2016). Preparation and characterization of high lux PES nanoiltration membrane using hydrophilic nanoparticles by phase inversion method for application in advanced wastewater treatment. *Journal of Applied Research in Water and Wastewater* 3: 232–235.

Zulfikar, M.A., Mohammad, A.W., and Hilal, N. (2006). Preparation and characterization of novel porous PMMA- SiO_2 hybrid membranes. *Desalination* 192: 262–270.

14

Occurrence, Fate, and Remediation of Arsenic

Gurudatta Singh, Anubhuti Singh, Reetika Shukla, Jayant Karwadiya, Ankita Gupta, Anam Naheed, and Virendra Kumar Mishra

Institute of Environment and Sustainable Development, Banaras Hindu University, Varanasi, Uttar Pradesh, India

14.1 Introduction

Arsenic (As) affects an estimated population of over 100 million around the world, especially Southeast Asia. Arsenic contamination in humans occurs mainly through the consumption of food and drinking water containing a high concentration. Two regions, Bangladesh and the West Bengal province of India, have been found to be affected by arsenic contamination. These regions have shown a high rate of exposure toward arsenic, putting a large percentage of their inhabitants at high risk (Bundschuh et al. 2012; Chakraborti et al. 2010; Naujokas et al. 2013). These regions have an excess of arsenic in groundwater with concentrations of up to 1000 μg/l. Apart from West Bengal, this problem is spreading to the other states of India, like Bihar and some parts of Uttar Pradesh (Yadav et al. 2020). In areas affected by arsenic contamination, water resources require strong policy to alleviate the problem. In general, to combat arsenic contamination for any region, alternate water sources are arranged for arsenic-free drinking water rather than using effective arsenic removal technology. This kind of approach does not provide a permanent solution to the problem. The adoption of new approaches to remediate water resources from arsenic pollution is urgently required because, in some areas, it is not possible to provide an alternative source of drinking water. This compels the development of new ecologically sustainable approaches toward the remediation of arsenic. Therefore, the transportation of arsenic in the aqueous environment and manifestation in groundwater, health impacts, and promising technological alleviation strategies are of the utmost importance.

Arsenic is a crystalline noxious metalloid universally present in environmental premises. This trace earth element ranks in 20th position based on occurrence in the Earth's crust, it is ranked 14th in seawater, and it is ranked 12th in the human body (Mandal and Suzuki 2002). Arsenic exists in four different oxidation states, namely arsenate (As(V)), arsenite (As(II)), arsenic (As(0)), and arsine (As(III)), among these forms, arsenate is the most stable (Gupta et al. 2011; Qu et al. 2009; Sharma and Sohn 2009; Zhao et al. 2010).

Pollutants and Water Management: Resources, Strategies and Scarcity, First Edition. Edited by Pardeep Singh, Rishikesh Singh, Vipin Kumar Singh, and Rahul Bhadouria.

Arsenate is present in oxygenic conditions, whereas arsine is generally present in anaerobic conditions. Among all the forms, the toxicity of arsenite is presumed to be the highest. Under reducing conditions, soil fauna can convert arsenite to arsenate (Hao et al. 2018). There are numerous biochemical forms of arsenic in the environment such as methylarsonic acid (MMA), dimethylarsinic acid (DMA), trimethylarsineoxide (TMAO), arsenobetaine (AsB), arsenolipid, arsenocholine (AsC), arsenosugars (AsS), etc. (Guérin et al. 2011). The use of MMA and DMA is extensive in pesticide and herbicide productions. DMA is also used as a cotton defoliant. Arsenobetaine and arsenolipid are found in marine animals (Huang and Matzner 2006), the latter being fat-soluble, where a concentration may reach up to 16 mg/kg in fish oil (Sele et al. 2012). Generally, organic forms are less toxic than inorganic ones (Meharg and Hartley-Whitaker 2002). According to EFSA (2009), Kile et al. (2011), and Wen et al. (2011), the toxicity of arsenic compounds can be found in the following order MMA(III) > As(III) > As(V) > DMA(V) > MMA(V), where MMA(III) is the most toxic species and recent research suggested it is the main toxic arsenic compound.

Experimental investigations are constantly going on to clean arsenic from the aquatic environment and various studies have presented their ideas. In this chapter, we discuss the various methods that can be applied for the remediation of arsenic from the natural environment.

14.2 Sources of Arsenic Contamination in the Environment

Natural and anthropogenic activities are a prominent source of arsenic (Shukla and Srivastava 2017). Natural sources may include rock components, minerals, and water reservoirs with higher concentrations of arsenic. Anthropogenic sources include mining activities, mining associated with gold, silver, and other metals, metal smelting, wood preservatives (Bundschuh et al. 2011), fossil fuels combustion (Pandey et al. 2011), waste incineration, application of pesticides, herbicides, fungicides, and crops desiccants, and stuff for livestock feed all contain arsenic complexes that release arsenic into the surrounding environment through dissolution (Bundschuh et al. 2012; de Esparza 2009; McClintock et al. 2012; Smedley and Kinniburgh 2002). Arsenic is a major constituent in a number of minerals including elemental arsenic, arsenide, sulfide, oxide, arsenate, and arsenite. Some of the important arsenic minerals are arsenopyrite (FeAsS), orpiment (As_2S_3), and realgar (As_2S_2), which can possibly be considered as arsenic ores due to the high arsenic content. (Hossain et al. 2006). Some natural and anthropogenic sources of arsenic are presented in Figure 14.1.

14.2.1 Natural Sources

Arsenic has a universal presence on Earth, however, in a small concentration (Smedley and Kinniburgh 2002). Yet, many locations around the world are recognized for the hazardous concentrations of arsenic, mainly in groundwater and soils. In its biogeochemical cycle, bounded arsenic may be released from minerals and rocks and transferred to the groundwater zone, which leads to elevated arsenic levels. During mining activity, oxidizing arsenopyrites and suspension of arsenic-rich $Fe(OH)_2$ through reduction (Saha and Sahu 2016) are mainly responsible for regulating arsenic concentrations in groundwater in West Bengal and eastern parts of Uttar Pradesh, Bihar, and Assam. As arsenic is chemically similar to

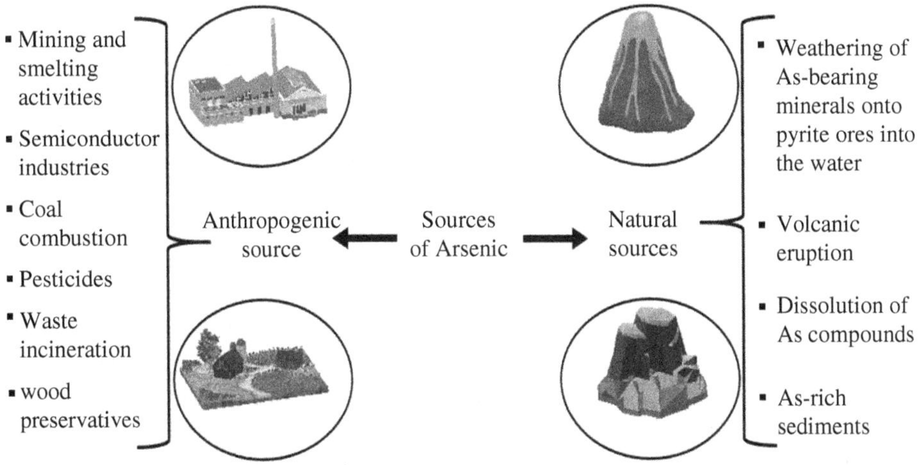

Figure 14.1 Natural and anthropogenic sources of arsenic.

Table 14.1 Arsenic concentration in various minerals/rocks.

Mineral/rock	Arsenic concentration (in mg/kg)	Reference
Phosphate mineral such as apatite	1000	Baur and Onishi (1969)
Carbonate minerals	10	Boyle and Jonasson (1973)
Silicate minerals	≤1	Baur and Onishi (1969)
Igneous rock (all types)	1.5	Onishi and Sandell (1955)
Sedimentary rocks	Ranges from 1.7 to 400	Smith et al. (1998)
Metamorphic rocks (phyllite)	18	Boyle and Jonasson (1973)
World soil	7.2 (average)	Boyle and Jonasson (1973)
Noncontaminated soil	≤ 10 (average)	Fitz and Wenzel (2002)

sulfur, it may replace sulfur from metal sulfide ores for metals like copper, gold, lead, silver, and uranium; metal sulfide may be converted into subsequent arsenite. Some of the sources of arsenic and its concentration are presented in Table 14.1.

Since iron oxyhydroxide has more probability of adsorbing arsenic, a higher concentration of arsenic is observed in groundwater because of reductive solubilization of iron oxyhydroxide, which breaks down to release arsenic into groundwater as seen in Bangladesh and India (Anawar et al. 2002; Chakraborti et al. 2001; Chowdhury et al. 2000; McArthur et al. 2001; Nickson et al. 1998). Due to volcanic activities, elements are redistributed in the Earth's crust, which builds up extremely high amounts of heavy metals, causing harm to the natural environment (Sriwana et al. 1998). These volcanic contaminants are accumulated in crater lakes, which are produced in volcano calderas after volcanic eruptions. Thus, water in the lakes, when dispersed on the surface, may potentially harm the surrounding ecosystem.

14.2.2 Anthropogenic Sources

Arsenic is found everywhere in the atmosphere, surface water, leachates, groundwater, soil, and sediments. Arsenic naturally occurs in fossil fuels (gasoline and coal), and therefore, vehicular emission and power plants emit arsenic into the atmosphere (Pacyna 1987). Ash released from power plants, often mixed with cement and other materials used in construction, releases arsenic into the soil. Also, the wood used in construction is usually treated with chromated copper arsenate (CCA), which is one of the major anthropogenic sources from which arsenic can readily leach out to adjacent lands (Sanders et al. 1994; Zagury et al. 2003). Soil application of organic pesticides accounts for approximately 80 to 90% of the arsenic introduced per annum (Nriagu and Pacyna 1988). Crop desiccants and food additives containing arsenic also contaminate the environment. The extraction of groundwater for agricultural purposes, on one hand, declines the groundwater table, while on the other hand, increases the arsenic proportion on the surface soil (Heikens 2006). The projected estimate shows an annual contribution of arsenic into the air, water, and soil ranging from 12000 to 25630 tons, 12000 to 72000 tons, and 52000 to 112000 tons, respectively (Nriagu and Pacyna 1988).

14.3 Health Effects of Arsenic Toxicity

Arsenic (As) naturally occurs in the environment within the soil, groundwater, and plants. According to Maity et al. (2012), several studies of arsenic-contaminated regions have shown the noxious consequence of arsenic on human well-being. Epidemiological signs stated that it can be cancerous for both humans and animals (WHO 2003). There are two pathways through that arsenic may enter into the human body, first, through direct uptake of arsenic-contaminated drinking water, and second, via consumption of arsenic-containing food, generally through vegetables accumulating arsenic in inorganic form (Fontcuberta et al. 2011). Arsenic toxicity in both acute and chronic forms may affect human beings. Arsenic toxicity may cause melanosis, leucomelanosis, keratosis, hyperkeratosis, dorsum, nonpetting edema, gangrene, and skin cancer; melanosis and keratosis occur most commonly (Karim 2000). Studies have shown the occurrence of a few cases of skin cancer patients extremely affected by As(III) and As(V) (Hsueh et al. 1995). Long-term exposure to inorganic arsenic disrupts the functioning of various organ systems like the respiratory system, digestive system, cardiovascular system, endocrine system, hematopoietic system, renal system, and reproductive system, which eventually leads to metastasis (Maharjan et al. 2005; Mandal et al. 1996). The World Health Organization (WHO) estimated that 5–10 years of exposure to inorganic arsenic may pose severe health effects in human beings (Hong et al. 2014). The health effects of arsenic on humans are presented in Figure 14.2.

Arsenicosis is a chronic contamination subsequence from drinking water with a high level of arsenic over an extensive period of time having concentration >0.05 mg/l that shows health effects including skin problems, skin cancers, internal cancers (bladder, kidney, lung), blackfoot disease, and possibly diabetes, high blood pressure, and reproductive disorders (WHO 2011). Arsenic has been recognized as a worldwide supreme ecological threat (Bundschuh et al. 2012; Chakraborti et al. 2010; Naujokas et al. 2013; Sohel et al. 2009).

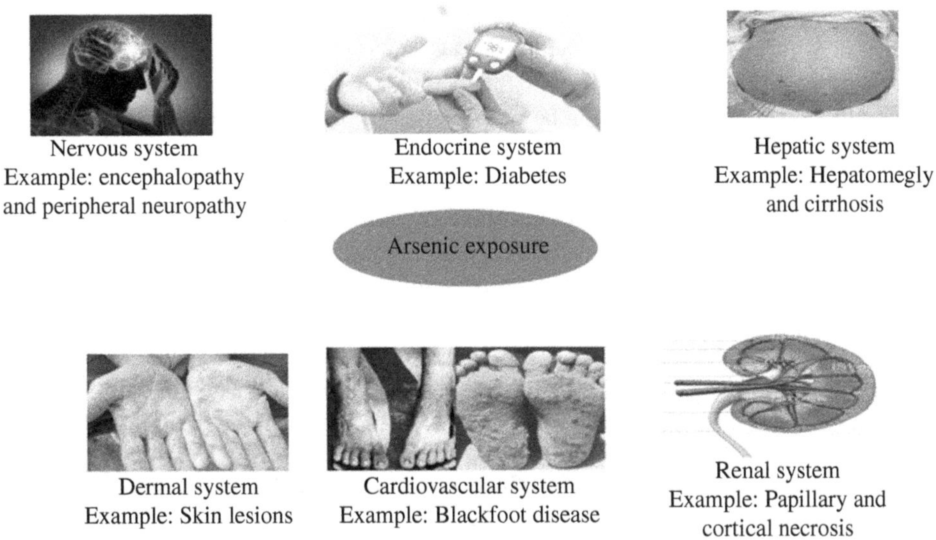

Figure 14.2 Health effects of arsenic on humans.

Additionally, arsenic has the capability to cross the placental barriers in mammals, and hence, it can affect both fetus and mother (Andrews 1989; Ramos-Peralonso 2014; WHO 2003).

14.4 Remediation Techniques for Arsenic Contamination

Remediation of arsenic from contaminated sites such as water and soil is necessary to reduce its effects on human health. Most of the physical, chemical, and biological tools have been applied for the remediation of arsenic from drinking water and soil. However, biological techniques include bacteria, fungi, and plants that have the potential to remediate arsenic contamination by various mechanisms (Su et al. 2010). These remediation techniques are discussed in the following sections.

14.4.1 Oxidation

The process of oxidation for arsenic removal from water involves the conversion of As(III) to As(V), i.e. from a soluble form to insoluble form or more toxic to less toxic, due to which it is precipitated and filtered out from the system (Johnston and Heijnen 2002; Masscheleyn et al. 1991). This process is crucial for groundwater removal of arsenic because of anaerobic conditions and predominance of As(III) (Masscheleyn et al. 1991; Singh et al. 2015) in the groundwater. As(III) can be oxidized either by natural oxidants like atmospheric O_2 and O_3 or by chemical oxidants like chlorine, chlorine dioxide, hypochlorite, hydrogen peroxide, etc. (Du et al. 2014; Johnston et al. 2001; Lee et al. 2003; Mondal et al. 2013; Sharma 2007; Vasudevan et al. 2006). Chemical oxidation has a greater pace of process when compared with natural oxidation (Ahmed 2001). But the efficiency of chemical oxidation is invariably affected by other substances present in the water, so the choice of oxidants preferred is critical. For example, the anions and organic matter present in groundwater greatly affect the use of ultraviolet (UV)/titanium dioxide (TiO_2) in arsenic oxidation (Guan et al. 2012). For the removal of arsenic from water that contains excess ammonia, it is better to use ozonation rather than chlorination because ozone reacts slowly with ammonia. It is also noted that Fe(VI) acts simultaneously as chemical oxidant, disinfectant, and coagulant (Sharma 2007).

An alternative process to oxidation is photochemical oxidation that utilizes ultraviolet irradiation in the presence of oxygen for the conversion of As(III) to its more soluble form, i.e. As(V), by generating hydroxyl ions from Fe(III) species (Yoon and Lee 2005; Sharma 2007; Ryu et al. 2013). Experiments revealed that it has an added benefit over oxidation due to low to no toxic sludge generation and nonenergy intensive treatment (Martinez-Huitle and Ferro 2006). Efficient oxidation can also be carried out by the mechanism of photocatalytic oxidation of soluble form to insoluble form followed by adsorption on titanium dioxide (TiO_2) (Bissen et al. 2001; Dutta et al. 2004, 2005; Miller et al. 2011). The adsorption of As(V) onto titanium dioxide (TiO_2) is influenced by pH, original concentration of the arsenic form of titanium dioxide (TiO_2), the occurrence of anions (e.g. CO_3^{2-} and PO_4^{3-}), and natural organic matter (Appelo et al. 1999; Bang et al. 2005; Bissen et al. 2001; Ferguson and Hering 2006; Miller et al. 2011). The outcome of reactions between anions and organic matter that are usually present in the groundwater is a significant drawback of using UV/titanium dioxide (TiO_2).

Biological oxidation of arsenic involves remediation via plants and microbes present in soil and groundwater. It is generally known that arsenic polluted groundwater usually contains concentrations of Fe and Mn. During treatment, the following reactions take place:

i) Oxidation of Mn(II) to Mn(IV) and Fe(II) to Fe(III)
ii) Oxidation of As(III) to As(V)
iii) Precipitation of manganese oxides
iv) Abiotic oxidation of As(III) by manganese oxides
v) Sorption of As(V) by manganese oxide

In situ oxidation is generally practiced for the elimination of iron from groundwater by pumping the water containing oxygen into the groundwater for the reduction of arsenic proportions (Liu et al. 2008). The aerated tube well water is deposited in a tank and released back into the groundwater. The dissolved oxygen content in tube well water oxidizes As(III) to As(V) and Fe(II) to Fe(III), thereby reducing the arsenic content. Different oxidation techniques and oxidants utilized are presented in Table 14.2 (Ghurye and Clifford 2004).

14.4.2 Adsorption

This is a technique by which a solid or a liquid surface holds gaseous or solid or solute particles on its surface. This process is mainly divided into two categories, i.e. physical or van der Waal's adsorption and chemical adsorption. Arsenic species are adsorbed on the upper layer of the adsorbent through van der Waal force and the factors which affect this adsorption are an initial concentration of arsenic, adsorbent dose, pH, temperature, availability of other chemicals, exposure time, and solubilization of arsenic in water (EPA 2000). This process is popularly used due to its high efficiency of removing arsenic from a region, fewer

Table 14.2 Different methods and oxidants used in the oxidation of As(III) to As(V).

S. no.	Type of oxidation	Oxidants	Working pH	Arsenic concentration (μg/l)	Reference
1	Chemical oxidation	Ozone and oxygen	7.5–8.5	46–62	Kim and Nriagu (2000)
		Chlorine	8.2–8.3	300	Hu et al. (2012)
		Hypochlorite	7	500	Viet et al. (2003)
		Potassium permanganate	8	50	Sorlini and Gialdini (2010)
		Hydrogen peroxide	7.5–10	50	Pettine et al. (1999)
2	Photocatalytic oxidation	UV/H_2O_2	8	100	Sorlini et al. (2014)
3	Biological oxidation	Chemoautotrophic-arsenite oxidizing bacteria	-	–	Katsoyiannis and Zouboulis (2004)

Source: Based on Ghurye and Clifford (2004).

adverse effects on the environment, low cost, and it also easily applied in rural areas (Maiti et al. 2008; Yazdani et al. 2016). Various kinds of absorbents can be utilized for the removal of arsenic from the environment but the chief absorbents that are popularly used are activated alumina, activated carbon, and red mud. All these types of absorbents will be discussed further in this chapter.

14.4.2.1 Activated Alumina

Activated alumina is a type of adsorbent which is used in arsenic remediation and is manufactured from aluminum hydroxide through dihydroxylation in such a way that a highly porous material is produced with a very high surface area. Between pH 6.0 and 8.0, activated alumina outer sides are positively charged, and between these ranges As(V) is adsorbed on the activated alumina surface with very high efficiency. The adsorption of As(III) on activated alumina is highly reliant on pH and it occurs strongly at pH 7.6 (Singh and Pant, 2004). The conventional activated alumina was found to have slow kinetics and low adsorption efficiency, so mesoporous alumina was manufactured and used for arsenic remediation. Mesoporous alumina has a larger superficial area ($307\,m^2/g$) and even pore size (3.5 nm) than the traditional activated alumina (Kim et al. 2004). The main reason for enhanced adsorption ability of mesoporous activated alumina is its even pore size more than the high surface area.

14.4.2.2 Activated Carbon

Activated carbon is a type of solid adsorbent which is used in arsenic remediation. It is also identified as activated charcoal and is a method of carbon that is manufactured from various substances such as coconut shells, wood char, lignin, fertilizer waste, sawdust, peat, etc. The efficiency of adsorption on activated carbon rests on several features such as pH, temperature, physical and chemical properties, ionic strength, etc. If the carbon is initially treated with Cu(II) before using it as an adsorbent, then the efficiency of arsenic removal is seen to be increased at an optimal pH of ~6 (Mohan et al. 2007). The study conducted by Gu et al. (2005) produced Fe-containing coarse structured activated carbon adsorbents (As-granular activated carbon [GAC]) for arsenic remediation from potable water. Impregnation of ferric ions on the GAC was done with the help of aqueous ferrous chloride followed by NaClO chemical oxidation. Arsenic adsorption was maximum at a concentration of ~6% iron and the adsorption of arsenate on the As-GAC was seen to occur in a varied range of pH (4.4–11.0). Also according to Rahman et al. (2020), arsenic adsorption was seen to be enhanced in aqueous solutions when Fe-incorporated activated carbon (MAC) from biomass mixture was produced through $FeSO_4$ impregnation and pyrolysis and used in aqueous solutions. There are many ways by which the efficiency of activated carbon can be enhanced to make it a better adsorbent, but it is a complex decision to select a type of absorbent due to various factors present in the environment.

14.4.2.3 Red Mud

Red mud, also identified as bauxite tailings, consists mainly of iron oxide, which is a very alkaline waste product that is generated in the industrial production of alumina. It can also be regarded as an alternative absorbent for arsenic removal from the environment. Red mud adsorption efficiency can be increased by treating it with heat and acid. The adsorption of arsenic on red mud particles is extremely reliant on pH; the optimal range is 5.8–7.5 for As(III) and 1.8–3.5 for As(V) (Altundoğan et al. 2002). Bauxsol, a type of red mud organized by neutralizing the red mud with seawater, is also used as an adsorbent (Genç et al. 2003). Also, the adsorption efficiency of bauxsol is increased by the addition of ferric sulfate or aluminum sulfate, or by acid and heat treatment (Genç-Fuhrman et al. 2004). The acid and heat-treated bauxsol were found to be best for arsenic adsorption on red mud particles. This treatment removed approximately 100% of the arsenate (pH 4.5) at an initial arsenate concentration of ~2 mg/l.

14.4.2.4 Ion Exchange

Ion exchange is another successful method by which arsenic remediation can be performed and contaminants can be removed from the environment. It is a physicochemical procedure in which a solid phase is utilized to exchange the equivalent number of ions from the contaminated water to the solid surface. The basic principle used in ion exchange is absorption. The solid surface is selected such that it can absorb the arsenic ions and remove the arsenic ions from polluted water. Synthetic resins are used as solid surfaces to exchange the arsenic ions in contaminated water from the resin ions. A bed of resin beads is made and the polluted water is allowed to flow through this bed. During this process, the arsenic ions are detached from the polluted water through ion exchange. Many types of resins can be made such as cation-exchange resins, cerium (Ce-IV), iron (Fe-III), copper (Cu-II), lanthanum (La-III), and yttrium (Y-III) resins; and anion-exchange resins like titanium dioxide (TiO_2) loaded on Amberlite XAD-7 resin and anion exchanger prepared from coir pith (Mohan and Pittman 2007). The features that affect ion exchange efficacy are pH, arsenic concentration, resin used, presence of ions such as carbonate, bicarbonate, sulfate, and chloride in water which can decrease the efficiency of arsenite, and arsenate removal via the ion exchange process (Sarkar and Paul 2016).

14.5 Coagulation and Flocculation

The coagulation and flocculation technique of removing arsenic, turbidity, and color is acknowledged because it is economical and highly efficient. The process of coagulation for arsenic removal is convenient and highly documented when used for large volumes of water (Johnston et al. 2001; Johnston and Heijnen 2002; Choong et al. 2007). It is an important primary step in which the dissolved particles are destabilized and attached to other particles known as coagulants for removal. The addition of metal-based oppositely charged coagulant (positively charged) in the coagulation process for colloids present makes the particle larger and settle. Usually, coagulants utilized are ferric chloride ($FeCl_3$) and aluminum sulfate $Al_2\ (SO_4)_3$. $FeCl_3$ hydrolyses in water to form positive charged ferric hydroxide [$Fe(OH)_3$]. Flocculation involves the addition of negatively charged flocculants causing neutralization leading to the formation of flocs (Mondal et al. 2013). The use of a coagulation-flocculation based process for the treatment of arsenic from water leads to a change from the dissolved form of arsenic to insoluble solids, which can be separated by sedimentation and filtration. Parameters such as arsenic species, pH, coagulant doses, and agitation speed are crucial for the effectiveness of the process (Granada et al. 2003; Sancha and Ruiz 1984). Among the arsenic species, the removal of arsenate is more efficient related to arsenite. The efficiency of arsenic removal also depends on the pH. Below the pH of 7.6, $Al_2(SO_4)_3$, and $FeCl_3$ are equally competent. When pH is higher than 7.6, $FeCl_3$ has a better coagulating effect than $Al_2(SO_4)_3$ (Cheng et al. 1994; Garelick et al. 2005; Hering et al. 1996; Saha et al. 2001; Scott et al. 1995). A derivative of chitin, chitosan is an important coagulant biopolymer with a low dosage requirement, bulky flocs formation, rapid precipitation, high biodegradation, easier sludge treatment, highly efficient in removing suspended solids and ionic metals, anti-microbial, and with no secondary pollutants (Chen and Chung 2006; Zeng et al. 2008).

14.6 Bioremediation for the Treatment of Arsenic

The mechanism of bioremediation is based on the degradation, transformation, stabilization, or volatilization of pollutants from the location of the contaminant with the aid of microorganisms or plants. In situ bioremediation is widely used for the remediation of polluted sites, particularly where heavy metals are present as pollutants in wider regions. Generally, heavy metal pollutants are present in different components of the environment (water, soils, and sediments). Heavy metals can be available in the environment by various processes; they can be present either naturally or can be produced by humans through various industrial and commercial processes. As heavy metals are not a biodegradable entity, the chief purpose becomes the removal of heavy metals polluted sites. The heavy metals can be transformed over several mechanisms such as methylation, complexation, sorption, and changes in valence state. These alterations change the movement and bioavailability of heavy metals contaminants in a diverse environment (Kumar et al. 2020). Arsenic bioremediation can be done by either using plants (phytoremediation) or microorganisms (microbial bioremediation) or by using both plants and microorganisms together. The details of various mechanisms in arsenic bioremediation are discussed further in the following sections.

14.6.1 Phytoremediation

The phytoremediation process became more popular toward the end of the last century because of its cost-effectiveness, easy in situ applicability, eco-friendly character, and least negative effects on the nearby ecosystem (Kabata-Pendias and Pendias 2001; Salt et al. 1995; Vithanage et al. 2012; Watanabe 1997). Phytoremediation is a group of processes in which different types of plants are used to extract, absorb, contain, volatilize, or immobilize pollutants (inorganic and organic) from diverse parts of the ecological atmosphere, hydrosphere, or lithosphere (Kumar et al. 2020; Salt et al. 1998). Arsenic is not a necessary constituent for plant metabolism, on the other hand, inorganic arsenic is extremely toxic for plants. Arsenate correspondent to phosphate interrupts phosphate uptake in plants by replacing phosphate in various biochemical processes. For instance, it can disturb mitochondrial oxidative phosphorylation, and thereby, the formation of adenosine triphosphate (ATP), the main energy currency of cells (Fayiga and Saha 2016). Once arsenic enters the plant, it can then enter into the fruits or grains and is accumulated there, thereby affecting the edible parts of plants. There are various methods by which phytoremediation of arsenic-contaminated sites can be done such as phytoextraction, phytovolatilization, and phytoimmobilization. The details of these methods are discussed in the forthcoming sections. Different types of phytoremediation techniques are shown in Figure 14.3.

14.6.1.1 Phytoextraction

Phytoextraction is a method of arsenic phytoremediation in which a certain type of plant is grown to collect high levels of heavy metals from polluted soil in parts of the plant, mainly the stem, thereby making the removal of the pollutant possible by picking the shoot portions of the plant (Wan et al. 2020). *Pteris vittata* is an arsenic hyperaccumulator that can

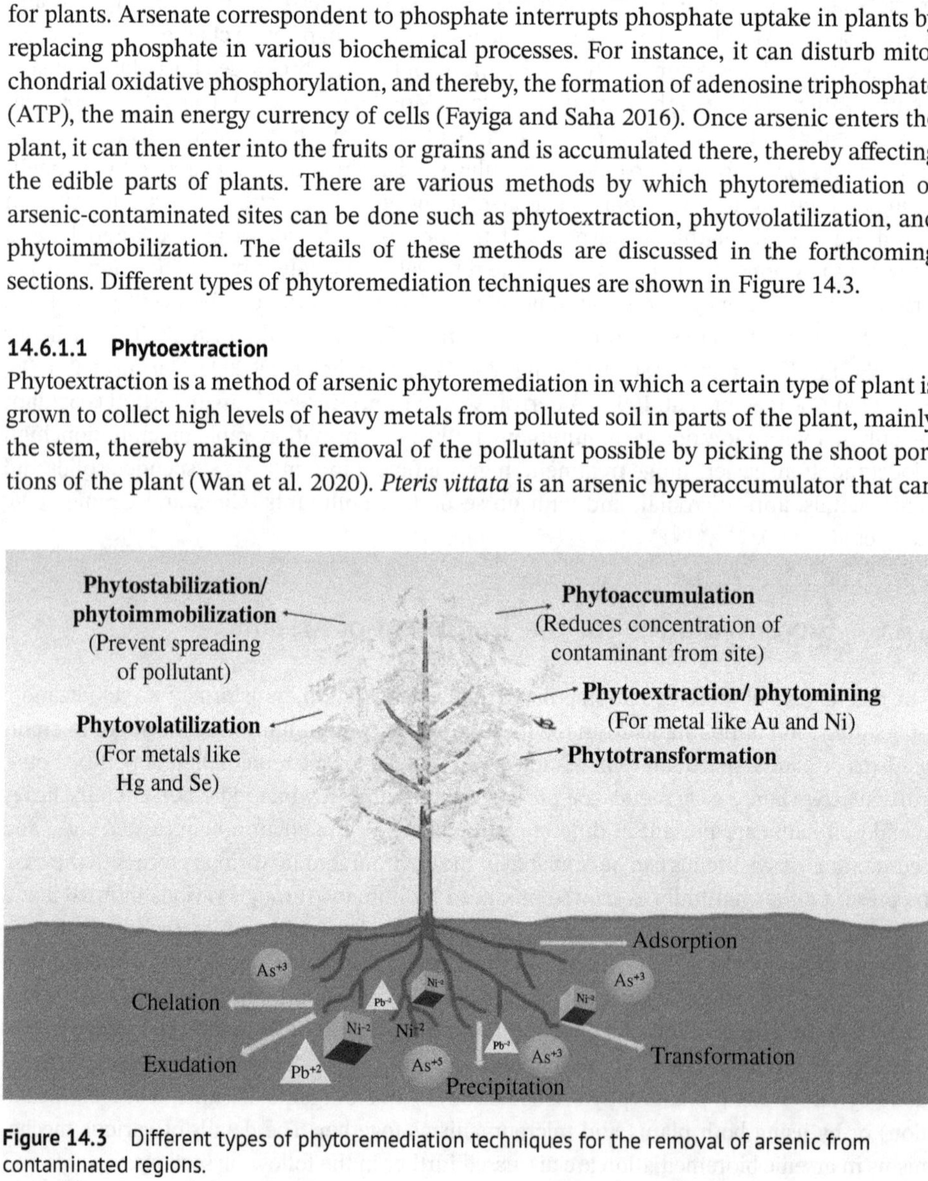

Figure 14.3 Different types of phytoremediation techniques for the removal of arsenic from contaminated regions.

attain an aboveground arsenic concentration of ~1% (dry weight) (Ma et al. 2001). It is a perennial plant, has high biomass, and can grow in variable environmental conditions. Using *P. vittata*, the highest arsenic removal efficiency of 18% (weight) per year was achieved in China in in situ phytoextraction projects (Chen and Chiou 2008). The arsenic hyperaccumulation capability of *P. vittata* is different among its various ecotypes/populations. The key defining feature for the varied property in its population is the amount of arsenic in the soil of the location where spores of *P. vittata* are accumulated – the higher the arsenic concentration in the soil, the lower amount of arsenic collecting capability. The main mechanisms for arsenic collection include the following points:

1) Root exudations that can solubilize arsenic in the soil
2) Application of effective phosphate transporter to transfer arsenic
3) As(III) must be reduced to As(V), which allows transport of arsenic upwards
4) Reclamation in the shoot zone mainly by compartmentation, precipitation, and efflux (Wan et al. 2020)

Reclamation of arsenic polluted soil can also be done by utilizing two other hyperaccumulators ferns, *P. cretica* and *Pityrogramma calomelanos* (Anh et al. 2018; Eze and Harvey 2018). They are less popular than *P. vittata* because of their lower biomass. Willow (*Salix* sp.) is not included in hyperaccumulating species but it is also relatively commonly utilized in the phytoextraction of arsenic polluted soil. Related with *P. vittata*, willow has massive biomass, widespread root organizations, fast-growing frequency, and possibly can also produce energy (Navazas et al. 2019). Phytoextraction is a profitable and eco-friendly technique for arsenic phytoremediation but the disposal of the hyperaccumulator plants or reusing everything valuable from the biomass should be the focus of forthcoming exploration in the field.

14.6.1.2 Phytovolatilization

Phytovolatilization is a method of arsenic phytoremediation in which the plant uptake the pollutants from the soil, sediment, or river, convert them to the volatilized form, and are transpired into the surroundings (EPA 2000). One of the first heavy metal contaminants in which this process was used was mercury (Erakhrumen and Agbontalor 2007). This process is generally possible with unstable organic complexes like trichloroethene, and inorganic chemicals that have volatile forms like Se and As (EPA 2000). The process of phytovolatilization occurs because of the phenomenon of water uptake by the plants from the soil for their growth, during this process, they can also uptake organic contaminants from the soil. Some of these organic contaminants at that point reach the leaves and are transpired to the atmosphere at relatively lower concentrations (Dushenkov et al. 1995).

14.6.1.3 Phytoimmobilization

Phytoimmobilization/phytostabilization is a method of arsenic phytoremediation in which either the heavy metal is precipitated within the rhizosphere or the heavy metal is absorbed on the roots of the plant. Both processes limit the movement of the contaminants in the soil and prevent the mobility of the contaminants to other regions of its surroundings (Awa and Hadibarata 2020). The process of phytoimmobilization occurs due to the changes developed by the plant being used in the soil chemistry, thereby facilitating the sorption and precipitate-forming processes of heavy metals in the lithosphere (García-Sánchez et al. 2018; Zarrin and Azra 2018; Zhang et al. 2009). The process of immobilization mainly occurs at contaminated heavy metal mining sites. This process is not a huge success because it limits the contaminant by inactivation and stabilization to only the contaminated site, and is not helpful in the permanent removal of the pollutant from the site. It is only a temporary solution for arsenic removal from the environment (Awa and Hadibarata 2020).

14.6.2 Microbial Bioremediation

Microorganisms are largely found in the soil and they possibly will reduce the biological availability of arsenic within the soil. Microbes possess mechanisms through which they can convert arsenic into different forms, we can, therefore, utilize this potential to remediate areas polluted by arsenic. Microbial bioremediation of arsenic polluted soil will be categorized into immobilization and arsenic mobilization. Oxidation of As(III) to As(V) lowers the mobility of arsenic, thereby decreasing its bioavailability, especially in rice fields (Mallick et al. 2015). Arsenic immobilization can be done by oxidation and reduction, chelation, pH change, biosorption, bioaccumulation, co-precipitation, co-dissolution, biomethylation, or altering the organic metallic complexes to radionuclides (Achal et al. 2012; Lee et al. 2012; Liu et al. 2011; Marwa et al. 2019; Pratush et al. 2018; Wan et al. 2020). Arsenate and arsenite cross the cells of prokaryotes and unicellular eukaryotes by the aid of phosphate transporters and aquaglyceroporins, respectively. Bacteria contain *arsRBC* operon which regulates arsenic resistance. This operon contains three genes, *arsR*, *arsB*, and *arsC*, which convert As(V) to As(III) (Rosen 2002; Tripathi et al. 2007).

Microbial immobilization of arsenic can be either directly or indirectly. *Aspergillus flavus*, rhizospheric fungi, have the ability to biotransform the moveable form of arsenic to the nonmobile form of arsenic particles, making the bioavailability of arsenic lower to plants (Afzal et al. 2019). In indirect immobilization, ureolytic bacteria may be used to secrete urease to cause the precipitation of calcite, thereby immobilizing arsenic by the adsorption of arsenic to calcite (Achal et al. 2012). The study conducted by Dey et al. (2016), isolated two rod-shaped Gram-positive bacteria, *Bacillus* sp. and *Aneurinibacillus aneurinilyticus* from arsenic polluted groundwater of West Bengal, India. These isolated bacteria can endure arsenate and arsenite concentrations of up to 4500 and 550 ppm, respectively. These bacteria have a removal capacity of 51.4–51.9% of As(III) and 50.3–53.0% of As(V), respectively, from a medium containing arsenic. The isolated bacterial species were arsenic resistant and can be used as a potential candidate for the bioremediation of arsenic.

Microalgae such as *Chlamydomonas reinhardtii* and *Scenedesmus obliquus* can also be utilized in the bioremediation of arsenic from the polluted site because of their bioaccumulation capability. In both these freshwater green algae species, it was seen that the biological accumulation of arsenic is higher in the media that is phosphorus limited in relation to the phosphorus enriched media. The reason for this is the structural similarity between the arsenate and phosphate results in the competitive binding to their transporter for entering into the cells due to the resemblance in their structures (Kumar et al. 2020; Wang et al. 2013).

Aspergillus niger also has a higher uptake and tolerant nature for As(V). It can be utilized for the removal of arsenic from contaminated water. Numerous different or varied species having arsenic removal capacity have been researched and reports suggest some of them have been applied for many years. Examples are *Scopulariopsis brevicaulis* (Gosio 1892), *Phaeolus schweinitzii* (Pearce et al. 1998), *Fusarium oxysporum* (Granchinho et al. 2002), *Sinorhizobium meliloti* (Carrasco et al. 2005), *Neosartorya fischeri, Aspergillus clavatus* (Čerňanský et al. 2007), *Aspergillus candidus* (Vala 2010), *A. niger* (Mukherjee et al. 2010), *Trichoderma* sp., *Neocosmospora* sp. and *Rhizopus* sp. (Srivastava et al. 2011). As microorganisms possess many mechanisms for arsenic resistance, they can be explored in many ways for arsenic bioremediation of arsenic-contaminated areas, and thus, can become a powerful tool against arsenic pollution in the different matrix of the environment. Examples of some plant growth-promoting microorganisms (PGPMs) and their use in regulating arsenic accumulation in plants are discussed in Table 14.3.

Table 14.3 Examples of plant growth-promoting microorganisms (PGPMs) used for regulating arsenic accumulation in plants and decreasing arsenic stress in the environment.

Plant growth-promoting microorganism	Variety of plant on which PGPMs are used	Effect of PGPMs	Reference
Bacteria			
Methylobacterium oryzae	*Acacia farnesiana*	The arsenic levels in the plant increased from 600 to 1700 mg/kg when *M. oryzae* was used during plant growth.	Alcántara-Martínez et al. (2018)
Bacillus flexus	*Oryza sativa*	Plants grown in the presence of arsenic; the plants with *B. flexus* had better grain yield than the plants not inoculated with *B. flexus*.	Das et al. (2016)
Bacillus megaterium	*Brassica juncea*	Positive effect on plant biomass.	Franchi et al. (2017)
Microbacterium paraoxydans	*Lupinus albus*	Assisted phytoextraction strategies.	Franchi et al. (2017)
Acinetobacter sp.	*Cicer arietinum* L.	Plant growth and production of the plant was increased. Decreases the toxic effect of arsenic.	Srivastava and Singh (2014)
Exiguobacterium	*Vigna radiata*	Plant growth and arsenic uptake increased. Increased shoot and root biomass.	Pandey and Bhatt (2016); Pandya et al. (2015)
Burkholderia sp. D54	*Spirulina platensis*	Increased in the growth of plants and uptake of metals.	Guo et al. (2011)
Pseudomonas fluorescens strain isolated in soil	*Helianthus annus var*	Enhanced antioxidant activities, proline, plant yield, physiology, growth, and reduced the malondialdehyde content.	Saleem et al. (2018)
Alphaproteobacteria, *Betaproteobacteria*, and *Gammaproteobacteria*	*Cirsium arvense*	Improves the phytoremediation process.	Cavalca et al. (2010)

(*Continued*)

Table 14.3 (Continued)

Plant growth-promoting microorganism	Variety of plant on which PGPMs are used	Effect of PGPMs	Reference
Staphylococcus arlettae	*Brassica juncea* (L.)	Increases arsenic accumulation in the root and increased biomass carotenoid, chlorophyll, and protein content.	Srivastava et al. (2013)
Fungi			
Rhizophagus intraradices (Arbuscular mycorrhiza)	*Glycine max*	The accumulation capacity of *Glycine max* was decreased from 7.8 mg As/kg to 6.0 mg As/kg	Li et al. (2016)
Trichoderma sp.	*Helianthus annuus*	The concentration of biomass was seen to be greater in arsenic-mixed soil inoculated with *Trichoderma* in comparison with uninoculated Arsenic-mixed soil.	Govarthanan et al. (2018)
Humicola sp.	*Bacopa monnieri*	The arsenic bioaccumulation in fungal biomass ranged between 0.146 and 11.36 g/kg biomass.	Tripathi et al. (2020)
Debaryomyces hansenii	*Oryza sativa*	*D. hansenii* showed arsenate reductase activity and improved detoxification mechanisms in plants.	Kaur et al. (2020)
Brevundimonas diminuta	*Oryza sativa*	The rhizoinoculation of bacterial strain reduced As(V) accumulation in aerial parts, especially in edible parts when grown in soil with As(Vs) (10 and 50 mg/kg) and also enhanced plant growth.	Singh et al. (2016)
Glomus geosporum, G. versiforme, G. mosseae	*Oryza sativa*	Arsenic concentration in the grains reduced up to 50% in the mycorrhiza-inoculated plants.	Wu et al. (2015)
Piriformospora indica	*Oryza sativa*	Arsenic accumulation increased from 0.65 mg/g dw in noncolonized arsenic-treated plant to 26.22 mg/g dw in pre-colonized plant root.	Mohd et al. (2017)
Rhizoglomus intraradices, Glomus etunicatum	*Triticum aestivum*	P deficiency due to As was removed and also a favorable P : As ratio was maintained.	Sharma et al. (2017)
Trichoderma asperellum	*Ipomoea aquatica*	The shoot dry weight increased by 21.6%, height by 35%, root dry weight by 87% at 5% inoculation level.	Su et al. (2017)

Table 14.3 (Continued)

Plant growth-promoting microorganism	Variety of plant on which PGPMs are used	Effect of PGPMs	Reference
Algae			
Anabaena sp.	*Oryza sativa*	Better growth of plants was seen when *Anabaena* was grown alongside *O. sativa* in the presence of arsenic	Ranjan et al. (2018)
Pseudomonas putida and *Chlorella vulgaris* consortium	*Oryza sativa*	The arsenic levels were significantly reduced in plants inoculated with this consortium in comparison with noninoculated plants. Also, better growth was seen in the inoculated plants.	Awasthi et al. (2018)
Chlorella vulgaris and *Nannochloropsis* sp.	*Oryza sativa*	The plants inoculated with this consortia showed reduced arsenic accumulation in roots and shoots	Upadhyay et al. (2016)

14.6.3 Microbially Assisted Phytoremediation/Phytobial Remediation

Phytobial remediation is a kind of arsenic remediation in which both plants and microbes are used together for a reduction in the number of contaminants in the environment. Microorganisms show various vital roles in the environment. They take part in various biogeochemical cycles and also assist in plant growth. The microorganisms occupy the rhizosphere and form a mutualistic relationship with the plants as free-living, root symbionts, and root endophytes (Hassan et al. 2020). High levels of heavy metal contamination are naturally endured by plant growth-promoting bacteria (PGPB). This property of PGPB assists equally to soil and plants. PGPB mutually interacts with plants, and through this mutual interaction, it takes part in the bioremediation of heavy metal contaminants, thereby increasing plant productivity (de Andrade et al. 2019).

Microorganisms convert arsenate to arsenite in the plant for adsorbing arsenite in their aerial parts. PGPB is used to raise the competence of plant remediation by removing metal build-up in plants or by enhancing plant growth or plant tolerance. It is done by phytohormone production in which ethylene levels are reduced, which leads to the production of a lengthier root system, thus facilitating the skill of the plants to withstand the contaminants (Glick 2010; Kong and Glick 2017). PGPB is classified into two categories of bacteria, i.e. plant growth-promoting endophytic bacteria (PGPE), which colonize the internal plant tissues, and plant growth-promoting rhizospheric bacteria (PGPR), which are present in the rhizospheric regions around the plant roots. Both types of PGPB enhance plant growth by helping plants use the nutrition from their surroundings (Afzal et al. 2019; Vurukonda et al. 2016).

According to Franchi et al. (2017), plants such as *Brassica juncea* and *Lupinus albus* when used for phytoremediation of arsenic-contaminated fields (arsenic in a concentration of 41.1 mg/kg) assisted by PGPB such as Actinomycetales (*Gordonia alkanivorans*,

Microbacterium paraoxydans, and *Rhodococcus equi)*, Betaproteobacteria *(Cupriavidus necator* and *Achromobacter denitrificans*), and Bacilli (*Bacillus megaterium*, *Lysinibacillus macroides*, and *Sporosarcina luteola*) isolated from contaminated soil had an encouraging outcome on plant biomass and aided phytoextraction approaches. To overcome the limitations of bioremediation, phytobial remediation can be utilized for the production of additional inexpensive, operative, and reduced time-taking methods to eliminate heavy metal contaminants from the environment.

14.6.4 Consortia-Based Approach

The consortia-based approach is a way of phytoremediation of regions contaminated with heavy metals in which two or more microorganisms are used together to remediate arsenic from the polluted areas. The efficiency is seen to be increased by the microbes when used in consortia. According to Liu et al. (2013), when a consortium containing two bacteria, *Bacillus* sp. PY1 and *Sphingomonas* sp. *PY2* and a fungus, *Fusarium* sp. PY3 (isolated from contaminated soils) were used in polluted regions, it was remarked that all these consortia were capable of pyrene and arsenic as carbon and energy sources, and in combination, they caused degradation of pyrene and volatilize arsenic. They also have the self-reliant ability for the degradation of pyrene and volatilization of arsenic as well. According to Hassan et al. (2020), when a consortium of isolated fungi from the polluted landfill soil was used, containing Ascomycota, all isolated fungi, and Basidiomycota, for mycoremediation of metal and metalloids like Cr, Cu, As, Fe, and Mn, they proved to be effective in the elimination of arsenic (77%), Mn (71%), Cr (60%), and Cu (52%). From the previous examples, it is seen that a consortia-based approach for the bioremediation of arsenic-contaminated fields is successful and further research should be done under this approach.

14.7 The Fate of Arsenic in the Environment

Arsenic (As) is an eminently possible lethal metalloid, which is found in all the compartments of soils, sediments, and aquatic environments in nature. Subsequently, significant studies have been reported on the circulation of arsenic in diverse ecological systems (Hughes 2002; Tabassum et al. 2019; Xie et al. 2017). Weathering and erosion of arsenic bearing minerals and rocks are well thought out as the chief and most prevalent sources of arsenic in water and soil (Abbas et al. 2018). Arsenic pollution is a consequence of an excessive volume of arsenic released from polluted aquifers to various components of the Earth's crust (Govil and Krishna 2018; Khalid et al. 2017; Neidhardt et al. 2012). Perhaps the use of arsenic-rich groundwater is a key contribution to soil and plants (Arco-Lázaro et al. 2018; Aysha et al. 2017). The build-up of arsenic in plants is one of the main possible pathways of arsenic transmission to humans. The major path of arsenic exposure to humans is utilizing arsenic-rich drinking water (Shahid et al. 2017; Tabassum et al. 2019). It can result in food chain contamination and dietary intake of arsenic by the use of arsenic-rich groundwater for crop irrigation. Overexploitation of minerals, ore extraction, mining activities, and smelting has resulted in the discharges of arsenic in agricultural/farmland soil. Arsenic from soil to plants is a key step of arsenic entering the food chain.

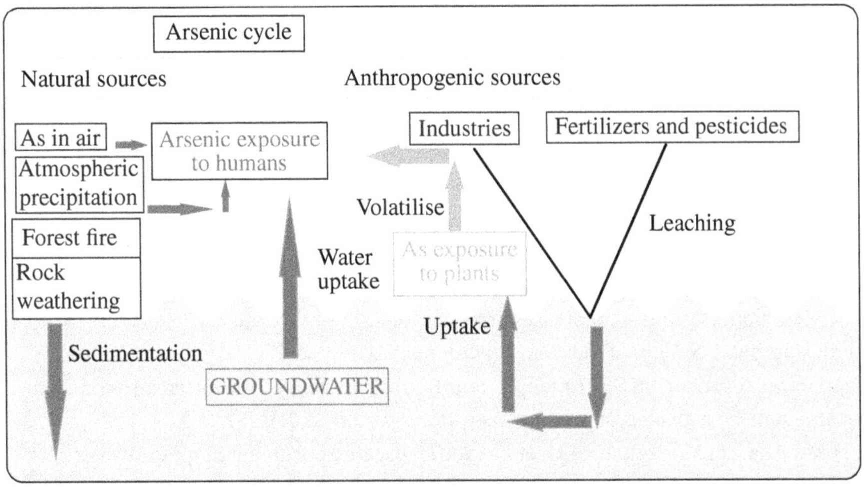

Figure 14.4 The fate of arsenic in the environment.

Earlier, Liu et al. (2005) described arsenic transmission from soil to vegetables. It has also been determined that plants can uptake arsenic through roots and leaves while taking arsenic from the soil through their roots (Shahid et al. 2017). A few studies also focused on the foliar absorption of arsenic from the atmosphere (Bondada et al. 2004; Shahid et al. 2019). The fate of arsenic in the environment is presented in Figure 14.4.

14.8 Conclusion

Arsenic contamination poses a global threat for all living organisms, for this reason, different strategies have been developed to cope with this serious challenge. In this chapter, we have provided a brief overview of the sources of arsenic contamination and the status of arsenic removal by different remediation strategies from aquatic systems (surface and groundwater). There are many remediation mechanisms applied to remediate arsenic toxicity from contaminated sites (surface and groundwater). They transform more toxic to less toxic forms of sustainability at minimum cost. Conversion and release of volatile arsenic species into the environment are very safe due to the dilution effect. Arsenic contamination poses a global threat for all living organisms, for this reason, different strategies have been developed to cope with this serious challenge. Still, most of the arsenic remediation techniques used are under experimental and pilot stages and these studies need to be implanted on a field scale.

References

Abbas, G., Murtaza, B., Bibi, I. et al. (2018). Arsenic uptake, toxicity, detoxification, and speciation in plants: physiological, biochemical, and molecular aspects. *International Journal of Environmental Research and Public Health* 15 (1): 59.

Achal, V., Pan, X., Fu, Q., and Zhang, D. (2012). Biomineralization based remediation of As (III) contaminated soil by *Sporosarcina ginsengisoli*. *Journal of Hazardous Materials* 201: 178–184.

Afzal, I., Shinwari, Z.K., Sikandar, S., and Shahzad, S. (2019). Plant beneficial endophytic bacteria: mechanisms, diversity, host range and genetic determinants. *Microbiological Research* 221: 36–49.

Ahmed, M.F. (2001). An overview of arsenic removal technologies in Bangladesh and India. Proceedings of BUET-UNU International Workshop on Technologies for Arsenic Removal from Drinking Water (pp. 5–7), Dhaka, Bangladesh.

Alcántara-Martínez, N., Figueroa-Martínez, F., Rivera-Cabrera, F. et al. (2018). An endophytic strain of Methylobacterium sp. increases arsenate tolerance in *Acacia farnesiana* (L.) Wild: a proteomic approach. *Science of the Total Environment* 625: 762–774.

Altundoğan, H.S., Altundoğan, S., Tümen, F., and Bildik, M. (2002). Arsenic adsorption from queous solutions by activated red mud. *Waste Management* 22 (3): 357–363.

Anawar, H.M., Akai, J., Mostofa, K.M.G. et al. (2002). Arsenic poisoning in groundwater: health risk and geochemical sources in Bangladesh. *Environment International* 27 (7): 597–604.

de Andrade, F.M., de Assis Pereira, T., Souza, T.P. et al. (2019). Beneficial effects of inoculation of growth-promoting bacteria in strawberry. *Microbiological Research* 223: 120–128.

Andrews, A.H. (1989). *Veterinary Medicine: A Textbook of the Diseases of Cattle, Sheep, Pigs, Goats and Horses: DC Blood & OM Radostits*. London: Baillière Tindall.

Anh, B.T.K., Minh, N.N., Ha, N.T.H. et al. (2018). Field survey and comparative study of *Pteris vittata* and *Pityrogramma calomelanos* grown on arsenic contaminated lands with different soil pH. *Bulletin of Environmental Contamination and Toxicology* 100 (5): 720–726.

Appelo, C.A.J., Drijver, B., Hekkenberg, R., and De Jonge, M. (1999). Modeling in situ iron removal from ground water. *Groundwater* 37 (6): 811–817.

Arco-Lázaro, E., Pardo, T., Clemente, R., and Bernal, M.P. (2018). Arsenic adsorption and plant availability in an agricultural soil irrigated with As-rich water: effects of Fe-rich amendments and organic and inorganic fertilisers. *Journal of Environmental Management* 209: 262–272.

Awa, S.H. and Hadibarata, T. (2020). Removal of heavy metals in contaminated soil by phytoremediation mechanism: a review. *Water, Air, & Soil Pollution* 231 (2): 47.

Awasthi, S., Chauhan, R., Dwivedi, S. et al. (2018). A consortium of alga (*Chlorella vulgaris*) and bacterium (*Pseudomonas putida*) for amelioration of arsenic toxicity in rice: a promising and feasible approach. *Environmental and Experimental Botany* 150: 115–126.

Aysha, M.I.J., Zakir, H.M., Haque, R. et al. (2017). Health risk assessment for population via consumption of vegetables grown in soils artificially contaminated with arsenic. *Archives of Current Research International*: 1–12.

Bang, S., Patel, M., Lippincott, L., and Meng, X. (2005). Removal of arsenic from groundwater by granular titanium dioxide adsorbent. *Chemosphere* 60 (3): 389–397.

Baur, W.H. and Onishi, B.-M.H. (1969). Arsenic. In: *Handbook of Geochemistry* (ed. K.H. Wedepohl), 33-A-1–33-0-5. Berlin: Springer-Verlag.

Bissen, M., Vieillard-Baron, M.M., Schindelin, A.J., and Frimmel, F.H. (2001). TiO_2-catalyzed photooxidation of arsenite to arsenate in aqueous samples. *Chemosphere* 44 (4): 751–757.

Bondada, B.R., Tu, S., and Ma, L.Q. (2004). Absorption of foliar-applied arsenic by the arsenic hyperaccumulating fern (*Pteris vittata* L.). *Science of the Total Environment* 332 (1–3): 61–70.

Boyle, R.W. and Jonasson, I.R. (1973). The geochemistry of arsenic and its use as an indicator element in geochemical prospecting. *Journal of Geochemical Exploration* 2 (3): 251–296.

Bundschuh, J., Bhattacharya, P., Sracek, O. et al. (2011). Arsenic removal from groundwater of the Chaco-Pampean plain (Argentina) using natural geological materials as adsorbents. *Journal of Environmental Science and Health, Part A* 46 (11): 1297–1310.

Bundschuh, J., Litter, M.I., Parvez, F. et al. (2012). One century of arsenic exposure in Latin America: a review of history and occurrence from 14 countries. *Science of the Total Environment* 429: 2–35.

Carrasco, J.A., Armario, P., Pajuelo, E. et al. (2005). Isolation and characterisation of symbiotically effective Rhizobium resistant to arsenic and heavy metals after the toxic spill at the Aznalcollar pyrite mine. *Soil Biology and Biochemistry* 37 (6): 1131–1140.

Cavalca, L., Zanchi, R., Corsini, A. et al. (2010). Arsenic-resistant bacteria associated with roots of the wild *Cirsium arvense* (L.) plant from an arsenic polluted soil, and screening of potential plant growth-promoting characteristics. *Systematic and Applied Microbiology* 33 (3): 154–164.

Čerňanský, S., Urík, M., Ševc, J., and Khun, M. (2007). Biosorption and biovolatilization of arsenic by heat-resistant fungi (5 pp). *Environmental Science and Pollution Research* 14 (1): 31–35.

Chakraborti, D., Basu, G.K., Biswas, B.K. et al. (2001). Characterization of arsenic bearing sediments in Gangetic delta of West Bengal-India. *Arsenic Exposure and Health Effects* 4: 27–52.

Chakraborti, D., Rahman, M.M., Das, B. et al. (2010). Status of groundwater arsenic contamination in Bangladesh: a 14-year study report. *Water Research* 44 (19): 5789–5802.

Chen, C.H. and Chiou, I.J. (2008). Remediation of heavy metal-contaminated farm soil using turnover and attenuation method guided with a sustainable management framework. *Environmental Engineering Science* 25 (1): 11–32.

Chen, C.C. and Chung, Y.C. (2006). Arsenic removal using a biopolymer chitosan sorbent. *Journal of Environmental Science and Health, Part A* 41 (4): 645–658.

Cheng, R.C., Liang, S., Wang, H.C., and Beuhler, M.D. (1994). Enhanced coagulation for arsenic removal. *Journal-American Water Works Association* 86 (9): 79–90.

Choong, T.S., Chuah, T.G., Robiah, Y. et al. (2007). Arsenic toxicity, health hazards and removal techniques from water: an overview. *Desalination* 217 (1–3): 139–166.

Chowdhury, U.K., Biswas, B.K., Chowdhury, T.R. et al. (2000). Groundwater arsenic contamination in Bangladesh and West Bengal, India. *Environmental Health Perspectives* 108 (5): 393–397.

Das, S., Jean, J.S., Chou, M.L. et al. (2016). Arsenite-oxidizing bacteria exhibiting plant growth promoting traits isolated from the rhizosphere of *Oryza sativa* L.: implications for mitigation of arsenic contamination in paddies. *Journal of Hazardous Materials* 302: 10–18.

Dey, U., Chatterjee, S., and Mondal, N.K. (2016). Isolation and characterization of arsenic-resistant bacteria and possible application in bioremediation. *Biotechnology Reports* 10: 1–7.

Du, J., Jing, C., Duan, J. et al. (2014). Removal of arsenate with hydrous ferric oxide coprecipitation: effect of humic acid. *Journal of Environmental Sciences* 26 (2): 240–247.

Dushenkov, V., Kumar, P.N., Motto, H., and Raskin, I. (1995). Rhizofiltration: the use of plants to remove heavy metals from aqueous streams. *Environmental Science & Technology* 29 (5): 1239–1245.

Dutta, P.K., Ray, A.K., Sharma, V.K., and Millero, F.J. (2004). Adsorption of arsenate and arsenite on titanium dioxide suspensions. *Journal of Colloid and Interface Science* 278 (2): 270–275.

Dutta, P.K., Pehkonen, S.O., Sharma, V.K., and Ray, A.K. (2005). Photocatalytic oxidation of arsenic (III): evidence of hydroxyl radicals. *Environmental Science & Technology* 39 (6): 1827–1834.

EPA (2000). Technologies and Costs for Removal of Arsenic from Drinking Water. National Washington, DC: Service Center for Environmental Publications.

Erakhrumen, A.A. and Agbontalor, A. (2007). Phytoremediation: an environmentally sound technology for pollution prevention, control and remediation in developing countries. *Educational Research Review* 2 (7): 151–156.

de Esparza, M.C. (2009). The presence of arsenic in drinking water in Latin America and its effect on public health. In: *Natural Arsenic in Groundwater of Latin America*, vol. 1, 17–29. The Netherlands: CRC Press/Balkema Publisher Leiden.

European Food Safety Authority (2009). Scientific opinion on arsenic in food. *EFSA Journal* 7 (10): 1351.

Eze, V.C. and Harvey, A.P. (2018). Extractive recovery and valorisation of arsenic from contaminated soil through phytoremediation using *Pteris cretica*. *Chemosphere* 208: 484–492.

Fayiga, A.O. and Saha, U.K. (2016). Arsenic hyperaccumulating fern: implications for remediation of arsenic contaminated soils. *Geoderma* 284: 132–143.

Ferguson, M.A. and Hering, J.G. (2006). TiO_2-photocatalyzed As (III) oxidation in a fixed-bed, flow-through reactor. *Environmental Science & Technology* 40 (13): 4261–4267.

Fitz, W.J. and Wenzel, W.W. (2002). Arsenic transformations in the soil–rhizosphere–plant system: fundamentals and potential application to phytoremediation. *Journal of Biotechnology* 99 (3): 259–278.

Fontcuberta, M., Calderon, J., Villalbí, J.R. et al. (2011). Total and inorganic arsenic in marketed food and associated health risks for the Catalan (Spain) population. *Journal of Agricultural and Food Chemistry* 59 (18): 10013–10022.

Franchi, E., Rolli, E., Marasco, R. et al. (2017). Phytoremediation of a multi contaminated soil: mercury and arsenic phytoextraction assisted by mobilizing agent and plant growth promoting bacteria. *Journal of Soils and Sediments* 17 (5): 1224–1236.

García-Sánchez, M., Košnář, Z., Mercl, F. et al. (2018). A comparative study to evaluate natural attenuation, mycoaugmentation, phytoremediation, and microbial-assisted phytoremediation strategies for the bioremediation of an aged PAH-polluted soil. *Ecotoxicology and Environmental Safety* 147: 165–174.

Garelick, H., Dybowska, A., Valsami-Jones, E., and Priest, N. (2005). Remediation technologies for arsenic contaminated drinking waters (9 pp). *Journal of Soils and Sediments* 5 (3): 182–190.

Genç, H., Tjell, J.C., McConchie, D., and Schuiling, O. (2003). Adsorption of arsenate from water using neutralized red mud. *Journal of Colloid and Interface Science* 264 (2): 327–334.

Genç-Fuhrman, H., Tjell, J.C., and McConchie, D. (2004). Increasing the arsenate adsorption capacity of neutralized red mud (Bauxsol). *Journal of Colloid and Interface Science* 271 (2): 313–320.

Ghurye, G. and Clifford, D. (2004). As (III) oxidation using chemical and solid-phase oxidants. *Journal-American Water Works Association* 96 (1): 84–96.

Glick, B.R. (2010). Using soil bacteria to facilitate phytoremediation. *Biotechnology Advances* 28 (3): 367–374.

Gosio, B. (1892). Action of microphytes on solid compounds of arsenic: a recapitulation. Science, 19(472), pp.104–106.

Govarthanan, M., Mythili, R., Selvankumar, T. et al. (2018). Myco-phytoremediation of arsenic-and lead-contaminated soils by *Helianthus annuus* and wood rot fungi, Trichoderma sp. isolated from decayed wood. *Ecotoxicology and Environmental Safety* 151: 279–284.

Govil, P.K. and Krishna, A.K. (2018). Soil and water contamination by potentially hazardous elements: a case history from India. In: *Environmental Geochemistry* (eds. B. De Vivo, H.E. Belkin and A. Lima), 567–597. Oxford: Elsevier.

Granada, J., Cerda, W., and Godoy, D. (2003). ESSAN: the way to significantly reduce arsenic in drinking water. *AIDIS-Chile Magazine* 34: 44–49.

Granchinho, S.C., Franz, C.M., Polishchuk, E. et al. (2002). Transformation of arsenic (V) by the fungus *Fusarium oxysporum melonis* isolated from the alga *Fucus gardneri*. *Applied Organometallic Chemistry* 16 (12): 721–726.

Gu, Z., Fang, J., and Deng, B. (2005). Preparation and evaluation of GAC-based iron-containing adsorbents for arsenic removal. *Environmental Science & Technology* 39 (10): 3833–3843.

Guan, X., Du, J., Meng, X. et al. (2012). Application of titanium dioxide in arsenic removal from water: a review. *Journal of Hazardous Materials* 215: 1–16.

Guérin, T., Chekri, R., Vastel, C. et al. (2011). Determination of 20 trace elements in fish and other seafood from the French market. *Food Chemistry* 127 (3): 934–942.

Guo, J., Tang, S., Ju, X. et al. (2011). Effects of inoculation of a plant growth promoting rhizobacterium Burkholderia sp. D54 on plant growth and metal uptake by a hyperaccumulator *Sedum alfredii* Hance grown on multiple metal contaminated soil. *World Journal of Microbiology and Biotechnology* 27 (12): 2835–2844.

Gupta, D.K., Srivastava, S., Huang, H.G. et al. (2011). Arsenic tolerance and detoxification mechanisms in plants. In: *Detoxification of Heavy Metals* (eds. I. Sherameti and A. Varma), 169–179. Berlin, Heidelberg: Springer.

Hao, L., Liu, M., Wang, N., and Li, G. (2018). A critical review on arsenic removal from water using iron-based adsorbents. *RSC Advances* 8 (69): 39545–39560.

Hassan, A., Pariatamby, A., Ossai, I.C., and Hamid, F.S. (2020). Bioaugmentation assisted mycoremediation of heavy metal and/metalloid landfill contaminated soil using consortia of filamentous fungi. *Biochemical Engineering Journal* 157: 107550.

Heikens, A. (2006). *Arsenic Contamination of Irrigation Water, Soil and Crops in Bangladesh: Risk Implications for Sustainable Agriculture and Food Safety in Asia*. RAP Publication (FAO).

Hering, J.G., Chen, P.Y., Wilkie, J.A. et al. (1996). Arsenic removal by ferric chloride. *Journal-American Water Works Association* 88 (4): 155–167.

Hong, Y.S., Song, K.H., and Chung, J.Y. (2014). Health effects of chronic arsenic exposure. *Journal of Preventive Medicine and Public Health* 47 (5): 245.

Hossain, F., Bagtzoglou, A.C., Nahar, N., and Delawer Hossain, M. (2006). Statistical characterization of arsenic contamination in shallow tube wells of western Bangladesh. *Hydrological Processes: An International Journal* 20 (7): 1497–1510.

Hsueh, Y.M., Cheng, G.S., Wu, M.M. et al. (1995). Multiple risk factors associated with arsenic-induced skin cancer: effects of chronic liver disease and malnutritional status. *British Journal of Cancer* 71 (1): 109–114.

Hu, C., Liu, H., Chen, G. et al. (2012). As (III) oxidation by active chlorine and subsequent removal of as (V) by Al13 polymer coagulation using a novel dual function reagent. *Environmental Science & Technology* 46 (12): 6776–6782.

Huang, J.H. and Matzner, E. (2006). Dynamics of organic and inorganic arsenic in the solution phase of an acidic fen in Germany. *Geochimica et Cosmochimica Acta* 70 (8): 2023–2033.

Hughes, M.F. (2002). Arsenic toxicity and potential mechanisms of action. *Toxicology Letters* 133 (1): 1–16.

Johnston, R. and Heijnen, H. (2002). Safe Water Technology for Arsenic Removal. Report. Geneva: World Health Organization (WHO).

Johnston, R., Heijnen, H. and Wurzel, P. (2001). Safe Water Technology. United Nations Synthesis Report on Arsenic in Drinking Water. Geneva: World Health Organization, p. 390.

Kabata-Pendias, A. and Pendias, H. (2001). *Trace Elements in Soils and Plants*, 3e. Boca Raton, FL, USA: CRC Press.

Karim, M.M. (2000). Arsenic in groundwater and health problems in Bangladesh. *Water Research* 34 (1): 304–310.

Katsoyiannis, I.A. and Zouboulis, A.I. (2004). Application of biological processes for the removal of arsenic from groundwaters. *Water Research* 38 (1): 17–26.

Kaur, J., Anand, V., Srivastava, S. et al. (2020). Yeast strain *Debaryomyces hansenii* for amelioration of arsenic stress in rice. *Ecotoxicology and Environmental Safety* 195: 110480.

Khalid, S., Shahid, M., Niazi, N.K. et al. (2017). Arsenic behaviour in soil-plant system: biogeochemical reactions and chemical speciation influences. In: *Enhancing Cleanup of Environmental Pollutants* (eds. N.A. Anjum, S. Singh and G. Narendra Tuteja), 97–140. Cham: Springer.

Kile, M.L., Hoffman, E., Rodrigues, E.G. et al. (2011). A pathway-based analysis of urinary arsenic metabolites and skin lesions. *American Journal of Epidemiology* 173 (7): 778–786.

Kim, M.J. and Nriagu, J. (2000). Oxidation of arsenite in groundwater using ozone and oxygen. *Science of the Total Environment* 247 (1): 71–79.

Kim, Y., Kim, C., Choi, I. et al. (2004). Arsenic removal using mesoporous alumina prepared via a templating method. *Environmental Science & Technology* 38 (3): 924–931.

Kong, Z. and Glick, B.R. (2017). The role of plant growth-promoting bacteria in metal phytoremediation. In: *Advances in Microbial Physiology*, vol. 71 (ed. R.K. Poole), 97–132. Cambridge, MA: Academic Press.

Kumar, M., Yadav, A., and Ramanathan, A.L. (2020). Arsenic contamination in environment, ecotoxicological and health effects, and bioremediation strategies for its detoxification. In: *Bioremediation of Industrial Waste for Environmental Safety* (eds. G. Saxena and R.N. Bharagava), 245–264. Singapore: Springer.

Lee, Y., Um, I.H., and Yoon, J. (2003). Arsenic (III) oxidation by iron (VI)(ferrate) and subsequent removal of arsenic (V) by iron (III) coagulation. *Environmental Science & Technology* 37 (24): 5750–5756.

Lee, K.Y., Bosch, J., and Meckenstock, R.U. (2012). Use of metal-reducing bacteria for bioremediation of soil contaminated with mixed organic and inorganic pollutants. *Environmental Geochemistry and Health* 34 (1): 135–142.

Li, H., Chen, X.W., and Wong, M.H. (2016). Arbuscular mycorrhizal fungi reduced the ratios of inorganic/organic arsenic in rice grains. *Chemosphere* 145: 224–230.

Liu, H., Probst, A., and Liao, B. (2005). Metal contamination of soils and crops affected by the Chenzhou lead/zinc mine spill (Hunan, China). *Science of the Total Environment* 339 (1–3): 153–166.

Liu, G., Zhang, X., Talley, J.W. et al. (2008). Effect of NOM on arsenic adsorption by TiO_2 in simulated As (III)-contaminated raw waters. *Water Research* 42 (8–9): 2309–2319.

Liu, S., Zhang, F., Chen, J., and Sun, G. (2011). Arsenic removal from contaminated soil via biovolatilization by genetically engineered bacteria under laboratory conditions. *Journal of Environmental Sciences* 23 (9): 1544–1550.

Liu, S., Hou, Y., and Sun, G. (2013). Synergistic degradation of pyrene and volatilization of arsenic by cocultures of bacteria and a fungus. *Frontiers of Environmental Science & Engineering* 7 (2): 191–199.

Ma, L.Q., Komar, K.M., Tu, C. et al. (2001). A fern that hyperaccumulates arsenic. *Nature* 409 (6820): 579–579.

Maharjan, M., Watanabe, C., Ahmad, S.A., and Ohtsuka, R. (2005). Arsenic contamination in drinking water and skin manifestations in lowland Nepal: the first community-based survey. *The American Journal of Tropical Medicine and Hygiene* 73 (2): 477–479.

Maiti, A., DasGupta, S., Basu, J.K., and De, S. (2008). Batch and column study: adsorption of arsenate using untreated laterite as adsorbent. *Industrial & Engineering Chemistry Research* 47 (5): 1620–1629.

Maity, J.P., Nath, B., Kar, S. et al. (2012). Arsenic-induced health crisis in peri-urban Moyna and Ardebok villages, West Bengal, India: an exposure assessment study. *Environmental Geochemistry and Health* 34 (5): 563–574.

Mallick, I., Islam, E., and Kumar Mukherjee, S. (2015). Fundamentals and application potential of arsenic-resistant bacteria for bioremediation in rhizosphere: a review. *Soil and Sediment Contamination: An International Journal* 24 (6): 704–718.

Mandal, B.K. and Suzuki, K.T. (2002). Arsenic round the world: a review. *Talanta* 58 (1): 201–235.

Mandal, B.K., Chowdhury, T.R., Samanta, G. et al. (1996). Arsenic in groundwater in seven districts of West Bengal, India–the biggest arsenic calamity in the world. *Current Science* 70 (11): 976–986.

Martinez-Huitle, C.A. and Ferro, S. (2006). Electrochemical oxidation of organic pollutants for the wastewater treatment: direct and indirect processes. *Chemical Society Reviews* 35 (12): 1324–1340.

Marwa, N., Singh, N., Srivastava, S. et al. (2019). Characterizing the hypertolerance potential of two indigenous bacterial strains (*Bacillus flexus* and *Acinetobacter junii*) and their efficacy in arsenic bioremediation. *Journal of Applied Microbiology* 126 (4): 1117–1127.

Masscheleyn, P.H., Delaune, R.D., and Patrick, W.H. Jr. (1991). Effect of redox potential and pH on arsenic speciation and solubility in a contaminated soil. *Environmental Science & Technology* 25 (8): 1414–1419.

McArthur, J.M., Ravenscroft, P., Safiulla, S., and Thirlwall, M.F. (2001). Arsenic in groundwater: testing pollution mechanisms for sedimentary aquifers in Bangladesh. *Water Resources Research* 37 (1): 109–117.

McClintock, T.R., Chen, Y., Bundschuh, J. et al. (2012). Arsenic exposure in Latin America: biomarkers, risk assessments and related health effects. *Science of the Total Environment* 429: 76–91.

Meharg, A.A. and Hartley-Whitaker, J. (2002). Arsenic uptake and metabolism in arsenic resistant and nonresistant plant species. *New Phytologist* 154 (1): 29–43.

Miller, S.M., Spaulding, M.L., and Zimmerman, J.B. (2011). Optimization of capacity and kinetics for a novel bio-based arsenic sorbent, TiO_2-impregnated chitosan bead. *Water Research* 45 (17): 5745–5754.

Mohan, D. and Pittman, C.U. Jr. (2007). Arsenic removal from water/wastewater using adsorbents – a critical review. *Journal of Hazardous Materials* 142 (1–2): 1–53.

Mohan, D., Pittman, C.U. Jr., Bricka, M. et al. (2007). Sorption of arsenic, cadmium, and lead by chars produced from fast pyrolysis of wood and bark during bio-oil production. *Journal of Colloid and Interface Science* 310 (1): 57–73.

Mohd, S., Shukla, J., Kushwaha, A.S. et al. (2017). Endophytic fungi *Piriformospora indica* mediated protection of host from arsenic toxicity. *Frontiers in Microbiology* 8: 754.

Mondal, P., Bhowmick, S., Chatterjee, D. et al. (2013). Remediation of inorganic arsenic in groundwater for safe water supply: a critical assessment of technological solutions. *Chemosphere* 92 (2): 157–170.

Mukherjee, A., Das, D., Mondal, S.K. et al. (2010). Tolerance of arsenate-induced stress in *Aspergillus niger*, a possible candidate for bioremediation. *Ecotoxicology and Environmental Safety* 73 (2): 172–182.

Naujokas, M.F., Anderson, B., Ahsan, H. et al. (2013). The broad scope of health effects from chronic arsenic exposure: update on a worldwide public health problem. *Environmental Health Perspectives* 121 (3): 295–302.

Navazas, A., Hendrix, S., Cuypers, A., and González, A. (2019). Integrative response of arsenic uptake, speciation and detoxification by *Salix atrocinerea*. *Science of the Total Environment* 689: 422–433.

Neidhardt, H., Norra, S., Tang, X. et al. (2012). Impact of irrigation with high arsenic burdened groundwater on the soil–plant system: results from a case study in the Inner Mongolia, China. *Environmental Pollution* 163: 8–13.

Nickson, R., McArthur, J., Burgess, W. et al. (1998). Arsenic poisoning of Bangladesh groundwater. *Nature* 395 (6700): 338–338.

Nriagu, J.O. and Pacyna, J.M. (1988). Quantitative assessment of worldwide contamination of air, water and soils by trace metals. *Nature* 333 (6169): 134–139.

Onishi, H. and Sandell, E.B. (1955). Geochemistry of arsenic. *Geochimica et Cosmochimica Acta* 7 (1–2): 1–33.

Pacyna, J.M. (1987). Atmospheric emissions of arsenic, cadmium, lead and mercury from high temperature processes in power generation and industry. In: *Lead, Mercury, Cadmium and Arsenic in the Environment* (eds. T.C. Hutchinson and K.M. Meema), 69–87. Chichester: Wiley.

Pandey, N. and Bhatt, R. (2016). Role of soil associated Exiguobacterium in reducing arsenic toxicity and promoting plant growth in *Vigna radiata*. *European Journal of Soil Biology* 75: 142–150.

Pandey, V.C., Singh, J.S., Singh, R.P. et al. (2011). Arsenic hazards in coal fly ash and its fate in Indian scenario. *Resources, Conservation and Recycling* 55 (9–10): 819–835.

Pandya, M., Rajput, M., and Rajkumar, S. (2015). Exploring plant growth promoting potential of non rhizobial root nodules endophytes of *Vigna radiata*. *Microbiology* 84 (1): 80–89.

Pearce, R.B., Callow, M.E., and Macaskie, L.E. (1998). Fungal volatilization of arsenic and antimony and the sudden infant death syndrome. *FEMS Microbiology Letters* 158 (2): 261–265.

Pettine, M., Campanella, L., and Millero, F.J. (1999). Arsenite oxidation by H_2O_2 in aqueous solutions. *Geochimica et Cosmochimica Acta* 63 (18): 2727–2735.

Pratush, A., Kumar, A., and Hu, Z. (2018). Adverse effect of heavy metals (As, Pb, Hg, and Cr) on health and their bioremediation strategies: a review. *International Microbiology* 21 (3): 97–106.

Qu, D., Wang, J., Hou, D. et al. (2009). Experimental study of arsenic removal by direct contact membrane distillation. *Journal of Hazardous Materials* 163 (2–3): 874–879.

Rahman, H.L., Erdem, H., Sahin, M., and Erdem, M. (2020). Iron-incorporated activated carbon synthesis from biomass mixture for enhanced arsenic adsorption. *Water, Air, & Soil Pollution* 231 (1): 6.

Ramos-Peralonso, M.J. (2014). European food safety authority (EFSA) scientific opinion on dietary reference values for iodine. *EFSA Journal* 12: 1–57.

Ranjan, R., Kumar, N., Dubey, A.K. et al. (2018). Diminution of arsenic accumulation in rice seedlings co-cultured with Anabaena sp.: modulation in the expression of lower silicon transporters, two nitrogen dependent genes and lowering of antioxidants activity. *Ecotoxicology and Environmental Safety* 151: 109–117.

Rosen, B.P. (2002). Biochemistry of arsenic detoxification. *FEBS Letters* 529 (1): 86–92.

Ryu, J., Monllor-Satoca, D., Kim, D.H. et al. (2013). Photooxidation of arsenite under 254 nm irradiation with a quantum yield higher than unity. *Environmental Science & Technology* 47 (16): 9381–9387.

Saha, D. and Sahu, S. (2016). A decade of investigations on groundwater arsenic contamination in Middle Ganga Plain, India. *Environmental Geochemistry and Health* 38 (2): 315–337.

Saha, J.C., Dikshit, K. and Bandyopadhyay, M. (2001, May). Comparative studies for selection of technologies for arsenic removal from drinking water. BUET-UNU International Workshop on Technologies for Arsenic Removal from Drinking Water, Dhaka, Bangladesh.

Saleem, M., Asghar, H.N., Zahir, Z.A., and Shahid, M. (2018). Impact of lead tolerant plant growth promoting rhizobacteria on growth, physiology, antioxidant activities, yield and lead content in sunflower in lead contaminated soil. *Chemosphere* 195: 606–614.

Salt, D.E., Blaylock, M., Kumar, N.P. et al. (1995). Phytoremediation: a novel strategy for the removal of toxic metals from the environment using plants. *Bio/Technology* 13 (5): 468–474.

Salt, D.E., Smith, R.D., and Raskin, I. (1998). Phytoremediation. *Annual Review of Plant Biology* 49 (1): 643–668.

Sancha, A. and Ruiz, G. (1984). Study of the process of arsenic removal from drinking water sources using aluminum salts. *Proceedings XIX Inter-American Congress of Sanitary and Environmental Engineering (AIDIS) Chile, Theme I Potable Water* (Vol. 2, pp. 380–410), Santiago, Chile.

Sanders, J.G., Riedel, G.F., and Osman, R.W. (1994). Arsenic cycling and its impact in estuarine and coastal marine ecosystems. In: *Arsenic in the Environment, Part I: Cycling and Characterization*, J. O. Nriagu, 289–308. Hoboken, NJ: J. Wiley and Sons.

Sarkar, A. and Paul, B. (2016). The global menace of arsenic and its conventional remediation – a critical review. *Chemosphere* 158: 37–49.

Scott, K.N., Green, J.F., Do, H.D., and McLean, S.J. (1995). Arsenic removal by coagulation. *Journal-American Water Works Association* 87 (4): 114–126.

Sele, V., Sloth, J.J., Lundebye, A.K. et al. (2012). Arsenolipids in marine oils and fats: a review of occurrence, chemistry and future research needs. *Food Chemistry* 133 (3): 618–630.

Shahid, M., Dumat, C., Khalid, S. et al. (2017). Foliar heavy metal uptake, toxicity and detoxification in plants: a comparison of foliar and root metal uptake. *Journal of Hazardous Materials* 325: 36–58.

Shahid, M., Dumat, C., Khalid, S. et al. (2019). Foliar uptake of arsenic nanoparticles by spinach: an assessment of physiological and human health risk implications. *Environmental Science and Pollution Research* 26 (20): 20121–20131.

Sharma, V.K. (2007). Disinfection performance of Fe (VI) in water and wastewater: a review. *Water Science and Technology* 55 (1–2): 225–232.

Sharma, V.K. (2007). Ferrate studies for disinfection and treatment of drinking water. In: *Advances in Control of Disinfection By-products in Drinking Water Systems* (eds. A. Nikolaou, L. Rizzo and H. Selcuk), 373–380. New York: Nova Science Publishers.

Sharma, V.K. and Sohn, M. (2009). Aquatic arsenic: toxicity, speciation, transformations, and remediation. *Environment International* 35 (4): 743–759.

Sharma, S., Anand, G., Singh, N., and Kapoor, R. (2017). Arbuscular mycorrhiza augments arsenic tolerance in wheat (*Triticum aestivum* L.) by strengthening antioxidant defense system and thiol metabolism. *Frontiers in Plant Science* 8: 906.

Shukla, A. and Srivastava, S. (2017). Emerging aspects of bioremediation of arsenic. In: *Green Technologies and Environmental Sustainability* (eds. R. Singh and S. Kumar), 395–407. Cham: Springer.

Singh, T.S. and Pant, K.K. (2004). Equilibrium, kinetics and thermodynamic studies for adsorption of As (III) on activated alumina. *Separation and Purification Technology* 36 (2): 139–147.

Singh, R., Singh, S., Parihar, P. et al. (2015). Arsenic contamination, consequences and remediation techniques: a review. *Ecotoxicology and Environmental Safety* 112: 247–270.

Singh, N., Marwa, N., Mishra, J. et al. (2016). *Brevundimonas diminuta* mediated alleviation of arsenic toxicity and plant growth promotion in *Oryza sativa* L. *Ecotoxicology and Environmental Safety* 125: 25–34.

Smedley, P.L. and Kinniburgh, D.G. (2002). A review of the source, behaviour and distribution of arsenic in natural waters. *Applied Geochemistry* 17 (5): 517–568.

Smith, E.R.G., Naidu, R. and Alston, A.M. (1998). Arsenic in the soil environment. Doctoral dissertation. Cambridge, MA: Academic Press.

Sohel, N., Persson, L.Å., Rahman, M. et al. (2009). Arsenic in drinking water and adult mortality: a population-based cohort study in rural Bangladesh. *Epidemiology* 20: 824–830.

Sorlini, S. and Gialdini, F. (2010). Conventional oxidation treatments for the removal of arsenic with chlorine dioxide, hypochlorite, potassium permanganate and monochloramine. *Water Research* 44 (19): 5653–5659.

Sorlini, S., Gialdini, F., and Stefan, M. (2014). UV/H 2 O 2 oxidation of arsenic and terbuthylazine in drinking water. *Environmental Monitoring and Assessment* 186 (2): 1311–1316.

Srivastava, S. and Singh, N. (2014). Mitigation approach of arsenic toxicity in chickpea grown in arsenic amended soil with arsenic tolerant plant growth promoting Acinetobacter sp. *Ecological Engineering* 70: 146–153.

Srivastava, P.K., Vaish, A., Dwivedi, S. et al. (2011). Biological removal of arsenic pollution by soil fungi. *Science of the Total Environment* 409 (12): 2430–2442.

Srivastava, S., Verma, P.C., Chaudhry, V. et al. (2013). Influence of inoculation of arsenic-resistant Staphylococcus arlettae on growth and arsenic uptake in *Brassica juncea* (L.) Czern. Var. R-46. *Journal of Hazardous Materials* 262: 1039–1047.

Sriwana, T., Van Bergen, M.J., Sumarti, S. et al. (1998). Volcanogenic pollution by acid water discharges along Ciwidey River, West Java (Indonesia). *Journal of Geochemical Exploration* 62 (1–3): 161–182.

Su, Y.H., McGrath, S.P., and Zhao, F.J. (2010). Rice is more efficient in arsenite uptake and translocation than wheat and barley. *Plant and Soil* 328 (1–2): 27–34.

Su, S., Zeng, X., Bai, L. et al. (2017). Inoculating chlamydospores of *Trichoderma asperellum* SM-12F1 changes arsenic availability and enzyme activity in soils and improves water spinach growth. *Chemosphere* 175: 497–504.

Tabassum, R.A., Shahid, M., Dumat, C. et al. (2019). Health risk assessment of drinking arsenic-containing groundwater in Hasilpur, Pakistan: effect of sampling area, depth, and source. *Environmental Science and Pollution Research* 26 (20): 20018–20029.

Tripathi, R.D., Srivastava, S., Mishra, S. et al. (2007). Arsenic hazards: strategies for tolerance and remediation by plants. *Trends in Biotechnology* 25 (4): 158–165.

Tripathi, P., Khare, P., Barnawal, D. et al. (2020). Bioremediation of arsenic by soil methylating fungi: role of Humicola sp. strain 2WS1 in amelioration of arsenic phytotoxicity in *Bacopa monnieri* L. *Science of the Total Environment* 716: 136758.

Upadhyay, A.K., Singh, N.K., Singh, R., and Rai, U.N. (2016). Amelioration of arsenic toxicity in rice: comparative effect of inoculation of *Chlorella vulgaris* and Nannochloropsis sp. on growth, biochemical changes and arsenic uptake. *Ecotoxicology and Environmental Safety* 124: 68–73.

Vala, A.K. (2010). Tolerance and removal of arsenic by a facultative marine fungus *Aspergillus candidus*. *Bioresource Technology* 101 (7): 2565–2567.

Vasudevan, S., Mohan, S., Sozhan, G. et al. (2006). Studies on the oxidation of As (III) to As (V) by in-situ-generated hypochlorite. *Industrial & Engineering Chemistry Research* 45 (22): 7729–7732.

Viet, P.H., Con, T.H., Ha, C.T. et al. (2003). Investigation of arsenic removal technologies for drinking water in Vietnam. In: *Arsenic Exposure and Health Effects V* (eds. W.R. Chappell, C.O. Abernathy and D.J. Thomas), 459–469. Amsterdam: Elsevier Science BV.

Vithanage, M., Dabrowska, B.B., Mukherjee, A.B. et al. (2012). Arsenic uptake by plants and possible phytoremediation applications: a brief overview. *Environmental Chemistry Letters* 10 (3): 217–224.

Vurukonda, S.S.K.P., Vardharajula, S., Shrivastava, M., and SkZ, A. (2016). Enhancement of drought stress tolerance in crops by plant growth promoting rhizobacteria. *Microbiological Research* 184: 13–24.

Wan, X., Lei, M., and Chen, T. (2020). Review on remediation technologies for arsenic-contaminated soil. *Frontiers of Environmental Science & Engineering* 14 (2): 1–14.

Wang, Z., Luo, Z., and Yan, C. (2013). Accumulation, transformation, and release of inorganic arsenic by the freshwater cyanobacterium *Microcystis aeruginosa*. *Environmental Science and Pollution Research* 20 (10): 7286–7295.

Watanabe, M.E. (1997). Phytoremediation on the brink of commericialization. *Environmental Science & Technology* 31 (4): 182A–186A.

Wen, W., Wen, J., Lu, L. et al. (2011). Metabolites of arsenic and increased DNA damage of p53 gene in arsenic plant workers. *Toxicology and Applied Pharmacology* 254 (1): 41–47.

WHO (2011). *Guidelines for Drinking-Water Quality*, vol. 4, 315–318. Geneva: World Health Organisation.

World Health Organization (2003). Arsenic in Drinking-Water: Background Document for Development of WHO Guidelines for Drinking-Water Quality (No. WHO/SDE/WSH/03.04/75). Geneva: World Health Organization.

Wu, F., Hu, J., Wu, S., and Wong, M.H. (2015). Grain yield and arsenic uptake of upland rice inoculated with arbuscular mycorrhizal fungi in As-spiked soils. *Environmental Science and Pollution Research* 22 (12): 8919–8926.

Xie, J.J., Hu, X., Shen, Y.W. et al. (2017). Bioavailability and speciation of arsenic in urban street dusts from Baoding city, China. *Chemical Speciation & Bioavailability* 29 (1): 135–142.

Yadav, S.K., Ramanathan, A.L., Kumar, M. et al. (2020). Assessment of arsenic and uranium co-occurrences in groundwater of central Gangetic Plain, Uttar Pradesh, India. *Environmental Earth Sciences* 79 (6): 1–14.

Yazdani, M.R., Tuutijärvi, T., Bhatnagar, A., and Vahala, R. (2016). Adsorptive removal of arsenic (V) from aqueous phase by feldspars: kinetics, mechanism, and thermodynamic aspects of adsorption. *Journal of Molecular Liquids* 214: 149–156.

Yoon, S.H. and Lee, J.H. (2005). Oxidation mechanism of As (III) in the UV/TiO_2 system: evidence for a direct hole oxidation mechanism. *Environmental Science & Technology* 39 (24): 9695–9701.

Zagury, G.J., Samson, R., and Deschênes, L. (2003). Occurrence of metals in soil and ground water near chromated copper arsenate-treated utility poles. *Journal of Environmental Quality* 32 (2): 507–514.

Zarrin, B. and Azra, Y. (2018). Genome editing weds CRISPR: what is in it for phytoremediation? *Plants* 7 (3): 51.

Zeng, D., Wu, J., and Kennedy, J.F. (2008). Application of a chitosan flocculant to water treatment. *Carbohydrate Polymers* 71 (1): 135–139.

Zhang, H., Dang, Z., Zheng, L.C., and Yi, X.Y. (2009). Remediation of soil co-contaminated with pyrene and cadmium by growing maize (*Zea mays* L.). *International Journal of Environmental Science and Technology* 6 (2): 249–258.

Zhao, F.J., McGrath, S.P., and Meharg, A.A. (2010). Arsenic as a food chain contaminant: mechanisms of plant uptake and metabolism and mitigation strategies. *Annual Review of Plant Biology* 61: 535–559.

15

Physical and Chemical Methods for Heavy Metal Removal

Monika Yadav[1]*, Gurudatta Singh*[2]*, and R.N. Jadeja*[1]

[1] *Department of Environmental Studies, Faculty of Science, The Maharaja Sayajirao University of Baroda, Vadodara, Gujarat, India*
[2] *Institute of Environment and Sustainable Development, Banaras Hindu University, Varanasi, Uttar Pradesh, India*

15.1 Introduction

The group of metal and metalloids having a density greater than $4 \pm 1\,g/cm^3$ are collectively known as heavy metals and are broadly considered as global contaminants of the environment (Duffus 2002). These metals naturally occur in the Earth's crust and they are commonly found in sediments, soils, rocks, water, microorganisms, etc., which have natural background concentrations. They are persistent in nature and cannot be destroyed; some broadly discussed heavy metals are copper, silver, zinc, cadmium, gold, mercury, lead, chromium, iron, nickel, tin, arsenic, selenium, molybdenum, cobalt, manganese, and aluminum (Mohammed et al. 2011). Heavy metals are useful in many ways: cobalt acts as a central atom in the vitamin B12 complex and plays a crucial role in the reduction of propionic acid during fermentation (Stowers et al. 2014). Copper, zinc selenium, etc. are essential trace elements that carry out several biological processes and also operate human metabolism (Mertz 1981). Technologically useful metals are zinc, iron, tin, lead, tungsten, copper, etc. which have been utilized in industry. Some artificially designed bioinorganic catalysts utilize heavy metals as the central atom for chemical transformations (Terfassa et al. 2014). Most of the heavy metals represent the "dark side of chemistry," such as cadmium, mercury, arsenic, chromium, lead, and thallium, because they show destructive effects, even at lower concentrations (Duruibe et al. 2007). Globalization, industrialization, uncontrollable urbanization, and a rapidly increasing population will all cause trouble for humans.

Heavy metals like tin, lead, nickel, etc. are released from printed circuit board (PCB) manufacturing centers in considerable quantities. Cadmium, zinc, lead, chromium, nickel, copper, vanadium, platinum, silver, titanium, etc. heavy metals are produced by electroplating, conversion-coating, anodizing-cleaning, electrolysis depositions, milling, and etching industries. Chromate copper-arsenate wood treatment utilized in wood processing

Pollutants and Water Management: Resources, Strategies and Scarcity, First Edition. Edited by Pardeep Singh, Rishikesh Singh, Vipin Kumar Singh, and Rahul Bhadouria.

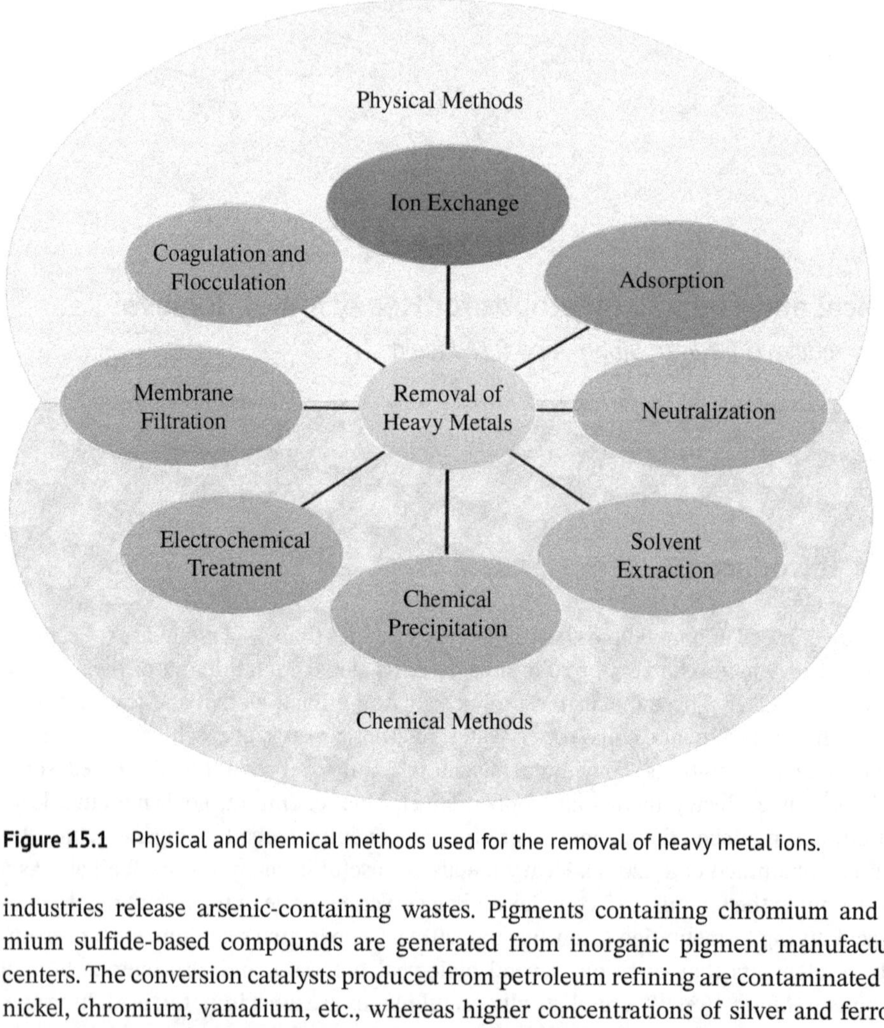

Figure 15.1 Physical and chemical methods used for the removal of heavy metal ions.

industries release arsenic-containing wastes. Pigments containing chromium and cadmium sulfide-based compounds are generated from inorganic pigment manufacturing centers. The conversion catalysts produced from petroleum refining are contaminated with nickel, chromium, vanadium, etc., whereas higher concentrations of silver and ferrocyanide containing films are released by photographic operations. The waste products generated by several industrial sectors produce a huge quantity of sludge, residues, etc. that can be hazardous in nature, thus, efficient wastewater treatment is required to eliminate them from the environment (Figure 15.1) (Sörme and Lagerkvist 2002).

15.2 Toxicity of Heavy Metals

Aquatic organisms are unfavorably affected by the presence of heavy metals in the environment. Water chemistry and ingredients deposited in surface water reflects its toxicity. Microorganisms mineralize metals and are taken up by plankton and enter into the phase of biomagnification, passing toward higher trophic levels of ecosystems and affecting aquatic organisms physiologically and morphologically. Heavy metals adversely affect human health including impairment in growth development, nervous system disorders, etc.

Table 15.1 Maximum contaminant level (MCL) standards for the most hazardous heavy metals.

S. no.	Heavy metal	MCL standard (mg/l)	Toxicities
1	As	0.05	Skin manifestations, visceral cancers, vascular disease
2	Cr	0.05	Headache, diarrhea, nausea, vomiting, carcinogenic
3	Cd	0.01	Kidney damage, renal disorder, human carcinogen
4	Ni	0.20	Dermatitis, nausea, chronic asthma, coughing, human carcinogen
5	Zn	0.80	Depression, lethargy, neurological signs, increased thirst
6	Pb	0.06	Damage to the fetal brain, diseases of the kidneys, circulatory system, and nervous system
7	Cu	0.25	Liver damage, Wilson disease, insomnia

Source: Modified from Barakat and Schmidt (2010).

Autoimmune disorders like rheumatoid arthritis, kidney diseases, etc. are caused by mercury and lead. Heavy metal ions can be carcinogenic (like hexavalent chromium, arsenic, cobalt, nickel, antimony, vanadium, mercury) as well as mutagenic (arsenic, vanadium) causing skin cancer, lung cancer, whitening of hairs, etc. Some of them are teratogenic (arsenic), endocrine-disrupting (silver, copper, zinc, selenium), and allergenic (nickel).

Thallium can cause neurological and behavioral alterations mainly in children and can also cause heart rhythm disturbances, central nervous system damage caused by mercury, lead, thallium, manganese, and tin, which can be irreversible. Cadmium causes bone marrow damage and osteoporosis, whereas hepatotoxic, nephrotoxic effects are generated by cadmium, mercury, and heptavalent manganese. Lead can adversely affect the immune system (Tchounwou et al. 2012). In order to minimize human and environmental exposure to hazardous chemicals, wastewater regulations have been established. The United States Environmental Protection Agency (USEPA) established maximum contaminant level (MCL) limits and toxicity generated due to the release of heavy metals, which are summarized in Table 15.1 (Barakat and Schmidt 2010).

The toxicity of lead is known in today's era and the environment should have a low allowable level because it is an accumulative poison. The report by Mohammed et al. (2011) showed that in Minamata Bay, Japan, sewage containing mercury was disposed of by Chisso Corporation. The mercury accumulates in sea creatures and via biomagnification eventually leads to mercury poisoning in the population. Unusual symptoms have been experienced by residents of Minamata like vision problems, numbness, and convulsions, which have led to the death of several hundred people – the reason was the consumption of mercury poisoned fish. According to the report of Rai and Pal (2002), an unusual and painful disease of a rheumatic nature was recorded in 1947 from a village on the bank of the Jintsu river, Japan, with 44 patients. Later on, it was termed as "itai-itai" disease due to the shrieking results from patients with painful skeletal deformities. The islands of Hokkaido witnessed fatal incidents of lung cancer due to the inhalation of dust containing Cr in higher oxidation states. This was associated with malignant growth in the respiratory tract and painless perforation in the nasal septum.

The greatest nonnuclear disaster of Europe occurred in 1986 when a huge chemical spill from a plant releases around 30 t of waste, including fungicides with mercury in the Rhine. It has killed almost 500 000 fish and governments had to close water systems in France, Germany, and the Netherlands. The Coto de Doñana nature reserve was contaminated with toxic chemicals due to the bursting of a dam that belonged to a mine in Spain. Around 5 million m^3 of mud with lead, copper, zinc, and cadmium flowed directly into the Rio Guadiamar. The groundwater of Ganges, Mekong, and Red River has been adversely affected due to arsenic contamination in a vast area of West Bengal (India) and Bangladesh (Mohammed et al. 2011).

15.3 Methods

Removal of heavy metals from inorganic effluents can be accomplished by several conventional treatment methods such as chemical precipitation, coagulation and flocculation, adsorption, ion exchange method, solvent extraction method, neutralization, membrane separation methods, and electrochemical methods (Figure 15.2). The various treatment strategies and methodologies exercised for heavy metal removal are discussed in the following sections.

Figure 15.2 Methods used for cleaning polluted water.

15.3.1 Physical Methods for Heavy Metal Removal

15.3.1.1 Membrane Filtration

In wastewater treatment, various types of porous and thin-film composite membranes are used. There are different types of driving forces regulating membrane processes such as low pressure (microfiltration, ultrafiltration, and distillation), high pressure (nanofiltration, reverse osmosis), and osmotic pressure (forward osmosis, electrodialysis, liquid membrane, etc.). Principle factors applied to the application of the membrane process are pore size, pore distribution, degree of hydrophilicity, surface charge, solution flow, and presence of functional groups. These factors play a crucial role in the interpretation of the overall membrane process of water production rate and heavy metal removal efficiency. The performance of the membrane is generally determined by its flux rate and selectivity. Cellulose acetate, polyamide, polysulfone, etc. are porous membranes used in this process where solution flow under pressure and withdrawing the membrane releases water at atmospheric pressure. For reverse osmosis (RO), the applied pressure must always be greater than osmotic pressure for sufficient water flux (Peters and Shem 1993).

Membrane separation technology is governed by steric hindrance mechanisms or size exclusion, Donnan exclusion (charge–charge repulsion) effect, and adsorptive ability toward the contaminants. Membrane filtration has received appreciable recognition for the treatment of inorganic effluents such as heavy metals, and it can also eliminate organic compounds and suspended solids (Abdullah et al. 2019). This technology has gathered more attention and has become a favorable approach to heavy metal removal due to its cost-effectiveness and high efficiency (Liu et al. 2001). For selective separation of heavy metals, multiple membrane processes have been developed to reduce the cost and pollution load. Metal ions like Cu^{2+}, Ni^{2+}, Cr(VI), and As(V) have been successfully removed by RO and nanofiltration (NF) (Chan and Dudeney 2008; Muthukrishnan and Guha 2008; Sudilovskiy et al. 2008; Taghizadeh et al. 2013).

The removal of metal ions through ultrafiltration (UF) can be accomplished with the help of chemicals during the separation process, they are known as micellar-enhanced UF (MEUF) and polymer-enhanced UF (PEUF) (Verbych et al. 2005; Barakat and Schmidt 2010; Huang et al. 2016). MEUF is a type of physicochemical membrane separation method that adds anionic surfactant into wastewater which agglomerates and forms micelles. The electrostatic force is the main factor through which heavy metals are trapped in the outer part of the micelle and remaining untrapped pass through the ultrafiltration membrane. The advantages of this method include high removal efficiency, high flux, and low energy cost, whereas the disadvantage is the high operating cost. Water-soluble polymers in wastewater are used in PEUF, as they help in the elimination of small solute molecules that cannot be removed by conventional ultrafiltration (Huang et al. 2010; Landaburu-Aguirre et al. 2010).

Ultrafiltration uses a permeable membrane to eliminate heavy metals, macromolecules, and suspended solids from an inorganic solution based on the molecular weight of eliminating compounds (1000–100 000 Da) and pore size of a membrane (5–20 nm). It can attain a removal efficiency of 90% (Vigneswaran et al. 2004). According to the study of Kosarek (1981), an integrated UF process with a chelating ligand successfully achieved the removal of As(V), Cd(II), Pb(II), Hg(II), Ni(II), and Se(II) from electroplating wastewater. Sato et al. (1977) showed the effective utilization of the reverse osmosis membrane to remove

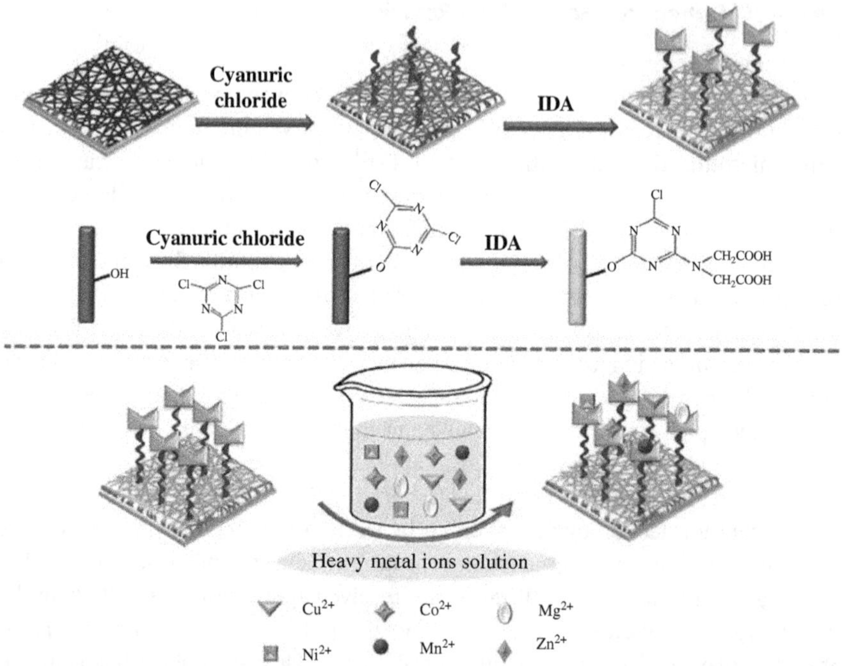

Figure 15.3 Membrane filtration method produced by Dutta and De (2017). (Used with permission from the Royal Society of Chemistry.)

chromium and cyanide from metal-plating wastewater (Sato et al. 1977). According to the study of Bhattacharyya et al. (1978), the charged ultrafiltration membrane was used for the elimination of Cu(II), Ni(II), and Zn(II) ions from electroplating water (Bhattacharyya et al. 1978). A nanofiber membrane based on poly(vinyl alcohol-*co*-ethylene) synthesized by activating with cyanuric chloride and covalently connecting with iminodiacetic acid demonstrated enhanced removal of copper, zinc, nickel, and cobalt (Figure 15.3) (Dutta and De 2017).

15.3.1.2 Coagulation and Flocculation

Coagulation/flocculation can remove heavy metals from solutions. Coagulation reduces the net surface charge of colloidal particles to stabilize through the electrostatic repulsion process; in short, it refers to the charge neutralization of particles. Coagulation requires coagulants and/or dissolved organic matter into larger forms. The coagulants used in this process are aluminum sulfate (alum), polyaluminium chloride (PACL), magnesium chloride (MgCl2), aluminum hydroxide oxides, polyethyleneimine (PEI), etc. (Pang et al. 2011). Flocculation slowly mixes the destabilized particles to increase particle size through additional collisions and interactions with inorganic polymers and dissolved organic polymers. Once the smaller particles are agglomerated into larger particles, they can be easily separated by filtration, flotation, or straining (Tripathy and De 2006). Chitosan is a biopolymer that can be easily biodegraded and act as a coagulating and flocculating agent (Yang and Zall 1984).

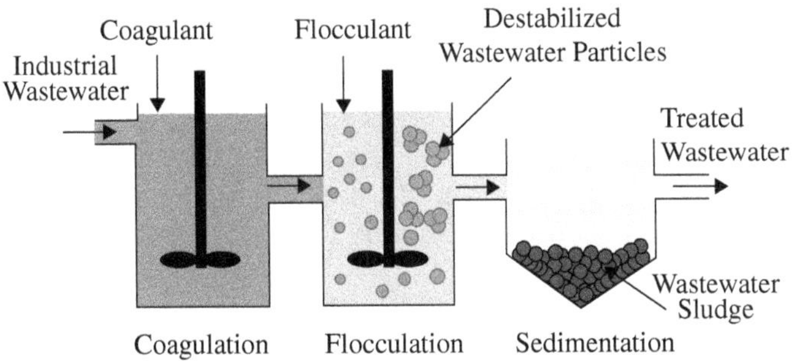

Figure 15.4 Coagulation and flocculation for wastewater treatment produced by Teh et al. (2016). (Used with permission from *Industrial & Engineering Chemistry Research.*)

The coagulation-flocculation process has been efficiently used in the treatment of drinking water and decontamination of industrial effluents (Figure 15.4) (Teh and Wu 2014; Teh et al. 2016). According to an investigation report of the US EPA (US Environmental Protection Agency 1978), utilizing lime softening and coagulation (ferric sulfate or alum) was successful in eliminating lead, chromium, and cadmium up to 98% in the pH range of 8.5–11.3. Removal of cadmium enhances as pH increases; 97% removal of lead has been investigated by using ferric sulfate and alum coagulation together from river water having 0.15 mg/l lead. Ferric sulfate coagulation accomplished 98% of chromium removal at the pH range 6.5–9.3 (Peters and Shem 1993). It was reported that 99% lead was removed at an optimum pH of 6.2–7.8 for alum, 8.7–10.9 for MgCl2, and 8–9.3 for PACL (Pang et al. 2011). $FeCl_3$ was used as a coagulant to increase the removal of metals like copper, lead, zinc, and nickel from synthetic wastewater (Johnson et al. 2008).

15.3.1.3 Ion Exchange

Ion exchange is a cost-effective method that involves inexpensive materials and is easy to operate, which attracts soluble ions from the liquid phase to the solid phase. It has been proved as an efficient method for eliminating heavy metals from aqueous solutions, mainly in low concentrations of metals. The method is sensitive to pH of the aqueous phase (Hamdaoui 2009). A special ion exchanger (naturally occurring inorganic zeolites or synthetic organic ion exchange resins) with cations or anions is used to eliminate heavy metal ions in the solution (Dizge et al. 2009). This is a reversible chemical reaction in which the exchange of ion occurs from a solution to an immobile solid particle. Organic ion exchange resins have high molecular weight poly-electrolytes to exchange the mobile ions sites that set the maximum amount of exchange per unit resin. Synthetic resins are generally used as they can be fabricated into specific applications, and they are manufactured by the polymerization of organic compounds into a porous 3D structure. These ion exchange resins are a solid substance that is insoluble in water and absorb positively or negatively charged ions from an electrolyte solution, which then releases other ions of the same charge into the solution in an equivalent amount (Yang et al. 2001).

Organic polymers are used for the production of anionic and cationic resins having a matching fundamental structure but different ionizable groups adhered to it. Resins are classified into strong or weak acid/base cation/anion exchangers. The ion exchange method

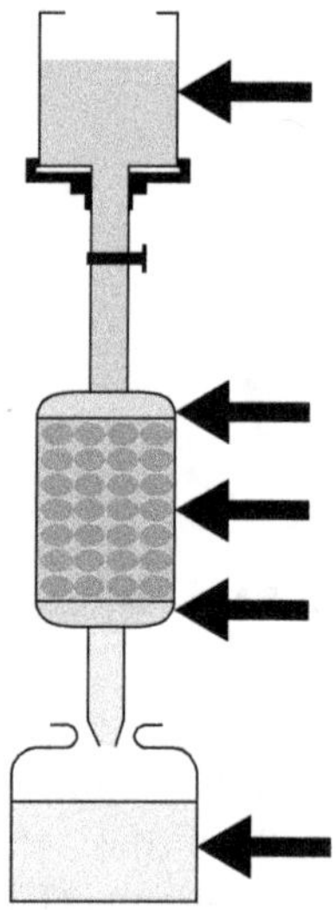

Figure 15.5 Column ion exchange used for heavy metal removal produced by Lalmi et al. (2018). (Used with permission from *Hydrometallurgy*.)

has been conveniently applied by industries for heavy metal removal from effluents. Since the matrix is sensitive, this method cannot be used in the elimination of high concentration metal ions (Kabata-Pendias and Pendias 1999). Column ion-exchange experiments have demonstrated the successful removal of lead by cation exchange resin in calcium (Figure 15.5) (Lalmi et al. 2018). Heavy metals like Cd, Cr, and Zn have been successfully removed by the liquid-ion exchange (LIE) method. The ion exchange process can eliminate or recover most of the metal ions effectively, but a higher cost input restricts its commercialization. Metal finishing operations utilize plating chemicals, acids, and bases that are generally ionized in a solution that can be easily removed by the ion exchange method. This method can also be used as the last stage method (end of the pipe) in pollution control in which metal cyanide complexes or toxic heavy metals are eliminated from the waste stream before discharging them for further use (Gupta et al. 1984).

Various types of resins are used for eliminating heavy metals with the help of the ion exchange method. According to the report of Badawy et al. (2009), Purolite C100 resin has been used for the successful elimination of lead ions from the binary mixture (Badawy et al. 2009). High-density ion exchange resin INDION 225H has been used by Thakare and Jana (2015) for the removal of copper from wastewater (Thakare and Jana 2015). Shaidan et al. (2012) removed 97% of nickel from wastewater using a strongly acidic cation exchange resin in a fixed bed (Shaidan et al. 2012). According to the study of Zewail and Yousef (2015), a spouted bed with AMBERJET 1200 Na resin has been used for the successful removal of lead (99%) and nickel (98%) (Zewail and Yousef 2015).

15.3.1.4 Adsorption

The load of heavy metal ions is increasing day by day and effluents released from industries contain more than the permissible limit of metal ions. Thus, there is an extreme need to eliminate these heavy metals from contaminated water using suitable methods. Adsorption is a surface phenomenon to remove contaminants/adsorbate from wastewater or solution. In this process, impurities stay adhered to the outer surface of the adsorbent rather than

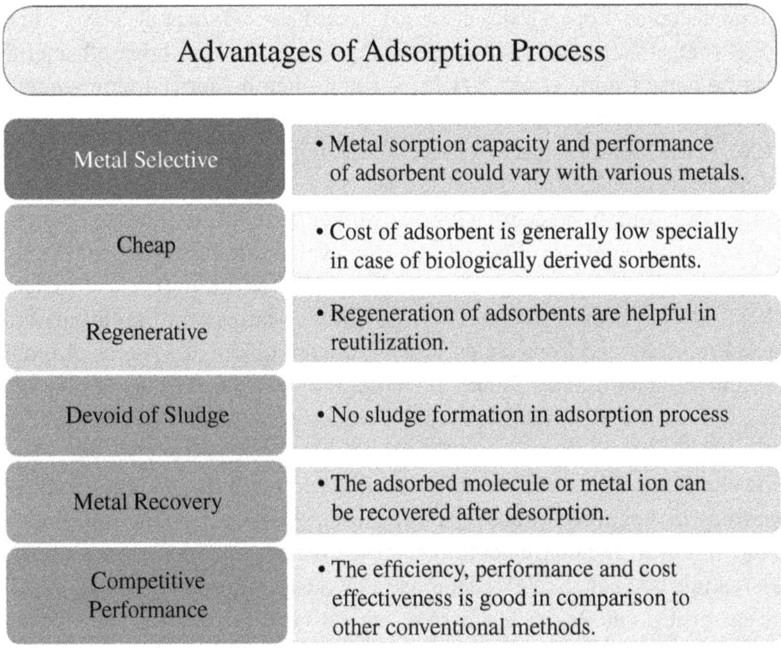

Figure 15.6 Advantages of the adsorption method.

entering into the internal crystal structure (Peters and Shem 1993). It is a feasible method that can be applied in most natural physical, chemical, and biological systems and its wide application has been observed in every domain of our daily life. In this process, the primarily adsorbed ion layer is held tightly, and the counter-ion layer is held a bit looser. Broadly, adsorption is a mass transfer method in which impurities are bounded to a solid surface with physical or chemical interactions, mainly van der Waals forces of attraction or hydrogen bonding (Abdel-Raouf and Abdul-Raheim 2017). Adsorption is a more acceptable approach in comparison with other methods like ozonation, ion exchange, membrane, and oxidative processes, among many others (Yadav et al. 2019). There are several advantages of adsorption over other conventional methods: they minimize biological and chemical sludge and they are cost-effective, eco-friendly, and highly efficient (Figure 15.6). Regeneration of adsorbents and recovery of metals is also possible in this process (Gunatilake 2015).

The mechanism of this method generally depends on solid–liquid equilibrium and mass transfer rates (Shah et al. 2009). Adsorbents are of various types and cost-effective in agriculture waste, natural material, industrial by-products, surface modified magnetic adsorbent, modified biopolymers, etc., and have been utilized for the elimination of heavy metals from metal-contaminated water (Sekhar et al. 2003). For the removal of some organic pollutants, activated carbon has been used in wastewater treatment (Silva et al. 2010). There are several studies revealing adsorption of metal ions using adsorbents such as photocatalyst beads (Idris et al. 2010), red mud (Gupta et al. 2001), coal (Wang et al. 2006), fertilizer industrial waste (Gupta et al. 2010), nanoparticles (Liu et al. 2010), activated sludge biomass (Wu et al. 2010), zeolite (Taamneh and Sharadqah 2017), etc. Some agricultural waste products like orange peel pith (López-Téllez et al. 2011), sunflower (Jain et al. 2010), water hyacinth, hazelnuts shells (Kobya 2004), coconut coir pith, fungal bioadsorbent, bacterial bioadsorbent, marine algae (Mane et al. 2011), plant and microbe derived biomass, etc. have been studied for the removal of heavy metals (Renu et al. 2017). According to a report by Kumar et al. (2014), the literature has investigated utilizing waste materials as an adsorbent for eliminating heavy metals (Kumar et al. 2014).

Since heavy metal contamination occurs with many other organic and inorganic pollutants, which can show an antagonistic effect on their removal, a suitable adsorbent should be selected with the multi-component system to facilitate the removal of heavy metals. Coal-fired fly ash (CFFA) has been investigated as a potential adsorbent for the elimination of cadmium, nickel, and copper from single and multi-contaminated systems (Kumar et al. 2019). Several investigations have shown the removal of heavy metals using sawdust. According to the study of Yu et al. (2003), maple sawdust was used for the removal of Cr(VI) with a removal efficiency of more than 80% (Yu et al. 2003). The study of Božić et al. (2009) showed the elimination of copper, nickel, zinc, and cadmium using beech sawdust, whereas another study of Božić et al. (2009) revealed that sawdust of poplar and linden trees exhibited the removal of zinc, cadmium, copper, nickel, and manganese (Božić et al. 2009, 2013). A study by Li et al. (2007) demonstrated the adsorption of Cr, Cu, and Pb using poplar tree sawdust (Li et al. 2007). The tree fern native to Taiwan was used for the removal of Zn (7.6 mg/g), Cu (10.6 mg/g), and Pb (39.8 mg/g) according to the report of Ho et al. (2002). These adsorption capacities were attained due to the negatively charged cellulose-based structure of tree fern that exhibits a strong affinity for metal ions (Ho et al. 2002).

Cellulose-graft-polyacrylamide/hydroxyapatite composite hydrogel was reported by Saber-Samandari et al. (2013), which successfully removed copper ion from an aqueous solution (Saber-Samandari et al. 2013). Disulfide-linked polymer networks have been used for the removal of lead, zinc, and copper via adsorption (Figure 15.7) (Ko et al. 2017). The study by Wang et al. (2011) demonstrated the amino-functionalized Fe_3O_4/SiO_2 core-shell

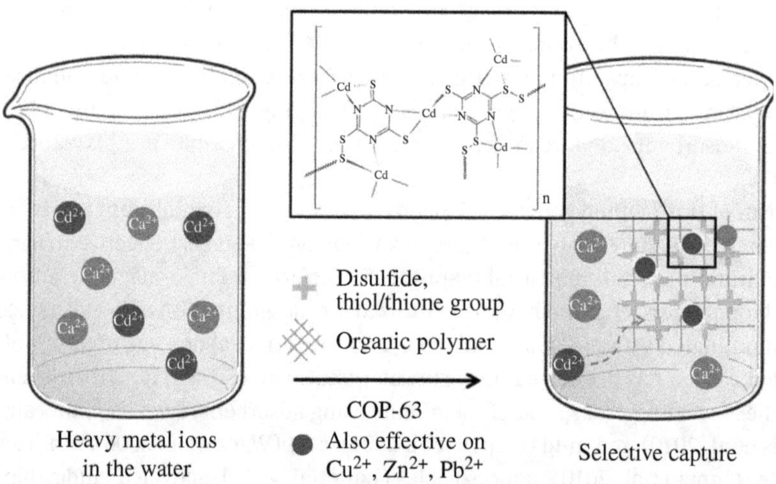

Figure 15.7 Adsorption method for removal of lead, zinc, and copper produced from Ko et al. (2017). (Used with permission from the *Journal of Hazardous Materials*.)

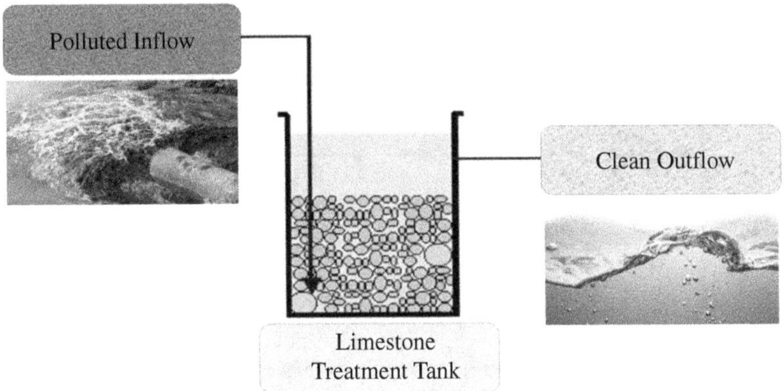

Figure 15.8 Neutralization method for the removal of contaminants.

nanoadsorbent used for the removal of copper, cadmium, and lead (Wang et al. 2011). The adsorption of Co, Ni, Zn, As, Ag, Cd, Hg, and TI was observed by Warner et al. (2012) using manganese dopant Fe_3O_4 nanoparticles (Warner et al. 2012).

15.4 Chemical Methods for Heavy Metal Removal

15.4.1 Neutralization

The removal of metals deposited in sediments has been accomplished by using the mechanism of precipitation-neutralization (Figure 15.8). Some of the common alkaline reagents used to increase the pH of acid water to basic for the removal of heavy metals via precipitation are limestone, ferrous salts, lime, and other compounds like $Mg(OH)_2$, $MgCO_3$, $BaCl_2$, and $CaCl_2$. Apatite has been used in the neutralization of acid mine wastewater for the complete removal of heavy metals, whereas the phosphorous-rich precipitate accumulated with microelements like iron, manganese, copper, magnesium, etc. is utilized as fertilizer. The water undergoes treatment with apatite, which can also be recycled (Ghirişan et al. 2007). Different kinds of reagents have been evaluated in the neutralization step of acid effluent, some of them are calcite (Papassiopi et al. 1996), soluble silicate and kiln dust (Conner and Hoeffner 1998), caustic soda, lime, limestone, FMBO (1 : 1)-diatomite1, Mn and Fe solutions, dolomite and fly ash, Portland cement, phosphate, and magnesium oxide. In comparison with other limes, calcium oxide (CaO) demonstrates the most economical, cost-effective, and efficient alternative use in the neutralization of cations in the form of hydroxides or complex anions. It is also insensitive toward temperature variation. The neutralization starts with the hydrolysis of lime then partial dissolution leads to the formation of hydroxyl ions due to the increase in pH. The results show a 97% removal of Pb and 98% removal of Sb and SO_4 (Balladares et al. 2018).

15.4.2 Solvent Extraction

One of the most acceptable and established processes in hydrometallurgy is solvent extraction, which has been used in the successful hydrometallurgical processing of cobalt, nickel, copper, zinc, uranium, molybdenum, vanadium, tungsten, zirconium, niobium, hafnium,

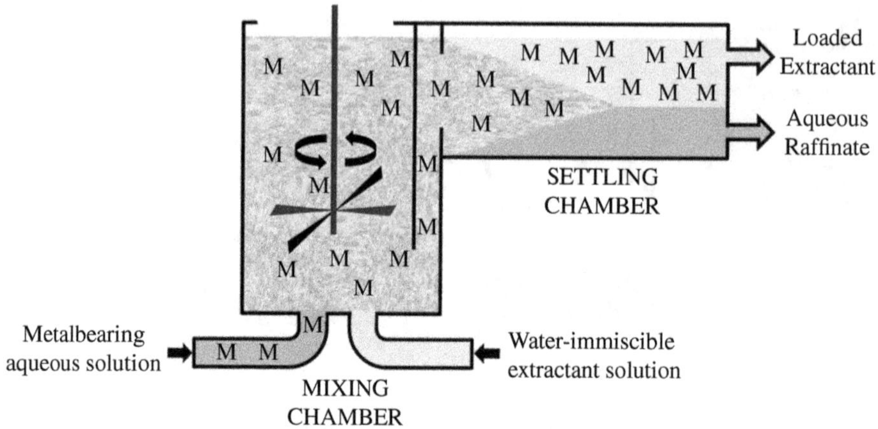

Figure 15.9 Schematic representation of a mixer-settler for continuous operation of solvent extraction produced by Wilson et al. (2014). (Used with permission from the Royal Society of Chemistry.)

indium, germanium, gallium, platinum group metals, boron, reprocessing nuclear fuels, decontamination of wet-process phosphoric acid, nitric acid recovery, etc. In the present scenario, several stable extractants are available that exhibit better selectivity for metal ions following eco-friendly processing routes (Figure 15.9) (Flett 1999; Wilson et al. 2014). In the processing of the metal separation method, organophosphorus has often been used as an extractant. The recovery of zinc and copper has been done successfully from sulfuric acid of low pH using di-(2-ethylhexyl)-phosphoric acid (D2EHPA) as an extractant. It also demonstrates the removal of nickel from aqueous sulfate solutions at pH range 6–7 and the extraction of chromium from aqueous sulfuric acid (Kongolo et al. 2003). Utilization of D2EHPA and Acorga M5640 (a hydroxyoxime derivative) effectively separates nickel from cobalt at lower pH (Sarma and Reddy 2002).

Using D2EHPA as an extractant, precipitation of copper and chromium was done at the initial level and later nickel sulfate was generated from pure raffinate solvent extraction aqueous solution via the salt crystallization method. For better stripping efficiency, the modifier has been used, in the case of chromium, the stripping mixture has the composition of NaOH and H_2O_2, while for other metals, sufficient H_2SO_4 or sulfate salt solution will be good enough for stripping (Reddy and Sarma 2001). Cyanex 272 is also an organophosphorus used for the extraction of Fe, Zn, Cr, Cu, and Ni from sulfuric and/or sulfate solutions. It also demonstrates the successful separation of cobalt from nickel around 99.4% at pH 6 (Flett 2004; Silva et al. 2005). According to the study of Juang and Huang 2003, zinc extraction has dominated the diffusion at the initial stage of extraction, whereas the limiting step for copper extraction depends on experimental conditions such as the prominent role of aqueous layer diffusion at lower metal concentration, high carrier concentrations, and high aqueous pH (Juang and Huang 2003). The Kelex 100® has been proved as a chelating extrant of Ga (III) and Ge (IV). The selectivity of Kelex 100 is good in a strongly basic solution for the separation of gallium/aluminum. It demonstrates successful metal recovery around 60% at pH 1.5 ± 0.1, 2.0 ± 0.1, and 3.1 ± 0.1 for cadmium, zinc, and chromium, respectively, indicating that cadmium has been extracted at lower pH in comparison with zinc and chromium (Mellah and Benachour 2006).

15.4.3 Chemical Precipitation

One of the commonly practiced industrial methods is chemical precipitation, mainly due to efficiency, cost-effectiveness, process control, and applicability at a wide range of temperatures. The effect of pH converts heavy metal ions into sulfide, hydroxide, carbonates, or any other less soluble compounds that can be eliminated by physical methods like sedimentation, filtration, or flotation. The method depends on the size, density, and surface charge of contaminants to be removed (Figure 15.10) (Chen et al. 2009; Demopoulos 2009). Some of the widely used inorganic precipitants for heavy metal precipitation are soda ash, lime, caustic soda, soda ash, sodium bicarbonate, sodium sulfide, and sodium hydrosulfide (Fu et al. 2007; Lim and Kim 2013). The pH greatly affects the solubility of metals, hence they exhibit various precipitation behaviors at different pH. It also depends on the type of contaminant and dosage of precipitants that have been continuously monitored during the precipitation process (Wang et al. 2018). Carbonate precipitation has been evolved as an effective alternative to hydroxide precipitation and dissolved heavy metals can be eliminated from wastewater using soda ash. Carbonate precipitation can occur at a lower range of pH around pH 9. Sulfide precipitation by adding Na_2S exhibit viable and efficient removal of zinc, copper, and lead from aqueous solutions (Chen et al. 2018). The treatment process used in acid mine drainage is precipitation through neutralization and peroxide addition. They also have some disadvantages such as large dosages of lime and peroxide and costly sludge fees. A highly potential bidentate or tetradentate ligand with two cystamine groups with benzene as a backbone, i.e. BDET, has been used to reduce metal pollutants from acidic water directly from mines. The precipitation of metals occurs due to the exceptional stability of BDET under oxidative and pH conditions, which can be used as an additive to lessen the metal concentration being discharged (Matlock et al. 2002). Precipitation of zinc, cadmium, magnesium, and manganese has been demonstrated by precipitation as hydroxide at high pH for effective removal from wastewater (Charerntanyarak 1999).

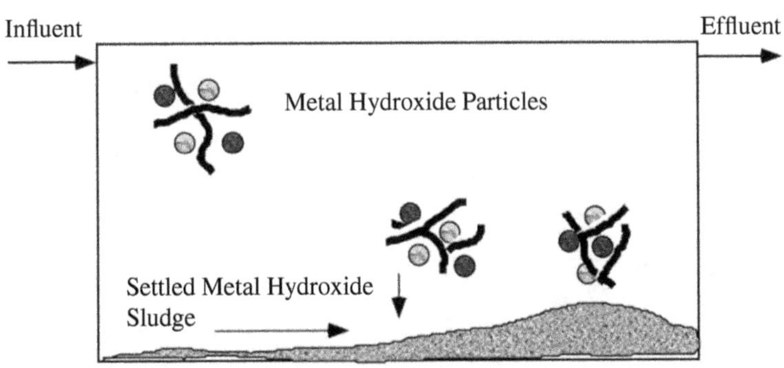

Figure 15.10 Chemical precipitation for metal ion removal produced by Chen et al. (2009).

15.4.4 Electrochemical Treatment

Much work has been done to find an effective and reasonable method to eliminate heavy metal from wastewater. The electrochemical process is one of the most efficient methods for the removal of hazardous soluble metal ions from industrial wastewater. Electrolysis gases provide aeration to enhance the reaction kinetics of the process. In this method, electricity was passed through an aqueous solution containing metal ions, a cathode plate, and an insoluble anode. This is based on the mechanism of applying an electric current to conjugate electrodes (Figure 15.11) (Sahu et al. 2017). Heavy metals precipitate as hydroxide in weakly acidic or neutralized electrolytes. Thus, the choice of electrode provides a specific application and also plays a crucial role in enhancing the process proficiency for contaminants (Tran et al. 2017).

Electrochemical systems mostly generate a kind of convenient to handle coalesced and dense water sludge, whereas electrical energy consumption per mass of contaminant is a drawback of the method along with passivation and electrode corrosion. The coupled technology of electrochemical reactor with membrane bioreactor demonstrated the successful removal of soluble microbial products (SMP), improved the physicochemical characteristics of sludge, enhanced the removal of phosphorus and heavy metal ions, and reduced membrane fouling (Giwa et al. 2019). Integrated electro-adsorption and electrochemistry arise due to electric field and functional electrodes enhancing heavy metal and pollutant removal. Some metals act as a precursor for the generation of electro-active catalytic species such as Cu^{2+}, Zn^{2+}, Ni^{2+}, Fe^{3+}, etc. Cu and Cu_2O metal and metal oxides are produced electrochemically via the in situ method showing higher electrochemical catalysis activity to oxidize organic pollutants (Gao et al. 2015). A conductive polymer like polypyrrole (PPy) has been used for the removal of lead, copper, chromium, etc. (Sun et al. 2020). According to the study of Solisio et al. (1999), copper ion was efficiently eliminated from dilute industrial effluent using an electrochemical reactor with a plate electrode (Solisio et al. 1999). Copper, nickel, and chromium were recovered from

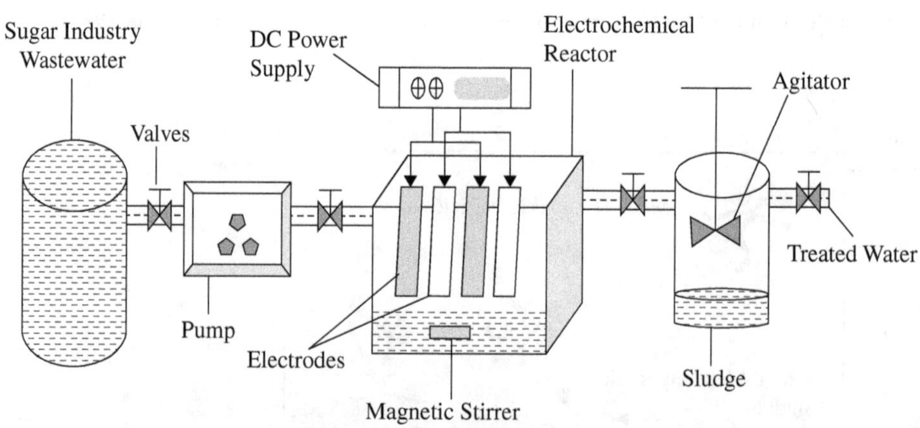

Figure 15.11 Electrochemical treatment for the removal of contaminants produced by Sahu et al. (2017). (Used with permission from the *Journal of Water Process Engineering*.)

industrial effluents utilizing a combination of electrochemical and biological processes. In the study, almost 80–90% of metals were recovered and the sludge was further used in agriculture (Hunsom et al. 2005).

15.5 Conclusion

The rapid increase of heavy metal load on the environment has become a severe threat to human civilization due to the dependency on nature. Thus, the elimination of heavy metals from the environment is of prime concern. Heavy metals can be eliminated by both physical and chemical methods; selection of the method depends on its technical applicability, economic factors, and simple approach. All methods are important in certain contexts such as eliminating heavy metals from wastewater via chemical precipitation is easy to operate and it is an inexpensive method, but it has the demerit of sludge formation. The method can be very helpful in low and high metal concentrations. Membrane filtration is also of great importance as it consumes fewer chemicals and generates less solid waste with high efficiency of heavy metal removal, but they have a lower flow rate. Sludge settling and dewatering are the advantages of the coagulation and flocculation method, but the method requires higher financial input and chemicals. In the case of the ion-exchange method, the pH of wastewater does not change, and there is chemical safety, stability, and process reliability, but the membrane costs are high and resin regeneration needs resin fouling. Adsorption is an efficient method of heavy metal removal from wastewater; the synthetic and natural adsorbents eliminate heavy metals based on surface phenomenon forming layers on the adsorbent. The utilization of waste products as adsorbent is also an economical and effective approach in the removal of heavy metals from aqueous solutions. The efficiency of adsorption depends on the type of adsorbent for eliminating heavy metals. The selection of method mainly depends on the concentration of heavy metal, environmental impact, financial investment, and ease of operation.

References

Abdel-Raouf, M.S. and Abdul-Raheim, A.R.M. (2017). Removal of heavy metals from industrial waste water by biomass-based materials: a review. *Journal of Pollution Effects & Control* 5: 180.

Abdullah, N., Yusof, N., Lau, W.J. et al. (2019). Recent trends of heavy metal removal from water/wastewater by membrane technologies. *Journal of Industrial and Engineering Chemistry* 76: 17–38.

Badawy, N.A., El-Bayaa, A.A., Abdel-Aal, A.Y., and Garamon, S.E. (2009). Chromatographic separations and recovery of lead ions from a synthetic binary mixtures of some heavy metal using cation exchange resin. *Journal of Hazardous Materials* 166 (2-3): 1266–1271.

Balladares, E., Jerez, O., Parada, F. et al. (2018). Neutralization and co-precipitation of heavy metals by lime addition to effluent from acid plant in a copper smelter. *Minerals Engineering* 122: 122–129.

Barakat, M.A. and Schmidt, E. (2010). Polymer-enhanced ultrafiltration process for heavy metals removal from industrial wastewater. *Desalination* 256 (1-3): 90–93.

Bhattacharyya, D., Moffitt, M., and Grieves, R.B. (1978). Charged membrane ultrafiltration of toxic metal oxyanions and cations from single-and multisalt aqueous solutions. *Separation Science and Technology* 13 (5): 449–463.

Božić, D., Stanković, V., Gorgievski, M. et al. (2009). Adsorption of heavy metal ions by sawdust of deciduous trees. *Journal of Hazardous Materials* 171 (1-3): 684–692.

Božić, D., Gorgievski, M., Stanković, V. et al. (2013). Adsorption of heavy metal ions by beech sawdust–Kinetics, mechanism and equilibrium of the process. *Ecological Engineering* 58: 202–206.

Chan, B.K.C. and Dudeney, A.W.L. (2008). Reverse osmosis removal of arsenic residues from bioleaching of refractory gold concentrates. *Minerals Engineering* 21 (4): 272–278.

Charerntanyarak, L. (1999). Heavy metals removal by chemical coagulation and precipitation. *Water Science and Technology* 39 (10-11): 135.

Chen, Q., Luo, Z., Hills, C. et al. (2009). Precipitation of heavy metals from wastewater using simulated flue gas: sequent additions of fly ash, lime and carbon dioxide. *Water Research* 43 (10): 2605–2614.

Chen, Q., Yao, Y., Li, X. et al. (2018). Comparison of heavy metal removals from aqueous solutions by chemical precipitation and characteristics of precipitates. *Journal of Water Process Engineering* 26: 289–300.

Conner, J.R. and Hoeffner, S.L. (1998). A critical review of stabilization/solidification technology. *Critical Reviews in Environmental Science and Technology* 28 (4): 397–462.

Demopoulos, G.P. (2009). Aqueous precipitation and crystallization for the production of particulate solids with desired properties. *Hydrometallurgy* 96 (3): 199–214.

Dizge, N., Keskinler, B., and Barlas, H. (2009). Sorption of Ni (II) ions from aqueous solution by Lewatit cation-exchange resin. *Journal of Hazardous Materials* 167 (1-3): 915–926.

Duffus, J.H. (2002). "Heavy metals" a meaningless term? (IUPAC Technical Report). *Pure and Applied Chemistry* 74 (5): 793–807.

Duruibe, J.O., Ogwuegbu, M.O.C., and Egwurugwu, J.N. (2007). Heavy metal pollution and human biotoxic effects. *International Journal of Physical Sciences* 2 (5): 112–118.

Dutta, K. and De, S. (2017). Aromatic conjugated polymers for removal of heavy metal ions from wastewater: a short review. *Environmental Science: Water Research & Technology* 3 (5): 793–805.

Flett, D.S. (1999). New reagents or new ways with old reagents. *Journal of Chemical Technology & Biotechnology: International Research in Process, Environmental & Clean Technology* 74 (2): 99–105.

Flett, D.S. (2004). Cobalt-nickel separation in hydrometallurgy: a review. *Chemistry for Sustainable Development* 12 (1): 81–91.

Fu, F., Zeng, H., Cai, Q. et al. (2007). Effective removal of coordinated copper from wastewater using a new dithiocarbamate-type supramolecular heavy metal precipitant. *Chemosphere* 69 (11): 1783–1789.

Gao, H., Zhang, J., Wang, R., and Wang, M. (2015). Highly efficient hydrogen production and formaldehyde degradation by Cu2O microcrystals. *Applied Catalysis B: Environmental* 172: 1–6.

Ghirişan, A.L., Drăgan, S., Pop, A. et al. (2007). Heavy metal removal and neutralization of acid mine waste water-kinetic study. *The Canadian Journal of Chemical Engineering* 85 (6): 900–905.

Giwa, A., Dindi, A., and Kujawa, J. (2019). Membrane bioreactors and electrochemical processes for treatment of wastewaters containing heavy metal ions, organics, micropollutants and dyes: recent developments. *Journal of Hazardous Materials* 370: 172–195.

Gunatilake, S.K. (2015). Methods of removing heavy metals from industrial wastewater. *Methods* 1 (1): 14.

Gupta, A., Johnson, E.F., Mindler, A.B., and Schlossel, R.H. (1984). The Use of Ion Exchange for Treating Wastewaters and Recovering Metal Values from Electroplating Operations, 67–74. Houston, Texas, Philadelphia, Pennsylvania, San Francisco: Separation of Heavy Metals and Other Trace Contaminants.

Gupta, V.K., Gupta, M., and Sharma, S. (2001). Process development for the removal of lead and chromium from aqueous solutions using red mud—an aluminium industry waste. *Water Research* 35 (5): 1125–1134.

Gupta, V.K., Rastogi, A., and Nayak, A. (2010). Adsorption studies on the removal of hexavalent chromium from aqueous solution using a low cost fertilizer industry waste material. *Journal of Colloid and Interface Science* 342 (1): 135–141.

Hamdaoui, O. (2009). Removal of copper (II) from aqueous phase by Purolite C100-MB cation exchange resin in fixed bed columns: modeling. *Journal of Hazardous Materials* 161 (2-3): 737–746.

Ho, Y.S., Huang, C.T., and Huang, H.W. (2002). Agricultural by-products as a metal sorbent: sorption of copper ions from aqueous solution onto tree fern. *Process Biochemistry* 37 (5): 1421–1430.

Huang, J.H., Zeng, G.M., Zhou, C.F. et al. (2010). Adsorption of surfactant micelles and Cd2+/Zn2+ in micellar-enhanced ultrafiltration. *Journal of Hazardous Materials* 183 (1-3): 287–293.

Huang, Y., Wu, D., Wang, X. et al. (2016). Removal of heavy metals from water using polyvinylamine by polymer-enhanced ultrafiltration and flocculation. *Separation and Purification Technology* 158: 124–136.

Hunsom, M., Pruksathorn, K., Damronglerd, S. et al. (2005). Electrochemical treatment of heavy metals (Cu2+, Cr6+, Ni2+) from industrial effluent and modeling of copper reduction. *Water Research* 39 (4): 610–616.

Idris, A., Hassan, N., Ismail, N.S.M. et al. (2010). Photocatalytic magnetic separable beads for chromium (VI) reduction. *Water Research* 44 (6): 1683–1688.

Jain, M., Garg, V.K., and Kadirvelu, K. (2010). Adsorption of hexavalent chromium from aqueous medium onto carbonaceous adsorbents prepared from waste biomass. *Journal of Environmental Management* 91 (4): 949–957.

Johnson, P.D., Girinathannair, P., Ohlinger, K.N. et al. (2008). Enhanced removal of heavy metals in primary treatment using coagulation and flocculation. *Water Environment Research* 80 (5): 472–479.

Juang, R.S. and Huang, H.L. (2003). Mechanistic analysis of solvent extraction of heavy metals in membrane contactors. *Journal of Membrane Science* 213 (1-2): 125–135.

Kabata-Pendias, A. and Pendias, H. (1999). Biogeochemistry of Trace Elements, 400. Warsaw: Polish Scientific Publishing Company.

Ko, D., Lee, J.S., Patel, H.A. et al. (2017). Selective removal of heavy metal ions by disulfide linked polymer networks. *Journal of Hazardous Materials* 332: 140–148.

Kobya, M. (2004). Removal of Cr (VI) from aqueous solutions by adsorption onto hazelnut shell activated carbon: kinetic and equilibrium studies. *Bioresource Technology* 91 (3): 317–321.

Kongolo, K., Mwema, M.D., Banza, A.N., and Gock, E. (2003). Cobalt and zinc recovery from copper sulphate solution by solvent extraction. *Minerals Engineering* 16 (12): 1371–1374.

Kosarek, L.J. (1981). Removal of various toxic heavy metals and cyanide from water by membrane processes. In: Chemistry in Water Reuse (ed. W.J. Cooper), 261. Ann Arbor, Michigan: Ann Arbor Science.

Kumar, M., Kumar, D., and Singh, B. (2014). Utilization of agro residue corncob for production of acetone-butanol using Clostridium acetobutylicum and process optimization through RSM. *Journal of Microbial & Biochemical Technology* 8 (2): 1–5.

Kumar, M., Goswami, L., Singh, A.K., and Sikandar, M. (2019). Valorization of coal fired-fly ash for potential heavy metal removal from the single and multi-contaminated system. *Heliyon* 5 (10): e02562.

Lalmi, A., Bouhidel, K.E., Sahraoui, B., and el Houda Anfif, C. (2018). Removal of lead from polluted waters using ion exchange resin with Ca (NO3) 2 for elution. *Hydrometallurgy* 178: 287–293.

Landaburu-Aguirre, J., Pongrácz, E., Perämäki, P., and Keiski, R.L. (2010). Micellar-enhanced ultrafiltration for the removal of cadmium and zinc: use of response surface methodology to improve understanding of process performance and optimisation. *Journal of Hazardous Materials* 180 (1-3): 524–534.

Li, Q., Zhai, J., Zhang, W. et al. (2007). Kinetic studies of adsorption of Pb (II), Cr (III) and Cu (II) from aqueous solution by sawdust and modified peanut husk. *Journal of Hazardous Materials* 141 (1): 163–167.

Lim, M. and Kim, M.J. (2013). Reuse of washing effluent containing oxalic acid by a combined precipitation–acidification process. *Chemosphere* 90 (4): 1526–1532.

Liu, M., Zhang, H., Zhang, X. et al. (2001). Removal and recovery of chromium (III) from aqueous solutions by a spheroidal cellulose adsorbent. *Water Environment Research* 73 (3): 322–328.

Liu, K., Nie, Z., Zhao, N. et al. (2010). Step-growth polymerization of inorganic nanoparticles. *Science* 329 (5988): 197–200.

López-Téllez, G., Barrera-Díaz, C.E., Balderas-Hernández, P. et al. (2011). Removal of hexavalent chromium in aquatic solutions by iron nanoparticles embedded in orange peel pith. *Chemical Engineering Journal* 173 (2): 480–485.

Mane, P.C., Bhosle, A.B., Jangam, C.M., and Vishwakarma, C.V. (2011). Bioadsorption of selenium by pretreated algal biomass. *Advances in Applied Science Research* 2 (2): 202–207.

Matlock, M.M., Howerton, B.S., and Atwood, D.A. (2002). Chemical precipitation of heavy metals from acid mine drainage. *Water Research* 36 (19): 4757–4764.

Mellah, A. and Benachour, D. (2006). Solvent extraction of heavy metals contained in phosphoric acid solutions by 7-(4-ethyl-1-methyloctyl)-8-hydroxyquinoline in kerosene diluent. *Hydrometallurgy* 81 (2): 100–103.

Mertz, W. (1981). The essential trace elements. *Science* 213 (4514): 1332–1338.

Mohammed, A.S., Kapri, A., and Goel, R. (2011). Heavy metal pollution: source, impact, and remedies. In: Biomanagement of metal-contaminated soils (eds. M.S. Khan, A. Zaidi, R. Goel and J. Musarrat), 1–28. Dordrecht: Springer.

Muthukrishnan, M. and Guha, B.K. (2008). Effect of pH on rejection of hexavalent chromium by nanofiltration. *Desalination* 219 (1-3): 171–178.

Pang, F.M., Kumar, P., Teng, T.T. et al. (2011). Removal of lead, zinc and iron by coagulation–flocculation. *Journal of the Taiwan Institute of Chemical Engineers* 42 (5): 809–815.

Papassiopi, N., Virčiková, E., Nenov, V. et al. (1996). Removal and fixation of arsenic in the form of ferric arsenates. Three parallel experimental studies. *Hydrometallurgy* 41 (2-3): 243–253.

Peters, R.W. and Shem, L. (1993). Separation of heavy metals: removal from industrial wastewaters and contaminated soil (No. ANL/ES/CP-78514; CONF-9303107-1). Argonne National Laboratory, Energy Systems Division, Lemont, IL, USA.

Rai, U.N. and Pal, A. (2002). Health hazards of heavy metals. *Environ News Newsletter* 8 (1) http://isebindia.com/01_04/02-01-2.html.

Reddy, B.R. and Sarma, P.B. (2001). Transfer of nickel from sodium sulphate solutions to the spent electrolyte through solvent extraction and stripping. *Hydrometallurgy* 60 (2): 123–128.

Renu, B., Agarwal, M., and Singh, K. (2017). Methodologies for removal of heavy metal ions from wastewater: an overview. *Interdisciplinary Environmental Review* 18 (2): 124–142.

Saber-Samandari, S., Saber-Samandari, S., and Gazi, M. (2013). Cellulose-graft-polyacrylamide/hydroxyapatite composite hydrogel with possible application in removal of Cu (II) ions. *Reactive and Functional Polymers* 73 (11): 1523–1530.

Sahu, O., Rao, D.G., Gopal, R. et al. (2017). Treatment of wastewater from sugarcane process industry by electrochemical and chemical process: aluminum (metal and salt). *Journal of Water Process Engineering* 17: 50–62.

Sarma, P.B. and Reddy, B.R. (2002). Liquid–liquid extraction of nickel at macro-level concentration from sulphate/chloride solutions using phosphoric acid based extractants. *Minerals Engineering* 15 (6): 461–464.

Sato, T., Imaizumi, M., Kato, O., and Taniguchi, Y. (1977). RO applications in wastewater reclamation for re-use. *Desalination* 23 (1-3): 65–76.

Sekhar, K.C., Kamala, C.T., Chary, N.S., and Anjaneyulu, Y. (2003). Removal of heavy metals using a plant biomass with reference to environmental control. *International Journal of Mineral Processing* 68 (1-4): 37–45.

Shah, B.A., Shah, A.V., and Singh, R.R. (2009). Sorption isotherms and kinetics of chromium uptake from wastewater using natural sorbent material. *International Journal of Environmental Science & Technology* 6 (1): 77–90.

Shaidan, N.H., Eldemerdash, U., and Awad, S. (2012). Removal of Ni (II) ions from aqueous solutions using fixed-bed ion exchange column technique. *Journal of the Taiwan Institute of Chemical Engineers* 43 (1): 40–45.

Silva, J.E.D., Paiva, A.P., Soares, D. et al. (2005). Solvent extraction applied to the recovery of heavy metals from galvanic sludge. *Journal of Hazardous Materials* 120 (1-3): 113–118.

Silva, H.S., Ruiz, S.V., Granados, D.L., and Santángelo, J.M. (2010). Adsorption of mercury (II) from liquid solutions using modified activated carbons. *Materials Research* 13 (2): 129–134.

Solisio, C., Panizza, M., Paganelli, P., and Cerisola, G. (1999). Electrochemical remediation of copper (II) from an industrial effluent: part I: monopolar plate electrodes. *Resources, Conservation and Recycling* 26 (2): 115–124.

Sörme, L. and Lagerkvist, R. (2002). Sources of heavy metals in urban wastewater in Stockholm. *Science of the Total Environment* 298 (1-3): 131–145.

Stowers, C.C., Cox, B.M., and Rodriguez, B.A. (2014). Development of an industrializable fermentation process for propionic acid production. *Journal of Industrial Microbiology & Biotechnology* 41 (5): 837–852.

Sudilovskiy, P.S., Kagramanov, G.G., and Kolesnikov, V.A. (2008). Use of RO and NF for treatment of copper containing wastewaters in combination with flotation. *Desalination* 221 (1-3): 192–201.

Sun, J., Liu, L., and Yang, F. (2020). A WO3/PPy/ACF modified electrode in electrochemical system for simultaneous removal of heavy metal ion Cu2+ and organic acid. *Journal of Hazardous Materials* 394: 122534.

Taamneh, Y. and Sharadqah, S. (2017). The removal of heavy metals from aqueous solution using natural Jordanian zeolite. *Applied Water Science* 7 (4): 2021–2028.

Taghizadeh, M., Asgharinezhad, A.A., Pooladi, M. et al. (2013). A novel magnetic metal organic framework nanocomposite for extraction and preconcentration of heavy metal ions, and its optimization via experimental design methodology. *Microchimica Acta* 180 (11-12): 1073–1084.

Tchounwou, P.B., Yedjou, C.G., Patlolla, A.K. et al. (2012). Molecular, clinical and environmental toxicology. *Molecular, Clinical and Environmental Toxicology* 3: 133–164.

Teh, C.Y. and Wu, T.Y. (2014). The potential use of natural coagulants and flocculants in the treatment of urban waters. *Chemical Engineering Transactions* 39 (6): 1603–1608.

Teh, C.Y., Budiman, P.M., Shak, K.P.Y., and Wu, T.Y. (2016). Recent advancement of coagulation–flocculation and its application in wastewater treatment. *Industrial & Engineering Chemistry Research* 55 (16): 4363–4389.

Terfassa, B., Schachner, J.A., Traar, P. et al. (2014). Oxorhenium (V) complexes with naphtholate-oxazoline ligands in the catalytic epoxidation of olefins. *Polyhedron* 75: 141–145.

Thakare, Y.N. and Jana, A.K. (2015). Performance of high density ion exchange resin (INDION225H) for removal of Cu (II) from waste water. *Journal of Environmental Chemical Engineering* 3 (2): 1393–1398.

Tran, T.K., Chiu, K.F., Lin, C.Y., and Leu, H.J. (2017). Electrochemical treatment of wastewater: selectivity of the heavy metals removal process. *International Journal of Hydrogen Energy* 42 (45): 27741–27748.

Tripathy, T. and De, B.R. (2006). Flocculation: a new way to treat the waste water. *Journal of Physical Sciences* 10 (December): 93–127.

US Environmental Protection Agency (1978). Manual of Treatment Techniques for Meeting the Interim Primary Drinking Water Regulations, EPA 600/8-77-005. Washington, DC: United States Environmental Protection Agency.

Verbych, S., Bryk, M., Alpatova, A., and Chornokur, G. (2005). Ground water treatment by enhanced ultrafiltration. *Desalination* 179 (1-3): 237–244.

Vigneswaran, R., Aitchison, S.J., McDonald, H.M. et al. (2004). Cerebral palsy and placental infection: a case-cohort study. *BMC Pregnancy and Childbirth* 4 (1): 1.

Wang, S., Soudi, M., Li, L., and Zhu, Z.H. (2006). Coal ash conversion into effective adsorbents for removal of heavy metals and dyes from wastewater. *Journal of Hazardous Materials* 133 (1-3): 243–251.

Wang, H., Zhang, H., and Cai, G. (2011). An application of phytoremediation to river pollution remediation. *Procedia Environmental Sciences* 10: 1904–1907.

Wang, D., Ye, Y., Liu, H. et al. (2018). Effect of alkaline precipitation on Cr species of Cr (III)-bearing complexes typically used in the tannery industry. *Chemosphere* 193: 42–49.

Warner, C.L., Chouyyok, W., Mackie, K.E. et al. (2012). Manganese doping of magnetic iron oxide nanoparticles: tailoring surface reactivity for a regenerable heavy metal sorbent. *Langmuir* 28 (8): 3931–3937.

Wilson, A.M., Bailey, P.J., Tasker, P.A. et al. (2014). Solvent extraction: the coordination chemistry behind extractive metallurgy. *Chemical Society Reviews* 43 (1): 123–134.

Wu, J., Zhang, H., He, P.J. et al. (2010). Cr (VI) removal from aqueous solution by dried activated sludge biomass. *Journal of Hazardous Materials* 176 (1-3): 697–703.

Yadav, M., Das, M., Savani, C. et al. (2019). Maleic anhydride cross-linked β-Cyclodextrin-conjugated magnetic nanoadsorbent: an ecofriendly approach for simultaneous adsorption of hydrophilic and hydrophobic dyes. *ACS Omega* 4 (7): 11993–12003.

Yang, T.C. and Zall, R.R. (1984). Absorption of metals by natural polymers generated from seafood processing wastes. *Industrial & Engineering Chemistry Product Research and Development* 23 (1): 168–172.

Yang, X.J., Fane, A.G., and MacNaughton, S. (2001). Removal and recovery of heavy metals from wastewaters by supported liquid membranes. *Water Science and Technology* 43 (2): 341–348.

Yu, L.J., Shukla, S.S., Dorris, K.L. et al. (2003). Adsorption of chromium from aqueous solutions by maple sawdust. *Journal of Hazardous Materials* 100 (1-3): 53–63.

Zewail, T.M. and Yousef, N.S. (2015). Kinetic study of heavy metal ions removal by ion exchange in batch conical air spouted bed. *Alexandria Engineering Journal* 54 (1): 83–90.

Part VI

Policy Dimensions on Water Security

Part I

Polyolefin Blends of Waste [illegible]

16

The Role of Government and the Public in Water Resource Management in India

Jitesh Narottam Vyas and Supriya Nath

Central Water and Power Research Station, Pune, Maharashtra, India

16.1 Introduction

Humans have disturbed hydrological cycles, which has affected ecosystem service and has had a negative impact on economic and social development (Shen and Chen 2010). The global economy is directly affected by the misuse of water and pollution at a local level (Hoekstra 2014). We need to manage our natural resources sustainably; particularly water requires to be managed in an integrated manner through integrated water resource management (Biswas 2004; Mukhtarov 2008; Rahaman and Varis 2005). For sustainable development, we require an eco-hydrological approach that brings synchronization between ecological needs and economic demand with respect to water resources (Shen and Chen 2010).

Water became a concern at international level when the first global conference, the United Nations Water Conference, devoted to water was held at Mar del Plata, Argentina, in 1977 (Savenije and Hoekstra 2002). Agenda 21 was adopted at the UN Conference on Environment and Development in Rio de Janeiro in 1992 (Savenije and Hoekstra 2002). An International Water Management Institute (IWMI) study anticipated that China, India, the Middle East, and North Africa may face "severe water shortage" due to excessive dependency on groundwater for irrigation (Barker et al. 2000).

Water resource management involves many activities such as planning, developing, assessment, monitoring, modeling, exploration, design of measures and strategies, implementation of policy, operation and maintenance, and evaluation in terms of both water quantity and quality (Savenije and Hoekstra 2002). It requires systematic, technological, institutional, administrative, lawful, and operational activities (Savenije and Hoekstra 2002). Management can only succeed if all stakeholders (both formal and informal) are involved and dedicated in the process of planning, decision making, and implementation (Savenije and Hoekstra 2002), otherwise, it is likely to fail.

We have focused our study on India as despite being naturally bestowed with many perennial rivers and good rainfall during monsoon, it is facing severe water scarcity in many regions (Singh et al. 2017). There is a lacuna in water resource management as some areas are facing

Pollutants and Water Management: Resources, Strategies and Scarcity, First Edition. Edited by Pardeep Singh, Rishikesh Singh, Vipin Kumar Singh, and Rahul Bhadouria.

flooding while some face drought. Even though it has rich traditional knowledge of water conservation, there is a lack of rainwater harvesting on a large scale (Borthakur and Singh 2016).

India has 4% of the world's freshwater resources with which it requires to meet the water requirements of 17% of the world's population (Kumar 2018; Guidelines for Urban Water Conservation Jal Shakti Abhiyan 2019). Currently, India is designated as a water-stressed nation, storing less than 1/10th of annual precipitation. Fluctuating monsoon, excessive water extraction from the ground, and water-intensive methods used in agriculture have imbalanced the demand–supply balance (Guidelines for Urban Water Conservation Jal Shakti Abhiyan 2019). Therefore, there is an urgent need to focus on water resource management in India. According to the National Institution for Transforming India (NITI) Aayog, around 50% of the population of India is confronting acute water scarcity (Guidelines for Urban Water Conservation Jal Shakti Abhiyan 2019).

In this chapter, we discuss the distribution of water in India, challenges faced in water resource management, traditional ways that were prevalent for water conservation, and present management measures taken by the Government of India and initiatives taken by the public for water resource management.

16.2 Distribution of Water in India

Water is unevenly distributed in India with some areas rich in water resources in the form of surface water (rivers and lakes) and high groundwater level as they receive abundant supply from rainfall. While some are facing severe water deficit as they are prone to drought and their groundwater level is low. Uneven distribution of rainfall due to geography and meteorological factors has aggravated this situation, making some areas prone to flood while some face severe drought, bringing the issue of the flood and drought syndrome (Kumar et al. 2005).

16.3 Challenges in Water Resource Management

16.3.1 Water Pollution

The major challenge in water resource management is the level of pollution in existing water resources. Our major river systems are facing high levels of pollution from industrial, agricultural, and domestic wastes. Major water pollutants are oils, pesticides, plastics, metal wastes, suspended solids, toxins, acids, dyes, medical wastes, etc.; they are nondegradable, and thus, cause serious pollution problems (Aggarwal et al. 2010). Groundwater is contaminated with excess fluoride, arsenic, iron, nitrate, and salinity (Maria 2003). This contamination changes water quality (Aggarwal et al. 2010) causing problems like reduction in dissolved oxygen, light penetration, death of aquatic organisms, and health problems.

16.3.2 Water Stress

Increasing urbanization, population growth, and changed lifestyles (Kumar et al. 2005) with an increase in water-intensive techniques in farming and modern industrial water needs have led to high demand for per capita water. The large quantity of water consumption has proven detrimental for river and groundwater (Jain 2019), as their water level has gone down drastically. Agriculture has the maximum consumption (80% of water) and 15% is used by industry (Kumar 2018). With an increasing focus on promoting industries, this ratio may change in the future. Excessive consumption has led to water scarcity and water stress.

16.3.3 Interstate Conflicts for Water

Our administrative requirements have governed our water resource management plan depending on administrative boundaries without considering our river basins as the hydrological unit (Kumar 2018). Since rivers do not follow any boundaries, it has caused interstate conflicts with rising water requirements in every sphere, i.e. domestic, industrial, and agricultural needs (Kumar 2018).

16.4 Traditional Ways of Water Resource Management in India

Various water harvesting systems have been in use by local people of different states for many centuries (Bhattacharya 2015). These traditional rainwater systems were simple yet efficient (Bhattacharya 2015). Ancient temples used to build large tanks that collected rainwater or tapped underground springs (Bhattacharya 2015). Even forts had some provisions to store rainwater (Bhattacharya 2015). Examples of rainwater harvesting structures are Zabo of Nagaland; Dongs, Garh, and Dara of Brahmaputra valley; Kund, Khadin, Talabs, Beri, Johad, Baoli, etc. of Thar Desert and Gujrat; Eri of Tamil Nadu; and Ahar pynes of Bihar (Table 16.1). These traditional water-knowledges should be applied in our modern society to deal with water scarcity (Bhattacharya 2015) without intervening with nature.

16.5 Present Management Measures

16.5.1 Government Schemes and Policies

Water scarce and rainfed regions need to focus on the implementation of policies and the development of techniques for farmers that depend on rain (Barker et al. 2000). Government policies play a major role in deciding the way any resources are being managed. For effective and pragmatic policy planning, we require comprehensive knowledge of the climate, spatial and temporal distribution of rainfall, and distribution of water resources (Surendran et al. 2017). Developing countries like India face twin problems of maintaining a rapid growth rate and balancing the effects of climate change resulting from this growth and modern lifestyle (Singh et al. 2019).

In 2004, the Ministry of Environment and Forests, Government of India, published a report on "India's Initial National Communication to the United Nations Framework Convention on Climate Change" which identified the following impacts of climate change on water (National Water Mission, Ministry of Jal Shakti, Department of Water Resources, RD & GR, Government of India):[1]

1 National Water Mission, http://nwm.gov.in.

Table 16.1 Traditional rainwater harvesting techniques used across different states in India.

Type of structure	Location in India	Use
Tanka	Rajasthan, Gujarat	Household-level water harvesting
Jhalara/Vav	Karnataka, Odisha	Water conservation
Bawdi	Rajasthan	Stepwell for water collection
Panam Keni	Kerala	Storage of water at household level
Zabo	Nagaland	Water conservation, agriculture, animal care
Eri	Tamil Nadu	Irrigation, groundwater recharge, and drinking water
Ahar pynes	South Bihar	Artificial rivulets for collection of water

Source: Jal Shakti Abhiyan Booklet (2019), Dept. of Drinking Water and Sanitation, Ministry of Jal Shakti). © 2020 Government of India (Ministry of Drinking Water and Sanitation).

- *Decline in the glaciers and the snowfields in the Himalayas*
- *Increased drought like situations due to overall decrease in the number of rainy days over a major part of the country*
- *Increased flood events due to overall increase in the rainy day intensity*
- *Effect on groundwater quality in alluvial aquifers due to increased flood and drought events*
- *Influence on groundwater recharge due to changes in precipitation and evapo-transpiration*
- *Increased saline intrusion of coastal and island aquifers due to rising sea levels*

In order to deal with projected impacts of climate change on water resources, the National Action Plan on Climate Change made principles and has given solutions for the impact of climate change in the form of eight National Missions among which the National Water Mission is included that is focused on water resource management (National Water Mission, Ministry of Jal Shakti, Department of Water Resources, RD & GR, Government of India).[1]

16.5.1.1 National Water Mission

As per the Ministry of Jal Shakti, Department of Water Resources, RD and GR, Government of India, the "National Water Mission" has recognized the ways for achieving the objectives of:[1]

a) *Comprehensive water database in public domain and assessment of the impact of climate change on water resource*
b) *Promotion of citizen and state actions for water conservation, augmentation and preservation*
c) *Focused attention to vulnerable areas including over-exploited areas*
d) *Increasing water use efficiency by 20%*
e) *Promotion of basin level integrated water resources management.*

The major objectives are ensuring integrated water resource management by:

- *conservation of water,*
- *reduction in wastage of water,*
- *equitable distribution both across and within states.*

The Mission aims to increase water use efficiency by regulatory processes like the difference in entitlements and price. It aims to recycle wastewater for meeting the daily requirement of water in urban areas. It also aims to adopt advanced technologies such as low-temperature desalination technologies which will solve water requirements in coastal areas (Comprehensive Mission Document of National Water Mission ministry of Jal Shakti 2011).[2]

16.5.1.2 Integrated Water Resources Management

The Government of India set-up the Central Water Commission (CWC) in 1945, which organizes awareness programs for the control, conservation, and utilization of water resources and coordinates with state governments. To achieve the goal of water management, Integrated Water Resources Management (IWRM) should be implemented.

CWC under the National Water Mission in March 2010 prepared guidelines on IWRM. As per IWRM Guidelines, IWRM has been defined as "a process which promotes the coordinated development and management of water, land and related resources, in order to maximize the resultant economic and social welfare in an equitable manner, without compromising the sustainability of vital ecosystems" (Guidelines for Integrated Water Resources Development and Management 2016). It aims at achieving water security, mitigating disasters, and managing risks of disasters at the river basin level.

The concept of IWRM emerged around 1980. According to the IWRM Guidelines, "IWRM addresses the 'three Es': **E**conomic efficiency, **E**nvironmental sustainability and social **E**quity. The three basic 'pillars' of IWRM are the **enabling environment** of appropriate policies and laws, the **institutional roles** and framework, and the **management instruments** for these institutions to apply on a daily basis" (Figure 16.1) (Guidelines for Integrated Water Resources Development and Management 2016).

Various components of IWRM and their relationship and interaction with one another are represented in Figure 16.2.

IWRM takes account of:

- Every normal aspect of water resources
- All sectoral interests and stakeholders

2 http://jalshakti-dowr.gov.in/sites/default/files/Document_of_NWM_Vol_I_April%2020117821020996_0.pdf

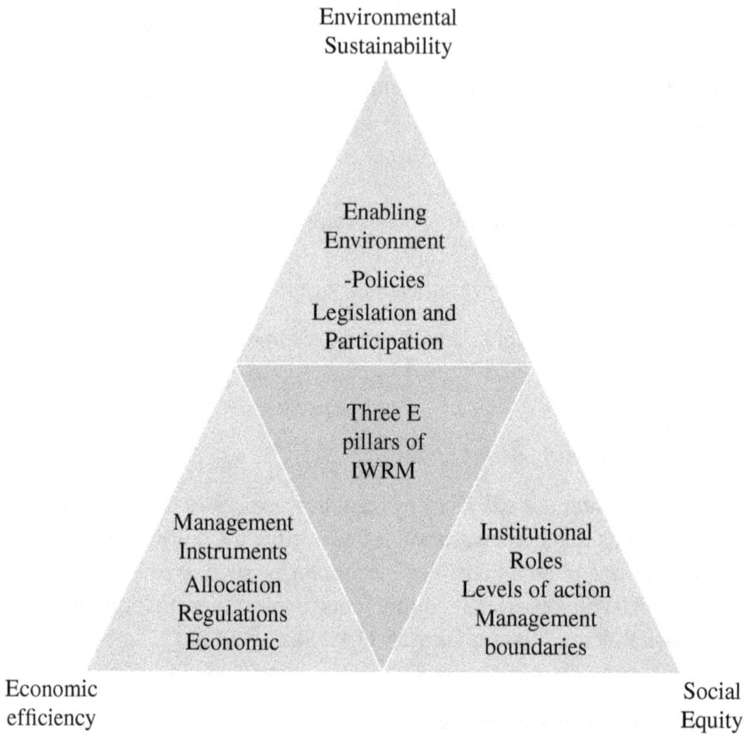

Figure 16.1 The Integrated Water Resources Management (IWRM) general framework explaining the concept of the three E pillars of IWRM. (*Source:* Based on Guidelines for Integrated Water Resources Development and Management 2016.)

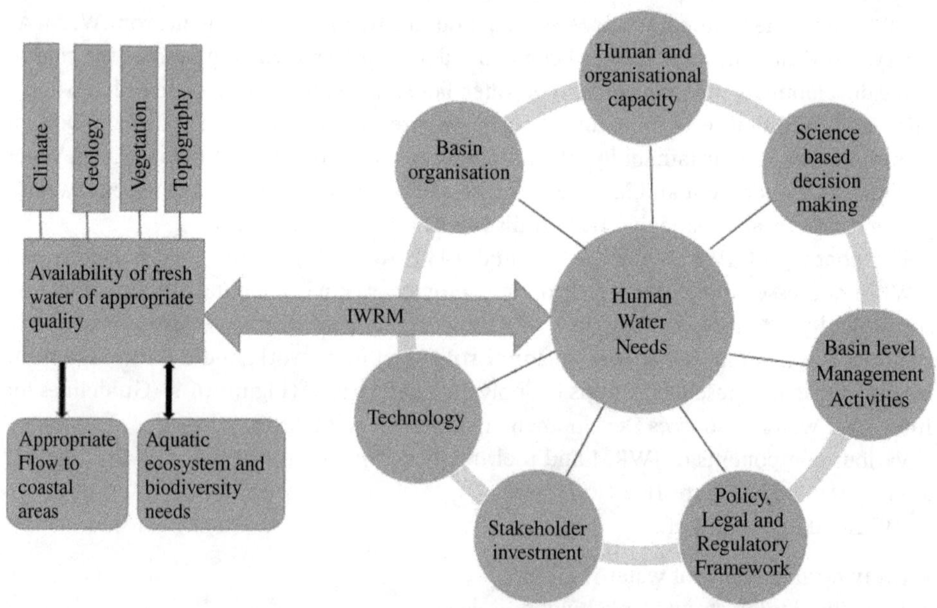

Figure 16.2 Various components of Integrated Water Resources Management (IWRM). (*Source:* Modified from Guidelines for Integrated Water Resources Development and Management 2016. © 2016 Government of India (CWC).)

- The spatial variation of resources and demands
- Relevant policy frameworks (national objectives and constraints)
- All institutional levels

Water security can be achieved only when a proper balance between the needs of various sectors and governance mechanisms are established. Well-developed, well-tested, scientific, socially acceptable, and efficient approaches to execute IWRM are still not extensively present (Guidelines for Integrated Water Resources Development and Management 2016).

16.5.1.3 National Water Policy

According to the Review of Water Policies and Programmes with Special Reference to Urban Water Supply, the direction of government water resource management programs is governed by national water management policies. They have legal and institutional significance. Public consultation plays a significant role in making these guidelines helpful and gaining acceptance by all the stakeholders. To resolve the drinking water security issues, central government has introduced various policies and programs as five-year plans. They asked states to recognize problems related to drinking water in villages and towns and to lay out special plans for resolving them (Review of Water Policies and Programmes with Special Reference to Urban Water Supply).[3]

The First National Water Policy (NWP) was formed in 1987, since then, India has faced various challenges in water resources management. The Ministry of Water Resources of the Government of India reviewed and updated the National Water Policy 1987 in 2002. The 2002 policy addressed and updated the lacunas that were in the 1987 policy. But this policy too was not sufficient for the management of water resources of a country in the twenty-first century.

The Indian Government formed national water policies in 1987, 2002, and 2012. Every time policies were upgraded and improved to meet the existing problems and to apply the latest technology and management methods for better water resource utilization (Review of Water Policies and Programmes with Special Reference to Urban Water Supply).[3]

16.5.2 Government Schemes

For the management of water resources, the government has to play a very important role, without government interest/support, it is not possible to manage water resources as it requires lots of funds, innovative technology, and clearance for various aspects. There are various schemes undertaken/run by the central government for the management of water resources in India, some of them are listed in the following sections.

16.5.2.1 Ganga Cleaning Programs

The Ganga Action Plan or GAP was a scheme/plan started by the Government of India in April 1985 so that pollution in the river Ganga reduces.[4] The Ganga Action plan was released in two stages; Stage I in 1985 covering 25 Ganga towns in three states and Stage II included 59 towns in five states.[3]

---

3 Review of Water Policies and Programmes with Special Reference to Urban Water Supply, https://shodhganga.inflibnet.ac.in/bitstream/10603/36784/11/11_chepter%203.pdf

4 https://www.gktoday.in/gk/namami-gange-project

At present, Ganga cleaning is underway from a central government scheme called the Namami Gange Project. This project, officially known as the *Integrated Ganga Conservation Mission Project*, has combined the efforts of ongoing activities and future concrete plans to rejuvenate the Ganga River in an inclusive manner.[3]

As per the Ministry of Jal Shakti, the aims and objectives of the National Mission for Clean Ganga (NMCG) is to achieve the directive of the National Ganga River Basin Authority (NGRBA), which are:

- *To ensure effective abatement of pollution and rejuvenation of the river Ganga by adopting a river basin approach to promote inter-sectoral co-ordination for comprehensive planning and management*; and
- *To maintain minimum ecological flows in the river Ganga with the aim of ensuring water quality and environmentally sustainable development.*

The objective of the Namami Gange mission is to successfully conserve, rejuvenate, and abate pollution of River Ganga.[5] The main focus areas under the Namami Gange program are creating sewerage treatment capacity, creating river-front development, river surface cleaning, biodiversity conservation, afforestation, public awareness, industrial effluent monitoring, and Ganga Gram.[4]

16.5.2.2 Jal Shakti Abhiyan

Jal Shakti Abhiyan (JSA) is a time-bound campaign with a task mode method meant for the improvement of conditions for around 1500 blocks that are facing problems of drought, water stress, or are over-exploited, and 254 districts with water conservation-related Government of India programs; 23 of these districts are aspirational districts. Designed on the lines of Gram Swaraj Abhiyan, JSA was planned to be carried out in two phases. Phase I was carried out between July 1 and September 15, 2019, for all states. For states/union territories (UTs) with receding monsoon (Andhra Pradesh, Tamil Nadu, Karnataka, and Puducherry), the campaign was carried out between October 2 and November 30, 2019.[6] During this campaign, senior officers, groundwater experts, and scientists from the Government of India were working together with state and district officials in India's most water-stressed districts. JSA objectives are to make water conservation and advancement of irrigation efficiency a jan andolan through asset making and communication campaigns (Jal Shakti Abhiyan Booklet 2019).

In this Abhiyan, central teams organized many community meetings at Gram-panchayat level and interacted with the village Sarpanch and local people on water conservation. The major focus was on the conservation of water by innovative techniques used for water conservation like spring water tapping, snow water harvesting/artificial glaciers, repaired/renovation of preexisting tanks, ponds, snow-check dams, etc. and impact assessment study on groundwater.

16.5.2.3 Interlinking of Rivers

This assignment is being controlled under the National Water Development Authority of India (NWDA), Ministry of Water Resources, Government of India. The interlinking of

5 https://nmcg.nic.in
6 sujal-swachhsangraha.gov.in

river project aims at connecting Indian rivers through reservoirs and canals. This project aims at resolving flood and drought syndrome as excess water will be diverted to places that lack water supply. The dependency of farmers on rain will reduce. This project will connect 60 rivers of India, including river Ganga.[7]

16.5.2.4 Command Area Development and Water Management Program

The Command Area Development and Water Management (CADWM) program was put into effect in 1976–1977 by the central government. The focus of Command Area Development (CAD) was to minimize the difference between irrigation capacity created and utilized by constructing various on-farm developmental activities like the creation of a field channel for the uniform supply of existing irrigation water.

It aims to:

> bridge the gap between irrigation potential created and utilized, to ensure timely and equitable distribution of irrigation water in the command, to enhance the irrigated area by means of efficient water management, to boost agriculture production and productivity in the command, to involve the beneficiary farmers at grass roots level for management of water resources through Participatory Irrigation Management(PIM) and organization of Pani Panchayats.

Command Area Development and Water Management program report 2011[8]

16.5.2.5 Atal Bhujal Yojana (Atal Jal)

The aim of this Yojana is to improve and manage the groundwater resources in the water-stressed areas of states viz. Gujarat, Haryana, Karnataka, Madhya Pradesh, Maharashtra, Rajasthan, and Uttar Pradesh. The success of this scheme depends on the efficiency of the community in using this scheme with the support of the state. The main objective of this Yojana is convergence among various ongoing schemes with the active involvement of local communities and stakeholders for the management of sustainable groundwater. The scheme will start functioning from 2020 to 2021 for five years until 2024–2025 (Atal Bhujal Yojana, Ministry of Jal Shakti, Department of Water Resources, RD & GR, Government of India).

16.5.2.6 National Hydrology Project

The objective of the National Hydrology Project is to "improve the extent, quality, and accessibility of water resources information and to strengthen the capacity of targeted water resources management institutions in India."[9]

The Ministry of Jal Shakti/Department of Water Resources, River Development and Ganga Rejuvenation/Central Water Commission have already applied the Hydrology Project. The project has been implemented in two phases, phase I is Hydrology Project

---

7 http://mowr.gov.in/schemes-projects-programmes/programmes/interlinking-rivers

8 https://archive.india.gov.in/sectors/water_resources/index.php?id=10#:~:text=The%20Centrally%20sponsored%20Command%20Area,team%20under%20an%20Area%20Development.

9 World Bank, https://projects.worldbank.org/en/projects-operations/project-detail/P152698

I (1995–2003) and phase II is Hydrology Project II (2004–2014) with the assistance of the World Bank.[10] Through this project, the Government of India is improving the planning, development, and management of water resources along with flood forecasting and reservoir operations in real-time. The project has led to a significant change in the availability and reliability of hydro-meteorological data in India. The Ministry of Jal Shakti/ Department of Water Resources, River Development and Ganga Rejuvenation have undertaken a National Hydrology Project (NHP) with World Bank assistance. The objective of NHP is to "improve the extent, quality, and accessibility of water resources information, decision support system for floods and basin level resource assessment/planning and to bolster the ability of targeted water resources specialist and management establishment in India."[11] The project concept is modernizing the monitoring network, transforming knowledge access, enhancing analytical tools, and modernizing institutions.[9] The various components of the project are water resource monitoring systems, water resource information systems, water resource operations and planning systems, and institutional capacity enhancement.[9]

16.5.2.7 Central Government Schemes Having Potential to Finance Renovation of Traditional and Other Water Bodies/Tanks

Under the Pradhan Mantri Krishi SinchayeeYojana (PMKSY), the Repair, Renovation and Restoration (RRR) of Water Bodies Scheme of the Department of Water Resources River Development and Ganga Rejuvenation is an important activity to ensure the utilization of the full capacity. The scheme of RRR of water bodies has become a part of the Har Khet Ko Pani (HKKP) component of PMKSY. The scheme emphasizes development work respect of water bodies such as the command area, catchment area, and de-siltation (Guidelines for the Scheme on Repair, Renovation and Restoration (RRR) of Water Bodies under PMKSY (HKKP) 2017).

16.6 The Role of the Public in Water Resource Management

The public plays an important role in water resource management as various nongovernmental organizations (NGOs) and stakeholders are involved. Some people have taken the initiative at their individual level and involved the community in their conservation practices, mobilizing local NGOs, industries, college, and school students and other volunteers for undertaking conservation works. Rainwater harvesting structures have been constructed in panchayat buildings, public schools, public health centers, government buildings, and household levels.

There are various steps through which communities can play a vital role in water resource management. Some existing ways are:

1) Collection of filtered rainwater in wells and cleaning the well
2) Taking necessary steps for water percolation to the soil such as contour bunding and trenching in slopes, etc.

---

10 http://cwc.gov.in/national-hydrology-project-nhp
11 National Hydrology Project, http://nhp.mowr.gov.in

3) Regular checking of subbase flow in rivers and streams, identify suitable sites for construction of subsurface dykes
4) Focusing on improving canal water distribution (warabandi) management and its efficiency
5) Use of appropriate/advanced water conservation and rainwater harvesting techniques for effective watershed management such as contour trenching and nala plugging in hills (Jal Shakti Abhiyan Booklet 2019).

There are various ways/techniques that the public can also easily adopt and implement at home. With the help of these techniques/methods, the public can contribute to the effective management of water which will be of great help to the nation. Some techniques are described in the following sections (Jal Shakti Abhiyan Booklet 2019).

16.6.1 Kitchen Gardens

We can use gray water for kitchen gardens and it can be used with a few precautions. The structure consists of a Nahani trap, a silt chamber, and a storage tank, which can easily be set-up with little effort (Jal Shakti Abhiyan Booklet 2019).

16.6.2 Soak Pit/Magic Pit

To manage gray water effectively, the soak pit/magic pit is a well-run alternative that may be used in every house and at community water sources. Pre-settled sewage from a collection and storage treatment is sent to the underground chamber from which it penetrates into the adjacent soil. An advantage of this technology is that it recharges the groundwater table and keeps the environment clean and safe (Jal Shakti Abhiyan Booklet 2019).

16.6.3 Afforestation

The process/initiative by humans to plant more trees on barren lands or at any place in order to create a forest is known as afforestation. It reduces wastage of water in the form of a reduction in annual rainwater runoff, helps in soil conservation, controls unexpected flooding, and it also improves water quality. Trees are responsible for lowering surface temperatures as trees provide a cooling effect on land, hence, loss of water due to evaporation is also reduced. The public should take the initiative to plant more trees and spread awareness about the importance of trees in surrounding areas (Jal Shakti Abhiyan Booklet 2019).

16.6.4 Rainwater Harvesting

Rainwater harvesting (RWH) is the collection of rainwater from rooftops, roadside, open areas, etc. and storing it for different uses and groundwater recharge. In a rooftop rainwater harvesting system, rainwater is captured from the roof catchments and stored in reservoirs or stored in a tank which is then used for household works. Collected rainwater is also stored in subsurface groundwater aquifers by adopting artificial recharge techniques. It can be easily adopted by public/community buildings. With some effort, households can also take up rooftop rainwater harvesting structures for their houses (Jal Shakti Abhiyan Booklet 2019).

16.6.4.1 Best Practices in Rainwater Harvesting

As per the Guidelines for Urban Water Conservation, Jal Shakti Abhiyan, best practices in rainwater harvesting are discussed which succeeded due to public support.

16.6.4.1.1 Garden Estate, Gurgaon In April 2013, the Garden Estate colony in Gurgaon adopted rainwater harvesting and raised the groundwater level by 1.7 m (Guidelines for Urban Water Conservation Jal Shakti Abhiyan 2019).

16.6.4.1.2 Tihar Jail, Delhi Tihar jail is another stellar example of a rainwater harvesting system that addressed their water scarcity issue by collecting rooftop rainwater of 1280 m^3 and a considerable increase in groundwater level was recorded in a shorter period of time since November 2002 (Guidelines for Urban Water Conservation Jal Shakti Abhiyan 2019).

16.6.4.1.3 Mother Dairy F&V Units, Delhi Mother Dairy used rooftop rainwater harvesting that diverted rainwater into recharge wells, collecting a total volume of 32 250 m^3 since April 2004 (Guidelines for Urban Water Conservation Jal Shakti Abhiyan 2019).

16.6.4.1.4 Rainwater Harvesting Measures in Chennai Water scarcity in Chennai caused indiscriminate groundwater extraction that caused saline water intrusion into fresh groundwater sources. The result of hard work by the public caused a considerable rise in the groundwater table in the past ten years (Guidelines for Urban Water Conservation Jal Shakti Abhiyan 2019).

16.6.4.1.5 Successful Endeavors of Public in Water Resources Management The participation and complete involvement of all stakeholders in water management has proven very effective in providing good results. Many success stories can explain the power of the public in changing the scenario of water resources management such as Hiware Bazar Panchayat, Vruksha Prem Seva Trust, etc. (Comprehensive Mission Document of National Water Mission ministry of Jal Shakti, Vol.1, April 2011).

The Government of India has been running various schemes like the interlinking of rivers, which will resolve flood drought syndrome, Namami Gange for the cleaning of River Ganga, Atal Bhujal Yojana to resolve groundwater-related problems, the National Hydrology Project for dissipating water resource information, CADWM to teach farmers better irrigation practices, and Jal Shakti Abhiyan for awareness and to make water conservation a public campaign. These government schemes have brought changes in many lives as the public has been involved and their active participation in the conservation of water through rainwater harvesting, cleaning of wells, contour bunding, warabandi management, afforestation, and gray water management through kitchen gardens and soak pits has resulted in an increase in groundwater level.

16.7 Recommendations and Conclusion

In order to utilize the available water resources systematically and productively, the role of government and the public is very important. If awareness of the public regarding conservation/effective utilization, as well as the collection and storage of water, increases, then only we will achieve effective results of all the government schemes.

We recommend the following measures for better water resource management:

1) The government should take the initiative to spread more awareness through advertisements in print, electronic media, or social media and by running campaigns throughout the country, especially in water-stressed areas.
2) The government should ask schools to organize workshops that explain the significance of freshwater and water-saving techniques to children at an early age so they become torchbearers for spreading further awareness.
3) The central government should form monitoring bodies that closely assess and ensure time-bound completion of all the water conservation-related schemes in India.
4) The government should also encourage NGOs and members of the public working in water conservation by rewarding and appreciating their noble work so that more people are associated with their work.

References

Aggarwal, A., Pandey, R.S., and Sharma, B. (2010). Water pollution with special reference to pesticide contamination in India. *Journal of Water Resource and Protection* 2 (5): 1–17. https://doi.org/10.4236/jwarp.2010.25050.

Barker, R., Van Koppen, B., and Shah, T. (2000). A global perspective on water scarcity and poverty: Achievements and challenges for water resource management (No. 613–2016-40702). IWMI Books, Reports H026191. Colombo: International Water Management Institute.

Bhattacharya, S. (2015). Traditional water harvesting structures and sustainable water management in India: a socio-hydrological review. *International Letters of Natural Sciences* 37: 30–38.

Biswas, A.K. (2004). Integrated water resources management: a reassessment: a water forum contribution. *Water International* 29 (2): 248–256. https://doi.org/10.1080/02508060408691775.

Borthakur, A. and Singh, P. (2016). India's lost rivers and rivulets. *Energy Ecology and Environment* 1: 310–314. https://doi.org/10.1007/s40974-016-0039-2.

Hoekstra, A.Y. (2014). Sustainable, efficient, and equitable water use: the three pillars under wise freshwater allocation. *WIREs Water* 1: 31–40. https://doi.org/10.1002/wat2.1000.

Jain, S.K. (2019). Water resources management in India–challenges and the way forward. *Current Science* 117 (4): 569–576. https://doi.org/10.18520/cs/v117/i4/569-576.

Kumar, C.P. (2018). Water resources issues and management in India. *Journal of Scientific and Engineering Research* 5 (9): 137–147.

Kumar, R., Singh, R.D., and Sharma, K.D. (2005). Water resources of India. *Current Science* 89 (5): 794–811.

Maria, A. (2003). The costs of water pollution in India. Paris: CERNA.

Mukhtarov, F.G. (2008). Intellectual history and current status of integrated water resources management: a global perspective. In: *Adaptive and Integrated Water Management* (eds. C.P. Wostl, P. Kabat and J. Möltgen), 167–185. Berlin, Heidelberg: Springer.

Rahaman, M.M. and Varis, O. (2005). Integrated water resources management: evolution, prospects and future challenges. *Sustainability: Science, Practice and Policy* 1 (1): 15–21. https://doi.org/10.1080/15487733.2005.11907961.

Savenije, H. H. G. and Hoekstra, A. Y. (2002). Sustainable Development: An insight into the Encyclopedia of Life Support Systems, Water Resources Management Vol. II, Oxford 2002: 155–180. UNESCO Publishing/EOLSS Publishers, Paris.

Shen, Y. and Chen, Y. (2010). Global perspective on hydrology, water balance, and water resources management in arid basins. *Hydrological Processes: An International Journal* 24 (2): 129–135. https://doi.org/10.1002/hyp.7428.

Singh, R., Singh, H., Singh, S. et al. (2017). Riparian land uses affect the dry season soil CO2 efflux under dry tropical ecosystems. *Ecological Engineering* 100: 291–300. https://doi.org/10.1016/j.ecoleng.2017.01.002.

Singh, R., Srivastava, P., Singh, P. et al. (2019). Human overpopulation and food security: challenges for the agriculture sustainability. In: *Urban Agriculture and Food Systems: Breakthroughs in Research and Practice* (ed. M. Khosrow-Pour), 439–467. Hershey, PA: IGI Global https://doi.org/10.4018/978-1-5225-8063-8.ch022.

Surendran, U., Sushanth, C.M., Mammen, G., and Joseph, E.J. (2017). FAO-CROPWAT model-based estimation of crop water need and appraisal of water resources for sustainable water resource management: pilot study for Kollam district-humid tropical region of Kerala, India. *Current Science* 112 (1): 76–86.

Web References/Online Sources

Atal Bhujal Yojana, Ministry of Jal Shakti, Department of Water Resources, RD & GR, Government of India. http://mowr.gov.in/schemes/atal-bhujal-yojana. Date of Access: 15 June 2020.

Command Area Development and Water Management (CADWM) Programme Report (2011) http://164.100.229.55/sites/default/files/msouser/Directorate%20of%20CAD-PIM.pdf.

Comprehensive Mission Document of National Water Mission, Ministry of Jal Shakti, Vol.1, April 2011. http://jalshakti-dowr.gov.in/sites/default/files/nwm16606419934.pdf.

http://cwc.gov.in/national-hydrology-project-nhp. Date of Access: 13 June 2020.

http://mowr.gov.in/schemes-projects-programmes/programmes/interlinking-rivers. Date of Access: 13 June 2020.

https://nmcg.nic.in. Date of Access: 24 June 2020.

http://sujal-swachhsangraha.gov.in. Date of Access: 24 June 2020.

https://www.gktoday.in/gk/namami-gange-project. Date of Access:15 June 2020.

Guidelines for Integrated Water Resources Development and Management, 2016. http://cwc.gov.in/sites/default/files/IWRM_Guidelines.pdf.

Guidelines for the Scheme on Repair, Renovation and Restoration (RRR) of Water Bodies under PMKSY (HKKP), 2017. http://pmksy-mowr.nic.in/documents/RRR_PMKSY_Guidelines_2017.pdf.

Guidelines for Urban Water Conservation Jal Shakti Abhiyan, 2019. https://ejalshakti.gov.in/jsa/jsaContent/GuidelinesforJalShaktiAbhiyan.pdf.

Jal Shakti Abhiyan Booklet, 2019. https://ejalshakti.gov.in/jsa/jsaContent/JSAbrochure.pdf.

National Hydrology Project, Ministry of Jal Shakti, Department of Water Resources, RD & GR, Government of India. http://nhp.mowr.gov.in. Date of Access: 14 June 2020.
National Water Mission, Ministry of Jal Shakti, Department of Water Resources, RD & GR, Government of India. http://nwm.gov.in. Date of Access: 13 June 2020.
Review of Water Policies and Programmes with Special Reference to Urban Water Supply (n.d.). https://shodhganga.inflibnet.ac.in/bitstream/10603/36784/11/11_chepter%203.pdf.
World Bank https://projects.worldbank.org/en/projects-operations/project-detail/P152698). Date of Access: 14 June 2020.

Index

Pollutants and Water Management: Resources, Strategies and Scarcity, First Edition. Edited by Pardeep Singh, Rishikesh Singh, Vipin Kumar. Singh, and Rahul Bhadouria.

b

d

i

j

k

l

n

t

www.ingramcontent.com/pod-product-compliance
Lightning Source LLC
LaVergne TN
LVHW062204200926
848484LV00006B/1314